Marine Rudders and Control Surfaces

Marine Rudders and Control Surfaces

Principles, Data, Design and Applications

Anthony F Molland

Emeritus Professor of Ship Design, University of Southampton, UK

and

Stephen R Turnock

Senior Lecturer in Ship Science, University of Southampton, UK

ELSEVIER

AMSTERDAM • BOSTON • HEIDELBERG • LONDON • NEW YORK • OXFORD
PARIS • SAN DIEGO • SAN FRANCISCO • SINGAPORE • SYDNEY • TOKYO
Butterworth-Heinemann is an imprint of Elsevier

B H

Butterworth-Heinemann is an imprint of Elsevier
Linacre House, Jordan Hill, Oxford OX2 8DP, UK
30 Corporate Drive, Suite 400, Burlington, MA 01803, USA

First edition 2007

British Library Cataloguing in Publication Data
A catalogue record for this book is available from the British Library

Library of Congress Cataloging-in-Publication Data
A catalog record for this book is available from the Library of Congress

ISBN: 978-0-75-066944-3

For information on all Butterworth-Heinemann publications
visit our web site at books.elsevier.com

Printed and bound by CPI Group (UK) Ltd, Croydon, CR0 4YY

Transferred to Digital Print 2011

Contents

Preface

Marine control surfaces are all pervasive and are used on a wide range of marine vehicles as rudders, stabilisers and for pitch control. They can range in size from a height of 40 cm for a control surface on an autonomous underwater vehicle to a height of 9 m and weighing over 80 tonnes for the rudder on a large container ship. An extensive amount of research and investigation into ship rudders and control surfaces has been carried out over a number of years, including wide ranging investigations by the authors into rudder–propeller interaction. The research has generally entailed experimental and theoretical investigations and the main results of many of these investigations have been published as contributions to Journals and at Conferences. It is considered, however, that there is a need to bring together the experimental and theoretical data sources and design methods into a book that will be suitable both for academic reference and as a practical design guide.

The book is aimed at a broad readership, including practising professional naval architects and marine engineers, small craft and yacht designers, undergraduate and postgraduate degree students and numerical analysts.

The book is arranged in four main parts. The first part (Chapters 1–4) covers basic principles, including the physics of control surface operation and a background to rudder and control surface types. The second part (Chapters 5 and 6) reviews and brings together design data sources including experimental data, together with theoretical and numerical methods. The review of available experimental data is extensive, covering various control surface types including all-movable, skeg and high-lift rudders. Performance data are presented for rudders working both in a free stream and downstream of a propeller. Theoretical and numerical methods applied to control surface design are described in some depth. Methods and interpretation of numerical analysis are described, together with applications to particular cases such as rudders in a free stream, downstream of a propeller and unsteady behaviour. In the third part (Chapters 7–10), rudder and control surface design is reviewed and described, covering design strategy and methodologies. The fourth part (Chapter 11) illustrates the use of the data and design methods through design applications with a number of worked practical examples of rudder and control surface design. The examples include design of a rudder for a container ship, including the use of a spade rudder and a semi-balanced skeg rudder, the design of a twisted rudder, a rudder cavitation check, structural analysis, design of cruising and racing yacht rudders and the design of a roll stabiliser fin. References are provided at the end of each chapter to facilitate access to the original sources of data and information and further depth of study where necessary.

The database of wind tunnel rudder–propeller interaction data gathered by the authors is an important and integral part of the book. The extensive existing database of experimental data has been restructured into a more readily useable form. Examples of the data are tabulated in an Appendix and the remainder are made accessible through the publisher's web site.

The authors acknowledge the help and support over many years of their colleagues in the School of Engineering Sciences at the University of Southampton. A.F.M. would wish, in particular, to acknowledge the support provided in the early days of his rudder research by the late Professor G. J. Goodrich. Thanks must also be conveyed to national and international colleagues for their continued support over the years.

Grateful thanks are due to the many undergraduate and M.Sc. students at the University of Southampton who, over many years, have contributed to a better understanding of the subject through project and assignment work. The work of postgraduate students who have contributed more formally to the Southampton research programme is much appreciated, including Dr. James Date, Dr. Sandy Wright and Dr. Jason Smithwick who is to be thanked also for his assistance in coordinating the database and plotting a number of the figures on rudder–propeller interaction.

The authors would extend their appreciation to the Engineering and Physical Sciences Research Council (EPSRC) for the substantial funding for the experimental rudder–propeller investigations and to the various industrial organisations who have provided incentives and funding for their research.

Finally, the authors wish especially to thank their wives and families for all their help and support.

A. F. Molland
S. R. Turnock
Southampton

Nomenclature

a	Axial induced inflow factor
a''	Rotational (circumferential) induced inflow factor
a_i	Lift curve slope for infinite aspect ratio
A	Total rudder area (m^2), or rudder axial force in direction of rudder axis (N)
A_L	Underwater lateral area (m^2)
Am	Movable rudder area (m^2)
A_n	Coefficients in Fourier series
A_R	Propeller slipstream area at rudder (m^2)
AR_G	Geometric aspect ratio ($S/\overline{c}$ or S^2/A)
AR	Effective aspect ratio (kAR_G)
A_x	Rudder axial force, in direction of rudder axis (N)
B	Ship breadth (m), or position of centre of buoyancy, or number of propeller blades
BM_E	Equivalent bending moment (Nm)
C	Courant number
C_A	Axial force coefficient ($A_x/\frac{1}{2}\rho AV^2$)
C_B	Block coefficient (∇/LBT)
C_D	Drag coefficient ($d/\frac{1}{2}\rho AV^2$)
C_{Do}	Profile drag coefficient, or minimum drag coefficient ($d_o/\frac{1}{2}\rho AV^2$)
C_{Di}	Induced drag coefficient ($d_i/\frac{1}{2}\rho AV^2$)
C_{Dc}	Cross flow drag coefficient
C_D^*	Drag coefficient ($d/\frac{1}{2}\rho AK_T n^2 D^2$)
C_D^{**}	Drag coefficient ($d/\frac{1}{2}\rho A(V^2 + K_T n^2 D^2)$)
C_F	Skin friction coefficient
C_f	Local skin friction coefficient
C_G	Correction factor
C_l	Local lift coefficient
C_{LL}	Local lift coefficient (Lifting line theory)
C_L	Lift coefficient ($L/\frac{1}{2}\rho AV^2$)
C_L^*	Lift coefficient ($L/\frac{1}{2}\rho AK_T n^2 D^2$)
C_L^{**}	Lift coefficient ($L/\frac{1}{2}\rho A(V^2 + K_T n^2 D^2)$)
C_{Lmax}	Maximum lift coefficient ($L_{max}/\frac{1}{2}\rho AV^2$)
C_M	Moment coefficient about rudder stock ($Q_R/\frac{1}{2}\rho AV^2\,\overline{c}$)
$C_{Mc/4}$	Moment coefficient about quarter chord
Cm	Movable chord (m)
Cmx	Moment about x-axis at root (Nm)
Cmy	Moment about y-axis at root (Nm)
Cmz	Moment about z-axis (centreline of stock) (Nm)
C_N	Normal force coefficient, normal to rudder ($N/\frac{1}{2}\rho AV^2$)
C_n	Normal force coefficient, normal to rudder ($N/\frac{1}{2}\rho AV^2$)
C_P	Pressure coefficient ($(P_L - P_O)/\frac{1}{2}\rho AV^2$)

$CP\bar{c}$	Centre of pressure chordwise, % $\bar{c}$ measured from leading edge
CPs	Centre of pressure spanwise, % S, measured from root
C_R	Root chord (m), or resultant force coefficient ($R/\frac{1}{2}\rho AV^2$)
Cs	Fixed chord (m)
C_T	Propeller thrust coefficient ($8T/\rho\pi D^2[V^2 + (0.7\pi nD)^2]$), or rudder tip chord (m)
c	Rudder chord (m), or speed of propagation [m/s]
$\bar{c}$	Rudder mean chord ($(C_R + C_T)/2$)
c_μ	Empirical coefficient in turbulent dissipation equation
D	Propeller diameter (m), or diameter of ship turning circle (m)
d	Rudder drag force, in direction of fluid free-stream flow (N)
d_i	Induced drag (N)
d_o	Rudder minimum drag force at zero incidence (N)
F	Total pressure force on a body (N)
Fc	Froude number, based on control surface chord ($V/\sqrt{gc}$)
F_H	Hull force (N)
F_R	Rudder force (N)
Fr	Froude number, based on ship length ($V/\sqrt{gL}$)
Fs	Froude number based on control surface span ($V/\sqrt{gS}$)
F_T	Total side force, ship axis (N), or total force acting on a body (N)
F_{visc}	Total viscous force on a body (N)
F_X	Rudder drag, ship axis (N)
F_Y	Rudder sideforce, ship axis (N)
G	Position of centre of gravity, or gap (m)
GM	Metacentric height (transverse) (m)
GM_L	Metacentric height (longitudinal) (m)
g	Acceleration due to gravity (m²/s)
H	Shape factor (δ^*/θ) for boundary layer
$H_1(\theta)$	General form of downwash variation (lifting line)
h	Depth of immersion of control surface (m)
J	Propeller advance ratio (V/nD)
K	Goldstein factor
K_Q	Propeller torque coefficient ($Q/\rho n^2D^5$)
K_R	Rudder drag/thrust coefficient ($d/\rho n^2D^4$), or Gutsche slipstream acceleration factor
K_T	Propeller thrust coefficient ($T/\rho n^2D^4$)
K_{To}	K_T at $J = 0$
k	Effective aspect ratio factor (AR/AR_G), or turbulent kinetic energy/unit mass, or taper ratio factor ($1 - TR$), or reduced frequency, or roughness height (m)
k_i	Induced drag factor
L	Rudder lift (sideforce), normal to direction of fluid free stream (N), or ship length (m)
LBP	Ship length between perpendiculars (m)
L_F	Stabiliser fin lift (N)
L_{FP}	Pitch-damping foil lift (N)

L_{max}	Maximum lift (N)
Lm	Lift on movable part of chord
Lt	Lift on total chord
l	Turbulent length scale, or length of body (m)
M	Root bending moment (Nm), or number of domains
m	Camber, or 2D lift curve slope (infinite aspect ratio)
N	Rudder normal force, normal to rudder axis (N), or number of panels on a body
n	Propeller rate of revolution (rps), or number of points in lifting line analysis, or number of panels in panel method
P	Propeller Pitch [m], or pressure (N/m^2)
P_{AT}	Atmospheric pressure (N/m^2)
P_G	Order of accuracy
P_O	Free stream or ambient pressure (N/m^2)
P_L	Local static pressure (N/m^2)
P_S	Stagnation pressure (N/m^2)
P_T	Total pressure (N/m^2)
P_V	Vapour pressure (N/m^2)
P/D	Pitch ratio
Q	Propeller torque (Nm), or volume flow rate (m^3/s)
Q_T	Total rudder torque (Nm)
Q_F	Frictional torque (Nm)
Q_H	Hydrodynamic torque (Nm)
Q_R	Rudder torque (Nm)
Q_{RE}	Equivalent rudder torque (Nm)
q	Velocity (u, v, w) (m/s)
R	Propeller radius (m), or total resultant rudder force (N), or lever of stabiliser fin (m), or ship resistance (N), or radius of turning circle (m)
RF	Reduction factor
Re	Reynolds number ($\rho Vc/\mu$ or Vc/ν)
R_G	Ratio of changes
R_P	Lever of pitch damping fins (m)
R_θ	Reynolds number based on momentum thickness ($V\theta/\nu$)
r	Radius of elemental section of propeller (m)
r_N	Nose radius of aerofoil
r_G	Refinement factor
S	Rudder or control surface span (m), or bounding surface
T	Propeller thrust (N), or ship draught (m)
T_i	Turbulence intensity
TR	Taper ratio (C_T/C_R)
t	Rudder or control surface section thickness (m), or thrust deduction fraction, or time (s)
U	Velocity (m/s)
U_d	Tangential disturbance velocity (m/s)
U_{G1}	Uncertainty
Ui	Normal component of relative body surface velocity (m/s)

U_0	Inflow velocity (m/s)
U_t	Total velocity at panel centroid (m/s)
U_∞	Velocity downstream at infinity (m/s)
U, V, W	Mean flow velocity components
u, v, w	Velocity components in x, y, z directions
u	Velocity (m/s)
u^+	Velocity based on wall shear stress
u_e	Velocity at edge of boundary layer (m/s)
V	Rudder or control surface inflow speed (m/s)
V_A	Wake flow speed ($V_S(1 - w_T)$) (m/s), or propeller induced axial velocity (m/s)
V_R	Velocity at rudder (lifting line) (m/s)
V_S	Ship speed (m/s)
V_T	Velocity at tangent to turning path (m/s), or propeller induced tangential velocity (m/s)
V_δ	Velocity at edge of boundary layer (m/s)
V_t	Transpiration velocity (m/s)
v	Velocity (m/s)
v_0	Amplitude of transverse gust (m/s)
w	Downwash (m/s)
w_L	Induced local velocity (lifting line) (m/s)
w_T	Wake fraction at propeller
X	Longitudinal distance, propeller plane to rudder leading edge (m)
X_1	Distance of rudder or control surface stock from leading edge (m)
x	position on propeller blade (r/R), or distance from leading edge of foil or body
$\bar{x}$	Distance of normal force from rudder or control surface stock (m)
x_V	Vertex node location
x, y, z	Cartesian (orthogonal) co-ordinate system
Y	Lateral separation of propeller axis and rudder stock (m)
$\bar{y}$	Distance of normal force from rudder root (m)
y	Distance perpendicular to body (m), or section thickness offset (m)
y^+	Non-dimensional distance from wall in boundary layer
Z	Vertical separation of propeller axis and rudder root (m)
z	Potential head (m)
α	Rudder or control surface incidence relative to incoming flow [deg.]
α_E	Effective rudder incidence [deg.]
$\bar{\alpha}$	Effective rudder incidence (lifting line) [deg.]
α_o	Rudder incidence for zero lift or sideforce [deg.]
α_{max}	Maximum rudder angle [deg.]
α_{stall}	Rudder stall angle [deg.]
$\alpha_T(\theta)$	Distribution of downwash due to tip vortex [deg.]
β	Drift angle [deg.]
β_R	Geometric local drift angle at rudder [deg.]

δ	Rudder incidence relative to ship or body axes [deg.], or boundary layer thickness (m)
δ^*	Boundary layer displacement thickness (m)
ε	Downwash angle [deg.], or rate of turbulent dissipation
ϕ	Heel angle [deg.], or velocity potential
Γ_L	Local circulation (m^2/s)
γ	Flow straightening factor (α_o/β_R)
η_0	Propeller open water efficiency ($JK_T/2\pi K_Q$)
η_H	Hull efficiency ($(1-t)/(1-w_T)$)
κ	Von Karman constant
λ	Proportion of propeller diameter D impinging on rudder span S, or inflow angle induced by upstream propeller [deg.]
μ	Dynamic viscosity (kg/ms), or normal dipole distribution per unit area, or coefficient in lifting line equations
μ_B	Coefficient of friction of bearing material
μ_W	Wake strength
ν	Kinematic viscosity (m^2/s)
ν_t	Eddy viscosity (m^2/s)
θ	Boundary layer momentum thickness (m), or pitch angle [deg.]
ρ	Density (kg/m^3)
σ	Cavitation number, or allowable stress (N/mm^2), or source distribution per unit area
τ_W	Surface or wall shear stress
Ω	Sweep of quarter chord [deg.], or propeller rate of revolution (rad/s)
ω	Circular passage frequency of propeller (rad/s)
ξ	Coverage, or proportion of rudder span covered by propeller race, ($\lambda D/S$)
ψ	Advance angle, $\tan^{-1}(J/\pi x)$
ζ	Vorticity (1/s)
∇	Ship displacement volume (m^3), or Grad operator
Δ	Ship displacement mass ($\nabla\rho$) [tonnes]

Abbreviations

ABS	American Bureau of Ships
BEM	Boundary element method, or blade element-momentum theory
CFD	Computational fluid dynamics
CLR	Centre of lateral resistance
CG	Centre of gravity
DNS	Direct numerical simulation
DNV	Det Norske Veritas
EPSRC	Engineering and Physical Sciences Research Council
FEA	Finite element analysis
FRP	Fibre reinforced plastic
GL	Germanischer Lloyd
IACC	International Americas Cup Class
IMO	International Maritime Organisation
INSEAN	Instituto di Architectura Navale (Rome)
ISO	International Standards Organisation
IESS	Institution of Engineers and Shipbuilders in Scotland.
ITTC	International Towing Tank Conference
JASNAOE	Japan Society of Naval Architects and Ocean Engineers
LCG	Longitudinal centre of gravity
LE	Leading edge of foil or fin
LES	Large eddy simulation
LR	Lloyd's Register of Shipping
NACA	National Advisory Council for Aeronautics (USA)
NECIES	North East Coast Institution of Engineers and Shipbuilders
N-S	Navier Stokes
RINA	Royal Institution of Naval Architects
RNG	Renormalisation group
SNAME	Society of Naval Architects and Marine Engineers (USA)
SNAJ	Society of Naval Architects of Japan
SNAK	Society of Naval Architects of Korea
TE	Trailing edge of foil or fin
VLCC	Very large crude carrier

Figure acknowledgements

The authors acknowledge with thanks the assistance given by the following companies and publishers in permitting the reproduction of illustrations from their publications:

Figure 1.6 reprinted courtesy of The Chapter of Winchester.

Figures 5.10, 5.11, 5.12, 5.27, 5.28, 5.29a, 5.29b, 5.29c, 5.51, 5.133, 5.134, 5.135, 5.136, 5.137, Table 5.7 reprinted courtesy of The Society of Naval Architects and Marine Engineers (SNAME), New York.

Figure 5.24 reprinted courtesy of Becker Marine Systems.

Figures 5.26 (a–c) reprinted courtesy of The Japan Society of Naval Architects and Ocean Engineers (JASNAOE), Tokyo (formerly The Society of Naval Architects of Japan (SNAJ), Tokyo).

Figures 5.25, 5.39, 5.41 (a–f), 5.42, 5.43, 5.45, 5.48, 5.49, 5.50, 5.55, 5.58, 5.67, 5.68, 5.74 (a–e), 5.78, 6.1, 6.31 (a–c), 6.32 reprinted courtesy of The Royal Institution of Naval Architects (RINA), London.

Figures 5.63, 5.64 and 5.65 reprinted courtesy of The Society of Naval Architects of Korea (SNAK), Seoul.

Figures 5.96, 5.97 5.98 and Table 5.16 reprinted courtesy of International Shipbuilding Progress, Delft University Press.

Figures 5.123(a–b) and 5.124 reprinted courtesy of VWS Berlin.

Figure 5.138 reprinted courtesy of The Institute of Mechanical Engineers (IMechE), London.

Visit http://books.elsevier.com/companions for further supporting materials.

PART ONE

Principles

PART ONE

Principles

1 Introduction

The fundamental concept of a movable device to steer a ship has been in use since ships were first conceived. The purpose of the device, or rudder, is to either maintain the ship on a particular course or direction, or to enable it to manoeuvre.

Although unaware of the mathematics associated with the dynamics of ship manoeuvring, people for many centuries have been aware of the role of the rudder in solely establishing an angle of attack on the hull, with the hydrodynamic forces developed on the hull largely turning the ship. For example, as James [1.1] observed nearly 2,000 years ago: "Or think of ships: large they may be, yet even when driven by strong gales they can be directed by a tiny rudder on whatever course the helmsman chooses." Although relatively small, submerged, and out of sight, the rudder is fundamental to the safe operation of the ship. The rudder has, therefore, always attracted considerable interest, research and development.

From the time of the early Egyptian ships onward, steering was carried out by means of a side-mounted steering oar over the after quarter, Figure 1.1. The oar rested in a notch cut in the gunwale, or passed through a hole cut in the gunwale, sometimes termed an 'oar port.' Later versions passed through a wooden bracket mounted on the side of the ship or were attached by a withy, or rope, forming a kind of hinge. The dominant use of the side-steering oar concept, sometimes termed a quarter rudder, continued until the twelfth century when there was a distinct change in the concept from a side-steering oar to a stern-mounted rudder using pintles and gudgeons (which may be broadly described as hinges), Figures 1.2 and 1.3. Although changes in ship propulsion from oars to sail to motor power have led to changes in detailed rudder–hull layout, the stern-mounted rudder remains the principal concept.

The seagoing ship basically evolved from the Egyptians from about 3000 BC. The same era also saw the development of the sail, although oars supplemented by sails, or vice versa, would be the custom for many centuries. Around 1000 BC, the Greek galley had evolved as the prevailing fighting ship. The Romans adopted the galley from the Greeks through the era from approximately 300 BC to 400 AD. From the mid-eighth to mid-eleventh centuries the Viking ship became the dominant type of craft. At about the same time, the standard vessel for northern Europe was the cog. The cog was descended from the Viking cargo vessel, although shorter and fatter with a straight stern and straight bow, a single mast and a square sail. It was the vessel used by the Hanseatic League of traders, established during the thirteenth century. Also prevalent at the time was the hulk, which was clinker built but with a bow and stern that curved upward. It was to the cog and hulk that the stern-hung

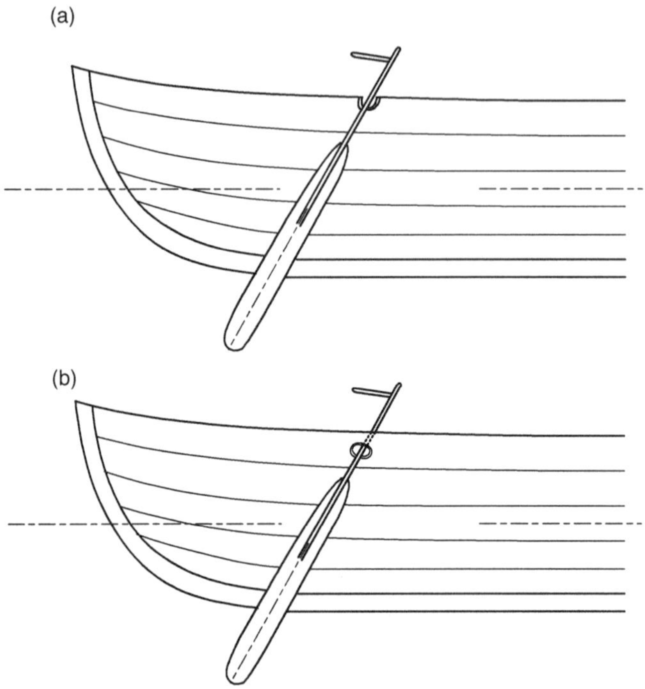

Figure 1.1 Early ships with side-mounted steering oars

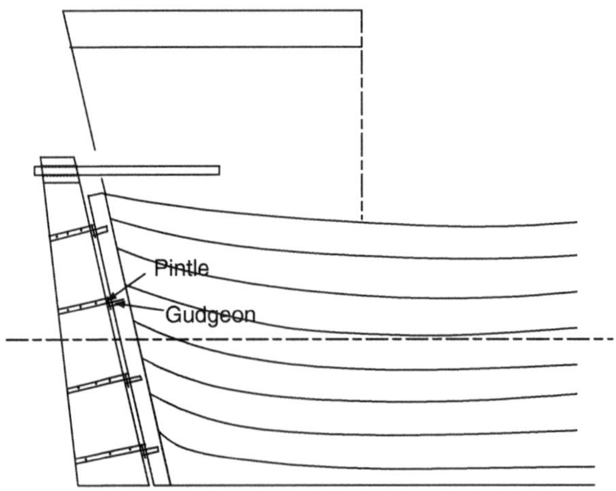

Figure 1.2 Early version of stern-mounted pintle and gudgeon rudder on a cog

pintle and gudgeon rudder was first fitted. Up to this era, the side-steering oar had been employed. By the end of the fourteenth century, the basic generic forms of the cog and hulk had merged together.

The side-mounted steering oar, or quarter rudder, had many variations over the years, including many shapes and attachment methods. As ships grew in size, many early arrangements had two-quarter rudders. Details of the development of the rudder are described by Mott [1.2] and descriptions of the various ship types through the ages, including their propulsion and steering methods, may be found in references [1.3–1.5].

It is important to note that the side-mounted steering oar was developed over the years to a very efficient level. In its most developed version, for example, on the Viking ships, it would have a form of rope hinge on the side of the ship and a tiller to assist the helmsman, Figure 1.4. The oar, or stock, would run through the centre of the blade, so the steering oar was effectively balanced, lowering torques and tiller forces.

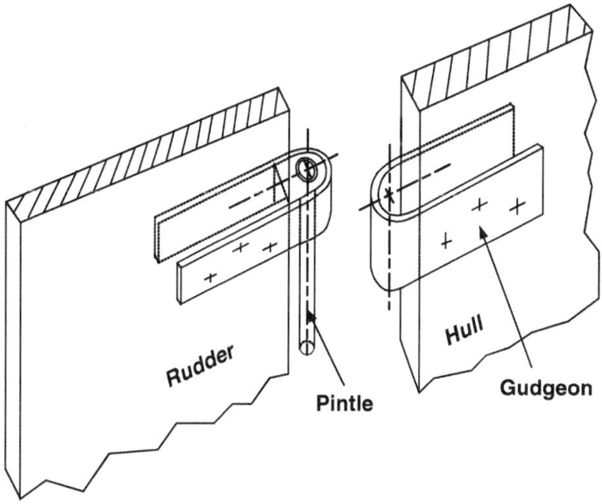

Figure 1.3 Pintle and gudgeon

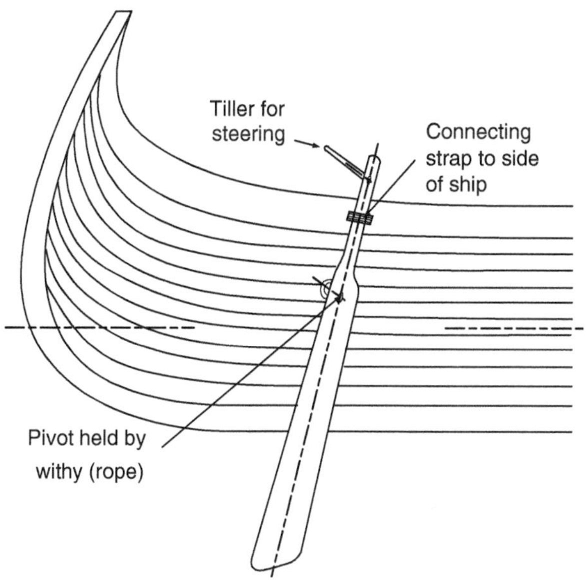

Figure 1.4 Viking steering oar

The reasons for the move from the side-steering oar to the stern-mounted rudder in the twelfth century are not clear [1.2]. It is likely to have been due to the growth in the size of ships and increase in freeboard and as a result, the side-steering oars were becoming too large to be manageable. Also, the stern rudder is less vulnerable to damage. It could also be partly due to the greater availability of iron to make the pintles and gudgeons (the rudder hinges), Figure 1.3.

The early days of the stern-mounted rudder were not without problems. The hull needed to be reshaped at its aft end to provide a reasonable flow into the rudder. More importantly, the stern-mounted rudder was not balanced, which led to very high torques and tiller forces. The initial response was to use long tillers to overcome the torque. Next, the whipstaff was introduced to provide extra torque by levers, Figure 1.5, but this restricted rudder angles. In the eighteenth century, the whipstaff was replaced by a geared steering wheel system with rope and pulleys, which overcame the torque problem. The geared steering wheel, although now mechanically aided, is still the main concept in use today.

The early historical development of ships and rudders can only be deduced from paintings, models, pottery, carvings and other iconic evidence. Identifying the dates of particular events can therefore be difficult and this includes the introduction of the stern rudder and the important move during the twelfth century from the side-steering oar, or quarter rudder, to the stern rudder mounted on pintles and gudgeons.

Some early Egyptian ships had centrally mounted steering oars over the stern. Iconic evidence would suggest that the structure was relatively weak and the oar was effectively unsupported. Hinges, or pintles and gudgeons, were not employed and

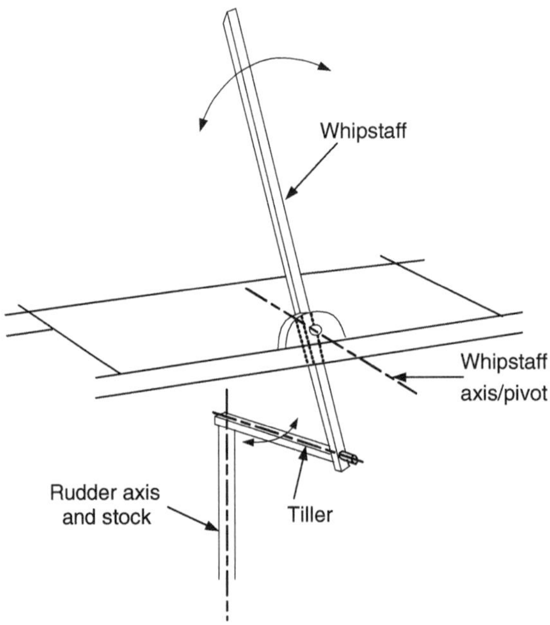

Figure 1.5 The whipstaff

these were not true stern-mounted rudders, as would be the case from the twelfth century onward.

The earliest depiction of a stern rudder can be traced to China (Needham [1.6], cited in [1.7]). It is shown on a detailed pottery model of a ship with a stern rudder, dating from the first century. Other evidence suggests that the rudder was held in place by a rope mechanism, rather than by a pintle and gudgeon mechanism. Paintings from the twelfth century show Chinese vessels with stern rudders of similar shape to the first European stern rudders, but attached above the waterline by nonmetallic parts. Xi and Chalmers [1.8], like Needham, consider that the concept of the sternpost rudders in Europe was derived from Chinese designs.

What is believed to be the earliest picture in the West of a stern-mounted rudder can be found in a relief on the baptismal font in Winchester Cathedral, Figure 1.6, which is considered to date from about 1160 to 1180 [1.2]. The picture is that of a hulk and the overall layout would suggest a stern rudder, with the arm of the helmsman clearly lying over the tiller. It should, however, be mentioned that some scholars consider that, as the leading edge of the rudder seems to project a little in front of the stern, the picture may in fact depict a side or quarter rudder mounted well aft, or possibly a stern rudder with its gudgeons mounted a little off the centreline to the starboard side [1.7,1.9]. The ships depicted on the seals of various towns such as Ipswich subsequently confirm the introduction of the stern rudder by about the end of the twelfth century.

Over the centuries, up to the advent of motor power, the layout of the pintle and gudgeon-hung rudder remained similar in concept to that adopted on the cog, Figure 1.2. Although there was little change in the fundamental concept of the rudder design through this period, improvements in the operation and design of rudders

Figure 1.6 Twelfth-century relief of a stern rudder: Winchester Cathedral. Photograph by the authors; published courtesy of The Chapter of Winchester

took place to meet the requirements of new ship types. The rudders on later large sailing ships were generally as shown in Figure 1.7 and on motor powered vessels up to the present time as in Figures 1.8–1.10.

There were relatively few changes in rudder design during the eighteenth century, although there was the introduction of the steering wheel and various topics of discussion on rudder operation. For example, Hutchinson [1.10] reports on discussions on a suitable maximum rudder angle. It had been found that helmsmen were putting the rudder hard over, even up to 40°, and the rudder was becoming inoperative. It was recommended that rules be made to limit rudder angle to 33°. Even after 250 years the debate and discussions on suitable maximum rudder angles remain ongoing.

A considerable amount of work was carried out on steering and rudders during the nineteenth century. Notable developments, such as the work of Jöessel [1.11]

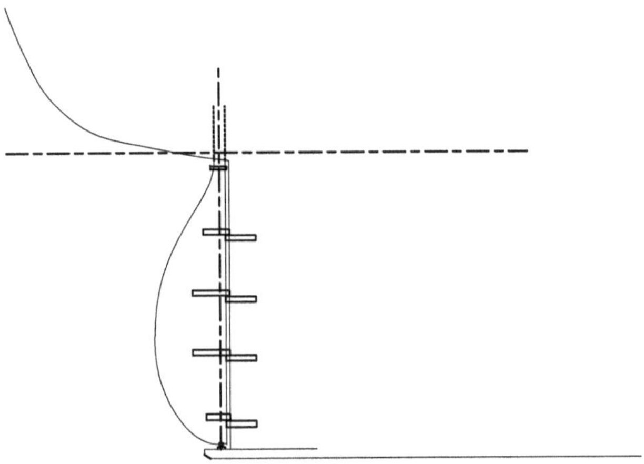

Figure 1.7 Large sailing ship

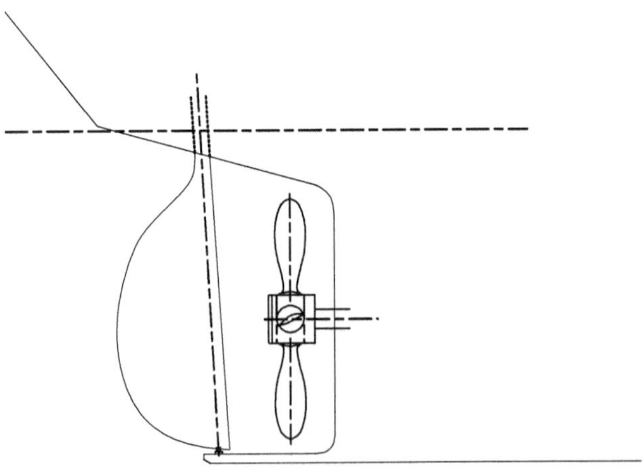

Figure 1.8 Early motor ship

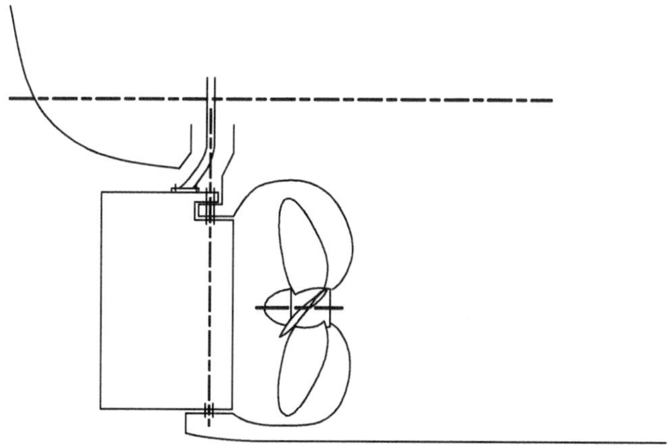

Figure 1.9 Motor ship

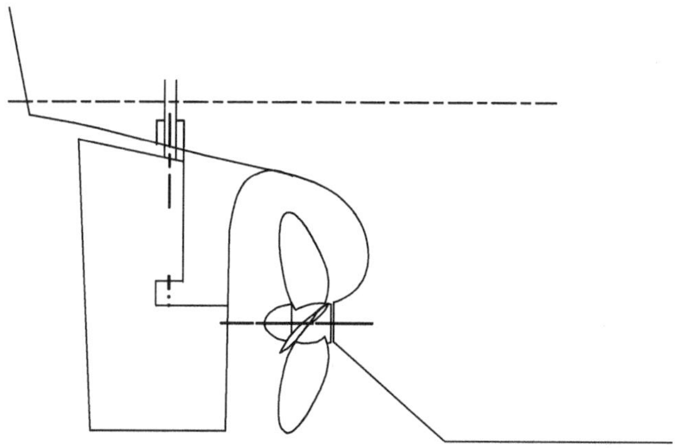

Figure 1.10 Recent motor ship

and Lumley [1.12] are worthy of mention. Jöessel carried out experiments on plate rudders in the Loire river in 1873 and developed empirical relationships for the torque on rectangular plates. These relationships were used for over 100 years before being superseded by more appropriate formulae. Lumley proposed designs in 1864 for flapped rudders, a concept still applied where high-lift forces are required to be developed. Typical other investigations into steering and rudders during that period, including balanced rudders, are reported in further papers to the Royal Institution of Naval Architects [1.13–1.15].

The early part of the twentieth century saw a marked increase in investigation into rudder performance, including the work of Denny [1.16], Bottomley [1.17] and others, particularly with the advent of new propulsion systems and the use of twin screws. By the 1950s and 1960s, significant progress on a more fundamental understanding of the physical performance of control surfaces had taken place, due partly to the transfer of information and data for lifting surfaces from the aeronautical industry and also in response to the needs of developing mathematical coursekeeping and manoeuvring simulations, such as the pioneering work of Nomoto et al. [1.18] and Eda and Crane [1.19]. At the same time, work was progressing on other marine control surfaces such as stabiliser fins for the reduction of roll and hydroplanes for the trim and pitch control of submarines.

An appreciation of safety has always existed, although this now tends to be on a more formal footing. Losses of rudders are not frequent, as investigations such as those reported in references [1.20] and [1.21] would indicate. Nevertheless, with changing ship types, size and speed and the introduction of new rudder types, it has been necessary to continuously review and update the design of rudders. At the same time, the international regulatory bodies have put the steering requirements for ships on a more formal basis, with requirements for manoeuvring standards that embody both the manoeuvring performance of the hull–propeller–rudder and an understanding by the operator of the ship's manoeuvring capabilities [1.22].

The last 30 years or so of the twentieth century saw further progress in experimental investigation into rudders and control surfaces, with deeper investigations into the physics of operation using laser Doppler velocimetry (LDV) and particle image velocimetry (PIV) techniques. This has been accompanied by the introduction and development of further theoretical modelling and the use of computational fluid dynamics (CFD), bringing us to the state of rudder and control surface design as we know it today.

References

1.1 James. A letter of James, Chapter 3, verse 4. *The New English Bible.* Oxford University Press, Oxford, 1961.

1.2 Mott, L.V. *The Development of the Rudder: A Technological Tale.* Chatham Publishing, London, 1997.

1.3 Landström, B. *The Ship.* Allen and Unwin, London, 1961.

1.4 Lavery, B. *Ship. 5000 years of Maritime Adventure.* National Maritime Museum. Dorling Kindersley Ltd., London, 2005.

1.5 Rougé, J. *Ships and Fleets of the Ancient Mediterranean.* Wesleyan University Press, CT, USA, 1981 (Translation by Susan Frazer).

1.6 Needham, J. *Science and Civilisation in China,* Vol. 4, Pt. 3. Cambridge University Press, Cambridge, 1971, pp. 627–656.

1.7 Sleeswyk, A. and Lehmann, L. Pintle and gudgeon and the development of the rudder: the two traditions. *Mariner's Mirror,* Vol. 68, 1982, 279–303.

1.8 Xi, L. and Chalmers, D.W. The rise and decline of Chinese shipbuilding in the Middle Ages. *Transactions of The Royal Institution of Naval Architects,* Vol. 146, 2004, pp. 137–147.

1.9 Brindley, H. Medieval rudders. *Mariner's Mirror,* Vol. 13, 1927, 85–88.

1.10 Hutchinson, W. *A Treatise on Naval Architecture,* 4th edition, 1794. New Impression 1969. The Conway Maritime Press, London.

1.11 Jöessel, Rapport sur des experiences relatives aux gouvernails. *Memorial du Genie Maritime*, Rapport 9, 1873.
1.12 Lumley, H. On the steering of ships. *Transactions of the Royal Institution of Naval Architects*, Vol. 5, 1864, pp. 128–134.
1.13 Barnaby, N. On the steering of ships. *Transactions of the Royal Institution of Naval Architects*, Vol. 4, 1863, pp. 56–78.
1.14 Shulden, M. On balanced rudders. *Transactions of the Royal Institution of Naval Architects*, Vol. 5, 1864, pp. 123–127.
1.15 Halsted, E.P. On screw ship steerage. *Transactions of the Royal Institution of Naval Architects*, Vol. 5, 1864, pp. 82–113.
1.16 Denny, M.E. The design of balanced rudders of the spade type. *Transactions of the Royal Institution of Naval Architects,* Vol. 63, 1921, pp. 117–130.
1.17 Bottomley, G.H. Manoeuvring of ships. Part II – Unbalanced rudders of twin screw ships. *Transactions of the Institute of Engineers and Shipbuilders in Scotland*, Vol. 67, 1923/24, pp. 509–559.
1.18 Nomoto, K., Taguchi, T., Honda, K. and Hirano, S. On the steering qualities of ships, (1) and (2). *Journal of the Society of Naval Architects of Japan*, Vols. 99 and 101, 1956, 1957.
1.19 Eda, H. and Crane, C.L. Steering characterstics of ships in calm water and in waves. *Transactions of SNAME*, Vol. 73, 1965, pp. 135–177.
1.20 Bunyan, T.W. A study of the cause of some rudder failures. *Transactions of The North East Coast Institution of Engineers and Shipbuilders*, Vol. 68, 1951–1952, pp. 313–330.
1.21 Ten year rudder analysis by Lloyd's Register. *The Naval Architect.* The Royal Institution of Naval Architects, London, March, 1976, p. 60.
1.22 IMO Resolution MSC 137(76). *Standards for ship manoeuvrability*, 2002.

2 Control surface types

2.1 Control surfaces and applications

The purpose of a control surface is to produce a force, which is used to control the motion of the vehicle. Control surfaces may be fixed or movable but, in the marine field, they are mainly movable with the prime example being the ship rudder.

Movable control surfaces are used on most marine vessels including boats, ships of all sizes, submarines and other underwater vehicles. Typical applications may be summarised as

Rudders: used to control horizontal motion of all types of marine vehicle.
Fin stabilisers: used to reduce roll motion.
Hydroplanes (or diving planes): used to control the vertical motion of submarines and other underwater vehicles.
Fins for pitch damping: used to control pitch motion in high-speed vessels.
Transom flaps: used to control running trim and/or to provide ride control.
Interceptors: used to control running trim and/or to provide ride control.

Examples of fixed control surfaces include *anti-pitching fins* on fast vessels and *keels* on sailing yachts.

In general, lift or sideforce on a control surface may be developed by applying incidence, Figure 2.1(a), introducing asymmetry by means of fixed camber, Figure 2.1(b), or introducing variable camber by means say of a flap, Figure 2.1(c). Further increases in lift may be achieved by the application of incidence to cases (b) and (c). Since movable control surfaces generally have to act in both directions, applications in the marine field tend to be symmetrical and confined to the use of (a) or (c) in Figure 2.1, or some variants of these two basic types. The cambered shape (b) is of course used extensively for aircraft lifting surfaces as well as marine applications such as sections for propeller blades and lifting foils on hydrofoil craft.

2.2 Rudder types

The choice of the rudder type will depend on factors such as ship or boat type and size, the shape of the stern, size of rudder required and whether there is a propeller upstream of the rudder.

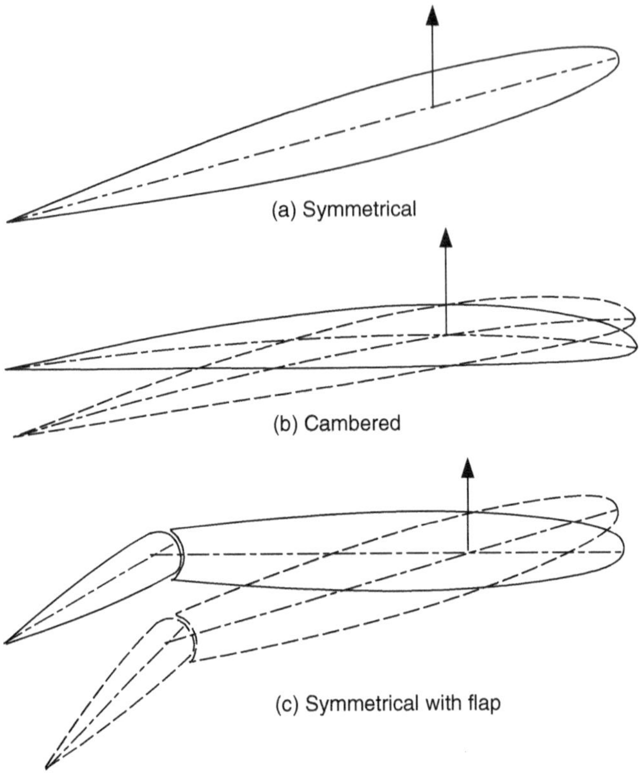

Figure 2.1 Sideforce on a control surface

The principal rudder types, or concepts, are summarised in Figure 2.2 and some comments on each are as follows:

(a) *Balanced rudder*: Open sternframe with a bottom pintle, which is a support bolt or pin with a bearing. The upper bearing is inside the hull. It has been applied to vessels such as tugs and trawlers and extensively to single-screw merchant ships. Tends to have been superseded by the use of the semi-balanced skeg rudder, type (d).
(b) *Spade rudder*: A balanced rudder. Both bearings are inside the hull. Bending moments as well as torque are carried by the stock, leading to larger stock diameters and rudder thickness. Applied extensively to single and twin-screw vessels, including small powercraft, yachts, ferries, warships and some large merchant ships. Also employed as control surfaces on submarines and other underwater vehicles.
(c) *Full skeg rudder*: An unbalanced rudder. The rudder is supported by a fixed skeg with a pintle at the bottom. Applied mainly to large sailing yachts, but also applied as hydroplanes on underwater vehicles.
(d) *Semi-balanced skeg rudder*: Also known as a horn rudder or a Mariner rudder, following its early application to a ship of that type [2.1]. The movable part of the rudder is supported by a fixed skeg with a pintle at the bottom of the skeg.

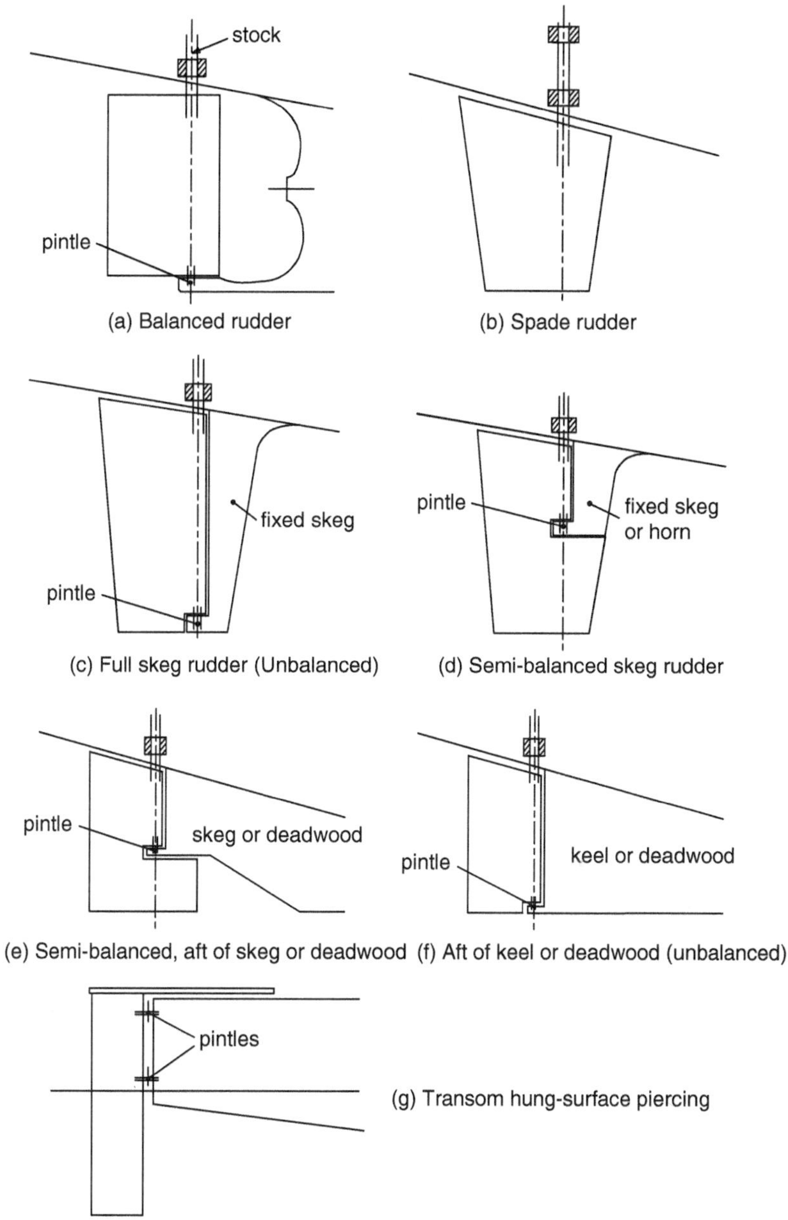

Figure 2.2 Rudder types

This pintle, at about half the rudder's vertical depth, is therefore usefully situated in the vicinity of the centre of pressure of the combined movable rudder plus skeg. Used extensively in single and twin-screw merchant ships of all sizes and some warships. In the single-screw application it is combined with an open, or Mariner, type stern arrangement.

(e) *Semi-balanced rudder, aft of skeg or deadwood*: Typically applied to twin-screw ships with a single rudder. Tends to have been superseded by the use of twin rudders of type (b) or (d).
(f) *Unbalanced, aft of keel or deadwood*: Typically applied to some older sailing craft.
(g) *Transom hung, surface piercing*: An unbalanced rudder. Typically applied to small sailing craft.

Other variants, such as twisted, flapped and high lift rudders may be considered as special cases of these principal rudder types.

In generic terms, rudders (a) and (b) in Figure 2.2 are balanced rudders, (d) and (e) are semi-balanced, whilst (c), (f) and (g) are unbalanced. An element of balance will reduce the rudder torque and reduce the size of the steering gear. It should, however, be noted that the centre of action of the rudder force tends to move with change in helm, or rudder angle and it is not possible to fully balance a rudder over a complete range of angles. "Balanced" is therefore only a broad generic term when used in the context of describing rudder types. Rudder balance is discussed in more detail in Chapter 7.

2.3 Other control surfaces

2.3.1 Fin stabilisers

Fin stabilisers are used to provide roll damping which leads to increased crew and passenger comfort and safety and reduced cargo damage. Excessive rolling tends to create more discomfort and danger than say pitching or other motions. As the energy involved in roll is much less than pitch, roll can be reduced more readily. Passive resistance to rolling is provided by bilge keels, fitted to many ships, Figure 2.3, but a more effective roll reduction can be achieved with fin stabilisers.

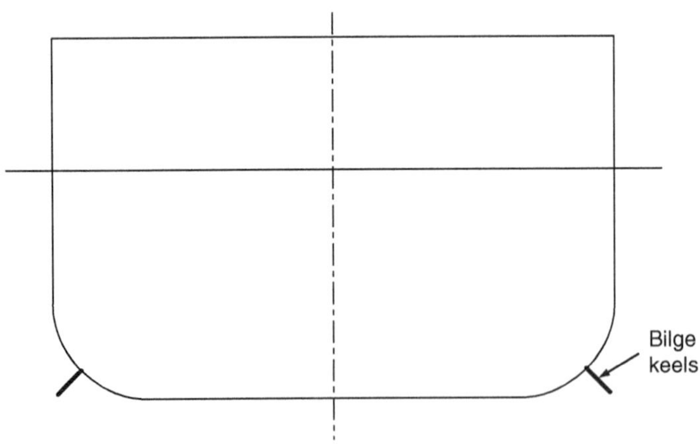

Figure 2.3 Bilge keels

The stabiliser fins are fitted at about the turn of bilge, Figure 2.4, and one or more pairs of fins may be fitted, depending on ship size and operational requirements. Aerofoil sections, flapped foils and foils with wedges at the trailing edge have been employed for fin stabilisers. The fins may be folding or retractable, or nonretractable on some smaller installations. The roll motion is sensed by a gyroscope, which feeds a controller whereby suitable incidence of the fins provides a restoring moment about the roll axis. Up to 70–80% roll reduction can be achieved depending on design and operational conditions. As they depend on forward speed to provide lift, and hence a restoring moment, fin stabilisers are most effective at higher speeds. An outline of the design procedure for stabiliser fins is given in Section 9.1.

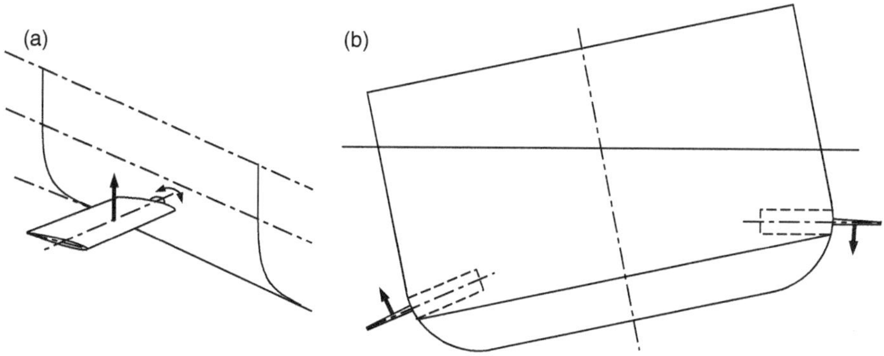

Figure 2.4 Fin stabilisers

2.3.2 Hydroplanes

Hydroplanes are control surfaces that are used to control the vertical motion of underwater vehicles and behave in the manner of a horizontal rudder, Figure 2.5. Aerofoil sections tend to be used for hydroplanes which will be all-movable or with a fixed skeg of 20–30% of chord from the leading edge, Figure 2.2(c). The design procedure tends to that of a control surface in a free stream and is discussed further in Section 9.2.

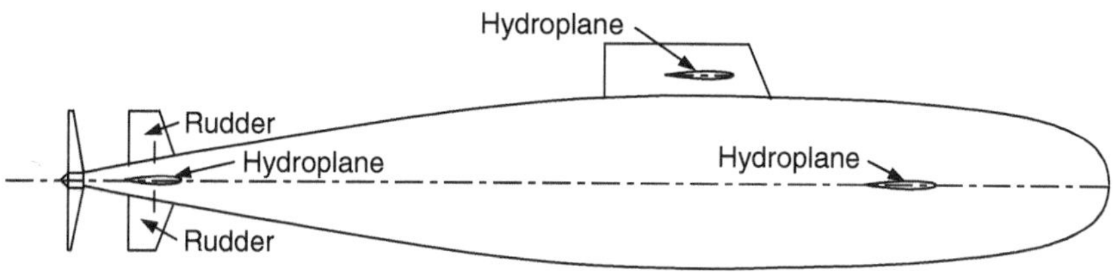

Figure 2.5 Hydroplanes: Submarine or underwater vehicle

2.3.3 Pitch damping fins

Fins are used to control the pitch motion of some higher speed commercial passenger ferries, Figure 2.6. These vessels are generally of semi-displacement form in monohull or multihull configuration. The fins tend to be of aerofoil section, all-movable and may be retractable. The design procedure is similar to that of a control surface in a free stream, taking into account the effective angle of attack due to the pitch motion, and is discussed further in Section 9.3.

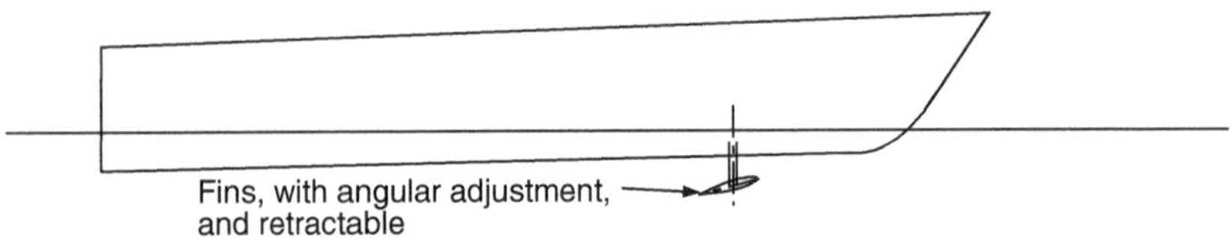

Figure 2.6 Pitch-damping fins

2.3.4 Transom flaps

Adjustable flaps, or trim tabs, at the bottom of the transom, Figure 2.7, are employed on high speed semi-displacement and planing craft to adjust the running trim and minimise the resistance to forward motion. They may also be used to provide pitch or ride control and may be used in conjunction with pitch-damping fins forward. Flap angle is typically changed using hydraulic actuators. Optimum flap angle will change with speed. Dissimilar flap angles port and starboard can be used to create a horizontal turning or steering moment on the hull. Further information on the design and applications of stern flaps is included in references [2.2–2.4].

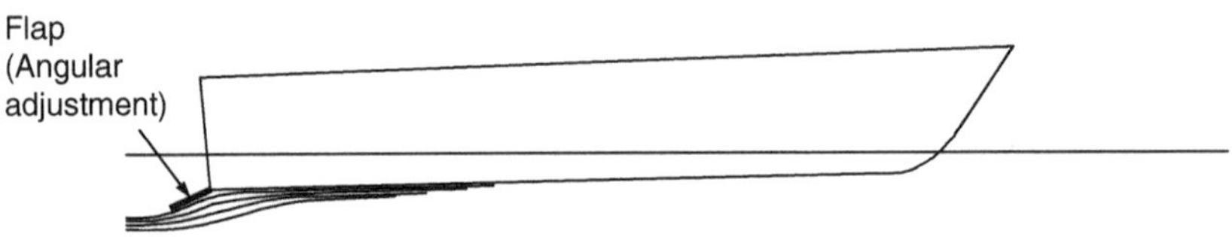

Figure 2.7 Transom flaps

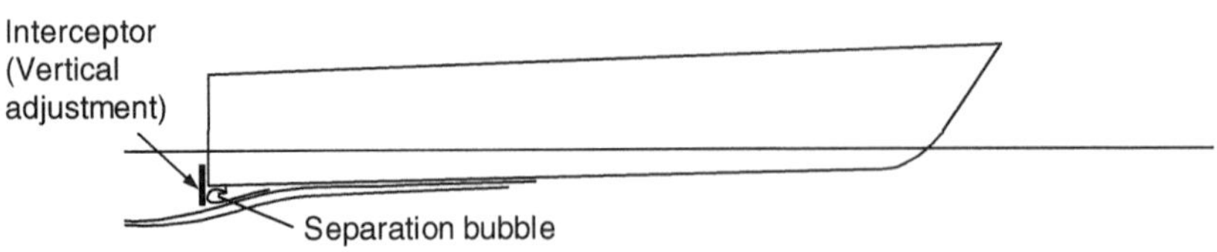

Figure 2.8 Interceptors

2.3.5 Interceptors

These are adjustable vertical plates at the bottom of the transom, normally one port and one starboard or one in each hull in the case of a catamaran, Figure 2.8.

Vertical movement is usually carried out using hydraulic actuators. Vertical adjustment can have an effect similar to a stern flap, providing adjustment to the running trim. They are also used on faster semi-displacement craft to provide pitch or ride control and may be used in conjunction with pitch-damping fins forward. Dissimilar vertical adjustment port and starboard leads to dissimilar horizontal forces on the interceptors and a horizontal turning or steering moment on the hull. Interceptors may be used as well as, or instead of, flaps. A typical application of interceptors has been to supplement the steering of vessels propelled by waterjets. Further information on the design and applications of interceptors is included in references [2.4–2.6].

References

2.1 Russo, V.L. and Sullivan, E.K. Design of the Mariner-Type ship. *Transactions of the Society of Naval Architects and Marine Engineers*, Vol. 61, 1953, pp. 98–151.

2.2 Cusanelli, D.S. and Karafiath, G. Advances in stern flap design and application. Proceedings of *Sixth International Conference on Fast Sea Transportation, FAST'2001,* Southampton, UK, September, 2001.

2.3 Cusanelli, D.S. Stern flap: An economic fuel-saving, go-faster, go-further device for the commercial vessel market. Proceedings of *Seventh International Conference on Fast Sea Transportation, FAST'2003,* Ischia, Italy, October, 2003.

2.4 Tsai, J.F., Hwang, J.L. and Chou, S.K. Study on the compound effects of interceptors with stern flap for two fast monohulls with transom stern. Proceedings of *Seventh International Conference on Fast Sea Transportation, FAST'2003,* Ischia, Italy, October, 2003.

2.5 Katayama, T., Suzuki, K. and Ikeda, Y. A new ship motion control system for high speed craft. Proceedings of *Seventh International Conference on Fast Sea Transportation, FAST'2003,* Ischia, Italy, October, 2003.

2.6 Brizzolara, S. Hydrodynamic analysis of interceptors with CFD methods. Proceedings of *Seventh International Conference on Fast Sea Transportation, FAST'2003,* Ischia, Italy, October, 2003.

3 Physics of control surface operation

3.1 Background

This chapter provides a background to basic flow patterns, terminology and definitions, together with the physics of control surface operation and rudder–propeller interaction. The depth of description is intended to provide an adequate background necessary to understand how a control surface works and to carry out practical design studies, assessments and investigations. Descriptions of fluid mechanics to a greater depth can be found in standard texts such as Massey and Ward-Smith [3.1] and Duncan *et al.* [3.2].

3.2 Basic flow patterns and terminology

3.2.1 Basic fluid properties

At an engineering level, it is sufficient to consider a fluid to be a continuous medium that will deform continuously to take up the shape of its container, being incapable of remaining in a fixed shape of its own accord.

Fluids are of two kinds:

Liquids, which are only slightly compressible and naturally occupy a fixed volume in the lowest available space within a container, and

Gases, which are easily compressed and expand to fill the whole space available within a container.

For flows at low speeds, it is frequently unnecessary to distinguish between these two types of fluid as the changes of pressure within the fluid are not large enough to cause a significant density change, even within a gas.

As with a solid material, the material within the fluid is in a state of stress involving two kinds of stress component:

(i) *Direct stress.* Direct stresses act normal to the surface of an element of material, Figure 3.1, and the local stress is defined as: stress = normal force per unit area of surface.

In a fluid at rest or in motion, the average direct stress acting over a small element of fluid is called the fluid *pressure* acting at that point in the fluid.

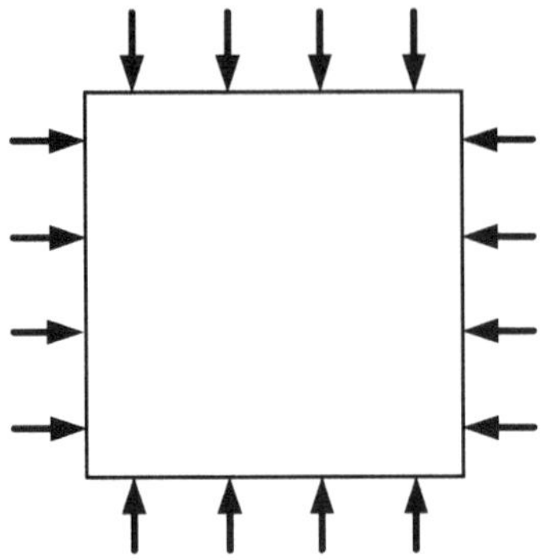

Figure 3.1 Direct stress

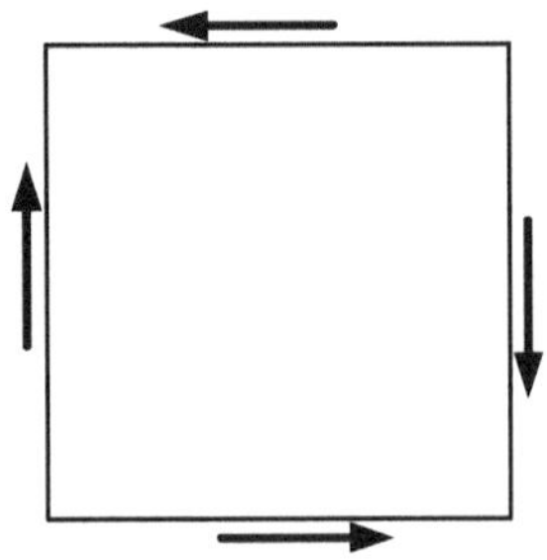

Figure 3.2 Shear stress

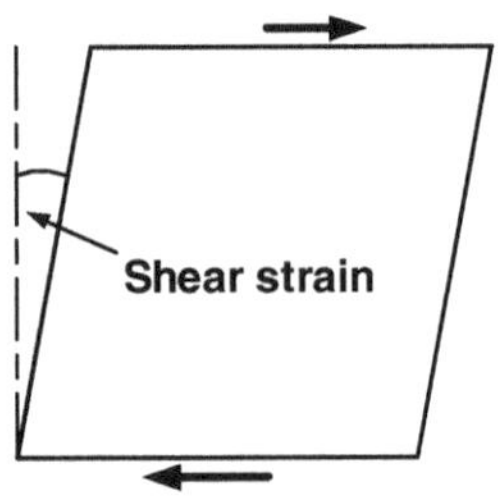

Figure 3.3 Shear strain

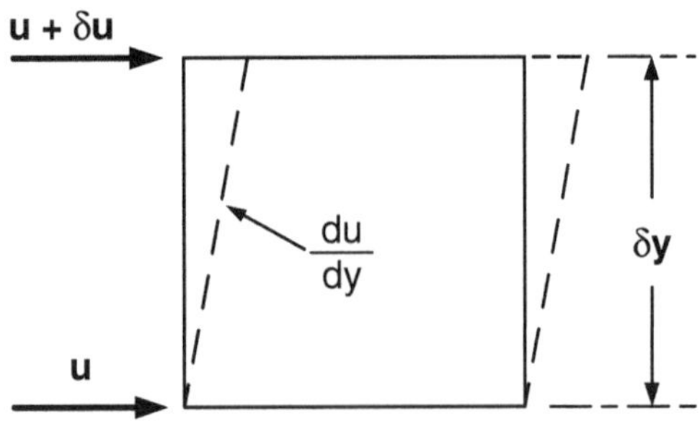

Figure 3.4 Shear stress

(ii) *Shear stress.* Shear stresses act tangentially to the surface of an element of material, Figure 3.2, and the local shear stress is defined as: shear stress = tangential force per unit area of surface.

In a fluid at rest there are no shear stresses. In a solid material, the shear stress is a function of the shear strain. In a fluid in motion, the shear stress is a function of the rate at which shear strain is occurring, Figure 3.3, that is, of the velocity gradient within the flow.

For most engineering fluids the relation, Figure 3.4, is a linear one and

$$\tau = \mu\left(\frac{\partial u}{\partial y}\right) \tag{3.1}$$

where τ = shear stress and μ = constant for that fluid.

Fluids that generate a shear stress due to shear flow are said to be viscous and the viscosity of the fluid is measured by

μ = coefficient of viscosity (or coefficient of *dynamic* viscosity)

or

$\nu = \frac{\mu}{\rho}$ = coefficient of *kinematic* viscosity

where ρ = fluid mass density

The most common fluids, for example, air and water, are only slightly viscous.

Values of kinematic viscosity and density suitable for engineering design applications are given in Tables 3.1 and 3.2. Kinematic viscosity values for fresh water, salt water and air are given in Table 3.1 and values of density for fresh water, salt water and air are given in Table 3.2.

Table 3.1 Viscosity of fresh water, sea water, and air

Temperature (°C)		10	15	20
Kinematic viscosity (m^2/s)	FW × 10^6	1.30	1.14	1.00
	SW × 10^6	1.35	1.19	1.05
(Pressure = 1 atm)	Air × 10^5	1.42	1.46	1.50

Note: FW, fresh water; SW, sea water.

Table 3.2 Density of fresh water, sea water, and air

Temperature (°C)		10	15	20
Density (kg/m^3)	FW	1,000	1,000	998
	SW	1,025	1,025	1,025
(Pressure = 1 atm)	Air	1.26	1.23	1.21

Note: FW, fresh water; SW, sea water.

3.2.2 Steady flow

In steady flow the various parameters such as velocity, pressure and density *at any point* in the flow do not change with time. In practice, this tends to be the exception rather than the rule. Velocity and pressure may vary from *point to point.*

3.2.3 Uniform flow

If the various parameters such as velocity, pressure and density do not change *from point to point* over a specified region, at a particular instant, then the flow is said to be uniform over that region. For example, in a constant section pipe (and neglecting the region close to the walls) the flow is steady and uniform. In a tapering pipe, the flow is steady and nonuniform. If the flow is accelerating in the constant section pipe then the flow will be nonsteady and uniform, and if accelerating in the tapering pipe the flow will be nonsteady and nonuniform.

3.2.4 Streamline

A streamline is an imaginary curve in the fluid across which, at that instant, no fluid is flowing. At that instant, the velocity of every particle on the streamline is in a direction tangential to the line, Figure 3.5. This gives a good indication of the flow, but only with steady flow is the pattern unchanging. The pattern should therefore be considered as instantaneous. Boundaries are always streamlines as there is no flow across them. If an indicator, such as a dye, is injected into the fluid then in steady flow the streamlines can be identified.

3.2.5 Forces due to fluids in motion

Forces occur on fluids due to accelerations in the flow. Applying Newton's Second Law:

$$\text{Force} = \text{Mass} \times \text{acceleration}$$

or

$$\text{Force} = \text{Rate of change of momentum}$$

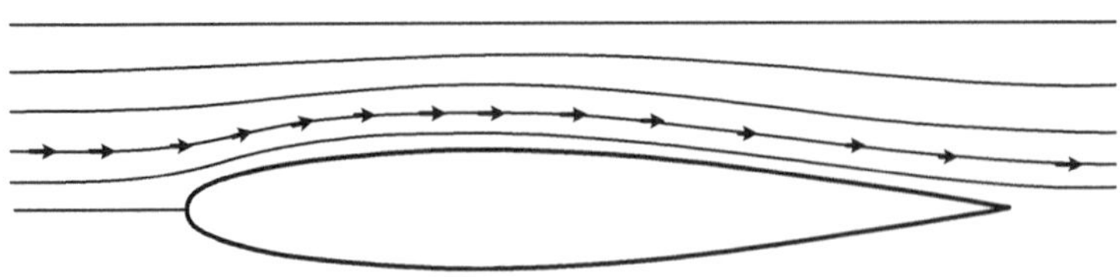

Figure 3.5 Streamlines

for a fixed mass m, all of which is moving at the same speed u, the law can be written as:

$$\text{Force} = \frac{\mathrm{d}(mu)}{\mathrm{d}t} \tag{3.2}$$

A typical application is a propeller where thrust (T) is produced by accelerating the fluid from velocity from V_1 to V_2:

$$T = \dot{m}(V_2 - V_1) \tag{3.3}$$

where $\dot{m}$ is the mass flow rate.

3.2.6 Continuity of fluid flow

Consider flow between (1) and (2) in Figure 3.6, whether in a tube, streamtube or slipstream, etc.

For no flow through the walls and constant flow rate, then for continuity:

$$\text{Mass flow rate} = \rho_1 A_1 V_1 = \rho_2 A_2 V_2 \text{ kg/s}$$

and if incompressible, $\rho_1 = \rho_2$ and

$$A_1 V_1 = A_2 V_2 = \text{volume flow rate} = Q \text{ m}^3/\text{s}$$

i.e.,

$$Q = A_1 V_1 = A_2 V_2 = \text{constant} \tag{3.4}$$

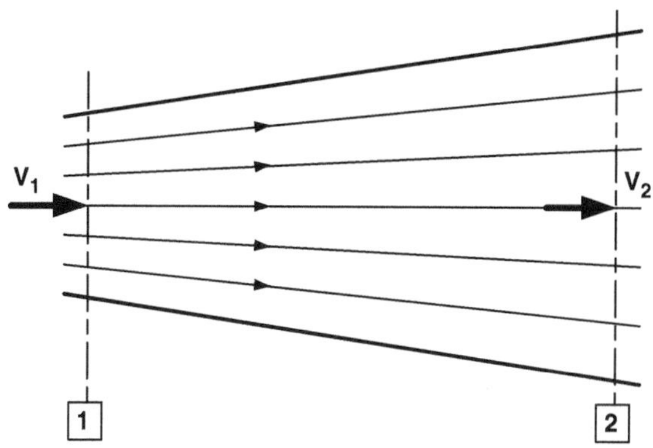

Figure 3.6 Continuity

3.2.7 Pressure and velocity changes in a moving fluid

The changes are described by Bernoulli's equation as

$$\frac{P}{\rho g} + \frac{u^2}{2g} + z = H = \text{constant (units of m)} \tag{3.5}$$

which is strictly valid when the flow is frictionless, termed inviscid, steady and of constant density.

H represents the total head, or total energy and, under these conditions, is constant for any one fluid particle throughout its motion along any one streamline.

In the equation, $P/\rho g$ represents the pressure head, $u^2/2g$ the velocity head (kinetic energy), and z the position or potential head (energy) due to gravity.

An alternative presentation of Bernoulli's equation in terms of pressure is

$$P + \frac{1}{2}\rho u^2 + \rho g z = P_T = \text{constant (units of pressure, N/m}^2\text{)} \tag{3.6}$$

and P_T = total pressure.

For example, when considering flow between two points on a streamline, Figure 3.7, then

$$P_0 + \frac{1}{2}\rho u_0^2 + \rho g z_0 = P_L + \frac{1}{2}\rho u_L^2 + \rho g z_L \tag{3.7}$$

where P_0 and u_0 are in the undisturbed flow upstream and P_L and u_L are local to the body.

Similarly, from Figure 3.7

$$P_0 + \frac{1}{2}\rho u_0^2 + \rho g z_0 = P_R + \frac{1}{2}\rho u_R^2 + \rho g z_R \tag{3.8}$$

In the case of air, its density is small relative to other quantities. Hence, the $\rho g z$ term becomes small and is often neglected.

General comment on Bernoulli's equation. Bernoulli's equation is strictly applicable to inviscid fluids. It should be noted that whilst in reality frictionless or inviscid fluids do not exist, it is a useful assumption that is often made in the description of fluid flows. If, however, Bernoulli's equation is applied to real fluids (with viscosity) it does not necessarily lead to significant errors, since the influence of viscosity in steady flow is usually confined to the immediate vicinity of solid boundaries and wakes behind solid bodies. The remainder of the flow, well clear of a solid body and termed the outer flow, behaves effectively as if it were inviscid, even if it is not so. The outer flow is discussed in more detail in Section 3.2.10.

Euler's dynamical equations for an inviscid fluid, including body forces, are described in Chapter 6. Similarly, the Navier–Stokes equations for a viscous incompressible fluid are described in Chapter 6.

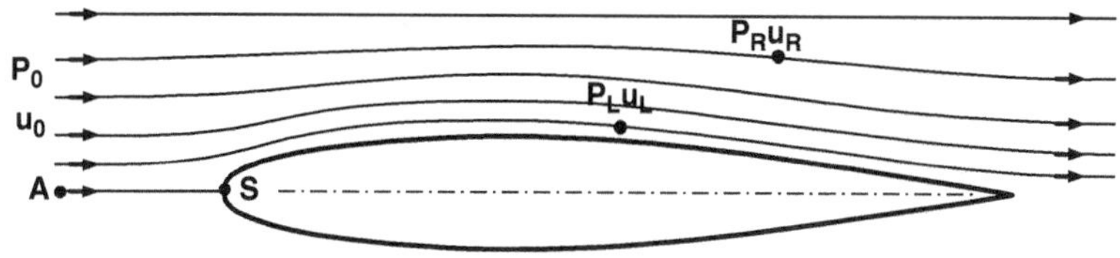

Figure 3.7 Pressure and velocity changes

3.2.8 Stagnation point

In Figure 3.7, the fluid is brought to rest at point *S*, which is known as the stagnation point. Applying Bernoulli's equation along streamline *A–S*, and assuming no change in head (z), then:

$$P_0 + \frac{1}{2}\rho u_0^2 = P_S + \frac{1}{2}\rho u_S^2$$

But $u_S = 0$
Hence,

$$P_S = P_0 + \frac{1}{2}\rho u_0^2 \tag{3.9}$$

P_0 is the background (or datum) pressure in the undisturbed flow and is referred to as the *static pressure*. P_S is the *stagnation pressure*, that is the pressure generated bringing the fluid to rest. The term $\frac{1}{2}\rho u_0^2$ is referred to as the *dynamic pressure*.

The dynamic pressure coefficient at any point in the flow with pressure P is defined by

$$C_P = \frac{P - P_0}{\frac{1}{2}\rho u_0^2} \tag{3.10}$$

and the pressure coefficient at the point *S* as

$$C_P = \frac{P_S - P_0}{\frac{1}{2}\rho u_0^2} = 1.0 \tag{3.11}$$

$C_P = 1.0$ only at the stagnation point *S*. Elsewhere, $C_P \leqslant 1.0$, noting that local u_L can exceed u_0, giving a negative C_P.

3.2.9 Boundary layer

When a slightly viscous fluid flows past a body, shear stresses are large only within a thin layer close to the body, called the boundary layer, and in the viscous wake formed by fluid within the boundary layer being swept downstream of the body, Figure 3.8. The boundary layer increases in thickness along the body length and is typically 2–3% of body length in thickness at the rear end of the body.

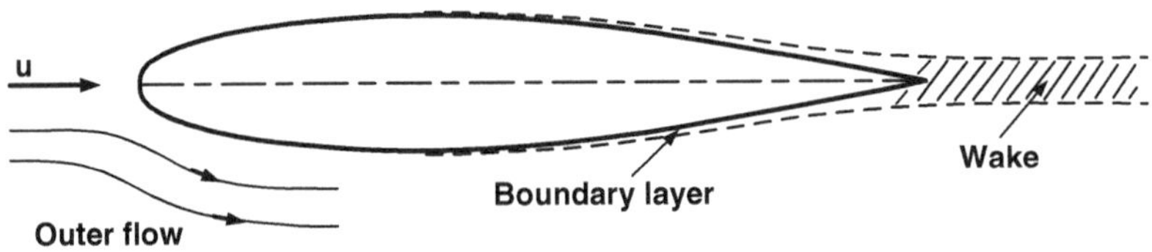

Figure 3.8 Boundary layer and outer flow

3.2.10 Outer flow

Outside the boundary layer, in the so-called outer flow in Figure 3.8, shear stresses are negligibly small and the fluid behaves as if it were totally inviscid, that is nonviscous or frictionless. In an inviscid fluid, the fluid elements are moving under the influence of pressure alone. Consideration of a spherical element of fluid shows that such pressures act through the centre of the sphere to produce a net force causing a translation motion, Figure 3.9. There is however no mechanism for producing a moment that can change the angular momentum of the element. Consequently, the angular momentum remains constant for all time and if, initially, flow started from rest, the angular momentum of all fluid elements is zero for all time. Thus the outer flow has no rotation and is termed irrotational.

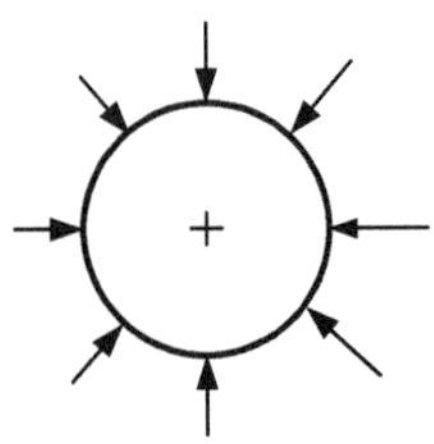

Figure 3.9 Pressure on a spherical element of fluid

3.2.11 Flow within the boundary layer

Flows within a boundary layer are unstable and a flow that is smooth and steady at the forward end of the boundary layer will break up into a highly unsteady flow, which can extend over most of the boundary layer.

Three regions can be distinguished, Figure 3.10:

1. *Laminar flow region*: In this region the flow within the boundary layer is smooth, orderly and steady, or varying only slowly with time.
2. *Transition region*: In this region the smooth flow breaks down.
3. *Turbulent flow region*: In this region the flow becomes erratic with a random motion and the boundary layer thickens. Within the turbulent region, the flow can be described by superimposing turbulence velocity components, having a zero mean averaged over a period of time, on top of a steady or slowly varying mean flow. The randomly distributed turbulence velocity components are typically ±20% of the mean velocity. The turbulent boundary layer also has a thin laminar sublayer close to the body surface. It should be noted that flow outside the turbulent boundary layer can still be smooth and steady and turbulent flow is not due to poor body streamlining as it can happen on a flat plate. Figure 3.11 shows typical velocity distributions for laminar and turbulent boundary layers. At the surface of the solid body, the fluid is at rest relative to the body. At the outer edge of the boundary layer, distance δ, the fluid effectively has the full free-stream velocity relative to the body.

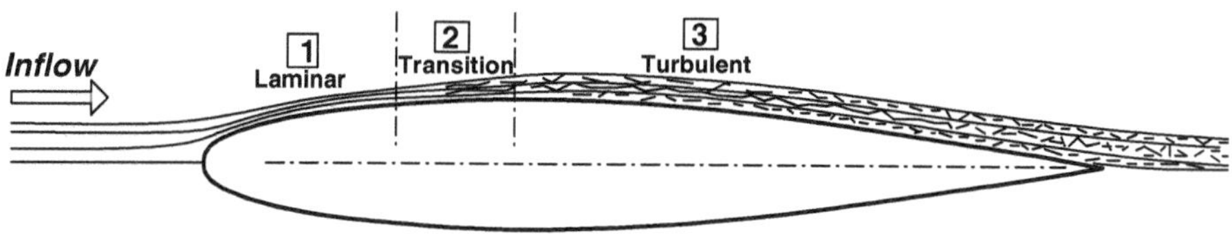

Figure 3.10 Boundary layer development

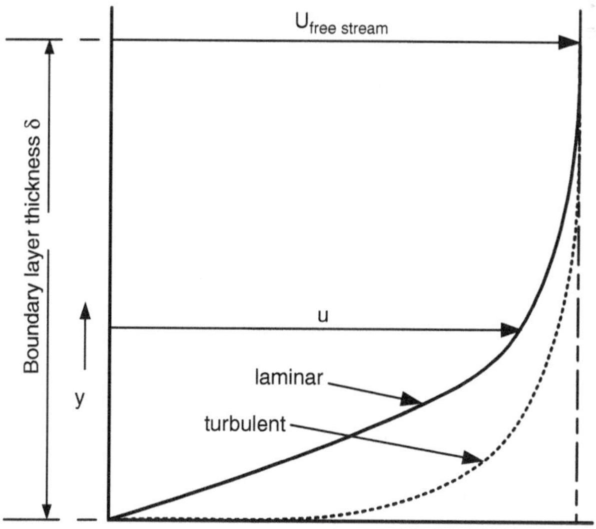

Figure 3.11 Typical boundary layer velocity profiles

The onset of the transition from laminar to turbulent flow will depend on the fluid velocity (v), the distance (l) it has travelled along the body and the fluid kinematic viscosity (ν). This is characterised by the Reynolds number (Re) of the flow, defined as

$$Re = \frac{vl}{\nu}$$

It is found that when Re exceeds about 0.5×10^6 then, even for a smooth body, the flow will become turbulent. At the same time, the surface finish of the body, for example, its level of roughness, will influence transition from laminar to turbulent flow.

Transition will also depend on the amount of turbulence already in the fluid through which the body travels. Owing to the actions of ocean waves, currents, shallow water and other local disruptions, marine rudders and control surfaces will be operating mainly in water with relatively high levels of turbulence. Consequently, their boundary layer will normally be turbulent. Only in relatively calm conditions and at low speed is a laminar boundary layer likely to be sustained.

3.2.12 Skin friction drag

The action of the frictional shear stresses on the surface of the body leads to a frictional drag force, opposing the motion of the body through the fluid. Typical boundary layer velocity profiles for laminar and turbulent flows are shown in Figure 3.11. The surface shear stress (τ_W) depends on the viscosity of the fluid (μ) and on the slope of the velocity profile, or velocity gradient, ($\partial u/\partial y$) at the surface of the body.

That is,

$$\tau_W = \mu \left(\frac{\partial u}{\partial y} \right)_{y=0} \tag{3.12}$$

and a local skin friction coefficient may be defined as

$$C_f = \frac{\tau_W}{\frac{1}{2}\rho U^2} \tag{3.13}$$

It is seen from Figure 3.11 that the velocity gradient ($\partial u/\partial y$) at the body surface is larger for the turbulent boundary layer and, consequently, the skin friction drag with a turbulent boundary layer is larger than that for a laminar boundary layer. This is illustrated for total C_F over a range of Re in Figure 3.12. At low Re laminar flow occurs with a lower C_F whilst at higher Re the flow is turbulent. In the region $Re = 10^5$–10^6 the flow may be laminar or turbulent, depending on the surface finish of the body and on the level of turbulence in the flow.

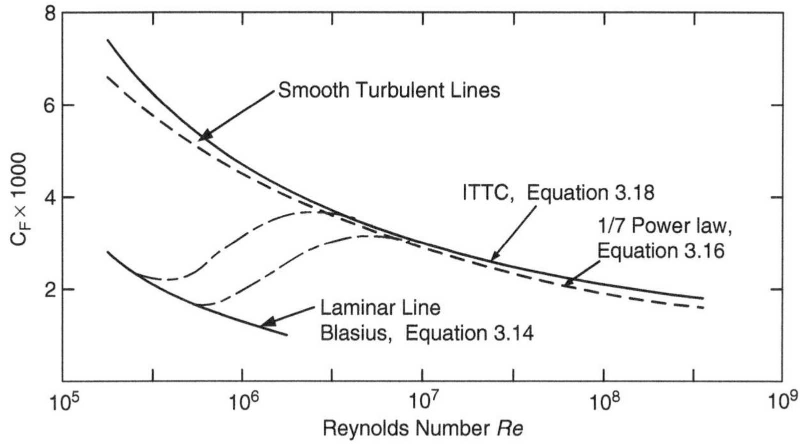

Figure 3.12 Skin friction lines

Estimates of skin friction coefficient C_F and boundary layer thickness δ may be made for laminar and turbulent boundary layers as follows:

For laminar flow, using the Blasius formulation

$$C_F = 1.328\ Re^{-1/2} \tag{3.14}$$

$$\frac{\delta}{x} = 4.91\ Re_x^{-1/2} \tag{3.15}$$

where x is distance from leading edge.

For turbulent flow, using a 1/7 power-law velocity distribution:

$$C_F = 0.072\ Re^{-1/5} \tag{3.16}$$

$$\frac{\delta}{x} = 0.370\ Re_x^{-1/5} \tag{3.17}$$

A further approximate formula, suitable for a wide range of *Re* is the ITTC skin friction correlation line:

$$C_F = \frac{0.075}{[\log Re - 2]^2} \tag{3.18}$$

3.2.13 Boundary layer displacement thickness

The boundary layer causes a reduction in flow, shown by the shaded area in Figure 3.13. The flow of an inviscid or frictionless fluid may be reduced by the same

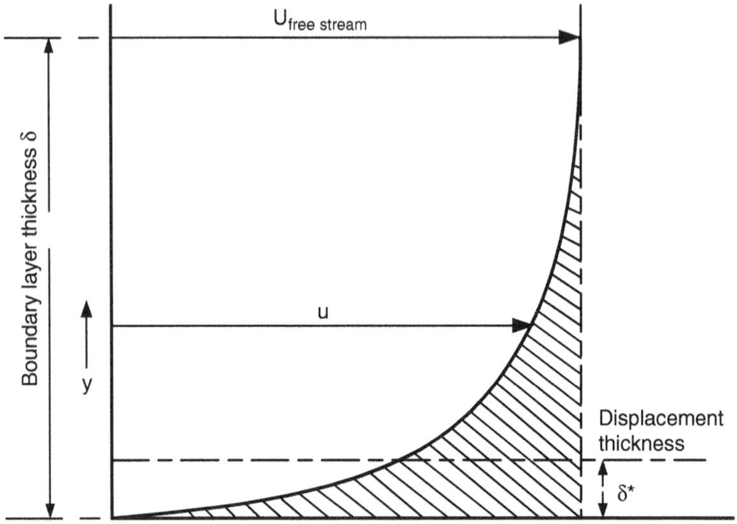

Figure 3.13 Boundary layer displacement thickness

amount if the surface is displaced outwards by the distance δ^*, where δ^* is termed the displacement thickness. The displacement thickness δ^* may be employed to reduce the effective span and effective aspect ratio of a control surface whose root area is operating in a boundary layer. This might include a rudder operating under a relatively flat hull, or a roll-stabiliser fin operating in the hull boundary layer near the bilge radius. Similarly, in theoretical simulations of fluid flow with assumed inviscid flow, and hence no boundary layer present, the surface of the body may be displaced outwards by δ^* to produce a body shape equivalent to that with no boundary layer. Approximate estimates of displacement thickness be made as follows:
Laminar flow:

$$\frac{\delta^*}{x} = 1.721 Re_x^{-1/2} \tag{3.19}$$

Turbulent flow, using a 1/7 power-law velocity distribution:

$$\frac{\delta^*}{x} = 0.0463 Re_x^{-1/5} \tag{3.20}$$

3.2.14 Drag due to surface roughness

Drag due to roughness on the surface is separation drag behind each item of roughness. With a turbulent boundary layer, the thin laminar sublayer close to the surface can smooth out roughness by flowing round small roughness without separating. Roughness only causes an increase in drag if it is large enough to project

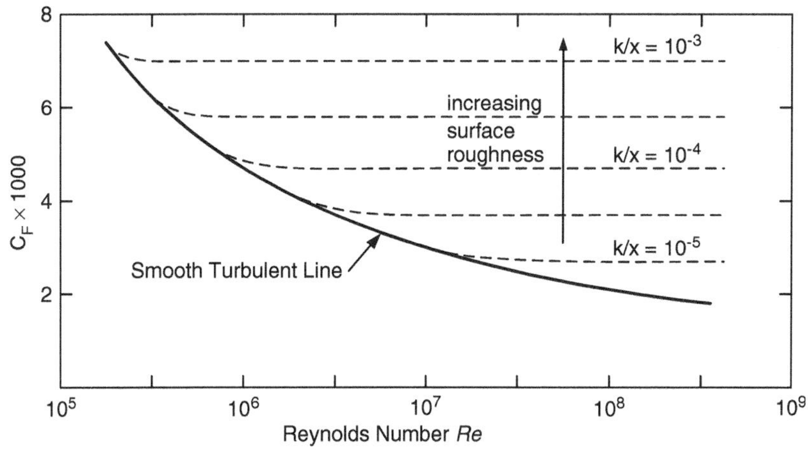

Figure 3.14 Effect of surface roughness

through the sublayer. If the surface is small enough to be entirely submerged in the laminar sublayer, then the surface is said to be hydrodynamically smooth. In this case, C_F is represented by the smooth turbulent line, Figure 3.14. As *Re* increases, say for increasing flow speed, the sublayer gets thinner and eventually a point is reached at which the drag coefficient ceases to follow the smooth turbulent line, Figure 3.14. From this point, increasing separation drag offsets falling C_F. The level of drag is now dependent only on the level of roughness, k/x, where k is the roughness height and x the distance from the leading edge. Approximate levels of drag for increase in roughness are shown in Figure 3.14. A detailed analysis would take into account the density and location of the roughness, such as that proposed by Grigson [3.3].

If a marine control surface is relatively free of corrosion and fouling, its skin friction drag can be estimated with adequate accuracy for design purposes using a smooth turbulent line such as in Figures 3.12 or 3.14.

3.2.15 Pressure drag

With flow around a curved surface, as well as skin friction drag, there is an additional drag due to the distribution of pressures over the body. The pressure drag arises from viscous effects and it is sometimes termed viscous pressure drag. Since it depends on the shape or form of the body, it is also known as form drag. Thus skin friction drag arises from the shear forces tangential to the surface τ_W and pressure drag from the resolution of the pressure forces P normal to the surface, Figure 3.15. Streamlined bodies will have a relatively small wake and the drag will be mainly skin friction, whilst a bluff body with a large wake will have a higher proportion of pressure drag, Figure 3.16. In the case of an aerofoil or control surface section, the sum of the frictional and form drag is termed the profile or section drag.

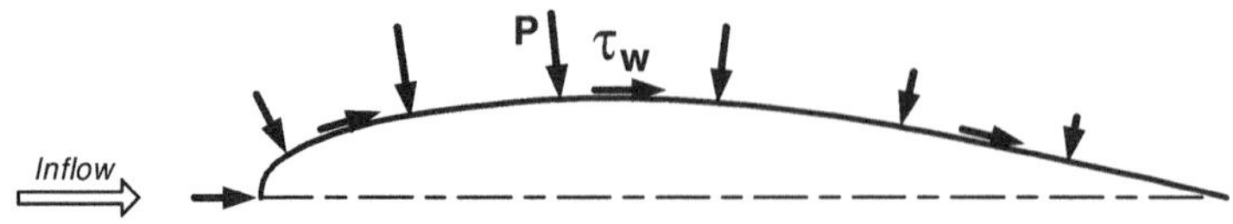

Figure 3.15 Forces on a body

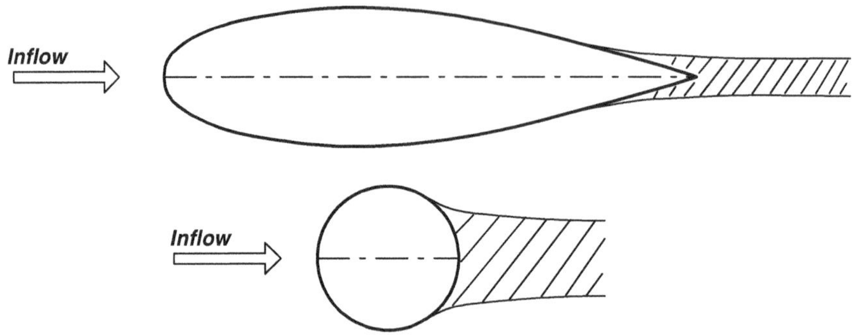

Figure 3.16 Pressure drag

3.2.16 Scale effect

Following the discussions of laminar flow, turbulent flow and Reynolds number (*Re*), it is appropriate to consider scale effect. That is, any effect as a result of scaling the forces on one size of body to another size of body, or from say a model to full scale.

For complete dynamic similarity of flow conditions, for a deeply submerged body with no surface wave effects, the *Re* has to be the same for both the model and full scale. Full-scale *Re* for control surfaces, based on chord length, range from about 5×10^5 on a small yacht rudder up to about 10^8 for a rudder on a large fast ship. An *Re* as high as 10^8 is not normally achievable at model scale. The precautions usually taken at model scale are to test at as high an *Re* as possible, that is to have a large model and high test speed, and to apply some form of turbulence stimulation near the leading edge of the body. This ensures that the model boundary layer is turbulent and hence replicates the full-scale turbulent flow. Any remaining difference in *Re* can lead to effects on the results when scaling from model to full scale. For example, it is apparent from Figure 3.12 that the full-scale skin friction drag would be overestimated if suitable corrections to the skin friction drag when going from a model *Re* of say 10^6 to a full-scale *Re* of say 10^8 are not applied. There are further possible implications for control surfaces such as a change in maximum lift and stall angle. More details of such changes due to scale effect, or effect of *Re*, are discussed later in Chapter 5.

3.2.17 Flow separation

For flow along a flat surface, with constant pressure in the direction of flow, the boundary layer grows in thickness with distance, but the flow will not separate from the surface. If the pressure is falling in the direction of the flow, termed a favourable pressure gradient, then the flow is not likely to separate. If, however, the pressure is increasing along the direction of flow, known as an adverse pressure gradient, then there is a relative loss of speed within the boundary layer. This process can reduce the velocity in the inner layers of the boundary layer to zero at some point along the body length, such as point A, Figure 3.17. At such a point, the characteristic mean flow within the boundary layer changes dramatically and the boundary layer starts to become much thicker. The flow is reversed on the body surface, the main boundary layer detaches from the body surface and a series of large vortices or eddies form behind the separation point A. Separated flows are usually unsteady, the vortices periodically breaking away into the wake downstream.

It should be noted that separation can occur in a laminar boundary layer as well as a turbulent one and indeed is more likely to occur in the laminar case. Inspection of the boundary-layer velocity profiles in Figure 3.11 indicates that the laminar layer has less momentum near the surface than the turbulent layer and is thus likely to separate earlier, Figure 3.18. Thus turbulent boundary layers are much more resistant to separation than laminar boundary layers. This leads to the result that drag due to separation is higher in laminar flow than in turbulent flow. This also explains why golf balls with dimples that promote turbulent flow have less drag and travel further than the original smooth golf balls.

It is also worth noting that a thick wake following separation should not be confused with the thickening of the boundary layer following transition from laminar to turbulent flow, described in Section 3.2.11.

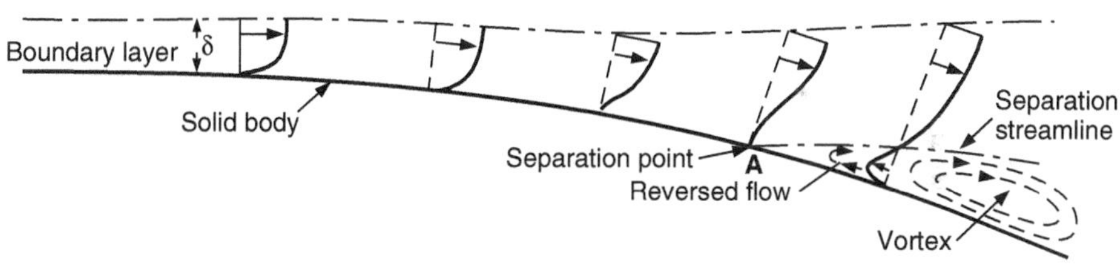

Figure 3.17 Separation

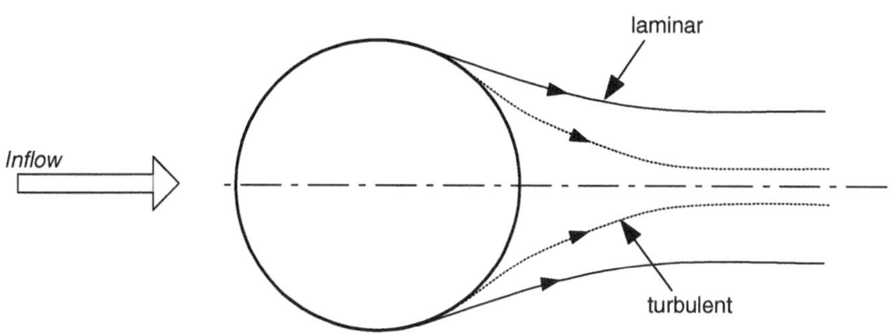

Figure 3.18 Separation drag

3.2.18 Cavitation

Devices immersed in a liquid may have their performance degraded by the formation of voids or cavities in the liquid. For example, this may occur when the suction (negative pressure) becomes high on a lifting device. Cavitation occurs when the local fluid pressure drops to the vapour pressure of the liquid, that is the pressure at which the liquid vapourises. Vapour pressure depends on temperature and the quality and content of the liquid. Cavitation can occur on marine rudders, control surfaces such as stabiliser fins, support struts and propellers. This is discussed further in Chapter 5.

3.2.19 Ventilation

Ventilation, or aeration, is a generally similar phenomenon to cavitation except that the void is filled with air drawn down from the free surface. The pressure in the void is therefore usually atmospheric, although the void may close after formation and the pressure then drops below atmospheric as the volume of the void increases. Ventilation typically may occur on surface piercing rudders, such as transom-hung rudders on yachts, and control surfaces on hydrofoil craft. Surface-piercing rudders are discussed further in Chapter 5.

3.3 Properties of lifting foils

A common fluid device is a foil designed to produce a lifting force acting across the direction of the incoming flow. The lifting action of the foil arises from the difference in the average pressure of the fluid over the upper and lower surfaces of the lifting foil, Figure 3.19. Typical applications of such lifting foils are aircraft wings, hydrofoils, rudders, stabiliser fins, propeller blades and sails.

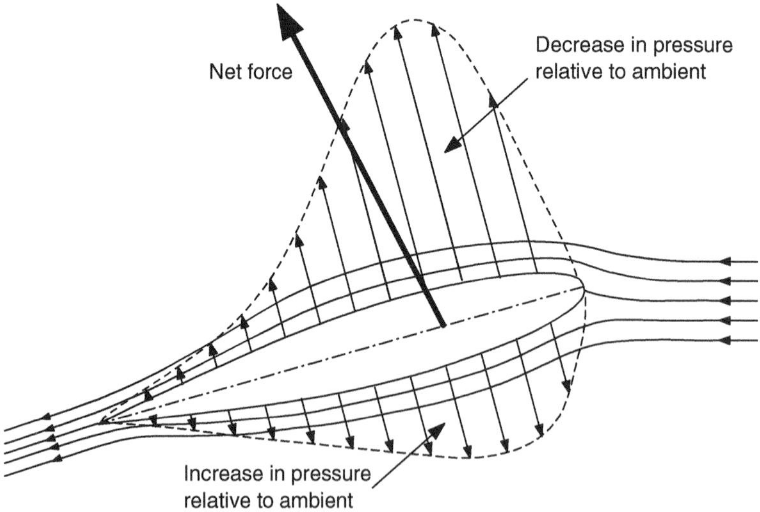

Figure 3.19 Pressures around a lifting foil

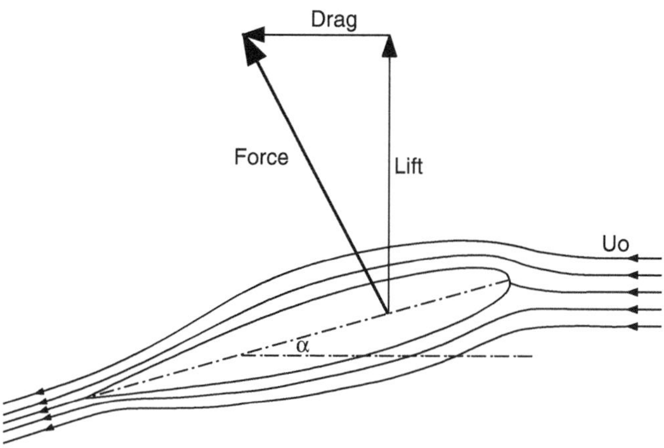

Figure 3.20 Forces on a lifting foil

Following normal practice, the total force is resolved into a *lift* component perpendicular to the fluid stream U_0 and a *drag* component parallel to U_0, Figure 3.20. These components are usually presented in terms of dimensionless coefficients for a given foil angle of attack or incidence α, where α is the angle between the foil and the direction of flow:

Lift coefficient

$$C_L = \frac{\text{Lift}(L)}{0.5\rho A U_0^2} \tag{3.21}$$

Drag coefficient

$$C_D = \frac{\text{Drag}(d)}{0.5\rho A U_0^2} \tag{3.22}$$

where A is the planform area, U_0 the fluid free-stream speed, and ρ the fluid density.

C_L and C_D depend on the foil geometry, the incidence to the incoming fluid flow and the Reynolds number (*Re*).

Geometric definitions, Figure 3.21:

Chord (c)	length of chord from leading edge to trailing edge
Span (S)	overall length of lifting surface
Plan area (A)	span × mean chord
Root chord (C_R)	length of chord at root
Tip chord (C_T)	length of chord at tip
Taper Ratio (TR)	C_T/C_R
Mean chord ($\bar{c}$)	$(C_T + C_R)/2$, or Area/Span, A/S
Aspect ratio (AR)	span/mean chord, $S/\bar{c}$, or span^2/plan area, S^2/A

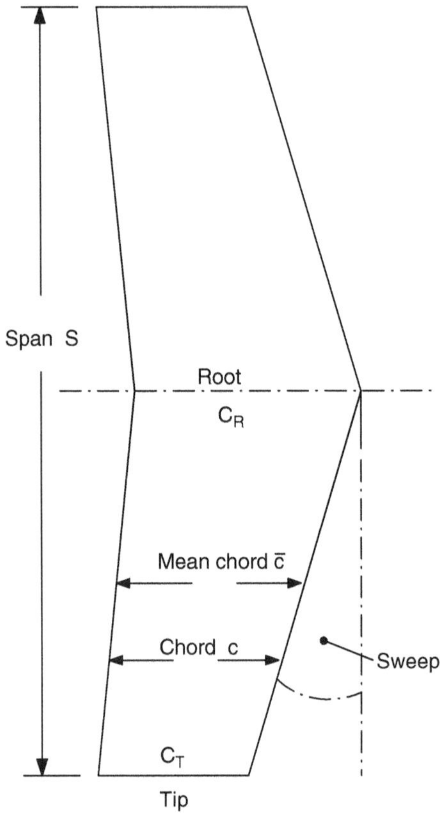

Figure 3.21 Planform of typical lifting wing

Aspect ratio for the case when a lifting surface is operating close to a flat surface, Figure 3.22. The geometric aspect ratio AR_G is S/c.

However, the flat surface behaves as a reflection plane and in this case the effective aspect ratio AR is two times the geometric aspect ratio AR_G, that is,

$$\text{Effective aspect ratio } AR = 2\,S/c, \text{ or } \quad = 2AR_G$$

It can be noted that this relationship is not strictly correct for movable foils such as a rudder below a shaped hull, or gaps between movable foils or sails and hull, although it may often be a reasonable approximation. This subject is described and discussed further in Section 5.5.

Section details: The geometry of the section is usually defined relative to the nose–tail datum line, Figure 3.23, and is generated by setting off a thickness distribution either side of a specified mean camber line. A symmetrical section, such as that used for rudders, stabiliser fins and other control surfaces, has zero camber. The section parameters are:

Camber ratio m/c, thickness ratio t/c, and nose radius ratio r_N/C.

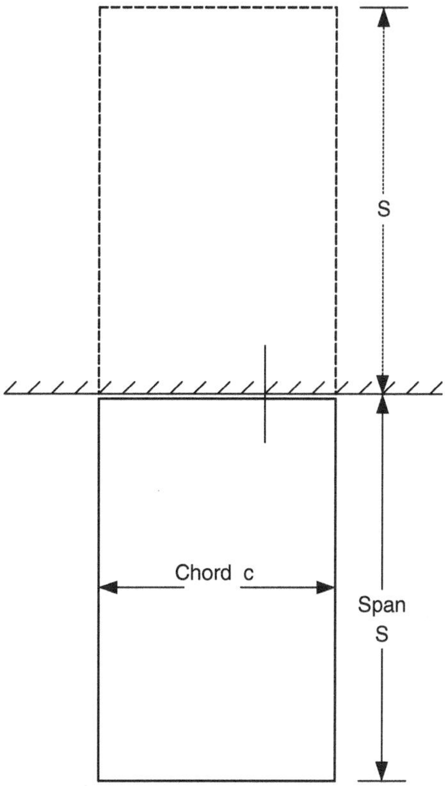

Figure 3.22 Effect of reflection plane

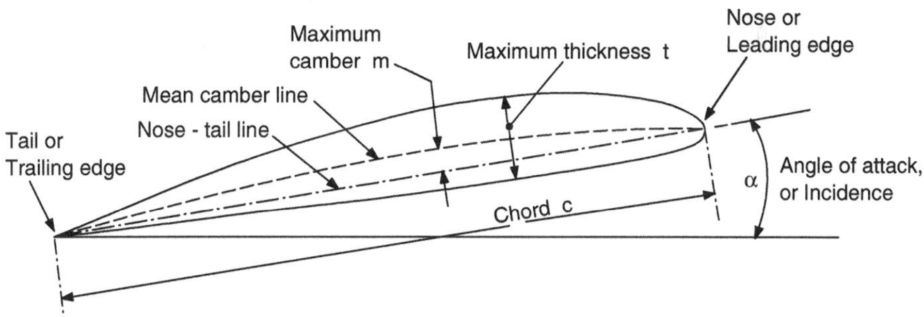

Figure 3.23 Section details

An extensive range of sections, for the National Advisory Council for Aeronautics (USA) (NACA) standard series, together with relevant two-dimensional (2-D) performance data, is available in Abbot and Von Doenhoff [3.4].

Lift and drag characteristics: Figure 3.24 shows C_L and C_D plotted against incidence α. Initially, C_L rises nearly linearly with incidence with a characteristic *lift slope* and

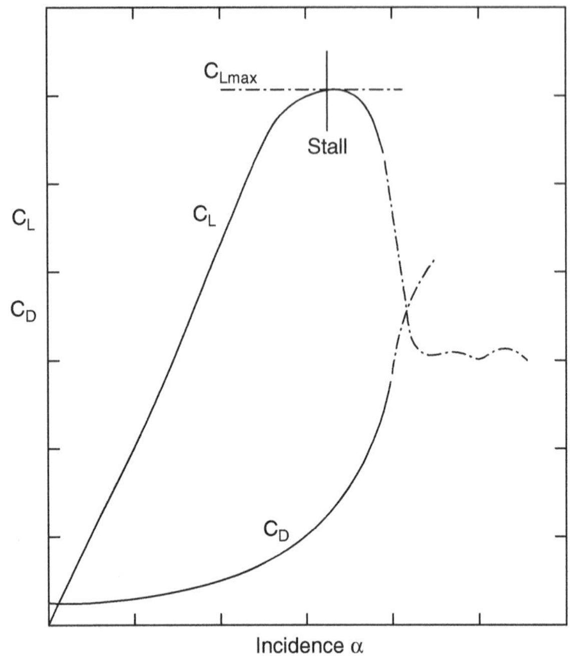

Figure 3.24 Lift and drag characteristics

then a little more steeply until a point is reached where the curve flattens and ultimately falls quite rapidly, after which the foil is said to be stalled. C_D rises approximately parabolically with incidence (see Section 3.4) until near stall when there is a significant increase in the rate of rise.

Stall: At small angles of attack or incidence, flow around the foil is smooth and attached to both the upper and lower surfaces, Figure 3.25(a). The points of separation (Section 3.2.17) of the flow on both the upper and lower surfaces are near the trailing edge. As incidence is increased, flow detaches from the upper surface, starting from the trailing edge. It can be noted, Figure 3.25(a), that the pressure on the upper side increases (is less negative) as the flow moves towards the trailing edge, known as an adverse pressure gradient as discussed under the mechanics of separation, Section 3.2.17. As incidence is increased there is a further increase in the adverse pressure gradient. The separation point moves rapidly further forward as incidence increases further, Figure 3.25(b), until the whole upper surface is in the separation zone. This process is called stall and it leads to a loss of lift and increase in drag, Figure 3.24. It should be noted that there is not a complete loss of lift after stall, as the foil is still driving off its lower pressure surface, Figure 3.19. At this stage, however, the level of lift is generally small and typically no more than about half of the maximum lift.

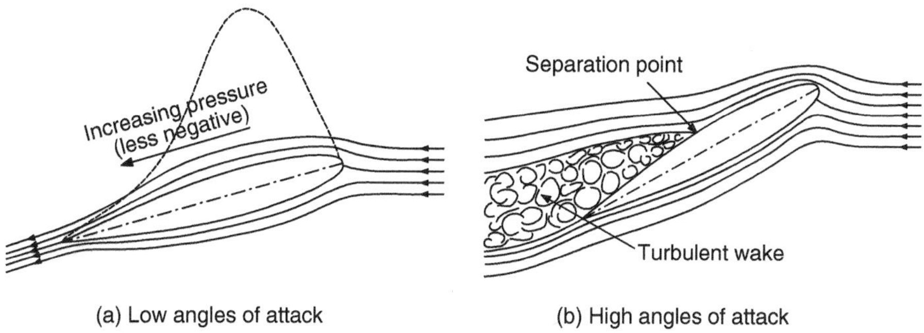

Figure 3.25 Flow over foil

3.4 Induced drag

3.4.1 Aerofoils of infinite span

With infinite span, fluid motion is 2-D and in the direction of flow perpendicular to the span. Infinite span can, for example, be simulated using a foil completely spanning a wind tunnel. Infinite aspect ratio characteristics are commonly called section characteristics, or 2-D section data. 2-D section data are associated with the shape of the section rather than the shape of the wing or control surface. Such an approach is useful for investigating the profile drag of the section, pressure distributions at various angles of attack, and for theoretical use with large aspect ratio foils when applying the characteristics of the component sections. Extensive experimental 2-D section data are available, such as those reported in Abbot and Von Doenhoff [3.4].

The profile drag of the section (C_{Do} in coefficient form) is made up of skin friction drag and form, or viscous pressure, drag. Both components originate from viscous effects.

3.4.2 Aerofoils of finite span

With infinite span, fluid motion is strictly 2-D. With finite span or finite aspect ratio, fluid motion also takes place in the spanwise direction and is 3-D. When lift is produced, there is an increase in pressure on the underside of the foil and a decrease on the upper side. Fluid pressure must equalise at the tip of a finite span foil, since a pressure difference between the upper and lower surfaces can only be sustained by the foil itself. The resulting spanwise pressure gradients force the lower surface flow out towards the tip and draw the upper surface flow in towards the centre of the span. The flow leaving the trailing edge forms a surface of discontinuity and a sheet of vortices is set up, Figure 3.26. This sheet of vortices is unstable and rolls up into a pair of concentrated trailing vortices close to the tips, as shown schematically in Figure 3.27. The core of these trailing vortices is at a reduced pressure which, when accompanied by suitable ambient conditions, can lead to visible vapour trails from the tips of aircraft wings and tip vortices in the case of marine propellers.

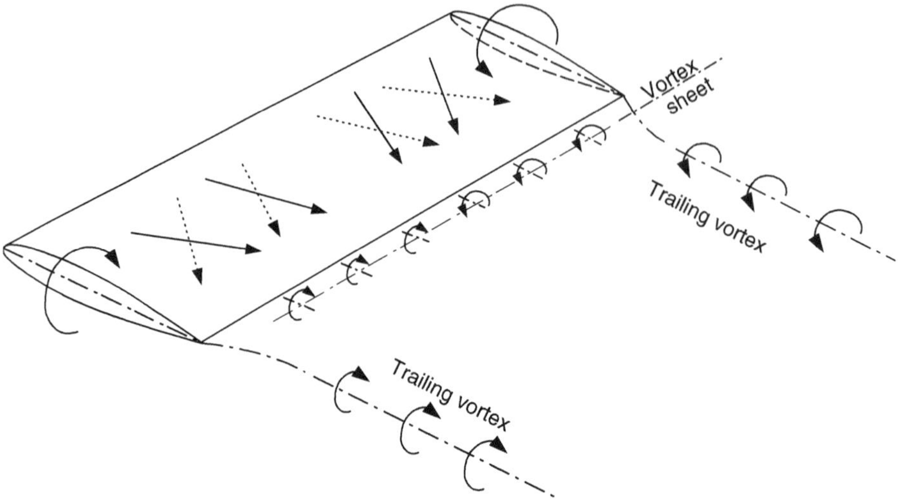

Figure 3.26 Vortex system

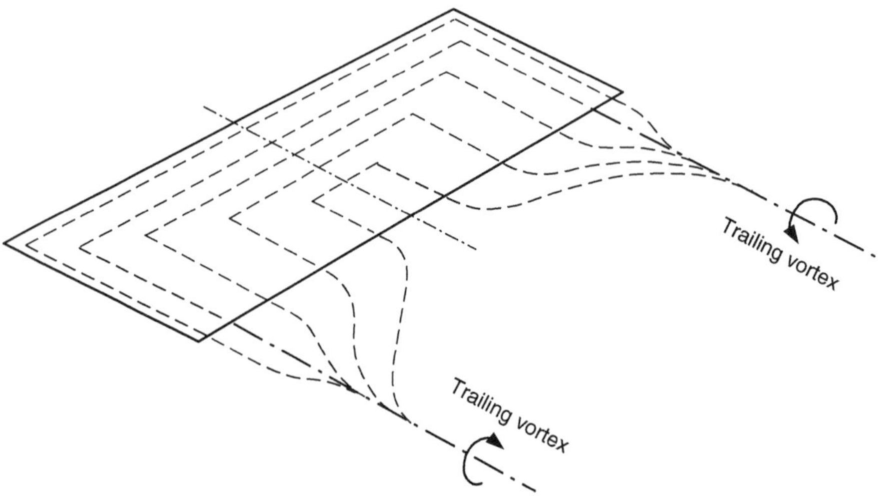

Figure 3.27 Trailing tip vortices

The general effect of the vortex system is to produce a downward flow along the foil and behind it. This induced downward flow is termed downwash. With increasing lift, the vortex grows stronger and the downwash velocity increases. When the downwash w is combined with the incoming flow V_0, the local velocity is rotated or inclined to the free stream by an angle ε. The effective angle of attack is altered and the geometric angle of attack is effectively reduced by an angle ε (Figure 3.28). The lift L_0 perpendicular to the effective angle can be resolved into lift L perpendicular to the incident flow V_0 and drag D_i parallel to V_0. D_i is termed the *induced drag* of the section and is a result of the finite span of the aerofoil.

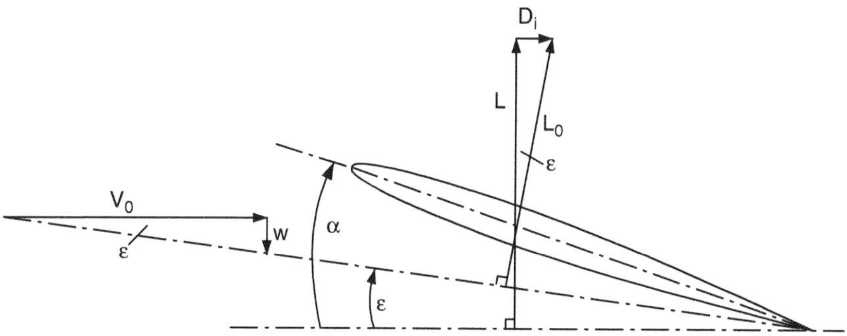

Figure 3.28 Derivation of induced drag

The aerofoil can also now be considered as one of infinite span but acting at an effective angle of attack (α–ε).

It was shown by Prandtl [3.5] using lifting line theory, that an elliptical distribution of lift leads to a constant downwash across the span and that this condition results in minimum values of downwash and induced drag as follows:

$$\varepsilon = \frac{C_L}{\pi AR} \tag{3.23}$$

and

$$C_{Di} = \frac{C_L^2}{\pi AR} \tag{3.24}$$

where C_{Di} is the induced drag coefficient.

The total drag experienced by the foil is obtained by adding C_{Di} to the 2-D section profile drag C_{Do}.

i.e.,

$$C_D = C_{Do} + k_i \frac{C_L^2}{AR} \tag{3.25}$$

where the minimum value of the induced drag factor $k_i = 1/\pi$. Actual values of k_i will be larger than $1/\pi$ and will depend on the shape of the planform characteristics, such as taper ratio and sweep. k_i values of 0.35–0.37 are more appropriate for practical design purposes and this is discussed further in Chapter 5. The total drag C_D, Figure 3.29, on a 3-D lifting surface is thus seen to be made up of profile drag C_{Do}, which is dependent on the section characteristics, and induced drag C_{Di} which is dependent on lift squared and aspect ratio. For the sake of description, this diagram has been idealised, suggesting that the profile drag is constant at all angles. In reality, other than at relatively small incidence, there will be some increase in profile drag with increase in incidence.

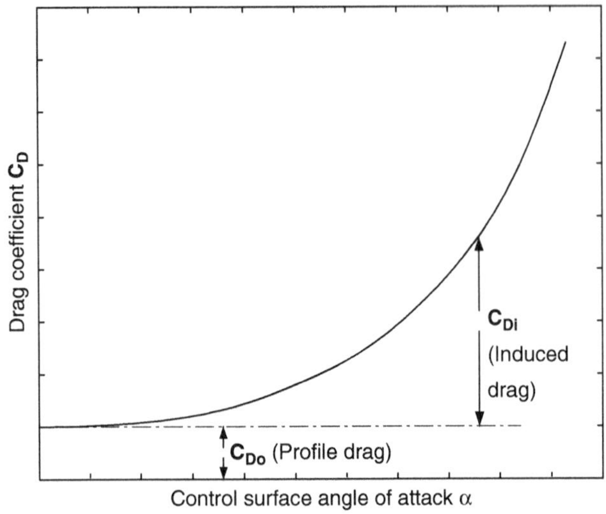

Figure 3.29 Components of drag

These results illustrate the importance of aspect ratio, in that induced drag (Equation (3.24)) will be reduced by an increase in aspect ratio. Also, the reduction in effective angle of attack (Equation (3.23)) indicates why foils of lower aspect ratio tend to stall at larger angles.

The aerofoil section behaves as if it formed part of an aerofoil of infinite span at an angle of incidence ($\alpha - \varepsilon$), and gives C_L and C_D values corresponding to 2-D motion at this incidence. Hence, if the lift curve slope a_i for a 2-D foil is written as

$$\left[\frac{dC_L}{d\alpha}\right]_{2D} = a_i \text{ (per degree)}$$

the lift coefficient C_L of a 3-D foil at incidence α is obtained as

$$C_L = a_i(\alpha - \varepsilon) = a_i\left[\alpha - \frac{57.3C_L}{\pi AR}\right] \tag{3.26}$$

and

$$\frac{dC_L}{d\alpha} = \frac{a_i}{[1 + (57.3a_i/\pi AR)]} \tag{3.27}$$

Using the theoretical 2-D lift curve slope $a_i = 2\pi$ (per rad.),

$$\frac{dC_L}{d\alpha} = \frac{2\pi}{[1 + (2/AR)]} \tag{3.28}$$

Equations (3.27) or (3.28) can be used to convert 2-D data into 3-D data. The theoretical 3-D lift curve slope tends to give values at a particular aspect ratio that are higher than those derived experimentally. Experimental lift curve slopes are discussed further in Chapter 5.

3.5 Rudder–propeller interaction

3.5.1 Physics of rudder–propeller interaction

When a ship is moving ahead the flow passing through the propeller is accelerated and rotated. The swirl and acceleration induced in the flow by the propeller alters the speed and incidence of the flow arriving at a rudder situated aft of the propeller. This controls the forces and moments developed by the rudder. These forces are important in determining the overall performance and manoeuvring characteristics of the vessel. The rudder itself both blocks and diverts the upstream flow onto and through the propeller, Figures 3.30 and 3.31, which in turn affects the thrust produced and torque developed by the propeller. In a similar manner, the rudder–propeller combination influences the sideforce developed by the hull.

The average flow generated by the propeller has axial and swirl components. The accelerated axial component increases the inflow velocity into, and the forces developed by, the rudder. The net effect of the swirl is an effective shift in local

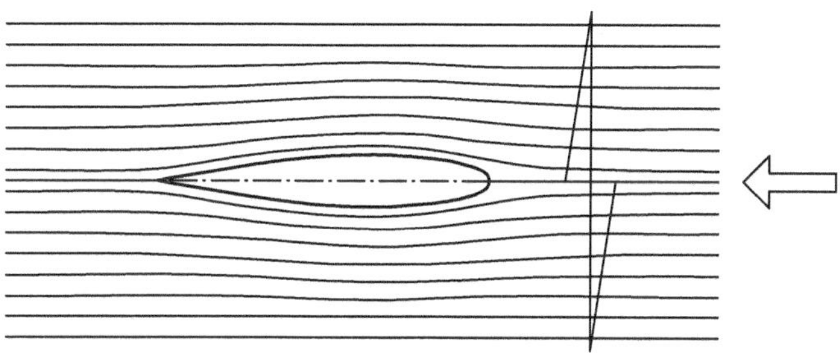

Figure 3.30 Blockage effect of rudder on propeller

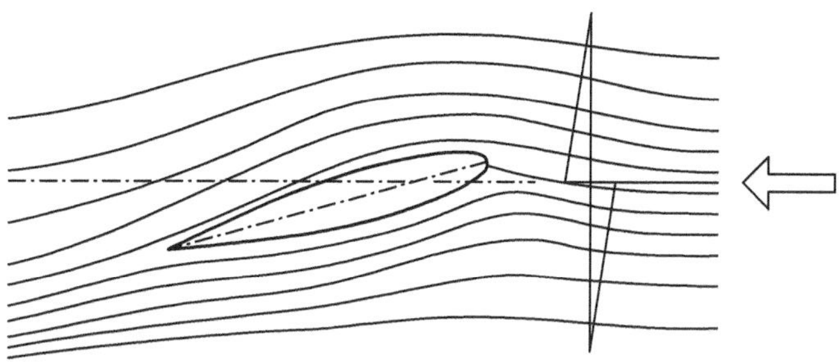

Figure 3.31 Rudder diverts flow leading to crossflow at propeller

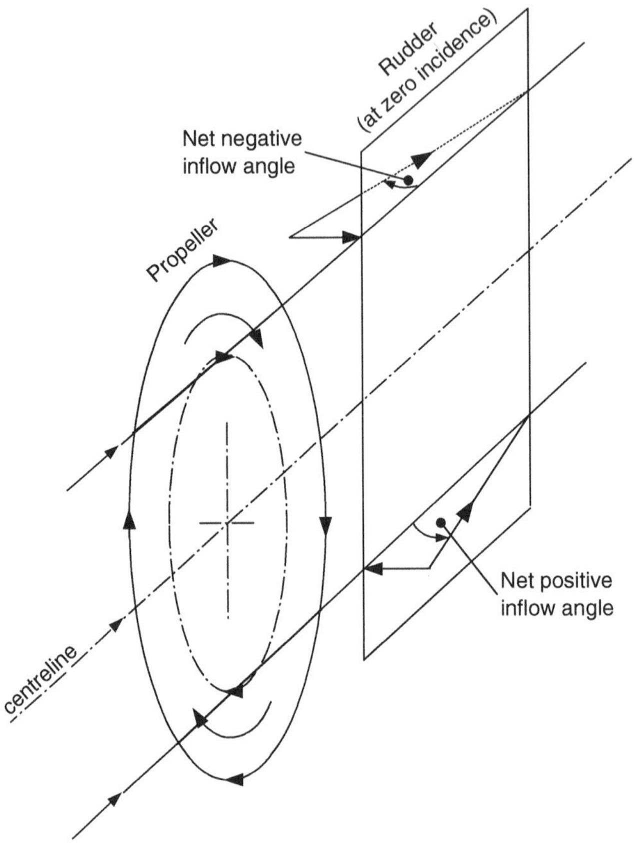

Figure 3.32 Propeller-induced rudder incidence

rudder incidence in one direction above the propeller axis and in the opposite direction below the axis, Figure 3.32. For a rudder with the same area above and below the propeller axis and at zero incidence, the normal forces will cancel, Figure 3.33(a), but the drag induced by the lift force on each side of the rudder is additional to the basic profile drag of the rudder. With high propeller thrust loadings, such as with low ship speed and high propeller revolutions, the induced angles can become large and the flow is vectored onto the rudder in such a way that the drag may become negative, or a thrust is produced by the rudder. The resolution of the forces in such a case is illustrated in Figure 3.33(b), showing the development of a negative rudder axial force. Such rudder propulsive effects are discussed further in Section 5.9. With say a tapered rudder, or a rudder for which the propeller tip protrudes below the tip of the rudder, the lift forces port and starboard will not necessarily cancel and the effect on the rudder is an angular offset in the rudder performance. Similar effects arise from variation in lateral and longitudinal rudder–propeller separation. These effects are discussed further in Section 5.4.

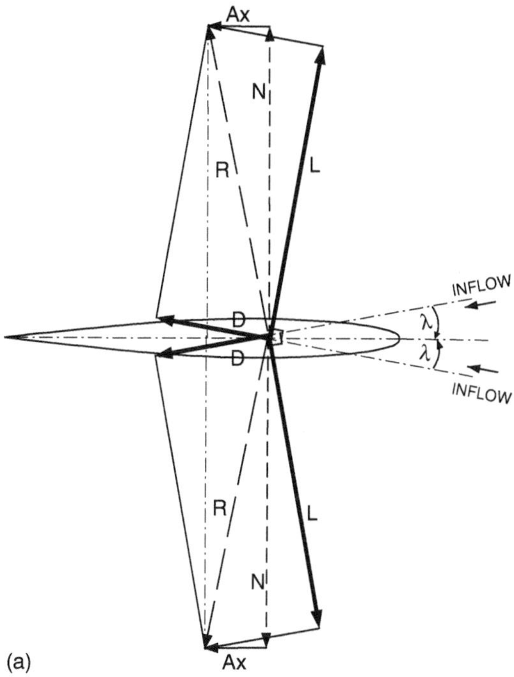

(a)

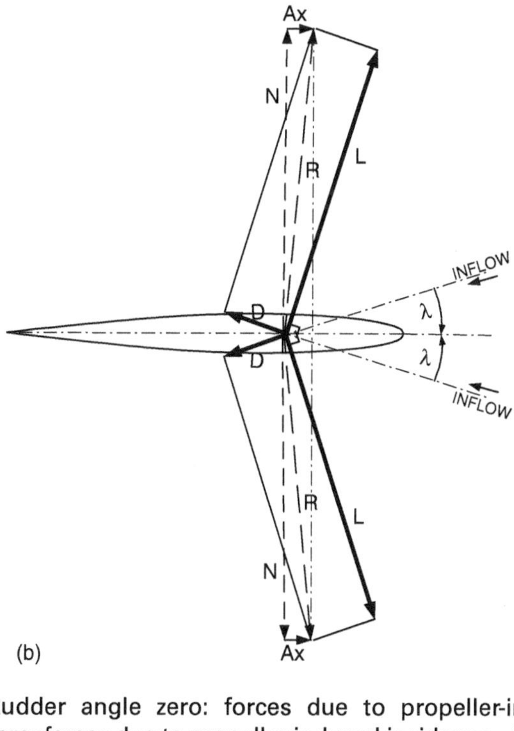

(b)

Figure 3.33 (a) Rudder angle zero: forces due to propeller-induced incidence; (b) Rudder angle zero: forces due to propeller-induced incidence – high thrust loading

3.5.2 Physical parameters governing rudder–propeller interaction

When addressing the rudder–propeller interaction problem it is necessary to identify the various independent parameters on which the rudder forces depend. Free surface effects and cavitation are not included at this stage as they are not considered fundamental in determining the normal operating conditions of the propeller and rudder. It is convenient to group the parameters into four categories and these groups of parameters can then be used to assess their effect on the rudder and propeller performance characteristics. This will include the effects on lift, drag, stall and centre of pressure in the case of detailed rudder design and manoeuvring and, for ship resistance and propulsion, the total thrust of the rudder–propeller combination (propeller thrust – rudder drag). The groups are as follows:

(i) *Flow variables*: These control the magnitude of the forces developed and include the time-dependent quantities V (free-stream velocity) and n (propeller rate of revolution) and the properties of the fluid, density (ρ) and dynamic viscosity (μ). Also included is the yaw angle β between the rudder–propeller combination and the free stream.

(ii) *Rudder geometric variables*: These determine how the flow passes over the rudder and hence the force developed. This is controlled by the rudder incidence α, span S, mean chord c, stock position X_1, thickness t, section shape, taper ratio, sweep and twist.

(iii) *Propeller geometric variables*: These control how the propeller imparts energy into the flow, accelerates the flow and generates thrust. This is determined by its diameter D, pitch P, blade area ratio, boss diameter, section shape, skew, pitch and thickness distributions and number of blades.

(iv) *Relative position and size of the rudder and propeller*: The two units can be separated longitudinally (X), laterally (Y) and vertically (Z). The relative size is defined as the coverage ξ and is equal to the proportion of the rudder span in way of the propeller race. This can be expressed as $\xi = \lambda\ D/S$ where λ is the fraction of the propeller diameter impinging on the rudder.

The flow parameters in nondimensional form become Reynolds number Re and advance ratio J. The remaining three categories consist solely of geometric properties.

If the basic geometric properties of the rudder and propeller are assumed fixed, the lift and other dependent variables can then be expressed as a function of the following nondimensional variables:

$$C_L = f\left\{[J, Re, \beta], \left[\alpha, AR, \frac{X_1}{c}\right], \left[\frac{X}{D}, \frac{Y}{D}, \frac{Z}{D}, \xi\right]\right\} \qquad (3.29)$$

And for a particular rudder–propeller combination, the parameters reduce to

$$C_L = f\left\{[J, Re, \beta], [\alpha], \left[\frac{P}{D}\right]\right\} \qquad (3.30)$$

It follows that, for a given ship, modelling of rudder–propeller interaction should consider the above parameters. From these, the two fundamental controlling parameters (for a fixed pitch propeller) are the propeller advance ratio J and rudder incidence α. *Re* and yaw angle will be of less importance. Axial momentum theory, discussed in Section 3.6, indicates that the propeller-induced velocities are a function of propeller thrust loading (K_T/J^2) which is independent of P/D and this a parameter commonly used in rudder–propeller interaction studies.

The geometrical groups of parameters (rudder, propeller and position) will determine the magnitude of the influence of J and α on the performance of the rudder and propeller.

3.6 Propeller-induced velocity upstream of rudder

An approximation to the velocity in a propeller slipstream may be made using axial momentum theory. The propeller is assumed to be many bladed, or termed an actuator disc, and be capable of imparting axial motion to a fluid, thus producing a thrust on the disc.

Consider an actuator disc of diameter D and area A_1 advancing through an undisturbed fluid at speed V_0, Figure 3.34. The motion of the fluid is imparted uniformly across the disc area and the flow is accelerated from V_0 upstream to V_1 at the disc and V_2 far downstream. Outside the slipstream and upstream and downstream the pressure is at P_0. The total energy is increased abruptly at the disc, with a pressure rise from P_1 to P_1'. The speed at a rudder whose leading edge is distance X downstream of the propeller disc may be developed as follows:

$$\text{Thrust } T = \text{rate of change of momentum} = \rho A_1 V_1 (V_2 - V_0) \qquad (3.31)$$

also,

$$T = \text{area of disc} \times \text{pressure change over disc} = A_1(P_1' - P_1)$$

using Bernoulli's equation, upstream, $P_0 + 0.5\rho V_0^2 = P_1 + 0.5\rho V_1^2$ and downstream,

$$P_0 + 0.5\rho V_2^2 = P_1' + 0.5\rho V_1^2$$

hence

$$T = 0.5\rho A_1(V_2^2 - V_0^2) \qquad (3.32)$$

from Equation (3.32),

$$V_2 = \left[V_0^2 + \frac{2T}{\rho A_1} \right]^{1/2} \qquad (3.33)$$

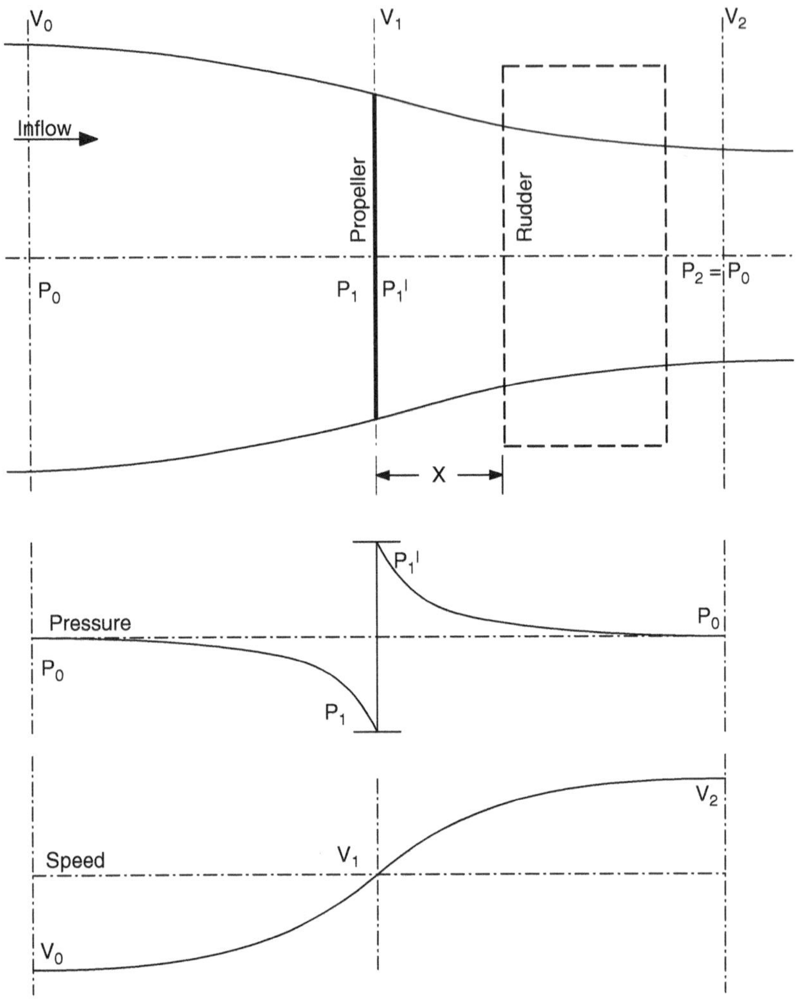

Figure 3.34 Actuator disc

and it can be shown that

$$\frac{2T}{\rho A_1} = \frac{8K_T}{\pi J^2} V_0^2 \tag{3.34}$$

where $K_T = T/\rho n^2 D^4$ and $J = V_0/nD$
hence,

$$V_2 = V_0 \left[1 + \frac{8K_T}{\pi J^2}\right]^{1/2} \tag{3.35}$$

also, when $V_0 = 0$ and $J = 0$, such as in the bollard pull (static thrust) condition,

$$V_2 = \left[\frac{8K_T}{\pi} n^2 D^2\right]^{1/2} \tag{3.35a}$$

Equating (3.31) and (3.32), $V_1(V_2 - V_0) = 0.5(V_2^2 - V_0^2) = 0.5(V_2 + V_0)(V_2 - V_0)$ and

$$V_1 = \frac{[V_2 + V_0]}{2} \tag{3.36}$$

(i.e., half of the acceleration has occurred at the propeller disc)

A Gutsche-type correction [3.6] can be applied to V_1 to account for the fluid acceleration between the propeller and rudder, leading to the velocity at the rudder V_R (rather than V_2, which is far downstream). This amounts to a simple correction based on the distance of the rudder from the propeller (X/D) and can be represented as

$$K_R = 0.5 + \frac{0.5}{[1 + (0.15/(X/D))]} \tag{3.37}$$

noting that when $X/D = 0$, $K_R = 0.5$ and when $X/D = \infty$, $K_R = 1.0$

the added increment of velocity $= K_R(V_2 - V_0)$

$$= K_R \left\{V_0\left[1 + \frac{8K_T}{\pi J^2}\right]^{1/2} - V_0\right\}$$

$$= K_R V_0 \left\{\left[1 + \frac{8K_T}{\pi J^2}\right]^{1/2} - 1\right\}$$

final velocity at rudder V_R = original velocity + added increment

$$= V_0 + K_R V_0 \left\{\left[1 + \frac{8K_T}{\pi J^2}\right]^{1/2} - 1\right\}$$

and, finally

$$V_R = V_0\left[1 + K_R\left\{\left[1 + \frac{8K_T}{\pi J^2}\right]^{1/2} - 1\right\}\right] \tag{3.38}$$

It is seen from Equation (3.38) that the propeller-induced velocity arriving at the rudder is a function of the propeller thrust loading K_T/J^2 (or thrust per unit area, Equation (3.34)) and distance X between the propeller and rudder (on which K_R depends). With increase in propeller thrust loading there is an increase in induced velocity.

As an example, if a rudder is situated 0.9 m downstream of a 6.0 m propeller that is operating at a J value of 0.7 and a $K_T = 0.1$, then $K_T/J^2 = 0.204$, $K_R = 0.75$ and $V_R = 1.175\ V_0$, that is a 17.5% increase over the speed of advance of the propeller.

It should be noted that Equation (3.38) is based solely on axial momentum changes and does not include frictional effects, the effect of a finite number of blades or rotational and radial velocities induced by the propeller, all of which will have an effect on the magnitude and incidence of the flow arriving at the rudder. Such theoretical predictions can overestimate the induced speed by as much as 30% compared with actual values, particularly at higher thrust loadings. Estimates of the differences are included in example application 4 in Chapter 11.

Approximate empirical propeller slipstream corrections for semi-displacement and planing vessels have been proposed by Hatch [3.7] and Du Cane [3.8]. These provide approximate predictions at the preliminary design stage when the detailed propeller characteristics and K_T/J^2 are not known.

Low rev./large diameter propeller up to 500 rpm:

$$V_R = 1.15 V_A \tag{3.39}$$

Faster, smaller propellers up to 1500 rpm:

$$V_R = 1.25 V_A \tag{3.40}$$

High-speed craft:

$$V_R = 1.20 V_S \tag{3.41}$$

where V_A is wake speed (Section 3.7), equivalent to V_0 in Figure 3.34, and V_S is ship speed.

Contraction of slipstream:
Continuity of flow requires that $A_1 V_1 = A_2 V_2 = A_R V_R$
hence,

$$A_R = \frac{A_1 V_1}{V_R} \tag{3.42}$$

and

$$D_R = D\left[\frac{V_1}{V_R}\right]^{1/2} \tag{3.43}$$

using Equations (3.35), (3.36) and (3.38),

$$\frac{D_R}{D} = \frac{\left[\left[1 + (8K_T/\pi J^2)\right]^{1/2} + 1\right]^{1/2}}{\left[2\left\{1 + K_R\left\{\left[1 + (8K_T/\pi J^2)\right]^{1/2} - 1\right\}\right\}\right]^{1/2}} \tag{3.44}$$

Equation (3.38) can be used as a first approximation to determine the mean velocity in the slipstream of the propeller and Equation (3.44) can be used as a first approximation to determine the diameter of the slipstream, hence the proportion of the rudder covered by the slipstream. The remaining portion(s) of the rudder outside the slipstream would be treated as if the speed is the upstream inflow velocity V_0. This is an idealised and approximate approach and is outlined in Figure 3.35(a). In reality, inflow to the propeller is nonuniform and the propeller-induced velocity is nonuniform over the blade radius as indicated in Figure 3.35(b).

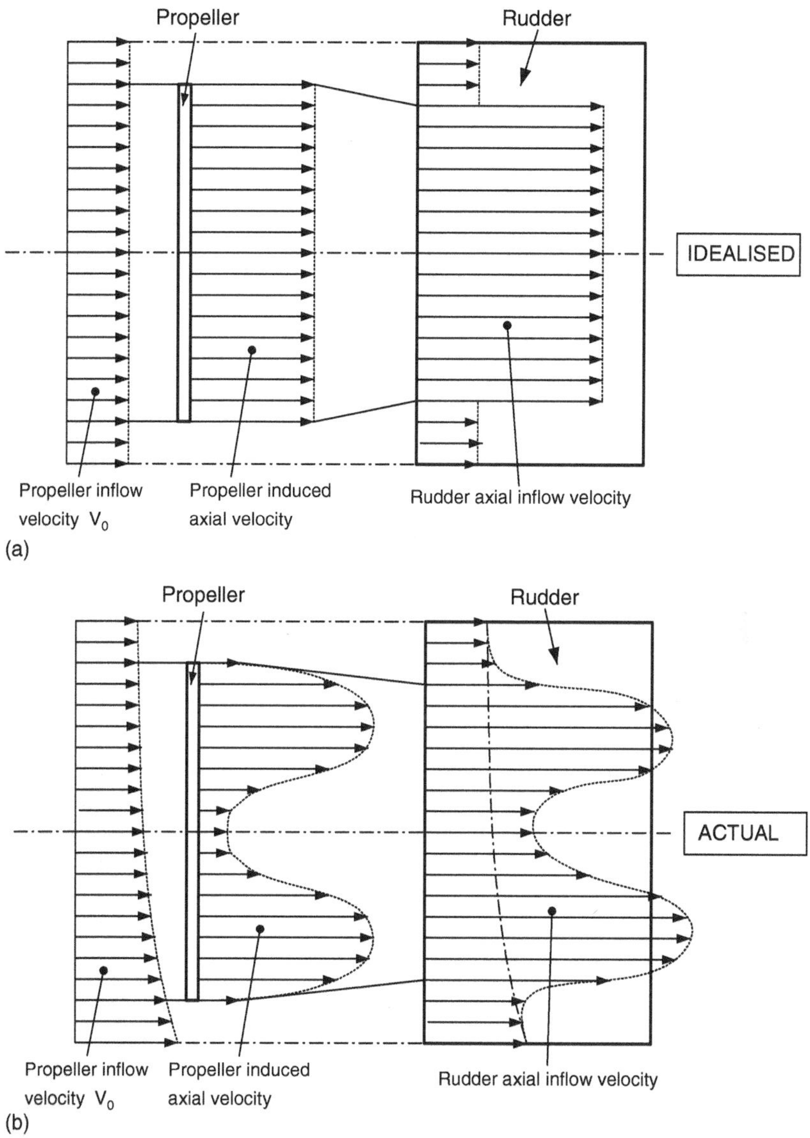

Figure 3.35 (a) Idealised and (b) Actual rudder axial inflow velocity

Variations on the above equations, incorporating suitable empirical modifications to Equation (3.38), are used in many mathematical models of ship manoeuvring, such as those described by Mikelis [3.9], Kose [3.10] and Inoue *et al.* [3.11].

3.7 Influence of hull on rudder–propeller performance

The hull upstream of an isolated rudder or propeller–rudder combination has two main effects:

(i) Owing to potential flow and boundary-layer effects, the hull slows down the inflow to the propeller and/or rudder to a speed V_A compared with the free stream or ship speed V_S, Figure 3.36.
(ii) The hull affects the inflow angle when on a turn; namely, it has flow-straightening effects. Flow-straightening effects on to the rudder from both the hull and propeller are discussed further in Section 5.4.2.7.

The wake or effective speed V_A into the propeller and/or rudder can be estimated using a suitable wake fraction w_T for that particular ship or boat type, where the wake or inflow speed as defined as

$$V_A = V_S(1 - w_T) \tag{3.45}$$

The value of w_T can be as low as 0.05 for fine high-speed forms and up to 0.450 for large full-form tankers. At the design stage, an average wake fraction (approximately in way of the propeller disc) will normally be estimated using empirical data. For example, for typical single-screw merchant ships, an approximate estimate of wake fraction may be made from:

$$w_T = 1.10 - 3.4C_B + 3.3C_B^2 \quad (C_B = 0.60 - 0.80) \tag{3.46}$$

where C_B is the block coefficient.

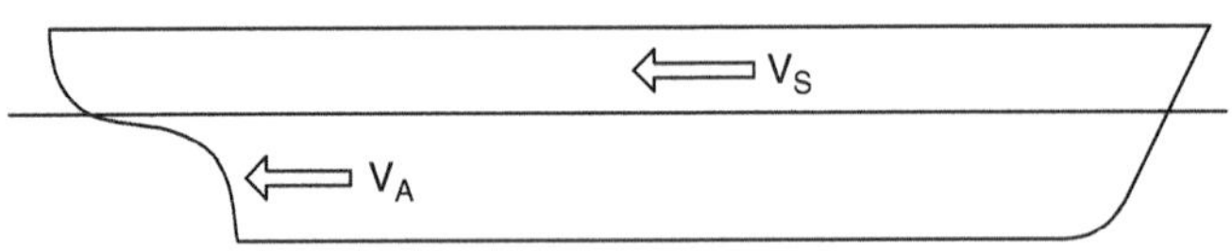

Figure 3.36 Wake speed V_A

For small single-screw ships and sailing craft, Barnaby [3.12] suggests:

$$w_T = 0.80C_B - 0.26 \quad (C_B = 0.40 - 0.60) \tag{3.47}$$

Values for typical twin-screw merchant and naval ships will normally be less than those for single-screw ships. For twin-screw ships, an approximate estimate of wake fraction may be made from:

$$w_T = 0.71 - 2.39C_B + 2.33C_B^2 \quad (C_B = 0.55 - 0.75) \tag{3.48}$$

For small twin-screw vessels, Barnaby [3.12] suggests:

$$w_T = 0.60C_B - 0.24 \quad (C_B = 0.40 - 0.55) \tag{3.49}$$

Further data and methods for estimating wake fraction may be obtained from Carlton [3.13] and Harvald [3.14] for a wide range of ship forms, and Bailey [3.15] for round bilge fast craft semi-displacement forms.

References

3.1 Massey, B.S. and Ward-Smith J. *Mechanics of Fluids*, 7th edition. Stanley Thornes Ltd., Cheltenham.

3.2 Duncan, W.J., Thom, A.S. and Young, A.D. *Mechanics of Fluids*. Edward Arnold, Australia.

3.3 Grigson, C.W.B. The drag coefficients of a range of ship surfaces. *Transactions of the Royal Institution of Naval Architects*, Vol. 123, 1981, pp. 195–208.

3.4 Abbot, I.H. and Von Doenhoff, A.E. *Theory of Wing Sections*. Dover Publications, New York, 1958.

3.5 Prandtl, L. Applications of modern hydrodynamics to aerodynamics. *NACA Report No. 116*, 1921.

3.6 Gutsche, F. Die induktion der axialen strahlzusatz-geschwindigheit in der umgebung der schraubenebene. *Schiffstecknik*, Heft 12/13, 1955.

3.7 Hatch, G.N. *Creative Naval Architecture*. Thomas Reed Publishing, London, 1971.

3.8 Du Cane, P. *High Speed Craft*. Temple Press Ltd., London, 1951.

3.9 Mikelis, N.E. A procedure for the prediction of ship manoeuvring response for initial design. *ICCAS '85*. North-Holland, Trieste, 1985.

3.10 Kose, D. On a new mathematical model of manoeuvring motions of a ship and its applications. *International Shipbuilding Progress*, Vol. 29, No. 336, August 1982, pp. 205–220.

3.11 Inoue, S., Hirano, M., Kijima, K. and Taskashina, J. A practical calculation method of ship manoeuvring motion. *International Shipbuilding Progress*, Vol. 28, No. 325, 1981, pp. 198–223.

3.12 Barnaby, K. *Basic Naval Architecture.* Hutchinson, London, 1963.

3.13 Carlton, J.S. *Marine Propellers and Propulsion.* Butterworth Heinemann, Oxford, 1994.

3.14 Harvald, Sv.A.A. *Resistance and Propulsion of Ships.* A. Wiley Interscience, New York, 1983.

3.15 Bailey, D. The NPL high speed round bilge displacement hull series. *Maritime Technology Monograph No. 4*, The Royal Institution of Naval Architects, London, 1976.

4 Control surface requirements

4.1 Rudder requirements

4.1.1 Requirements

The rudder has to be able to develop sufficient sideforce to maintain the ship on a straight course at typical service speeds, to change course at service speed, and to manoeuvre at slower speeds. In the case of coursekeeping, interest is centred on minimizing deviations from the set course. In the cases of changes in course or manoeuvring, interest centres on the ease and rapidity with which a ship takes up a new course. The effectiveness of the rudder in these two situations will depend on the directional stability of the ship.

4.1.2 Directional stability

A ship is said to be directionally stable if, when deflected from its straight-line path, by say wind or waves, it returns to a new straight-line path, although this will not necessarily be in the same direction as the original path. A high measure of directional stability will result in good coursekeeping but low ability to manoeuvre, whilst a measure of directional instability will result in poor coursekeeping but good ability to manoeuvre. A compromise will normally have to be accepted, depending on the duties of the vessel under consideration. For example, a service vessel working in a port will need good stopping and manoeuvring ability whilst a large sea-going ship that covers large distances without manoeuvring will primarily require good coursekeeping. Directional instability is not a desirable property, as repeated rudder corrections generally have to be applied to maintain a course. Also, a vessel with low directional stability will readily enter a turn but may be slow to respond to reversed rudder angle in order to leave the turn.

The directional stability of a vessel will depend on several features including the fineness or fullness of the form, fine forms having higher directional stability, the operating draught and trim with an increase in trim by the stern increasing directional stability, and the amount of deadwood or fixed skeg area aft, increases in such area leading to an increase in directional stability. Increases in directional stability lead to a decrease in manoeuvring performance and, if increases are needed, the best compromise is reached if directional stability is increased by increasing movable rudder area, if this is possible, rather than increasing fixed skeg or deadwood area [4.1].

4.1.3 Rudder action

To turn the ship, the rudder must be capable of placing and holding the ship at an angle of attack to the flow of water past the hull. Hydrodynamic forces then develop on the hull, which largely turn the ship, the rudder now providing only a small turning effect. The amount of turning effect due to the rudder depends on the ship type and hull form, with the rudder on a full-form tanker or bulk carrier with low directional stability contributing very little whilst on a fine-form vessel such as a container ship with larger directional stability, the rudder will continue to assist the turn by a small amount. The angle between the tangent to the path and the ship centreline is termed the drift angle, Figure 4.1.

The forces acting on the ship during a steady turn are shown in Figure 4.2 and comprise the propeller thrust T, the hull and rudder forces F_H and F_R and the centrifugal force on the ship, $\Delta V^2/R$, which opposes the centripetal acceleration. F_H is the resultant of the hull forces due to the angle of attack of the hull and acts at the centre of lateral resistance, CLR. In the steady state, the centrifugal and rudder forces are balanced by the hydrodynamic forces on the hull. As the steady state is reached, the hull force F_H moves aft of the centre of gravity G until the moments due to the rudder, $F_R \times x_R$ and the hull, $F_H \times x_H$, are in equilibrium. It is seen from Figure 4.2 that the ship pivots about the pivot point P, which is typically 15–35% of ship length from the bow. To an observer on board, the ship appears to be pivoting about P, the bow swinging inwards and the stern outwards. When first developing drift, F_H is well ahead of G, and the effective vertical turning axis of the ship is forward of amidships. This provides a larger lever for the rudder action and leads to a greater steering efficiency when the rudder is situated at the stern compared with, say, a rudder situated at the bow. Bow rudders as well as stern rudders may however be installed in some ships, such as ferries that have to operate and steer efficiently in both directions.

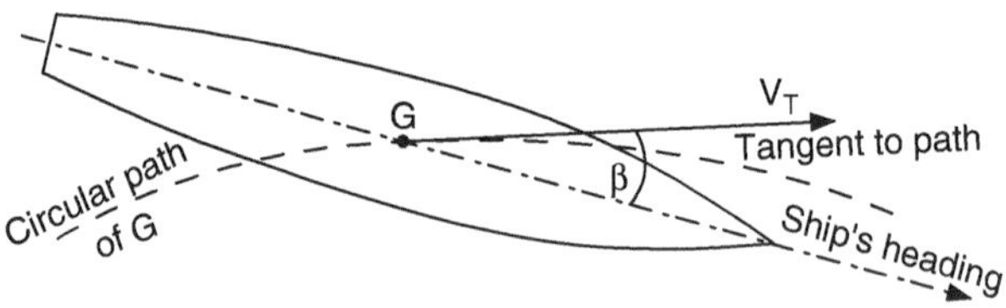

Figure 4.1 Drift angle β

4.1.4 Heel angle due to rudder

An undesirable effect of the rudder is that it produces heel. This arises because the transverse force on the rudder generally acts well below the roll centre and thus

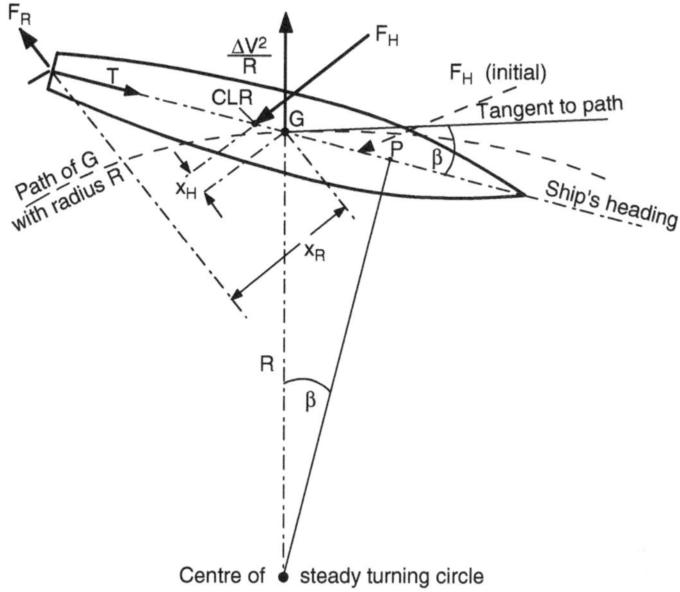

Figure 4.2 Forces acting on a turn

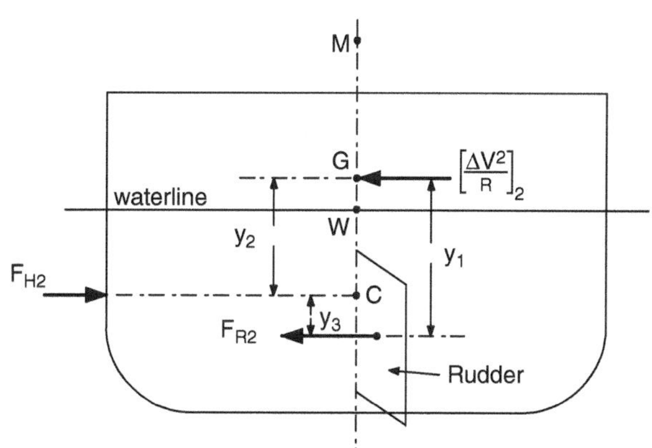

Figure 4.3 Heel due to rudder

imparts roll as well as yaw moments on the ship, Figure 4.3. Assume that the hull, rudder and centrifugal forces F_H, F_R and $\Delta V^2/R$ in Figure 4.2 are resolved perpendicular to the ship's longitudinal axis and are denoted F_{H2}, F_{R2} and $(\Delta V^2/R)_2$, as shown in Figure 4.3.

Initially, when the helm is first applied, say to starboard, $(\Delta V^2/R)_2$ and F_{H2} are small and the ship will heel inward to starboard due to the rudder moment $F_{R2} \times y_1$.

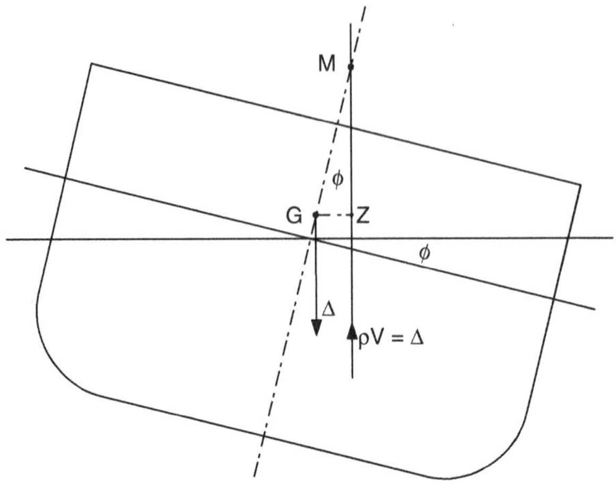

Figure 4.4 Hydrostatic restoring moment

Taking moments about G, then for an angle of heel ϕ, the resulting moments may be written as

$$F_{R2} \times y_1 = \Delta gGM \sin\phi + F_{H2} \times y_2 \tag{4.1}$$

where $\Delta gGM \sin\phi$ is the hydrostatic restoring moment at angle ϕ, Figure 4.4.

Since F_{H2} is initially small, a first estimate of inward heel due to the rudder may be made from

$$F_{R2} \times y_1 = \Delta gGM \sin\phi \tag{4.2}$$

When the transverse centrifugal force $(\Delta V^2/R)_2$ builds up and the steady turn has developed, there is an outward heeling moment to port. Taking moments about C, the centre of lateral resistance (CLR), through which the resultant hull force F_{H2} acts, and assuming relatively small angles of heel, the resulting moments may be written as

$$(\Delta V^2/R)_2 \times y_2 \cos\phi = \Delta gGM \sin\phi + F_{R2} \times y_3 \tag{4.3}$$

If the cosines of the rudder angle, drift angle and heel angle are taken as 1.0, then a first approximation to the steady-heel angle on the turn may be made from

$$[\Delta V^2/R] \times y_2 = \Delta gGM \sin\phi + F_{R2} \times y_3 \tag{4.4}$$

Approximate estimates of the heel due to the rudder are included in example application 18 in Chapter 11.

In the steady turn and for relatively small angles of heel, the horizontal forces are

$$F_{H2} - F_{R2} = \Delta V^2/R \tag{4.5}$$

It should be noted that whilst the above equations explain the physics of the roll moments, the position of the CLR changes with change in heel and drift angles, and any estimates of heel angle will be only approximate.

It can also be added that as a result of the heel moment produced by the rudder, dangerous combinations of heeling moments can occur if on a fast ship the helm is decreased or port helm is applied when in the process of a starboard turn.

The heeling moment produced by a rudder can also be seen and applied as a stabilising moment in roll and the rudder used as a roll stabiliser. The viability of such an application for the rudder is discussed in Section 9.1.7.

4.1.5 Measures of manoeuvrability

4.1.5.1 *Turning circle*

The turning circle, which is relatively easy to measure at model-scale or on full-scale trials, is used as a measure of a vessel's turning ability and an indication of the efficiency of the rudder. The geometry of the turning circle is shown in Figure 4.5. The rudder is put over to a specified angle and held whilst the ship completes a circle. If say starboard rudder is applied, the initial effect is for the ship to drift to port before turning about the vertical axis, after which the bow starts to swing to starboard, eventually settling on a steady curved path. At this stage, the centrifugal and hydrodynamic forces are balanced and the radius of curvature R becomes constant. Typical measurements include advance, transfer, steady turning radius,

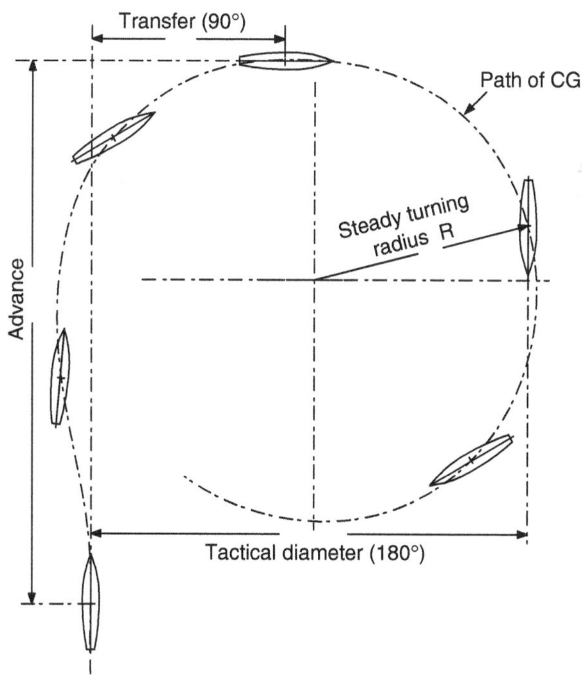

Figure 4.5 Turning circle

steady drift angle and tactical diameter, together with ship steady heel angle and speed. Tactical diameters may vary from about 4 to 7 ship lengths for merchant ships and down to 3.5 for naval vessels, whilst the steady speed on the turn may be as low as 60% of the approach speed.

Shallow water has an adverse effect on rudder effectiveness and steering efficiency. Turning ability is reduced and the diameter of the turning circle is increased.

4.1.5.2 *Zig–zag test*

This test is used to assess the rate of response of the ship to the rudder. On a steady approach, the rudder is put over to +20° and the ship allowed to turn. When the ship heading has changed by 20° the rudder is put over to –20° and held whilst the ship achieves a 20° heading in the other direction, when the rudder is again reversed and the procedure repeated. Alternatively, a 10°/10° zig–zag test may be used. Typical results from the procedure are shown in Figure 4.6. Interest is centred on the heading overshoot at helm reversal and the period of the cycle. These values can be compared with previous vessels, noting that larger values indicate an inferior response of the ship to change in helm.

4.1.5.3 *Spiral test*

This test is used to assess the directional stability or instability of the ship. On a steady approach, the rudder is put over to +15°, and the ship allowed to settle to a steady rate of change of heading and the rate recorded. The helm is then reduced to +10°, the ship allowed to settle, and the rate of change of heading again recorded. The process is repeated for +5°, −5°, −10° and −15° and the whole procedure then repeated back to +15°. The results obtained from the test are plotted as rate of change of heading against rudder angle, as shown in Figure 4.7. In the case of a directionally stable ship, the curve is a single function, but for an unstable ship the curve has the appearance of a hysteresis loop. The larger the loop the more unstable the vessel. For small rudder angles, the rate of change of heading depends on whether the rudder angle is increasing or decreasing.

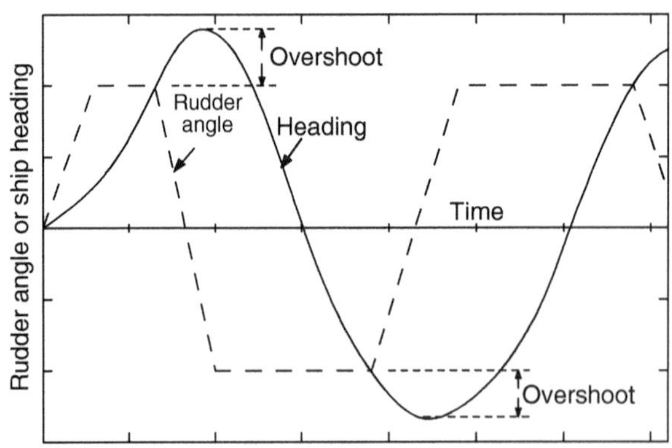

Figure 4.6 Zig–zag test

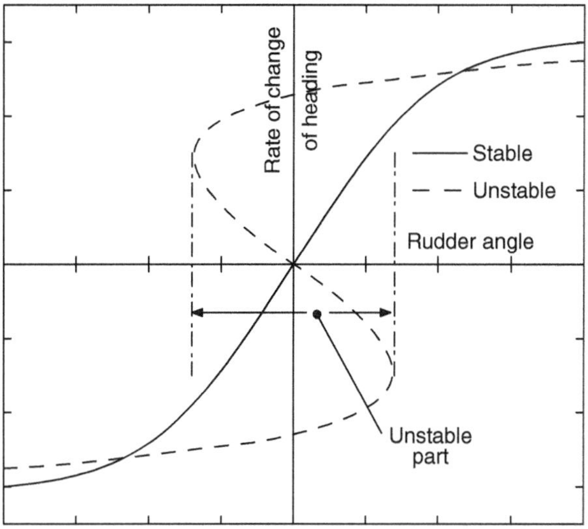

Figure 4.7 Spiral test

A thorough review of standard manoeuvring tests is presented by Burcher [4.1]. ITTC recommendations for ship manoeuvring trials can be found in reference [4.2]. International standards for ship manoeuvrability are contained in the IMO Standards for Ship Manoeuvrability [4.3, 4.4], and include requirements for turning ability, initial turning ability, yaw-checking and course-keeping ability, pull out and stopping ability. These are required to be demonstrated by means of a satisfactory performance in a turning circle, zig–zag test and a full astern stopping test. The mechanics of manoeuvring and the equations of motion may be pursued in more depth in standard naval architecture texts such as references [4.5, 4.6].

4.2 Rudder design within the ship design process

The previous section has described the action of the rudder and its role in course-keeping and manoeuvring. It is apparent that a fundamental requirement of the rudder is to produce sideforce in the most efficient manner, that is to produce the required lift with minimum drag. The rudder also has to fit into the practical layout of the aft end of the ship or boat with possible constraints in rudder size and shape, and operate under the influences of the upstream hull and, in many cases, propeller. It should be noted that it is advantageous to place the rudder in the slipstream of a propeller. The accelerating affect of the propeller on the flow leads to greater rudder inflow velocities and rudder forces, as described in Sections 3.5 and 3.6. Also, the propeller has a straightening effect on the cross flow at the stern on a turn, increasing rudder incidence and generating more lift.

As well as a limited space for the rudder, the draught may be limited which in turn will limit the span of the rudder. It was seen in Section 3.4 that a larger span and aspect ratio leads to a more efficient rudder in terms of reduced drag for a given lift. Good coursekeeping might favour a rudder with a high lift curve slope and rapid response resulting from a large aspect ratio, whilst manoeuvring performance might be enhanced with a large stall angle, which is more likely to be achieved with a small aspect ratio. Before starting the rudder design process a clear set of requirements for the rudder should be defined, but it is apparent that a compromise design is a likely outcome as a result of the various constraints.

The relative geometrical arrangement of the rudder, propeller and hull can have significant influences on both manoeuvring and propulsion performance. The effect of the hull is generally to slow down the flow into the propeller and that of the propeller is to accelerate and rotate the flow into the rudder, thus affecting its performance. The proximity of the rudder also influences the propeller upstream and the overall propulsive effect of the propeller–rudder combination. It is therefore necessary to devise an overall stern arrangement that satisfies the design requirements in terms of propulsion, speed and fuel consumption whilst ensuring the vessel is able to maintain its course and satisfy manoeuvring requirements at both low and service speeds. The position of the propeller relative to the hull is generally considered in terms of the stern tube or stern-frame design and minimum propeller to hull clearances. There is normally more freedom in the siting of the rudder relative to the propeller and hence hull.

The principal geometrical properties which affect rudder–propeller interaction relate to the relative positions of the rudder and propeller and may be summarised as the longitudinal separation (X/D), lateral separation (Y/D) and vertical position (Z/D), Figures 4.8–4.10.

4.2.1 Single screw

A typical single screw arrangement is shown in Figure 4.8, in this case employing a semi-balanced skeg rudder, Figure 2.2(d). In the single screw arrangement, the rudder is normally situated in line with the propeller, with the rudder tip approximately coincident with the propeller tip, leading to $Y/D = 0$ and typical values of Z/D of 1.0. The hull aperture is chosen to provide suitable clearances around the propeller to avoid propeller-excited vibration and, for example, the

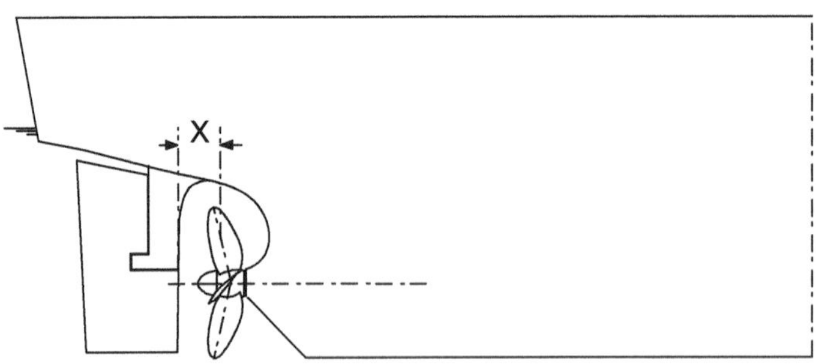

Figure 4.8 Typical single-screw layout

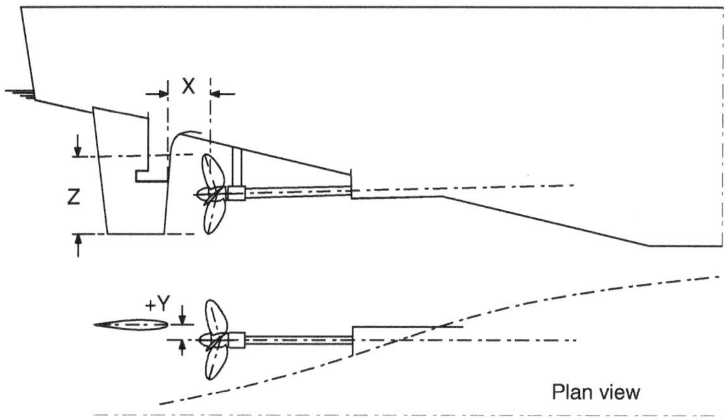

Figure 4.9 Typical twin-screw layout

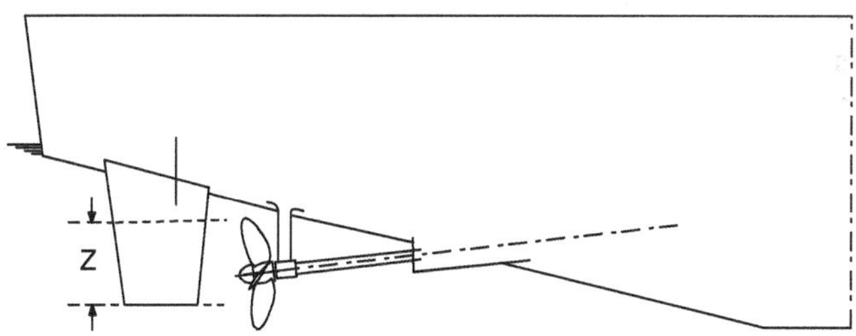

Figure 4.10 Typical twin-screw layout: smaller vessel

classification societies offer advice on minimum clearances. The only effective variable in such single screw arrangements is X/D.

X/D: A wide variation exists in the choice of X/D. A survey of existing ships indicates a range of X/D from 0.25 up to about 0.5, with some local variations occurring due to the amount of rake on the propeller, the use of controllable pitch propellers and the amount of taper on the rudder. It is generally not clear if the X/D value has been chosen on the grounds of manoeuvring (rudder forces), propulsion (thrust deduction), or whether these aspects have been considered. It should also be mentioned that in some small craft, such as motor sailing vessels, X/D can be very large and no significant interaction would be expected.

4.2.2 Twin screw

It is found that a wide range of combinations of X/D, Y/D and Z/D are employed on twin-screw vessels.

X/D: There is some freedom with the choice of X/D for twin-screw vessels, with values ranging from 0.25 up to as much as 0.70. As with the single screw cases, there does not appear to be an obvious reason why a particular value has been chosen, although the siting of the propeller for shaft support and clearance purposes and a suitable siting of the steering gear may have some influence.

Y/D: In the twin-screw case it is often the practice to offset the rudder laterally from the propeller shaft centreline, Figure 4.9. This enables the propeller tailshaft to be removed without removing the rudder. Such a lateral offset may also be used on smaller higher-speed craft to avoid the propeller hub core vortex impinging on the rudder, leading to rudder cavitation. Typical values of *Y/D* used in practice vary from 0.00 up to about 0.25.

Z/D: On large vessels *Z/D*, like the single-screw case, tends to a value of about 1.0, with the rudder tip approximately coincident with the propeller tip, Figure 4.9. On smaller vessels, the necessity of propeller shaft inclination can lead to *Z/D* values down to about 0.5, Figure 4.10. Limitations on spade rudder root bending moments and stock diameters will normally preclude extending the span of the rudder fully into the propeller race.

The proportion of the rudder that is within the propeller race is also termed the coverage. The coverage is a measure of the proportions of the rudder acted upon by the propeller race and by the free stream. It follows that a rudder with a larger coverage will be more strongly influenced by changes in propeller action.

4.3 Requirements of other control surfaces

Control surfaces such as roll-stabiliser fins or pitch-control foils behave in a similar manner to a rudder in that they have to develop sufficient lift to provide a satisfactory restoring moment. Such devices have to work in proximity of the hull and care must be taken in their siting and practical layout and attention paid to the influences of the adjacent hull on the hydrodynamic performance. This will include taking account of the boundary layer developed over the hull in way of the control surface, and designing and operating the control surface to avoid cavitation and its adverse effects on performance.

4.4 Rudder and control surface design strategy

A rudder and control surface design strategy is needed that will meet the design requirements and constraints described in the previous three sections. The contents of this book reflect these needs and proposes a suitable strategy that may be summarised as follows:

(1) A background understanding of the fundamental hydrodynamic characteristics of control surfaces is necessary, such as the development and behaviour of the boundary layer and its effect on separation and stall, and the implications of induced drag on a finite aspect ratio foil and the effect of the aspect ratio on induced drag. These principal hydrodynamic characteristics have been described in Chapter 3.

(2) The precise role of the rudder should be established, including likely modes of operation, choice of rudder type and section, such as those described in Sections 2.2 and 5.3.2, and the likely forces required to be developed.

In a similar manner, the precise requirements of control surfaces such as movable fins for roll or pitch control have to be established, including modes of operation, choice of section type and the forces required to be developed.

(3) Access should be available to a suitable database. In meeting this need, a wide range of data has been reviewed and tabulated in Chapter 5, and theoretical and numerical approaches to the estimates of such data are described in Chapter 6.

Examples of tabulated results for an extensive set of wind tunnel experiments on propeller–rudder interaction are given in Appendix 1, together with the means of access to a complete database of the results.

(4) A design methodology should be applied that will utilise the database and suitable design tools. Such a methodology for rudder design is described in Chapter 7. This includes the choice of rudder type, development of rudder size and estimates of the forces and moments developed by the rudder. This will take due account of the implications of the proximity of the rudder to the propeller and/or hull and operational features such as ship speed and drift angle, if present. The forces, moments and distribution of load on the rudder thus estimated may be used to verify the contribution of the rudder to coursekeeping and manoeuvring. They will also be used to complete the detailed design of the rudder in terms of rudder torque and sizing the steering gear and the structural design in terms of plating and stiffener thicknesses.

The methodology for the design of other control surfaces such as stabiliser fins is described in Chapter 9. This includes the estimation of the forces required to be developed to achieve the necessary restoring moments, the choice of suitable area and section type and the influence of local hull effects on the action of the control surface.

(5) In order to illustrate several of the rudder and control surface design problems that arise, examples of applications of the design process to various design problems are given in Chapter 11.

References

4.1 Burcher, R.K. Developments in ship manoeuvrability. *Transactions of The Royal Institution of Naval Architects*, Vol. 114, 1972, pp. 1–32.

4.2 ITTC 1975. ITTC 1975 Manoeuvring trial code. *14th International Towing Tank Conference*, 1975.

4.3 IMO Resolution A. 751(18). *Interim Standards for Ship Manoeuvrability*, 1993.

4.4 IMO Resolution MSC 137(76). *Standards for Ship Manoeuvrability*, 2002.

4.5 Lewis, E.V. (ed.) *Principles of Naval Architecture.* The Society of Naval Architects and Marine Engineers, New York, 1989.

4.6 Rawson, K.J. and Tupper, E.C. *Basic Ship Theory*, 5th edition. Butterworth Heinemann, Oxford, 2001.

PART TWO

Design Data Sources

5 Experimental data

5.1 Review of experimental data and performance prediction

The most common rudder and control surface design methods use the free-stream or open-water characteristics for the control surface being considered. Suitable corrections are applied to the inflow velocity and direction to take account of the influences of the adjacent hull and/or propeller. These methods are described and discussed further in Chapter 7. Extensive free-stream characteristics for the principal control surface types, derived mainly by experiment, have been published and these are reviewed, presented and discussed in Section 5.3.

In the case of a rudder working downstream of a propeller, an alternative and more physically correct approach is to treat and test the rudder plus propeller as a unit. This approach is discussed further in Chapter 7. Several experimental and theoretical investigations have been carried out and published experimental data for such cases are reviewed and presented in Section 5.4.

5.2 Presentation of experimental data

The notation and particulars of the control surface, forces and centre of pressure are given in Figures 5.1 and 5.2. The rudder and propeller coefficients are defined in the following manner:

Forces:

In the free stream, rudder lift (L) and drag (d) forces are nondimensionalised using the free-stream (ship wake) speed V:

$$C_L = \frac{L}{\frac{1}{2}\rho A V^2}, \qquad C_D = \frac{d}{\frac{1}{2}\rho A V^2} \tag{5.1}$$

These are presented in terms of the rudder incidence α.

Also,

$$C_N = C_L \cos\alpha + C_D \sin\alpha \tag{5.2}$$

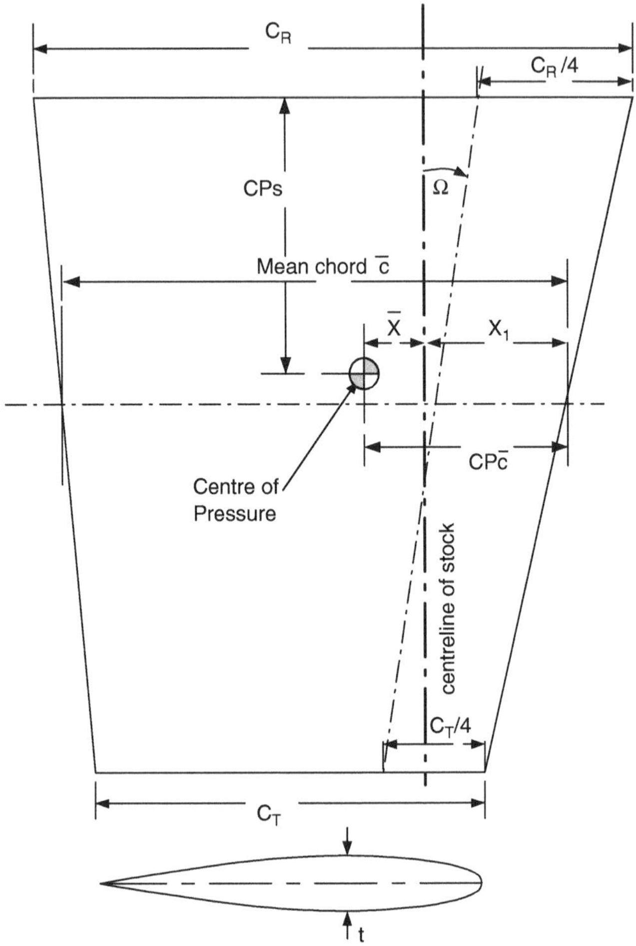

Figure 5.1 Notation for typical all-movable control surface

and

$$C_N = \frac{N}{\frac{1}{2}\rho A V^2} \tag{5.3}$$

In the case of the rudder downstream of a propeller, C_L and C_D are presented in terms of rudder incidence α and propeller thrust loading K_T/J^2 for particular rudder and propeller geometries, where

$$J = \frac{V}{nD} \tag{5.4}$$

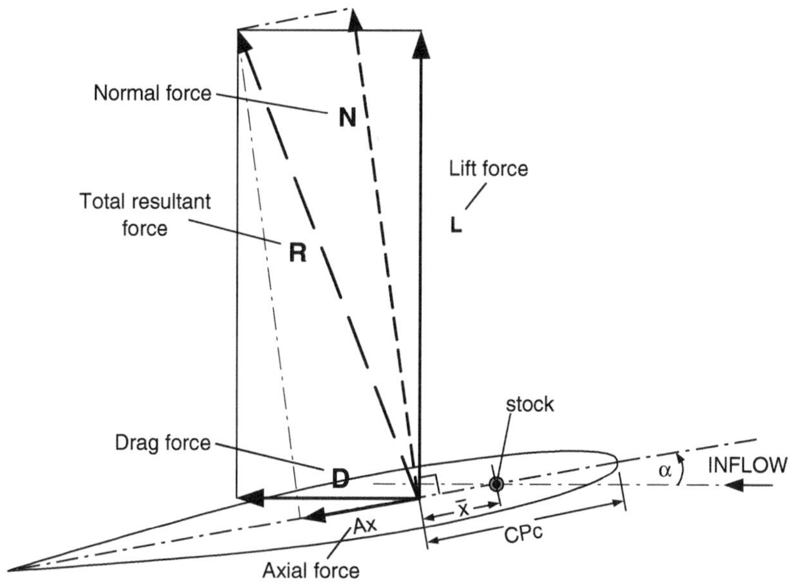

Figure 5.2 Notation of forces and angles for all-movable control surface

and propeller thrust and torque coefficients are defined as

$$K_T = \frac{T}{\rho n^2 D^4}, \qquad K_Q = \frac{Q}{\rho n^2 D^5} \tag{5.5}$$

Different presentations of the rudder and propeller forces are used for low speed and four-quadrant operation and these are discussed in Sections 5.4.2.5 and 5.4.2.6.

Centre of pressure, Figures 5.1 and 5.2:
Centre of pressure, chordwise, % chord from leading edge: *CPc*
Centre of pressure, spanwise, % span from root: *CPs*

For design purposes it is necessary to be able to estimate the forces and moments on a particular rudder at a given angle of attack to the flow and given inflow speed. The presentation of experimental data is usually in the form of lift, drag and moment characteristics in coefficient form over a range of rudder incidence, together with the location of the centre of pressure in the chordwise and spanwise directions, as illustrated in Figure 5.3. The presentation will also normally identify the stall angle α_{stall} and maximum lift coefficient C_{Lmax} together with the minimum profile drag coefficient, C_{Do}. The influence of a propeller upstream of a rudder on rudder performance characteristics will normally be depicted by curves resulting from different levels of propeller thrust loading, K_T/J^2, Figure 5.4, since the propeller induced velocity is a function of K_T/J^2 (Equation (3.38)). Where detailed pressure measurements have been carried out these will normally be presented as chordwise pressure distributions for 2-D data, Figure 5.5, and spanwise distributions for 3-D data, Figure 5.6, or combined chordwise and spanwise distributions, Figure 5.7.

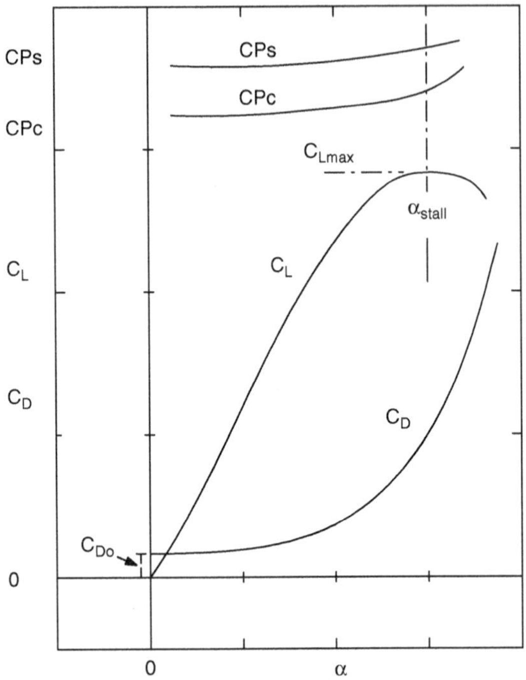

Figure 5.3 Lift, drag and CP data

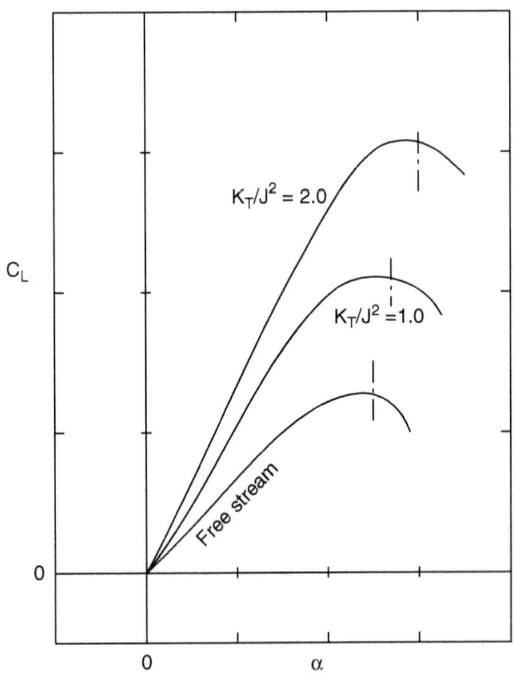

Figure 5.4 Lift data: influence of propeller loading

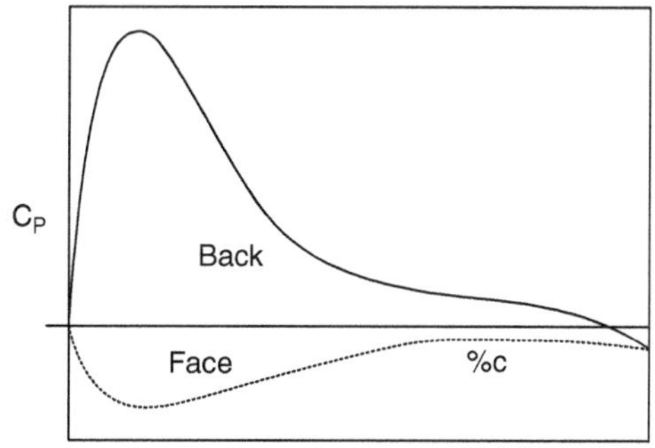

Figure 5.5 Chordwise pressure distribution

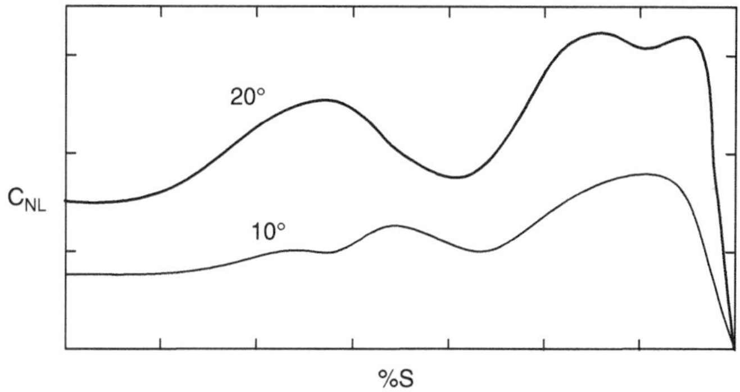

Figure 5.6 Spanwise pressure distribution

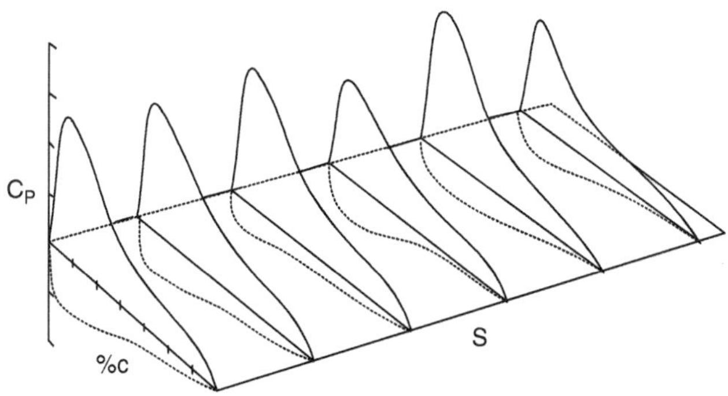

Figure 5.7 3-D presentation of surface pressures

5.3 Experimental data for rudder in free stream

5.3.1 General

Typical experimental free-stream control surface data, of relatively low aspect ratio and suitable for application in the marine field, include those of Jones [5.1], Whicker and Fehlner [5.2], Thieme [5.3] and Hagen [5.4] for all-movable control surfaces, Kato and Motora [5.5] and Kerwin *et al.* [5.6] for flapped rudders, Goodrich and Molland [5.7] and Molland [5.8] for semi-balanced skeg rudders. The results of some free-stream investigations on semi-balanced skeg rudders are also included in the work of Gong *et al.* [5.9] and Kang *et al.* [5.10]. Mention must also be made of earlier free-stream investigations for applications in the marine field, which include those of Jöessel [5.11], Johns [5.12], Denny [5.13], Darnell [5.14], Abell [5.15], Voepel [5.16], Flax and Lawrence [5.17], Bartlett and Vidal [5.18] and Windsor [5.19].

5.3.2 All-movable rudders

5.3.2.1 *Data*

The most comprehensive published set of freestream data for all-movable low aspect ratio control surfaces is that reported by Whicker and Fehlner [5.2]. A basic set of nine model foils was tested in a wind tunnel on a groundboard covering a range of aspect ratio, sweep, section shape and tip shape over a range of angles up to stall. The foils were tested in the ahead and astern conditions. A range of tunnel speeds was used, resulting in Reynolds numbers from 1×10^6 to 3×10^6.

A summary of the particulars of the test models and test conditions is given in Figure 5.8 and Table 5.1 and an example based on the results for one model is given in Figure 5.9. A major part of the investigation was carried out using models

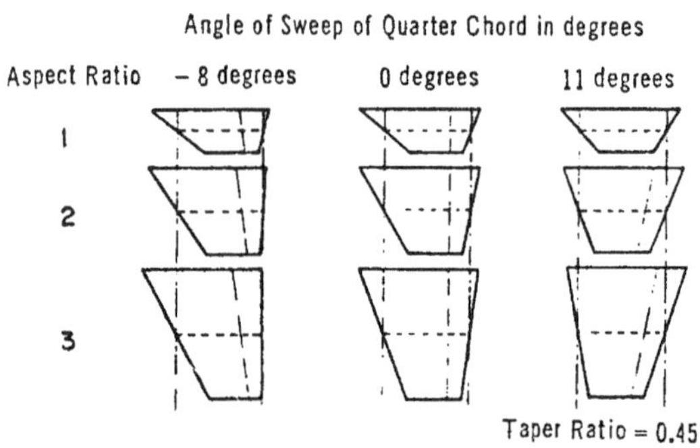

Figure 5.8 Planforms tested [5.2]

Table 5.1 Details of models and tests [5.2]

Effective aspect ratio: 1, 2, 3
Sweep of quarter chord: −8°, 0°, 11°
Taper ratio: 0.45
Tip shape: square and faired (Approximately, semi-circular)
Gap between control surface root and groundboard: $0.005\bar{c}$
Reynolds numbers for tests: 1×10^6 to 3.00×10^6
Basic section shape tested for major part of investigation: NACA0015

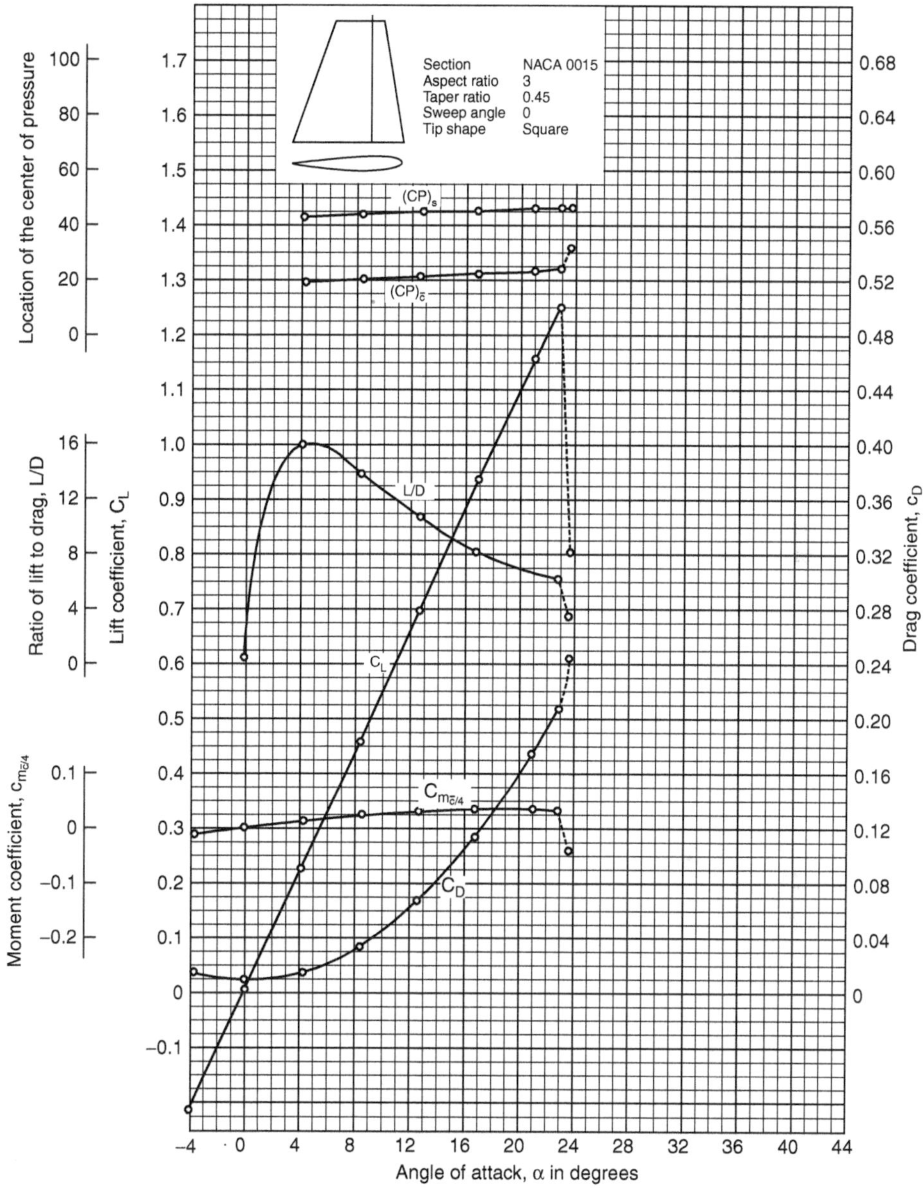

Figure 5.9 (a) Example of test results (ahead) [5.2]

with a NACA 0015 section and a summary of the results for all the models with this section is given in Table 5.2. Harrington [5.20] cross plotted the Whicker and Fehlner experimental lift, drag and chordwise centre of pressure data as useful design data curves and these are reproduced, with permission, in Figures 5.10, 5.11 and 5.12.

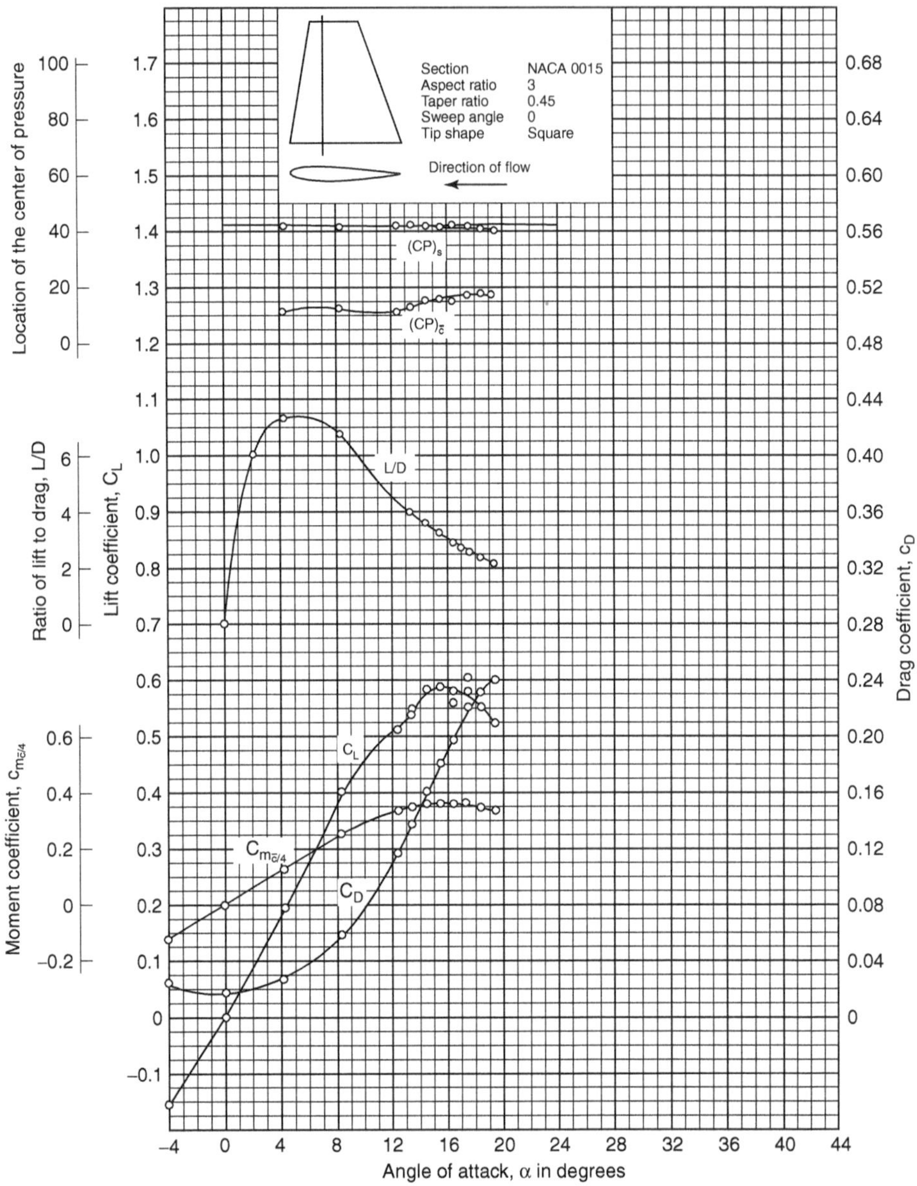

Figure 5.9 (b) Example of test results (astern) [5.2]

Whicker and Fehlner fitted equations to the lift, drag and moment data, which are very useful for predicting the characteristics of low aspect ratio foils. The equations are based on the theoretical work of Hembold [5.21], Polhamus [5.22] and Flax and Laurence [5.17]. A cross flow drag coefficient C_{Dc} is introduced to take account of the three-dimensional flow over low aspect ratio foils and resultant

Table 5.2 Tabulated free-stream test data, extracted from reference [5.2]

Test	*Fig. No.*	*Sweep (deg.)*	*AR*	$\frac{dC_L}{d\alpha}$	C_L *10°*	C_L *20°*	C_L *max*	α_{stall}	C_D *10°*	C_D *20°*	*CPc 10°*	*CPc stall*	*CPs 10°*	*CPs stall*
Ahead Square tips	39	−8	1	0.023	0.25	0.59	1.24	39.2	0.035	0.140	0.14	0.30	0.47	0.51
	49	−8	2	0.038	0.42	0.90	1.32	28.8	0.040	0.152	0.18	0.25	0.45	0.48
	61	−8	3	0.050	0.51	1.06	1.25	24.0	0.040	0.150	0.20	0.25	0.45	0.47
	44	0	1	0.023	0.27	0.59	1.26	38.5	0.037	0.140	0.16	0.32	0.47	0.50
	55	0	2	0.042	0.43	0.93	1.32	28.8	0.041	0.160	0.20	0.25	0.45	0.48
	66	0	3	0.053	0.55	1.10	1.25	23.0	0.045	0.155	0.20	0.23	0.45	0.47
	45	11	1	0.025	0.26	0.60	1.40	43.5	0.036	0.140	0.17	0.34	0.45	0.51
	60	11	2	0.045	0.45	0.94	1.32	28.8	0.043	0.162	0.19	0.25	0.46	0.48
	67	11	3	0.055	0.55	1.09	1.13	21.2	0.048	0.157	0.19	0.22	0.45	0.46
Ahead Faired tips	68	−8	1	0.020	0.23	0.53	1.03	36.5	0.030	0.122	0.10	0.30	0.46	0.48
	75	−8	2	0.040	0.40	0.86	1.23	29.0	0.037	0.142	0.18	0.24	0.44	0:47
	86	−8	3	0.049	0.51	1.02	1.22	24.0	0.040	0.147	0.20	0.22	0.43	0.46
	69	0	1	0.021	0.23	0.53	1.10	36.3	0.032	0.121	0.11	0.30	0.45	0.49
	79	0	2	0.041	0.41	0.86	1.14	26.5	0.040	0.145	0.18	0.23	0.45	0.46
	87	0	3	0.053	0.53	1.05	1.13	22.0	0.040	0.150	0.20	0.22	0.43	0.46
	70	11	1	0.022	0.24	0.54	1.21	39.2	0.030	0.124	0.12	0.32	0.45	0.50
	85	11	2	0.043	0.43	0.89	1.17	26.9	0.041	0.150	0.18	0.22	0.45	0.46
	88	11	3	0.051	0.52	1.05	1.08	21.0	0.041	0.150	0.18	0.20	0.44	0.46
Astern Square tips	89	−8	1	0.026	0.30	0.67	0.95	31.0	0.057	0.232	0.20	0.25	0.45	0.40
	92	−8	2	0.038	0.40	0.63	0.63	23.0	0.070	0.250	0.17	0.20	0.40	0.38
	95	−8	3	0.045	0.45	0.57	0.62	17.5	0.080	0.260	0.12	0.16	0.39	0.39
	90	0	1	0.026	0.30	0.68	0.93	29.5	0.059	0.230	0.20	0.25	0.43	0.40
	93	0	2	0.036	0.38	0.62	0.62	18.0	0.072	0.242	0.15	0.18	0.40	0.40
	96	0	3	0.040	0.45	0.52	0.58	16.0	0.077	0.240	0.12	0.17	0.42	0.40
	91	11	1	0.025	0.29	0.67	0.84	28.5	0.060	0.222	0.16	0.25	0.42	0.39
	94	11	2	0.030	0.38	0.60	0.65	18.3	0.060	0.240	0.11	0.17	0.42	0.43
	97	11	3	0.045	0.48	–	0.56	13.5	0.090	–	0.10	0.15	0.42	0.44

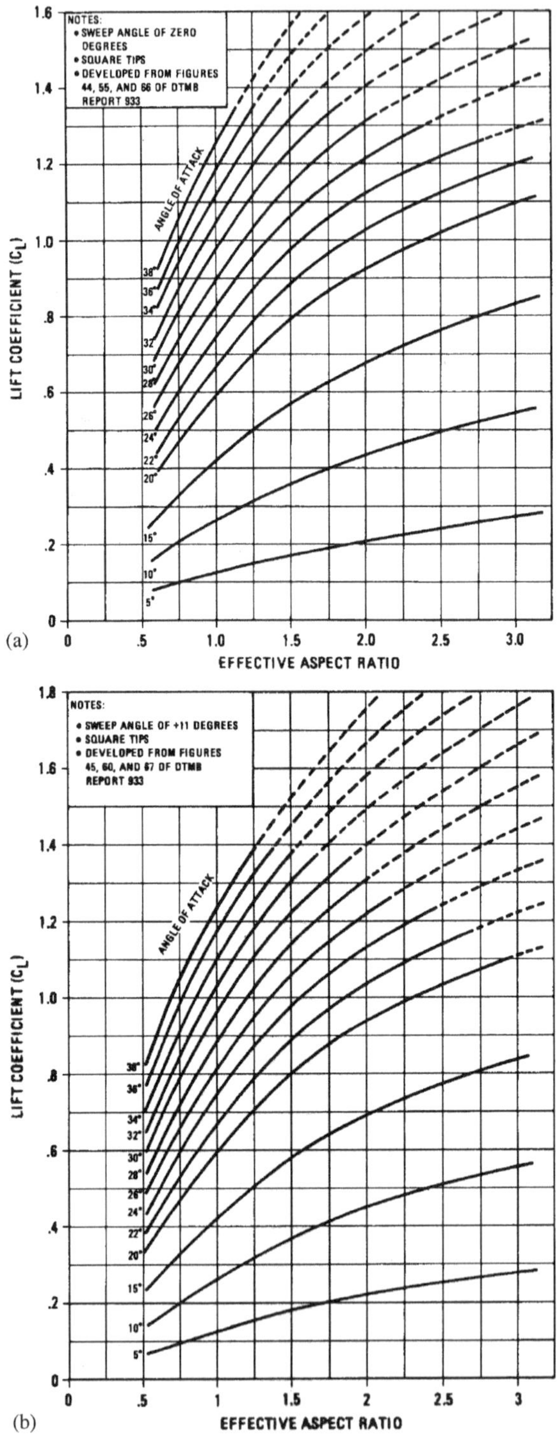

Figure 5.10 (a) Lift coefficient, sweep = 0°; (b) Lift coefficient, sweep = 11°

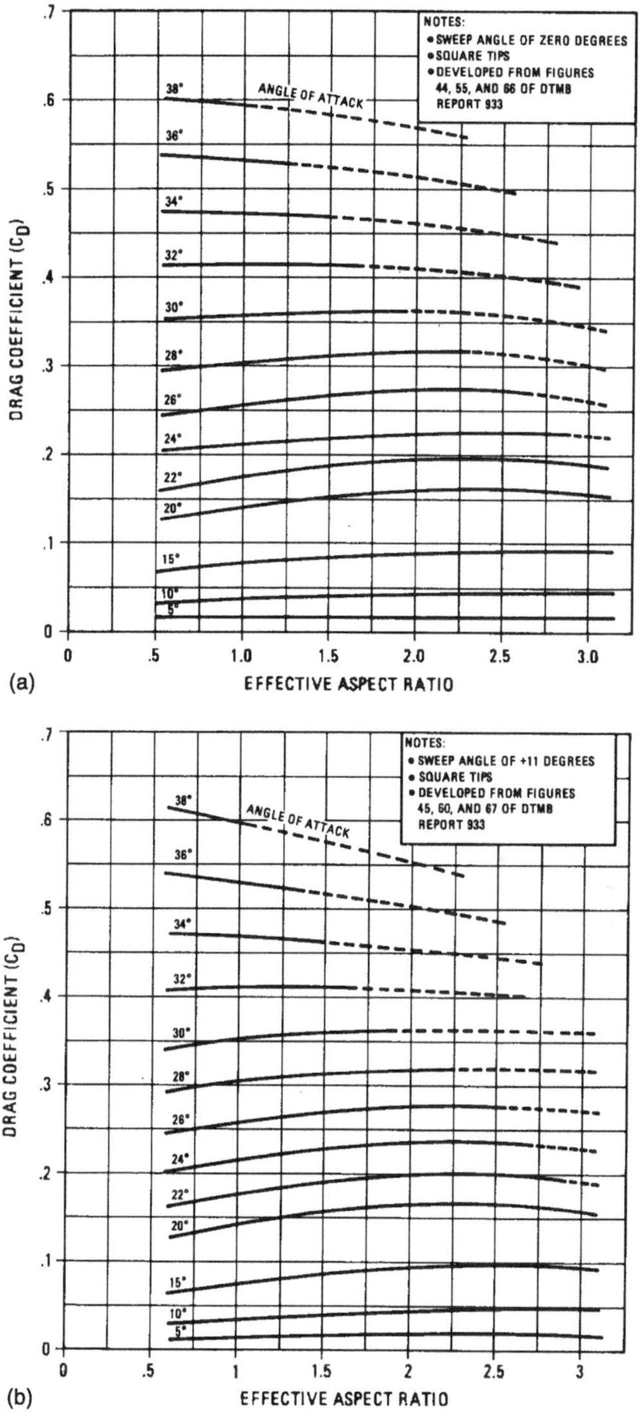

Figure 5.11 (a) Drag coefficient, sweep = 0°; (b) Drag coefficient, sweep = 11°

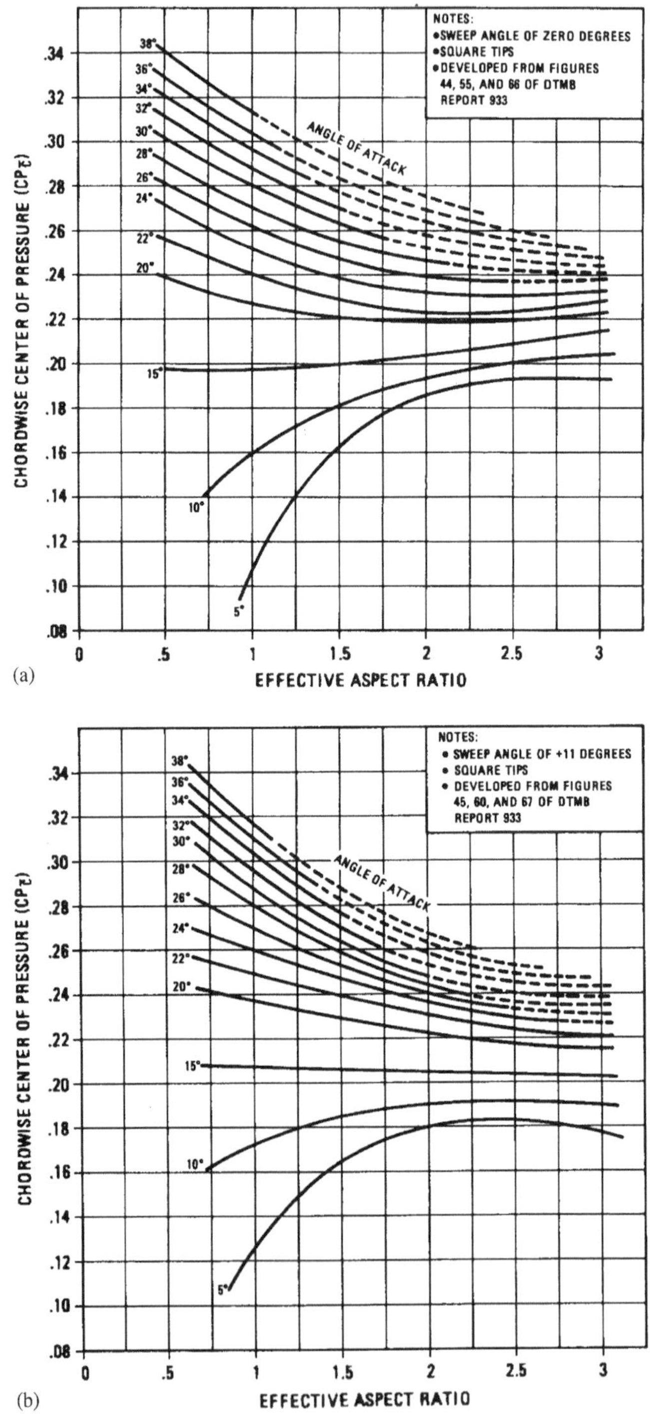

Figure 5.12 (a) Chordwise centre of pressure, sweep = 0°; (b) Chordwise centre of pressure, sweep = 11°

nonlinearities in the lift curve. An (Oswald) efficiency factor e is introduced into the induced drag term to bring predictions using the drag equation into line with the experimental data. A summary of the equations follows.

5.3.2.2 *Lift coefficient*

$$C_L = \left[\frac{dC_L}{d\alpha}\right]_{\alpha=0} \times \alpha + \frac{C_{Dc}}{AR}\left[\frac{\alpha}{57.3}\right]^2 \quad (5.6)$$

where
$$\left[\frac{dC_L}{d\alpha}\right]_{\alpha=0} = \frac{(0.9)(2\pi)AR}{57.3\left[\left[\cos\Omega\left(\frac{AR^2}{\cos^4\Omega}+4\right)^{0.5}\right]+1.8\right]} \quad (5.7a)$$

and
$$C_{Dc} = 0.1 + 1.6\frac{C_T}{C_R} \quad \text{Square tips} \quad (5.8)$$

$$C_{Dc} = 0.1 + 0.7\frac{C_T}{C_R} \quad \text{Faired tips} \quad (5.9)$$

The influence of sweep Ω on the lift curve slope $[dC_L/d\alpha]$ in Equation (5.7a) is found to be small. For example, changes in sweep of 10° lead to changes in the lift curve slope of less than 1%, although the values in Table 5.2 would suggest a little more. An alternative to Equation (5.7a) is Equation (5.7b), which represents a satisfactory mean line through the available data and is in a more readily usable form [5.8]. Its values are at about 98% of Equation (5.7a).

$$\left[\frac{dC_L}{d\alpha}\right]_{\alpha=0} = \frac{1.95\pi}{57.3\left(1+\frac{3}{AR}\right)} \quad (5.7b)$$

A comparison is given in Figure 5.13 between the lift curve slope derived from Equation (5.7b) and theory and with experimental data from references [5.1–5.3,5.6, 5.7 and 5.23].

5.3.2.3 *Drag coefficient*

$$C_D = C_{Do} + \frac{C_L^2}{\pi e AR} \quad (5.10a)$$

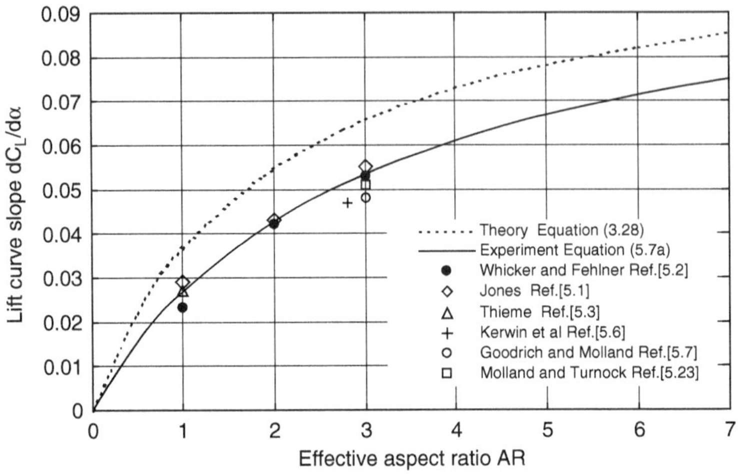

Figure 5.13 Effect of aspect ratio on lift curve slope: all-movable control surfaces

Using the notation of Equation (3.25):

$$C_D = C_{Do} + k_i \frac{C_L^2}{AR} \tag{5.10b}$$

where k_i is the induced drag factor.

It is assumed in reference [5.2] that $C_{Do} = 0.0065$ for a NACA 0015 section and that the efficiency factor e is taken as 0.9 (when $k_i = 1/\pi \times 0.9 = 0.35$). More appropriate values for design purposes, based on other published data such as references [5.6–5.8 and 5.23], are $C_{Do} = 0.010$–0.015 and $e = 0.86$ with $k_i = 0.37$.

Also,

$$C_N = C_L \cos\alpha + C_D \sin\alpha \tag{5.11}$$

Typical lift and drag data, based on these equations, are presented in conventional graphical form in Figure 5.14. The curves are based on Equations (5.6), (5.7b) and (5.10b) and $k_i = 0.37$.

5.3.2.4 *Centre of pressure, chordwise from leading edge*

$$CPc = \left[0.25 - \frac{Cm_{c/4}}{C_N}\right] \tag{5.12}$$

where $$Cm_{c/4} = \left(0.25 - \left[\frac{dCm_{c/4}}{dC_L}\right]_{C_L=0}\right)\left[\frac{dC_L}{d\alpha}\right]_{\alpha=0} \times \alpha - \frac{1}{2}\frac{C_{Dc}}{AR}\left[\frac{\alpha}{57.3}\right]^2 \tag{5.13}$$

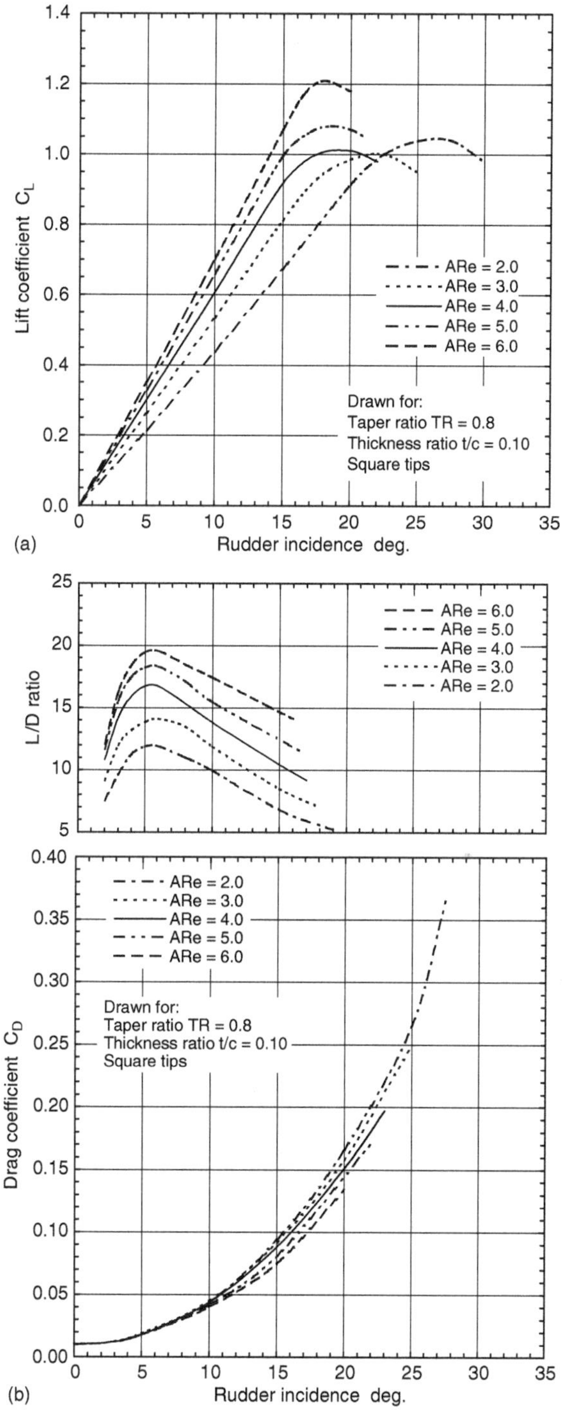

Figure 5.14 (a) Lift coefficient; all-movable rudders; (b) Drag coefficient and *L/D* ratio; all-movable rudders

and $$\left[\frac{dCm_{c/4}}{dC_L}\right]_{C_L=0} = \frac{1}{2} - \frac{1.11\left[(AR^2 + 4)^{1/2}\right] + 2}{4(AR + 2)} \tag{5.14a}$$

An alternative to Equation (5.14a) is (5.14b) which is in a more readily useable form has been proposed by Molland [5.24].

$$\left[\frac{dCm_{c/4}}{dC_L}\right]_{C_L=0} = \frac{0.25}{\left(1 + \frac{1}{AR}\right)} \tag{5.14b}$$

It was noted by Harrington [5.20] that the Whicker and Fehlner equations underestimate *CPc* at higher incidence for some of the control surfaces, particularly at low aspect ratio. The errors in the aspect ratio range of 2–3 are, however, relatively small. If precise figures are called for, then the *CPc* estimates should be checked using the faired cross plots of the experimental *CPc* values given in Figures 5.10–5.12.

In relation to control surface design, the forwardmost position of *CPc* (as incidence tends to zero) and its position at stall are important characteristics, and these two positions for different aspect ratios are shown in Figure 5.15. The forwardmost position of *CPc* shown in Figure 5.15 is a satisfactory fit to the available data and can be represented by Equation (5.14b).

i.e. since $$Cm = \frac{N \cdot CP\bar{c}}{\frac{1}{2}\rho A V^2 \bar{c}} = \frac{N}{\frac{1}{2}\rho A V^2} \times \frac{CP\bar{c}}{\bar{c}}$$

$$= C_N \times \frac{CP\bar{c}}{\bar{c}}$$

as $\alpha \to 0$, $C_N \to C_L$

Therefore

$$Cm = C_L \times \frac{CP\bar{c}}{\bar{c}}$$

and

$$\frac{CP\bar{c}}{\bar{c}} = \frac{Cm}{C_L} = \left[\frac{dCm_{c/4}}{dC_L}\right]_{\alpha=0=C_L} \tag{5.15}$$

Whilst available data indicate a movement of *CPc* forward by about 1–2% for a taper ratio of 0.60 and aft by about 1% for a taper ratio of 1.0, the scatter of the data is generally as large as this (up to ±1%); also, variation due to thickness ratio and sweep (say ±10°) tend to lie within this scatter.

CPc at stall, also shown in Figure 5.15, represents a mean curve through all the data; no specific trends were deduced for the influence of thickness, taper and sweep.

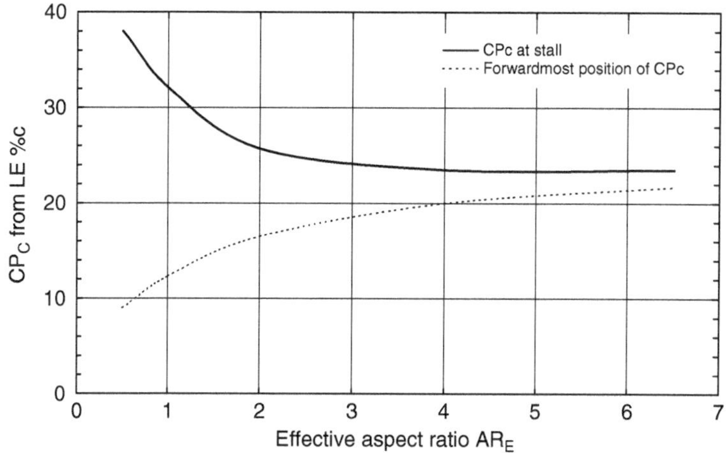

Figure 5.15 Centre of pressure chordwise; all-movable rudders

5.3.2.5 *Centre of pressure, spanwise from root*

A theoretical estimate, assuming an elliptical spanwise load distribution, is

$$CPs = \left[\frac{4}{3\pi}\right] S \tag{5.16a}$$

The centre of pressure spanwise tends to move towards the control surface tip (about 3% to 7% of *S*) as angle of attack is increased, reaching its final position at stall. *CPs* at stall is important in respect of estimating maximum forces. The position of *CPs* at stall, for variation in aspect ratio and taper ratio is shown in Figure 5.16. Data for these curves were derived from references [5.2, 5.8 and 5.23]. A suitable fit to the data for *CPs* at stall [5.24] is

$$CPs = \left[\frac{0.85}{(5 + AR)^{0.25}} \times \left(\frac{C_T}{C_R}\right)^{0.11}\right] S \tag{5.16b}$$

Equations (5.6)–(5.16b) define empirical fits to many of the data available and are valid for an aspect ratio range of about 2–6 and taper ratio range of about 0.6–1.0.

5.3.2.6 *Stall angle α_{stall}*

The principal influence on stall angle, α_{stall}, is aspect ratio and this effect, for different Reynolds numbers, is shown in Figure 5.17.

5.3.2.7 *Maximum lift coefficient C_{Lmax}*

The principal influences on C_{Lmax} of Reynolds number, thickness and aspect ratio are shown in Figure 5.18. It is noted that the C_{Lmax} value for $t/c = 0.10$ rises to a maximum at $t/c = 0.15$ and then decreases down to the value at $t/c = 0.20$.

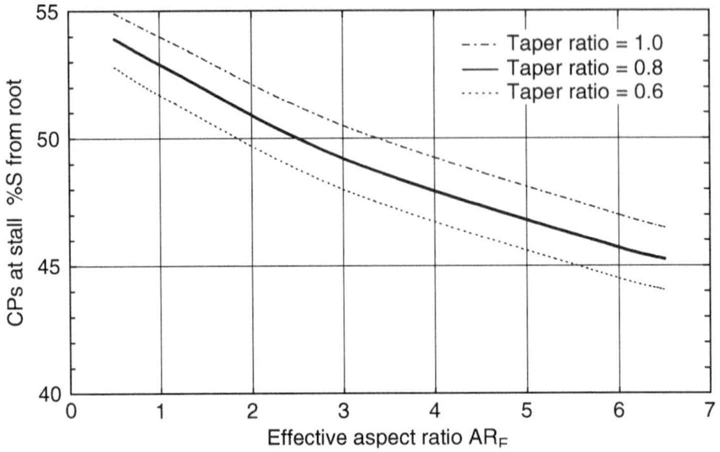

Figure 5.16 Centre of pressure spanwise; all-movable rudders

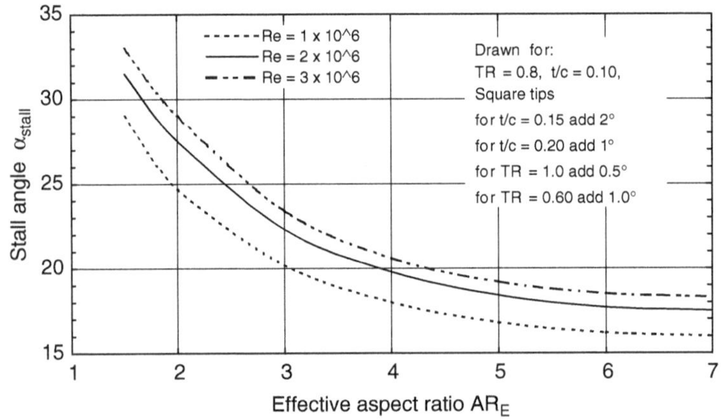

Figure 5.17 Stall angle

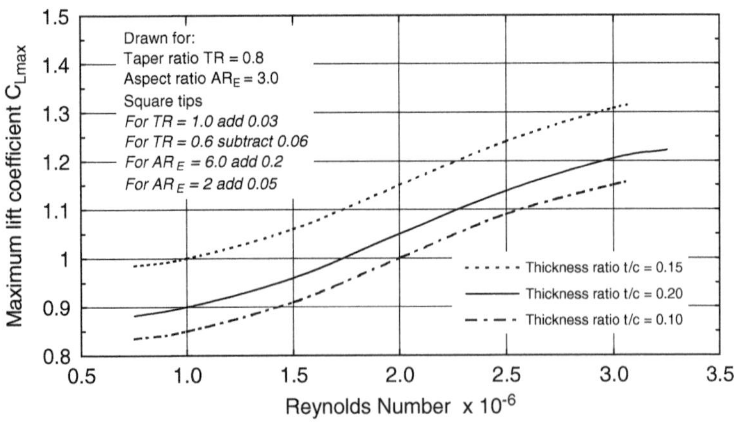

Figure 5.18 Maximum lift coefficient; all-movable rudders

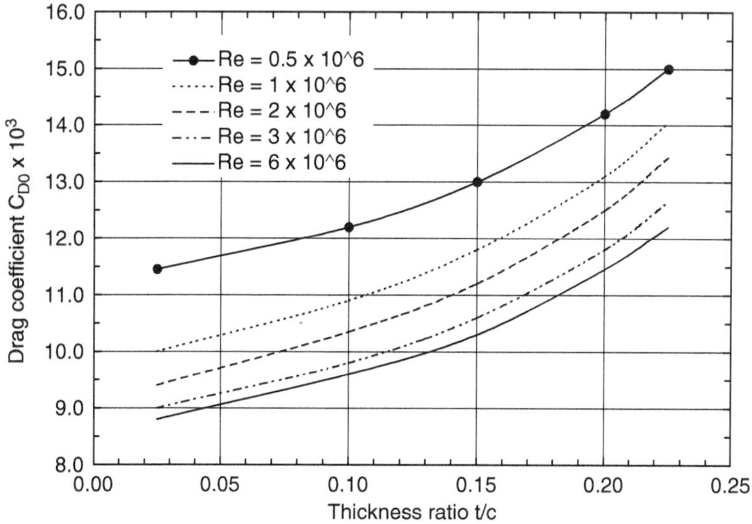

Figure 5.19 Drag coefficient C_{Do}

5.3.2.8 *Minimum drag coefficient C_{Do}*

Figure 5.19 shows mean curves for minimum drag coefficient C_{Do}, [or drag coefficient at $\alpha = 0$ for symmetrical sections] for various section thicknesses and Reynolds number. These curves were faired through data derived from references [5.2, 5.6, 5.23 and 5.25]. From the available data, the influence of taper ratio and sweep on C_{Do} is not conclusive and no attempt is made to account for these parameters in the presentation.

5.3.2.9 *Effect of Reynolds number*

Lift curve slope: The results of Jones [5.1] and Whicker and Fehlner [5.2] would indicate that Reynolds number has a relatively small influence on lift curve slope, particularly above a *Re* of about 1×10^6.

Minimum drag coefficient, C_{Do}: It is seen in Figure 5.19 that *Re* has a noticeable influence on C_{Do}, although the effect is diminishing at $Re = 3 \times 10^6$ and above.

Stall angle, α_{stall}: It is seen in Figure 5.17 that there is a noticeable influence on α_{stall} although the effect is diminishing at $Re = 3 \times 10^6$ and above.

Maximum lift coefficient, C_{Lmax}: The biggest influence of *Re* tends to be on C_{Lmax}, with significant influence from $Re = 1 \times 10^6$ up to about 3×10^6. Above $Re = 3 \times 10^6$ the influence is less significant.

5.3.2.10 *Planform parameters,* AR, TR *and sweep*

Aspect ratio, *AR*, has by far the most significant influence on performance characteristics, as indicated for lift curve slope in Figure 5.13, induced drag (Equation (5.10b)), stall angle in Figure 5.17, maximum lift coefficient in Figure 5.18 and centre of pressure in Figures 5.15 and 5.16. Based on a change in lift curve slope with

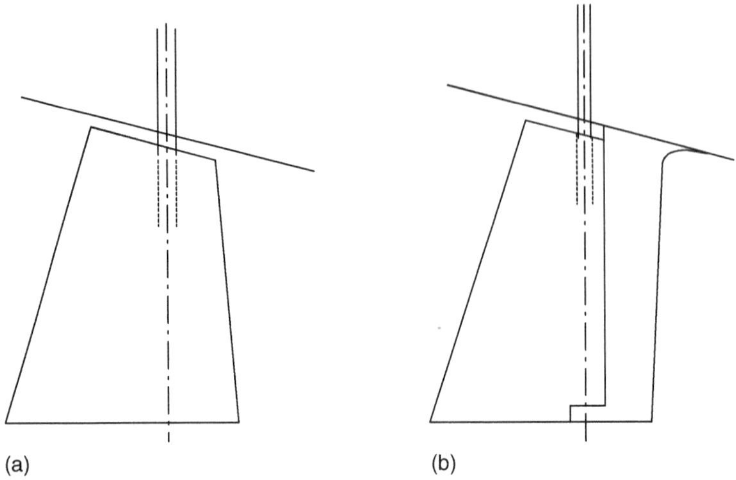

Figure 5.20 Inverse taper

change in aspect ratio described by Equation (5.7b), a satisfactory correction to lift for change in aspect ratio from AR_1 to AR_2 is of the form:

$$\frac{C_{L2}}{C_{L1}} = \frac{[1 + 3/AR_1]}{[1 + 3/AR_2]} \tag{5.17}$$

The fundamental planform should produce elliptical loading for minimum induced drag (Section 3.4). Moderate taper ratios, *TR*, of 0.6–0.8 go some way to meeting this requirement. The influence of taper ratio is summarised in Table 5.3, using $TR = 0.80$ as a datum. It is seen that lift curve slope, C_{Lmax} and α_{stall} all increase with increase in taper ratio. The effect is much greater when decreasing the taper ratio down to 0.6, compared with increasing it to 1.00. The influence of *TR* on *CPc* is very small, but it does tend to influence *CPs* at stall, *CPs* moving toward the tip with increase in *TR*.

The possible use of inverse taper ratio, $TR > 1.0$, Figure 5.20, has been proposed on a number of occasions. In particular, it has been used on yachts both for rudders and keels. Potential advantages of inverse taper tend to be negated by the disadvantages, Mandel [5.26] and Burcher [5.27]. With inverse taper, the chord is small near the hull where the slow moving boundary layer is dominant, with a bigger chord towards the tip, which is in clear flow. The inverse taper means that more of the rudder is in the propeller race. The possibility of ventilation is minimised. With a lower *CPs*, there is an increase in lever and righting moments for yacht keels and rudders. The advantage of keeping the rudder out of the boundary layer does, however, tend to be offset by the desirability to use the hull as an endplate reflection plane and to increase the effective aspect ratio. Also, the increased bending moments on an inverse taper ratio spade rudder are undesirable. It is clear that there are several conflicting properties that need careful consideration and quantification, involving say experimental and theoretical work, before using inverse taper in a particular application.

Table 5.3 Corrections for Taper Ratio, with $TR = 0.80$ as datum

	Taper Ratio			
Parameter	*0.45*	*0.60*	*0.80*	*1.00*
$dC_L/d\alpha$	0.921	0.962	1.000	1.005
C_{Lmax}	0.869	0.948	1.000	1.020
α_{stall}	0.911	0.956	1.000	1.022
CPc at 10°	0.864	0.955	1.000	1.023
CPc at stall	0.962	1.000	1.000	1.000
CPs at 10°	1.000	1.000	1.000	1.000
CPs at stall	0.923	0.981	1.000	1.020

The influence of sweep can be determined from Table 5.2 where it is seen that, in the main, sweep has a relatively small influence on the principal performance characteristics.

5.3.2.11 *Tip shape*

The data presented in Figures 5.9–5.19 are for square tips. A broad indication of the results of using faired tips for rudders is given by Whicker and Fehlner [5.2] where, as an alternative to square tips, faired tips (approximating to a semi-circle) were tested. Use of faired tips would indicate that a small saving in drag is made at low angles of attack. For example, using the Whicker and Fehlner results, Table 5.2, there is a decrease in C_{Do} of up to 0.001, which can amount to up to 10% decrease in minimum rudder drag at zero incidence. At maximum L/D, C_D is reduced by about 5% and L/D increased by about 3%, whilst at higher incidence L/D is about the same for both tip shapes. For the faired tip shape, stall angle is reduced by 2°–3°. There is little change in lift curve slope between the two tip shapes, but the maximum lift coefficient C_{Lmax} of the faired tip is reduced by up to 0.15.

Hoerner [5.28] reports a small decrease in C_{Do} with rounded tips compared with square tips. Data presented in Chapter 7, Figure 10 of that reference for lift against drag indicate no difference at low angles of attack and significant losses by the faired tip at larger angles of attack, which leads Hoerner to conclude that shapes having square tips are generally the ones exhibiting the least drag due to lift. The choice of tip shape therefore might well depend on whether ultimate rudder drag reduction at very small angles of attack is required, or whether square tips are used for maintaining large rudder angle performance for manoeuvring or resisting broaching in the case of sailing craft. The evidence available does, however, suggest that by using a square tip on the rudder, which lends itself to accurate and easy production, the hydrodynamic losses, if any, at low angles of attack are relatively small compared with the gains to be made at higher angles of attack. At the same time, for high-speed vessels, rounded tips at the leading edge may decrease the chances of tip and sole cavitation, as discussed in Section 5.8.

5.3.2.12 *Section shape*

The chordwise section shape of lifting surfaces has received much attention. For example, efficient low drag, high lift/drag ratio, wings have been developed in the case of aircraft and special section shapes have been designed for movable control surfaces in the marine field to delay the onset of cavitation and/or to help control separation and delay stall.

When considering section shape, a fundamental design factor is thickness, the minimum requirements for which will normally be governed by structural needs. Actual chosen thickness will also be influenced by hydrodynamic requirements, since thickness change will affect the minimum drag, separation tendencies and hence stall angle and maximum lift coefficient. The distribution of thickness chordwise, and the position of maximum thickness, also has an influence on these characteristics.

Application of NACA Sections: Abbot and Van Doenhoff [5.25] present test data for a wide range of section shapes based mainly on the NACA series; it should be emphasised that these results are for two-dimensional sections, or infinite aspect ratio, and require correction when applied to actual finite aspect ratios. Analysis of the data in reference [5.25] for different section shapes yields a general trend. For example, the common or so called 'normal' aerofoil shapes, such as the NACA00 series with maximum thickness 30% aft of leading edge, have very satisfactory characteristics for movable control surfaces such as rudders. The so called 'low drag' shapes, such as the NACA 65 or 66 series with maximum thickness 45–50% aft, have lower drag coefficients at low angles of attack, up to about 3°, yielding what is termed a drag bucket. These low drag forms, with maximum thickness further aft do, however, exhibit earlier separation tendencies as angle of attack is increased, with consequent increase in form drag and earlier stall. Since there is little significant difference in the induced drag at a given lift for the section shapes being discussed, this leads to higher drag for the 'low drag' shape at higher angles of attack. A comparison between these principal characteristics for the 'normal' and 'low drag' forms is shown schematically in Figure 5.21. It should be added that the data in reference [5.25] indicate that the drag bucket is only achieved for very smooth sections, and that the presence of even minimal roughness will eliminate this drag bucket effect.

The overall results suggest that, for applications in the marine field, the low drag shapes such as the NACA 65 and 66 series are possibly suitable for applications with limited range requirements for angle of attack, such as yacht keels, when the benefits of laminar flow and the drag bucket may be utilised. The normal aerofoil shapes, such as the NACA00 series, are more suitable for movable appendages such as rudders and other control surfaces which require efficient operation over a wider range of angles. The NACA64 series has, however, been proposed for rudders on higher speed ships where cavitation may be a problem. This is discussed further in Section 5.8.

The NACA00 series is well proven for both small craft and large ship rudders and other movable control surfaces; a further consideration is that its maximum thickness at 30% aft of leading edge satisfies the typical area of location of the rudder stock. The trailing edge thickness of many published aerofoil sections, which have been primarily evolved for aircraft use, is generally too small for practical marine applications. Typical methods of increasing the thickness to practical values, for example 2–3 mm say in the case of small craft and 15–25 mm in the case of larger ships, are to thicken the trailing edge locally and fair in, or remove the last 2–3% of the section to attain adequate thickness. Thickening the trailing edge for practical

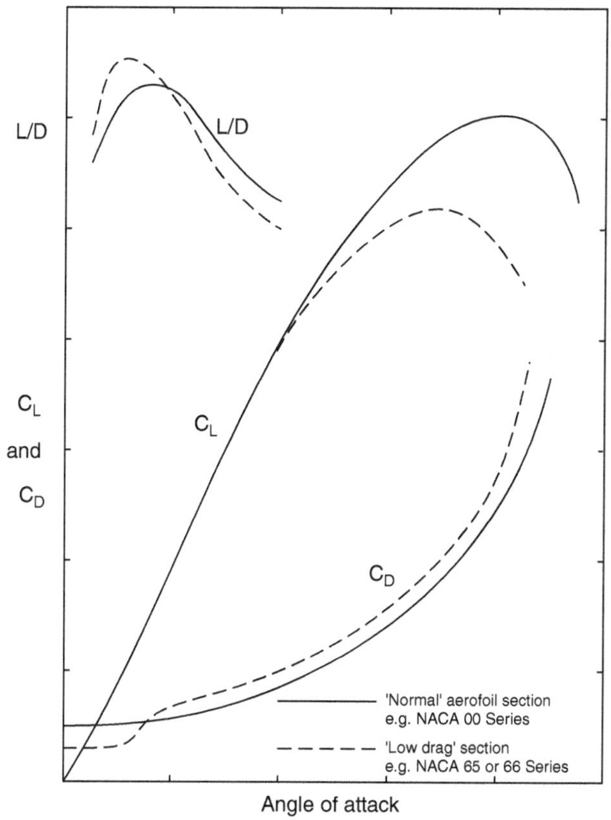

Figure 5.21 Comparison between NACA00 aerofoil section and low drag section characteristics

reasons by this order of magnitude has a small influence on drag and can in fact lead to a small increase in maximum lift coefficient, Hoerner [5.28] and Thiemann [5.29].

For practical design purposes, offsets as a percentage of maximum thickness, for the NACA00, 64 and 66 series, together with an HSVA section, are given in Table 5.4.

Application of other sections: Significant investigations into section shape, for application to movable control surfaces in the marine field, include those by Whicker and Fehlner [5.2], Thieme [5.3], Thiemann [5.29], Kwik [5.30–5.32], Landgraf [5.33], and Kracht [5.34]. Typical basic section shapes considered in the various investigations are summarised in Figure 5.22. Several variants of these basic shapes have also been investigated. The origins of the basic shapes, and sources of data for such shapes, are summarised as follows:

(a) *Flat plate*: Effectively where rudders began, being the simplest device to deflect a fluid. Thieme [5.3] presents results for flat plates with thickness/chord ratios (t/c) of 0.015, 0.03, 0.05 and 0.07.
(b) *NACA aerofoil sections*: Well documented and many data available in reference 5.25. In the marine field, the NACA00 series, of various thickness, tends to be the

Table 5.4 Offsets for control surface sections

	NACA 00	*NACA 64*	*NACA 66*	*HSVA-MP71*
x/c %c	$y/(t/2)_{max}$	$y/(t/2)_{max}$	$y/(t/2)_{max}$	$y/(t/2)_{max}$
0	0	0	0	0
1.25	0.316	0.241	0.215	0.230
2.50	0.436	0.334	0.292	0.306
5.00	0.592	0.466	0.409	0.419
7.50	0.700	0.575	0.500	0.502
10	0.780	0.646	0.578	0.583
15	0.891	0.774	0.703	0.706
20	0.956	0.868	0.799	0.801
25	0.990	0.936	0.872	0.881
30	1.000	0.980	0.928	0.939
35	0.988	1.000	0.967	0.972
40	0.967	0.996	0.991	0.996
45	0.933	0.959	1.000	1.000
50	0.882	0.901	0.994	0.965
60	0.761	0.738	0.924	0.766
70	0.611	0.534	0.729	0.546
80	0.437	0.318	0.462	0.335
90	0.241	0.121	0.186	0.140
100	0.021	0	0	0.054
LE rad. %*c* (Approx.)	$110(t/c)^2$	$66(t/c)^2$	$59(t/c)^2$	–

Note: Offsets have been normalised relative to maximum thickness
Leading edge radii for the NACA 64 and 66 sections are approximate.

standard shape or datum, and the one with which other section shapes are compared. The NACA64 series sections have also been proposed where vulnerability of the rudder to cavitation may be a problem.

(c) *IfS sections*: Developed by the Institute für Schiffbau, Hamburg, using theoretical and experimental investigations. Designed to achieve a pressure distribution that leads to a lower travel of *CPc*, increases lift curve slope, helps delay stall and increases C_{Lmax}. Descriptions and data are presented by Thieme [5.3], Kwik [5.30] and Landgraf [5.33].

(d) *HSVA sections*: Developed by HSVA, Hamburg, with the view to achieving lower drag, and pressure distributions that delay the onset of cavitation. The example shown has a maximum thickness 45% aft of LE. Descriptions and some data are presented by Kracht [5.34] and Brix [5.35].

(e) *Wedge at tail*: Tends to increase lift curve slope and maximum C_L, with some increases in drag. The wedge seems to have been applied originally as a retrofit to improve the poor performance of some rudders already in service, Thieme [5.3]. Tends not to work well going astern. Wedges are also being applied to stabiliser fins. Descriptions and data are presented by Thiemann [5.29] and Thieme [5.3].

(f) *Fish tail*: Developed by and applied to the high lift Schilling rudder. Designed to achieve a pressure distribution that delays stall.

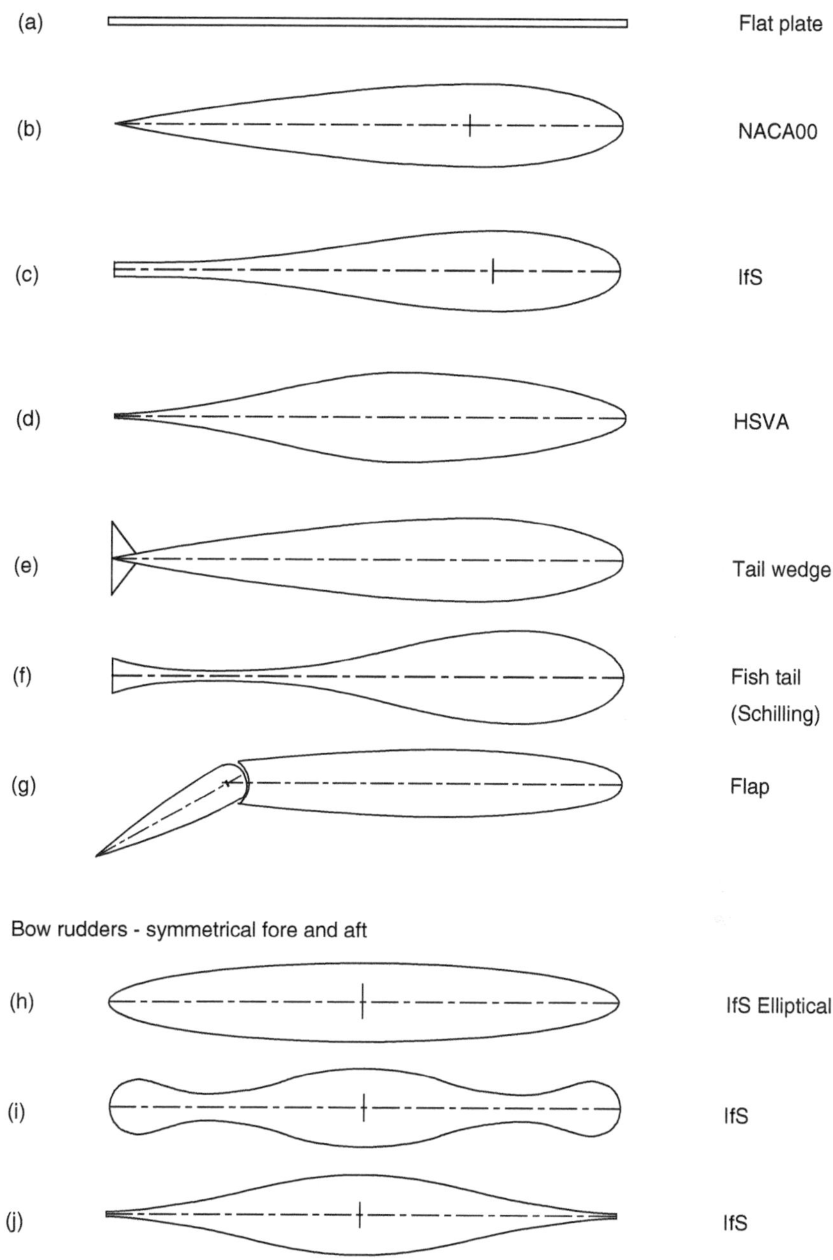

Figure 5.22 Typical basic section types

(g) *Flapped foil*: Effectively provides variable camber. For marine purposes, usually based on a symmetrical NACA section with 20–30% chord flap. Tends not to work so well going astern. Applied to high-lift rudders, stabiliser fins and pitch ride control fins. Design data are available from sources such as

Kerwin *et al.* [5.6] and Ames and Sears [5.36] and the concept is discussed further in Section 5.3.3.

Sections (h), (i) and (j), Figure 5.22, are examples of bow rudders that have been investigated, noting that the sections are symmetrical fore and aft:

(h) *Elliptical section*: Presented by Thieme [5.3].
(i) IfS section: Presented by Thieme [5.3].
(j) IfS section: Investigated by Thieme [5.3] and Kwik [5.31].

Tabulated performance data for control surfaces with section types (a)–(c) and (h)–(j) from Thieme [5.3], together with examples of three wedge sections from Thiemann [5.29], are given in Table 5.5. Details of sections for which data are listed in Table 5.5 are given in Figure 5.23. For the wedge sections, the wedges are defined by their overall breadth, B_W, as a percentage of chord, Figure 5.23. Thiemann tested other wedges sizes and wedge/foil combinations including truncating the aft end of the section. All the data in Table 5.5 are for control surfaces with aspect ratio $AR = 1.0$ and square tips.

Inspection of Table 5.5 enables the attributes of the various sections to be compared. For example, compared with say the NACA sections, the flat plates exhibit a higher lift curve slope, but are accompanied by a very high drag. Compared with the NACA sections, the IfS sections show an increase in lift curve slope, C_{Lmax} and

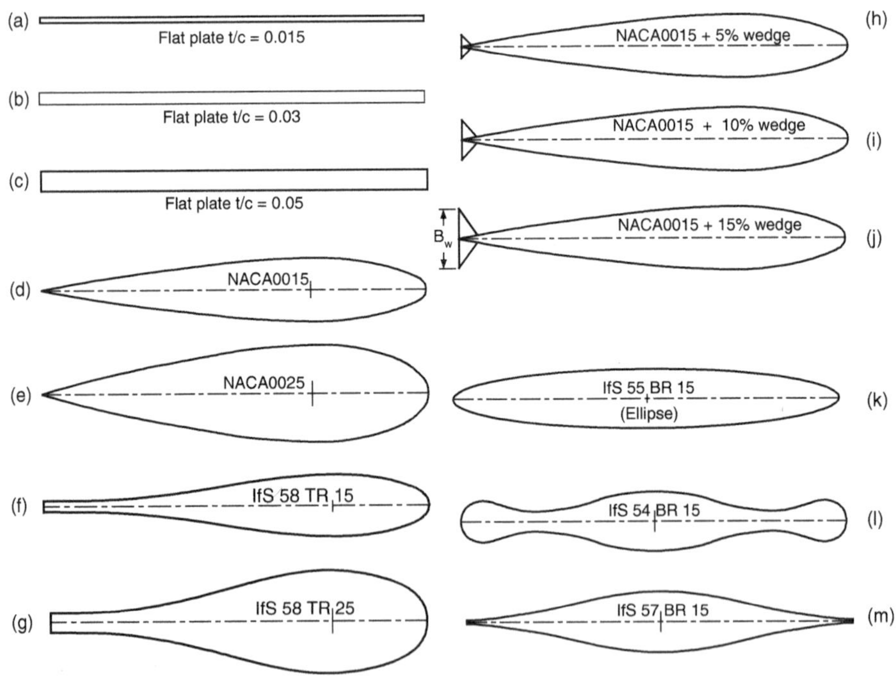

Figure 5.23 Details of section shapes for Table 5.5 (from references [5.3,5.29])

Table 5.5 Tabulated free-stream test data, different section shapes, effective aspect ratio = 1.0, square tips, [5.3] and [5.29] for the three wedge sections

Section Shape	t/c	*Fig. No.*	*Re* $\times 10^6$	*Ahead/ Astern*	$\frac{dC_L}{d\alpha}$	C_L *10°*	C_L *20°*	C_L *max*	α_{stall}	C_{Do}	C_D *10°*	C_D *20°*	*CPc 10°*	*CPc stall*
Flat plate	0.015	36	0.71	Either	0.030	0.349	0.749	1.18	40	0.018	0.072	0.271	0.26	0.40
Flat plate	0.030	37	0.71	Either	0.031	0.342	0.722	1.14	40	0.033	0.092	0.289	0.28	0.41
Flat plate	0.050	38	0.67	Either	0.033	0.345	0.697	1.13	40	0.054	0.117	0.315	0.31	0.43
NACA0015	0.15	20	0.79	Ahead	0.028	0.289	0.622	1.05	34	0.020	0.042	0.135	0.18	0.35
NACA0015	0.15	21	0.56	Astern	0.048	0.385	0.643	1.06	40	0.050	0.102	0.258		
NACA0025	0.25	22	0.78	Ahead	0.026	0.270	0.586	1.34	46	0.031	0.052	0.145	0.20	0.35
NACA0025	0.25	23	0.56	Astern	0.053	0.412	0.672	0.91	37	0.095	0.142	0.289		
IfS 58TR 15	0.15	24	0.78	Ahead	0.032	0.322	0.670	1.18	34	0.029	0.053	0.166	0.25	0.34
IfS 58TR 15	0.15	25	0.56	Astern	0.026	0.272	0.648	1.11	40	0.052	0.092	0.286		
IfS 58TR 25	0.25	26	0.78	Ahead	0.030	0.313	0.660	1.45	48	0.043	0.076	0.191	0.29	0.37
IfS 58TR 25	0.25	27	0.56	Astern	0.030	0.293	0.674	0.98	40	0.085	0.139	0.356		
NACA0015 + 5% wedge	0.15	T4	0.77	Ahead	0.033	0.327	0.670	1.26	36	0.058	0.091	0.210	0.27	0.35
NACA0015 + 10% wedge	0.15	T7	0.77	Ahead	0.033	0.327	0.653	1.25	35	0.088	0.142	0.257	0.28	0.36
NACA0015 + 15% wedge	0.15	T8	0.76	Ahead	0.031	0.319	0.661	1.20	35	0.108	0.159	0.294	0.29	0.36
IfS 55 BR 15	0.15	32	0.71	Either	0.037	0.335	0.601	0.721	30	0.028	0.059	0.182	0.30	0.36
IfS 54 BR 15	0.15	33	0.78	Either	0.030	0.341	0.630	0.990	38	0.060	0.121	0.284	0.34	0.42
IfS 57 BR 15	0.15	35	0.72	Either	0.030	0.293	0.572	1.038	40	0.028	0.068	0.217	0.24	0.41

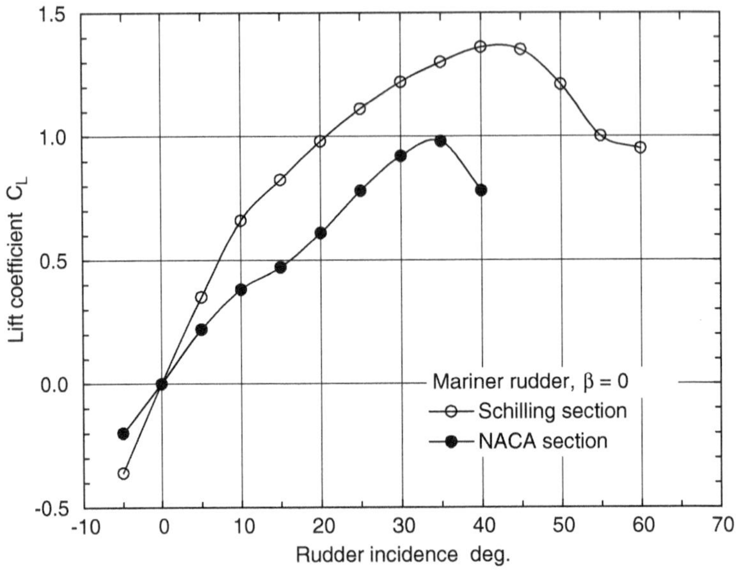

Figure 5.24 Wind tunnel free-stream tests on a Schilling Mariner rudder, courtesy of Becker Marine Systems

α_{stall}, together with a much smaller travel of *CPc*. The trailing edge wedge sections also show an increase in lift curve slope and C_{Lmax}, but are accompanied by a significant increase in drag.

From the results presented by Kracht [5.34], it is found that, compared with the NACA0020 section, the HSVA-MP-71-20 section shows some improvement in lift curve slope but has a slightly lower stall angle. There is some widening of the cavitation inception envelope for the HSVA sections and this is discussed in Section 5.8.

The main beneficial properties of the Schilling section and rudder are illustrated in Figure 5.24, which compares free-stream wind tunnel tests on a Schilling Mariner rudder with the test results for an equivalent rudder with a NACA section [5.37]. A significant increase in lift curve slope and a delay in stall are demonstrated.

5.3.2.13 *Astern condition*

With astern operation, the trailing edge of the rudder becomes the leading edge and vice versa. Rudder inflow speed tends to be simply the astern ship speed, with no help from the race of a propeller (if fitted in line of rudder). Most merchant ships tend not to go at high speed astern, and torques can be limited by possible limitations on speed and rudder angle. Exceptions tend to be warships and some ferries, where the need to steer at relatively high speeds astern may be important. Considering rudder types, it is generally accepted that flapped rudders, and rudders with thickened trailing edges or with trailing edge wedges do not perform as well going astern.

Data for the astern condition are relatively limited. Relevant all-movable rudder data are, however, reported by Whicker and Fehlner [5.2], Thieme [5.3] and Molland and Turnock [5.38]. The results of Whicker and Fehlner and Thieme are summarised in Tables 5.2 and 5.5. Complete sets of results based on the Whicker and Fehlner data for the ahead and astern conditions for an aspect ratio of 3.0 and $t/c = 0.15$ are shown in Figures 5.9(a) and 5.9(b).

Compared with the ahead condition, the results in Table 5.2 and Figures 5.9(a) and 5.9(b) indicate that in the astern condition, depending also on aspect ratio and sweep

- $\frac{dC_L}{d\alpha}$ is lower, being about 75–85% of the ahead condition.
- α_{stall} is much lower than the ahead condition by 5–10°.
- C_{Lmax} is about 45–75% of the ahead condition.
- *CPc* at stall is 5–8% aft of the ahead case.
- *CPs* at stall is 5–10% less than ahead case.

The results of Thieme (Table 5.5), broadly follow these trends for thicker ($t/c = 0.25$) sections but not for the sections with $t/c = 0.15$. In particular, Thieme finds an increase in lift curve slope with the astern condition. The results of Molland and Turnock (for $t/c = 0.20$) also show broad agreement with the above trends except for lift curve slope which, like the Thieme results, were some 30% higher than the ahead condition. These results suggest that, for thicker sections (say t/c 0.20–0.25), an increase in lift curve slope of the order of 30% should be assumed for the astern condition.

The above data indicate that, when going astern, *CPc* will be well ahead of the stock which could lead to large negative torques. The earlier stall and lower C_{Lmax} will, however, contribute to limiting the maximum astern torque to a certain extent. Positioning of the stock to take account of ahead and astern torques is discussed further in example application 1 in Chapter 11.

5.3.3 Flapped rudders

Flapped control surfaces, Figures 2.1(c) and 5.22(g), effectively work as variable camber devices and, for a given control surface incidence, can produce a significant increase in lift. Due to their facility to produce high lift, they tend to be applied in situations where space may be limited, such as for fin stabilisers, or for rudders required to produce high manoeuvring forces. The concept is not new and Lumley [5.39] describes the application of such rudders as early as 1862 on merchant and war ships. Lumley refers to the flap as the tail due to the analogy to a fish. Drawings of Lumley's system are shown in Figure 5.25 where a self-acting increased movement of the flap can be achieved in the same direction as the main rudder, (a) being the chain driven version and (b) the second system which, in concept, is the same as that applied to many modern flapped rudders and control surfaces. Further refinements to the second system are included in Lumley's paper.

Kato and Motora [5.5] present one of the first systematic sets of open water tests on flapped rudders. The rudders had an effective aspect ratio of 2.0 and a NACA0020 section. The tests were carried out in a recirculating water channel at a Reynolds

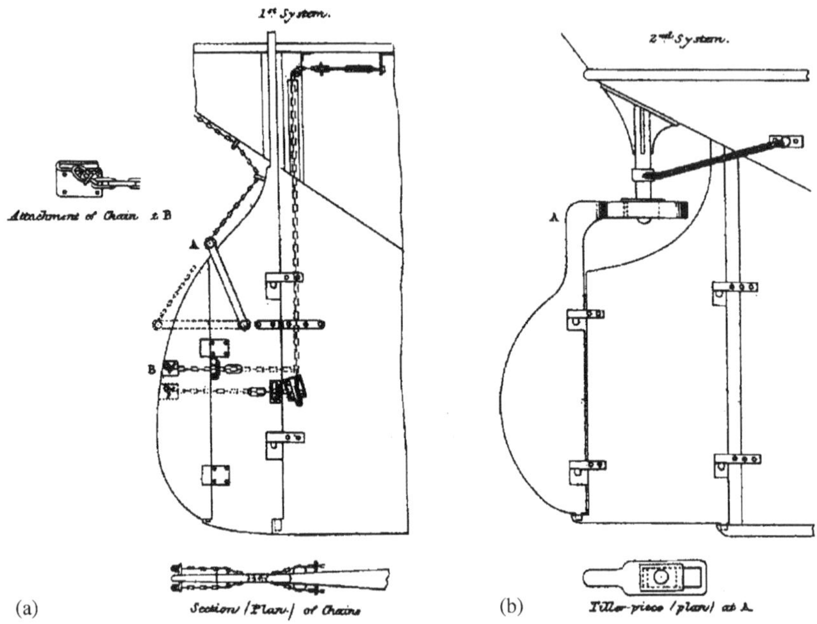

(a) (b)

Figure 5.25 The Lumley flapped rudder

number of about 0.125×10^6. Flap chord/total chord ratios, c_f/c, of 0.50, 0.25 and 0.165 were tested and the lift and moment on the main rudder and flap were measured. Drag measurements are not presented. Examples of the tests are reproduced (with permission) in Figure 5.26, where α is the rudder angle and β the flap angle. It is concluded from the tests that the best performance could be achieved by a rudder with a flap ratio of 0.25 and a flap angle/rudder angle ratio, β/α, of 2, when the lift is almost doubled.

Kerwin *et al.* [5.6] carried out free-stream tests on a series of twelve flapped rudders in a water tunnel. Tests were carried out for flap chord/total chord ratios of 0.2, 0.30, 0.40, 0.50 and 0.60. The basic particulars of the rudder series are given in Figure 5.27 and Table 5.6 and examples of three flaps are given in Figure 5.28. The principal hydrodynamic characteristics of the rudder series are given in Table 5.7 and examples of the results for 20%, 30% and 50% flap chords are reproduced, with permission, in Figures 5.29(a–c). A comparison between flapped and all-movable rudders is made in example application 5 in Chapter 11.

5.3.4 Skeg rudders (full-depth skeg)

Data for rudders with a full-depth skeg, Figure 2.2(c), are very limited and recourse normally has to be made to flapped foil data, treating the flap as the movable rudder and the 'rudder' as the fixed skeg. A restriction on the use of such data, for example the data in [5.6], is that the flap chord/total chord ratio is often lower than

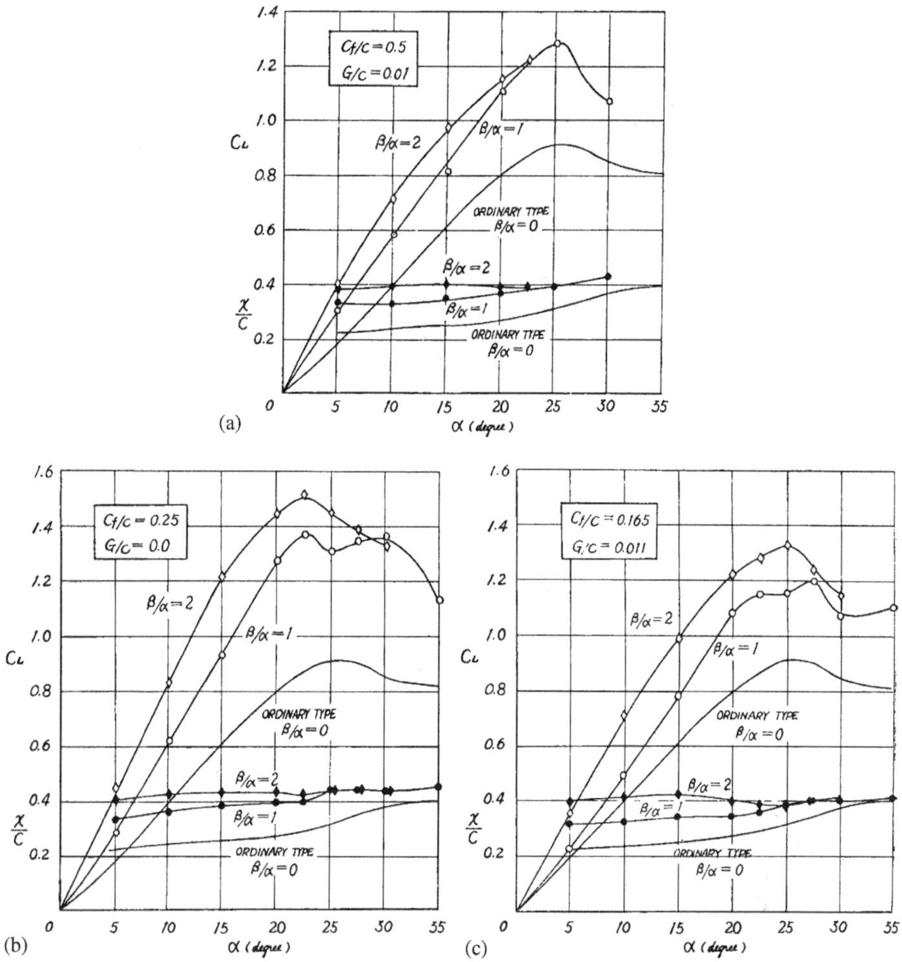

Figure 5.26 Characteristics of flapped rudders [5.5]

that suitable for skeg rudders. Data for movable rudder area/total area ratios *Am/A* of 0.40, 0.50 and 0.60, cross plotted from data in [5.6] are shown in Figure 5.30.

An extrapolation for *Am/A* of 0.70 and $AR = 4$ based on Figure 5.34 is given in Figure 5.31.

Figure 5.32 compares the lift coefficients and lift/drag ratios for all-movable and skeg rudders (with $AR_E = 2.8$ and for $\beta = 0$) using data from [5.6]. It is seen that whilst the rate of increase of lift is considerably less for the skeg rudder, its stall angle is delayed by about 10° and the maximum lift coefficient is about the same as the all-movable rudder. However, the lift/drag ratio, or measure of efficiency, for the skeg rudder is considerably less than the all-movable rudder. In other words, for the same developed lift at a particular angle, there is a significant increase in drag over the all-movable rudder. Example application 12 in Chapter 11 is used to illustrate the magnitude of these differences.

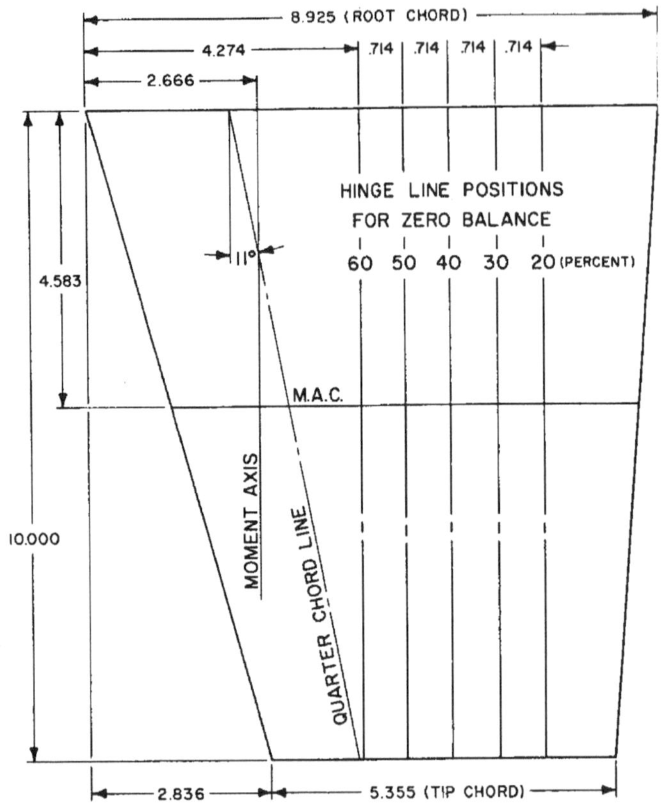

Figure 5.27 Details of model rudders [5.6]

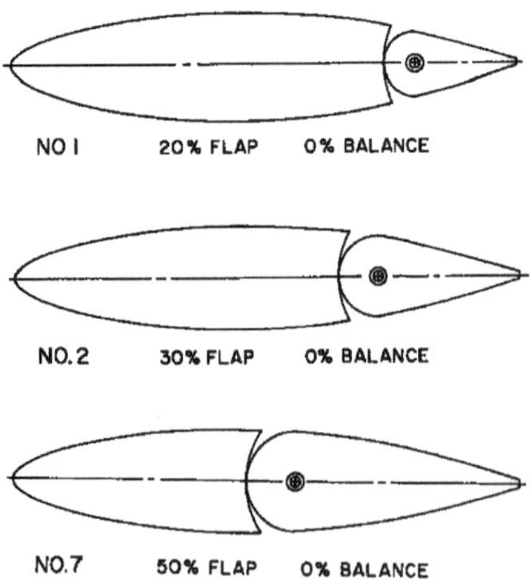

Figure 5.28 Sections of three of the models tested [5.6]

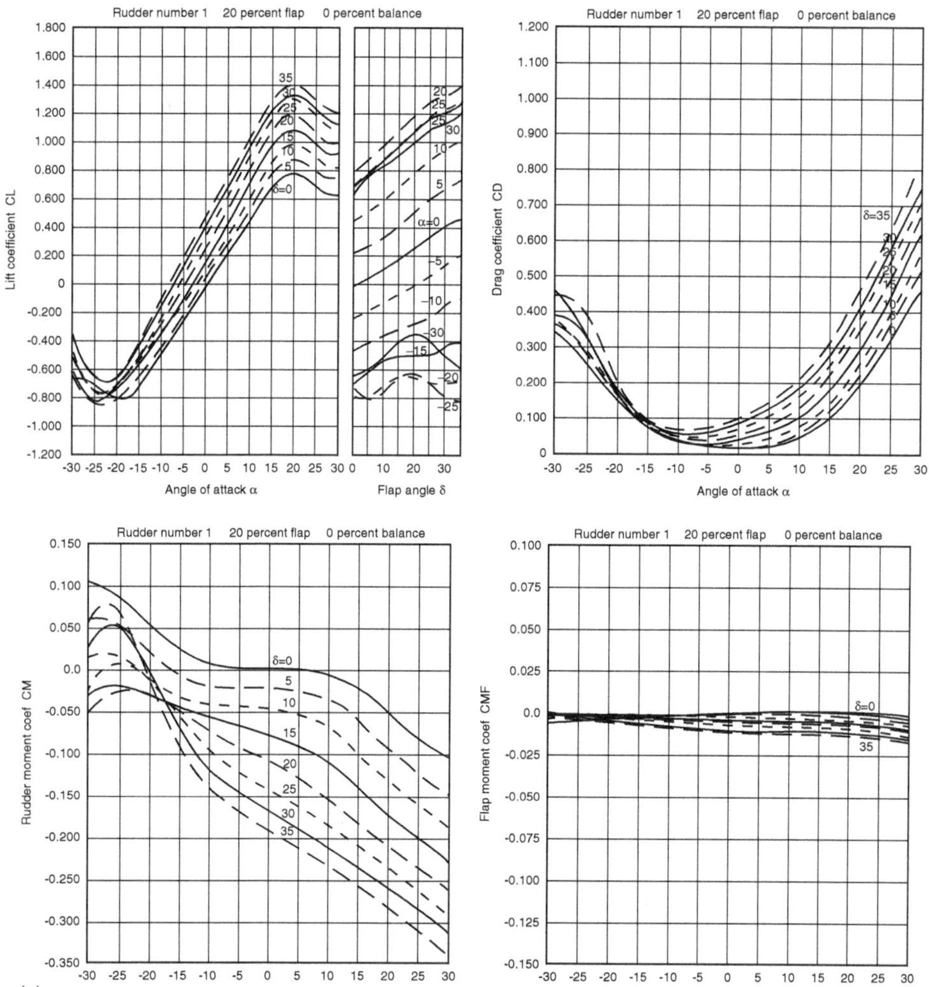

(a)

Figure 5.29 (a) Flapped rudder characteristics: 20% flap [5.6]

Figure 5.33 gives a further indication of the drag penalty incurred by the skeg rudder. Whilst the increase in drag with increasing angle of attack is primarily due to early separation of the flow over the movable part of the rudder (profile drag), it is assumed for comparative purposes to be a function of C_L^2 and hence can be included in the induced drag component. C_{Do} is assumed to remain approximately constant at about 0.016 for $\beta = 0$ and 0.025 for $\beta = \pm 5°$ for these skeg rudders (compared with about 0.013 for the equivalent all-movable rudder). The curves in Figure 5.33 were derived from Figures 5.30(b) and 5.30(c) and clearly show that as skeg size is increased, there is a marked increase in induced drag factor k_i (Equation (5.10b)), and hence induced drag. For example, with a movable chord 60% of total, the induced drag factor k_i has increased to 0.78 (for $\beta = 0$) compared

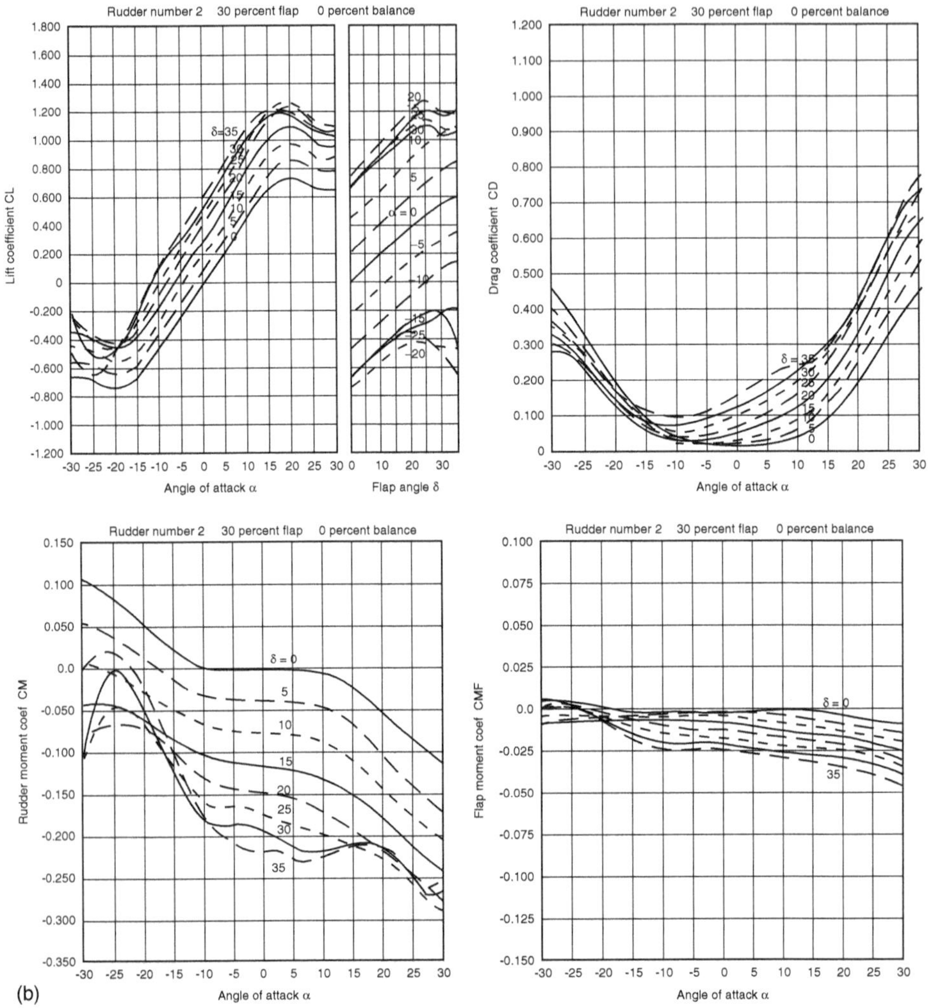

Figure 5.29 (b) Flapped rudder characteristics: 30% flap [5.6]

with about 0.37 for the all-movable case. There is a further increase in k_l with negative skeg angle β, but a slight improvement with positive β. The curves indicate a relatively rapid increase in k_l as movable chord is decreased below about 70%.

Figure 5.34 is based on data from references [5.6, 5.40–5.44] and all-movable data. It gives the relationship between the lift on a skeg rudder (movable part plus skeg) in the region of 10° of rudder incidence, relative to the lift on an all-movable equivalent, *Ls/L* and the movable area ratio, *Am/A*. Based on the various data, a design mean curve is suggested (Equation (5.18)). It should be noted that the high data are for two dimensional (infinite aspect ratio) sections with the gap between the rudder and skeg sealed. The lower results, from reference [5.6], are probably accounted for by the relatively large gap required to allow for experiments on flap balance.

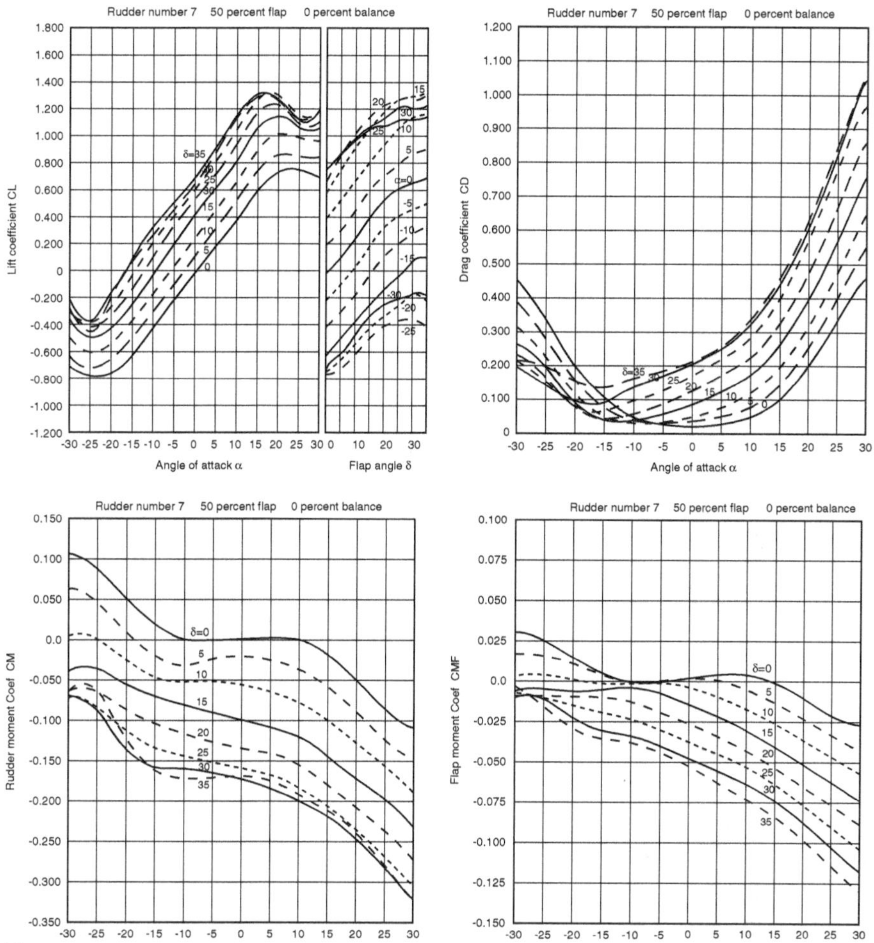

(c)

Figure 5.29 (c) Flapped rudder characteristics: 50% flap [5.6]

Table 5.6 Principal particulars of flapped rudders [5.6]

Effective aspect ratio: 2.8
Taper ratio: 0.60
Sweep: 11° aft
Section: Tip NACA66010, Root NACA66020
Flap/chord ratios: 20%, 30%, 40%, 50%, 60%
Rudder area forward of moment axis: 17% of rudder area
Rudder root gap: $0.009\bar{c}$
Approximate Re $= 1.2 \times 10^6$

Table 5.7 Hydrodynamic characteristics of flapped rudder series [5.6]

Rudder	% Flap	% Balance	$(C_L)_{max}$	$(C_L)_{max}$ $(\alpha = 0)$	$C_D \begin{Bmatrix} \delta = 0 \\ C_L = 0.6 \end{Bmatrix}$	C_D at $(C_L)_{max}$	$(C_D)_{min}$	$\frac{(C_L)_{max}}{(C_D)_{min}}$	$(C_{MF})_{max}$ (Pos, C_L)	$\frac{(C_L)_{max}}{(C_D)_{min}}$ $(\alpha = 0)$
1	20	0	1.40	0.47	0.07	0.47	0.013	108	−0.017	36
2	30	0	1.24	0.60	0.08	0.39	0.016	78	−0.042	38
3	30	39	1.18	0.29	0.11	0.58	0.018	66	−0.007	16
4	40	0	1.36	0.66	0.10	0.53	0.015	91	−0.080	44
5	40	19	1.63	0.69	0.09	0.69	0.016	102	−0.037	44
6	40	38	1.29	0.51	0.11	0.70	0.017	76	−0.010	30
7	50	0	1.35	0.70	0.11	0.54	0.017	80	−0.135	41
8	50	18	1.29	0.64	0.12	0.66	0.018	72	−0.067	35
9	50	36	1.26	0.61	0.13	0.93	0.020	63	−0.025	31
11	60	18	1.20	0.86	0.17	0.67	0.016	75	−0.105	54
12	60	36	1.17	0.76	0.18	0.83	0.025	47	−0.045	30
32	0[a]	0[a]	0.88	0.00	0.07	0.19	0.013	68	(−0.120)[a]	–

[a] Rudder 32 has no flap.

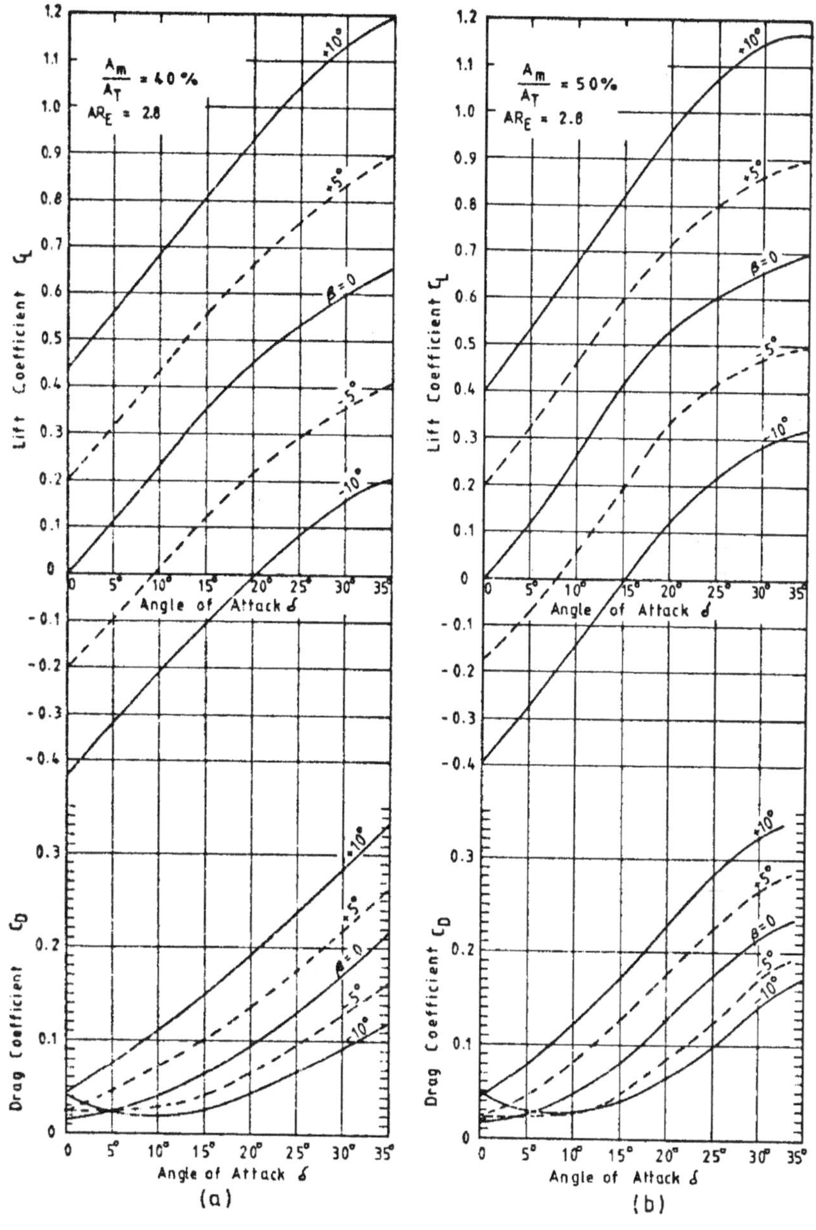

Figure 5.30 Skeg rudder characteristics

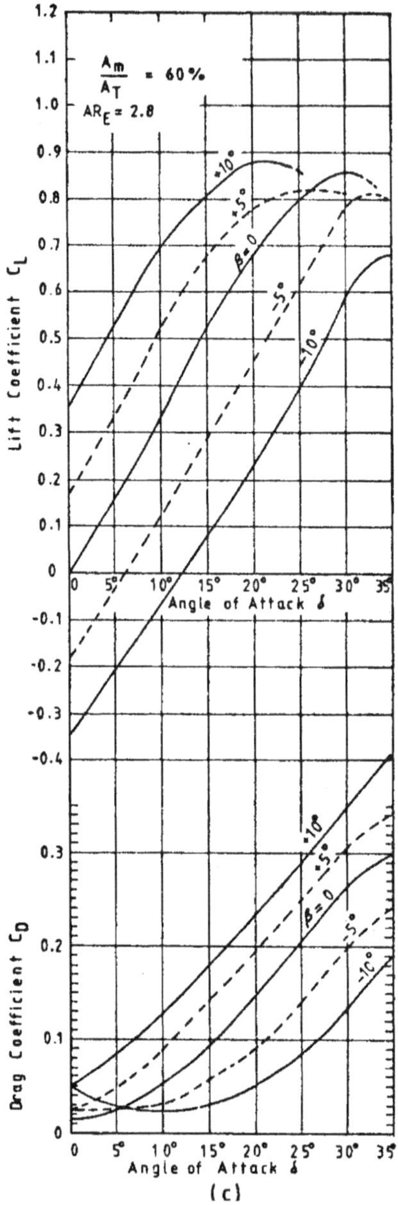

Figure 5.30 (Continued)

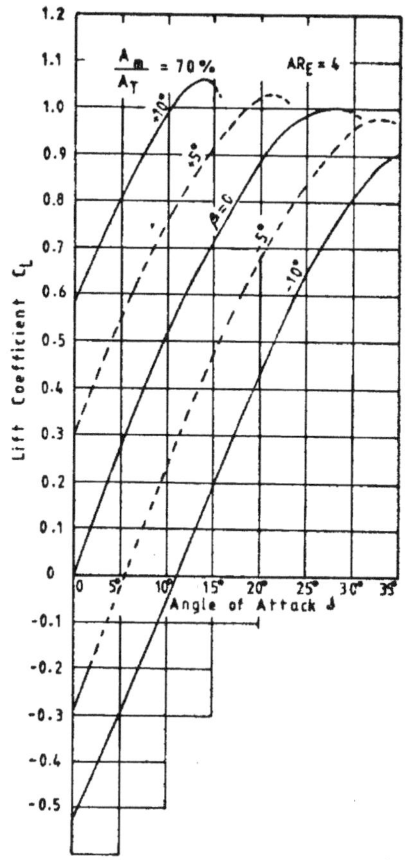

Figure 5.31 Skeg rudder characteristics; $A_m/A_T = 70\%$, $AR_E = 4.0$

$$\frac{Ls}{L} = \frac{2.4}{[1.4 + 1/(Am/A)]} \tag{5.18}$$

Figure 5.34 indicates that if the gap size between the rudder and skeg can be minimised an improvement in lift production follows. At the same time it is seen that even with the gap sealed, the skeg rudder still falls short of the rate of increase of lift attainable by the all-movable rudder.

If the skeg rudder is considered as a total movable plus skeg combination, then it is not unreasonable to assume that the influence of the overall rudder characteristics such as aspect ratio, section shape and tip shape discussed for the all-movable rudder, will also apply to skeg rudders.

In order to estimate torque, and hence stock size for the skeg rudder, the proportion of load carried by the movable part together with its centre of pressure is required. Figure 5.35, derived from the pressure measurements in references [5.41] and [5.42], offers guidance for these characteristics. Typical examples of these pressure distributions are shown in Figure 5.36. A comparison is made with the simple

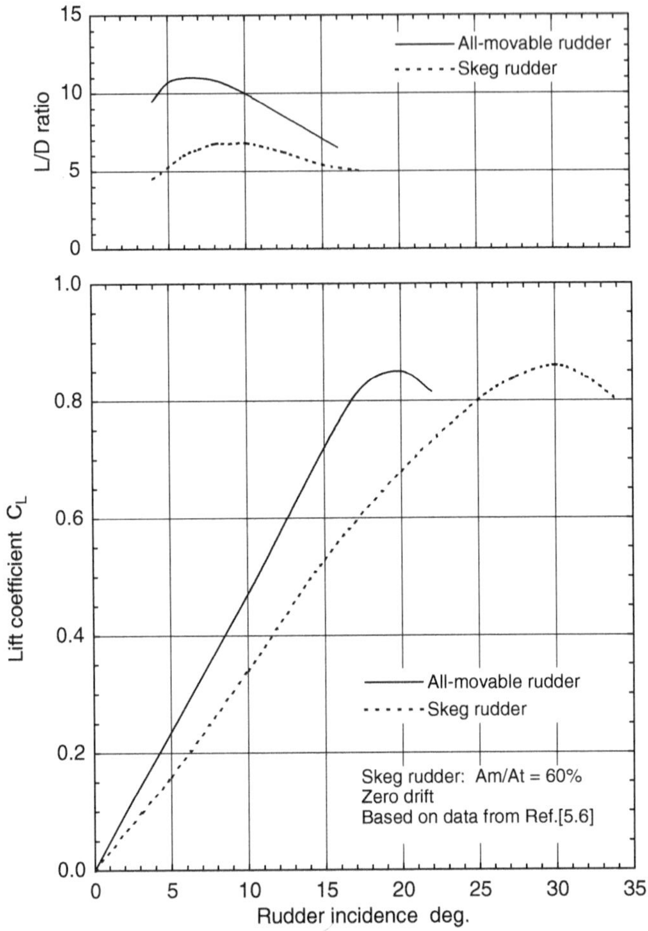

Figure 5.32 Comparison between all-movable and full-skeg rudder characteristics

assumption that the dotted distribution in Figure 5.36 is equivalent to the actual distribution. This suggestion was made, for example, by Jaeger [5.45]. Using this assumption it is found that the ratio of lift on the movable part to total lift is

$$\frac{L_m}{L_T} = \frac{C_m}{[2C_s + C_m]} \tag{5.19}$$

This relationship is included in Figure 5.35 and it is seen that whilst it errs a little on the high (safe) side for higher C_m/c ratios, it should prove satisfactory for general design purposes.

Analysis of the pressure data in references [5.41] and [5.42] indicated that, for both the 0.50 and 0.80 C_m/c ratios, the chordwise centre of pressure is approximately 33% of movable chord aft of the stock as rudder angle tends to zero, moving to approximately 40% aft of stock for rudder angles 20°–30°.

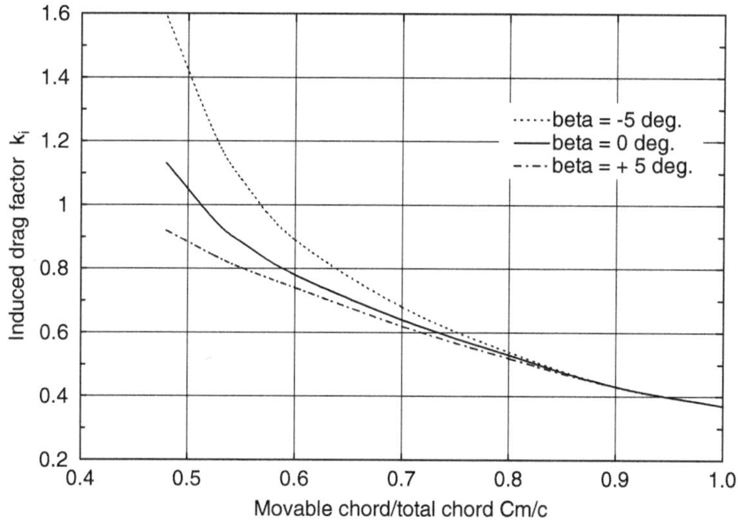

Figure 5.33 Induced drag factor: full-skeg rudders

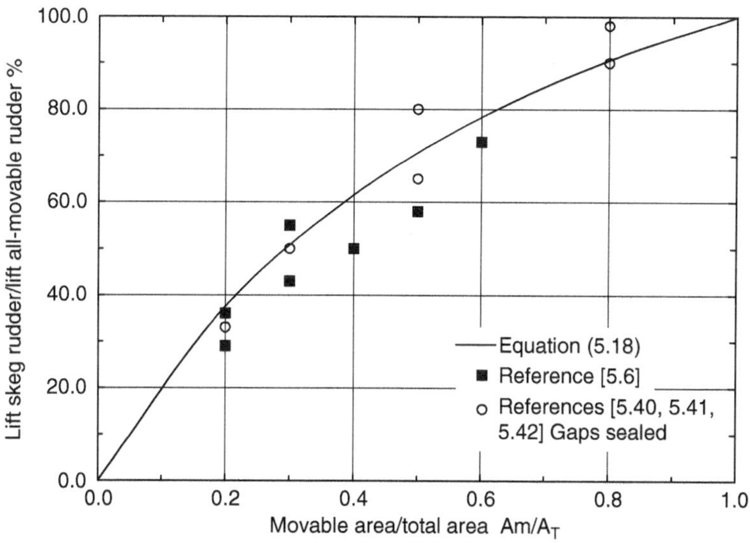

Figure 5.34 Lift of full-skeg rudder relative to all-movable rudder; $\beta = 0°$

Design procedure: Since the amount of skeg rudder data is limited, and all-movable rudder characteristics relatively well defined, the proposed design procedure is to relate the rudder plus skeg characteristics (for a particular skeg size), to the all-movable configuration by means of Figure 5.34. Having estimated the total (rudder plus skeg) force from Figure 5.34, the lift on the movable portion can be obtained from Figure 5.35 and the centre of pressure chordwise *CPc* assumed to be $0.33\bar{c}$ at small rudder

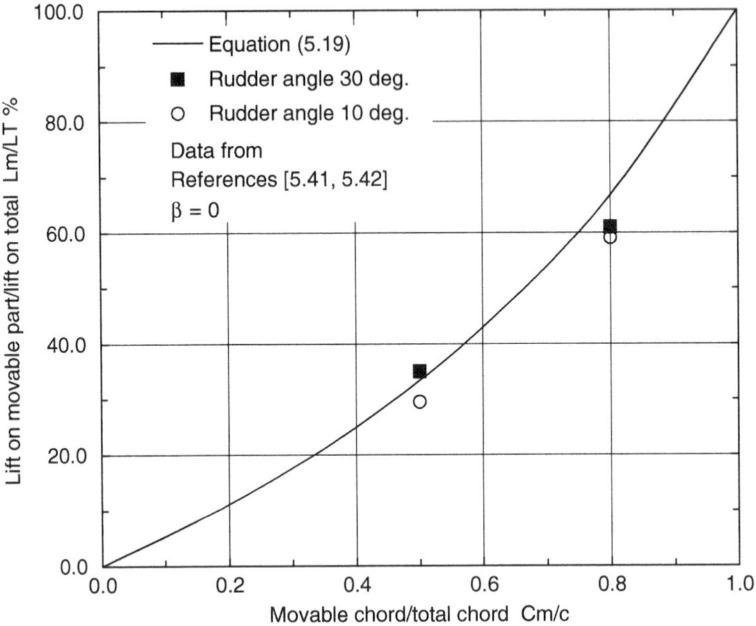

Figure 5.35 Lift of movable part of skeg rudder relative to total

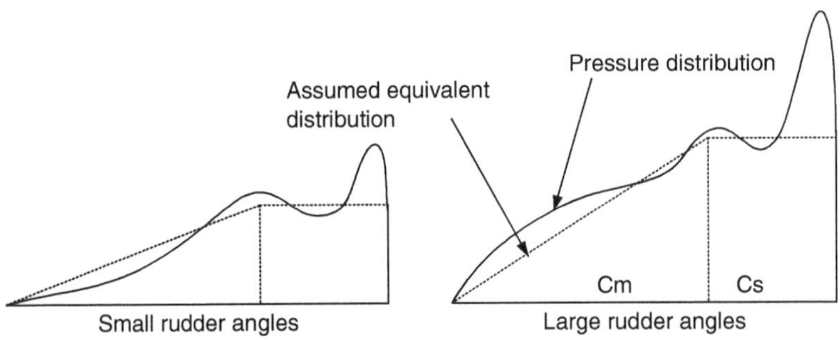

Figure 5.36 Typical chordwise pressure distributions for skeg rudder; $\beta = 0°$

angles and up to $0.40\bar{c}$ at large angles. The centre of pressure spanwise *CPs* is assumed to be in the same position as the equivalent all-movable rudder. In the absence of drag data for the movable part alone, or for a movable plus skeg combination for $C_m/c > 0.6$, Figures 5.30 and 5.33, it may be assumed that the maximum lift force is equivalent to the maximum normal force for torque and bending moment calculations. An example application to illustrate this design procedure for a sailing yacht is included as example application 10 in Chapter 11.

The fixed skeg may have adverse effects on manoeuvring and coursekeeping performance when negative drift angle on the fixed skeg occur, Figures 5.37 and 5.38.

This may be particularly significant for sailing yachts employing such skeg rudders. These characteristics are illustrated and discussed further in an application to a sailing yacht, example application 12 in Chapter 11.

5.3.5 Semi-balanced skeg rudders

The results of free-stream wind tunnel tests on semi-balanced skeg rudders are presented by Goodrich and Molland [5.7]. Tests were carried out on three rudders having taper ratios of 0.59, 0.80 and 1.00 and skeg and overall characteristics that were typical for the rudders fitted to many modern ships. Principal particulars of the three rudders are given in Table 5.8 and Figure 5.39. Leading edge roughness forming a turbulence strip was applied to both sides of the rudders. The rudders were mounted through the tunnel floor and the gap between the rudder and floor in each case was about $0.0055\bar{c}$. One purpose of the investigation was to provide a better understanding of the performance of the skeg rudder when the skeg is subjected to drift angle, which results from the cross-flow at the stern when the ship is on a turn, Figure 5.37. The influence of drift angle on the all-movable rudder is simply to decrease its effective incidence; in the case of the skeg rudder, drift angle also leads to negative inflow angles on the skeg as shown in Figure 5.38.

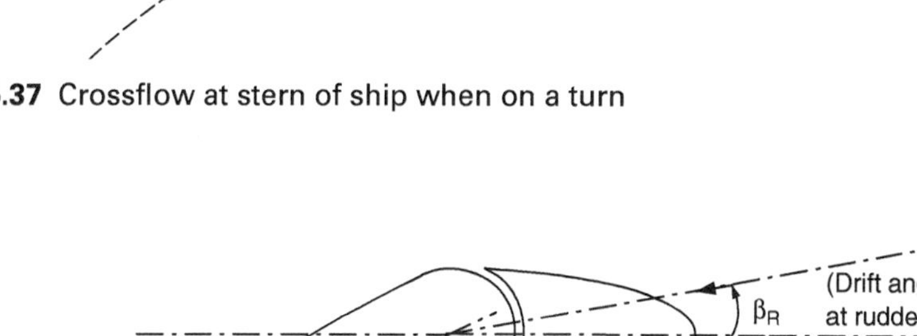

Figure 5.37 Crossflow at stern of ship when on a turn

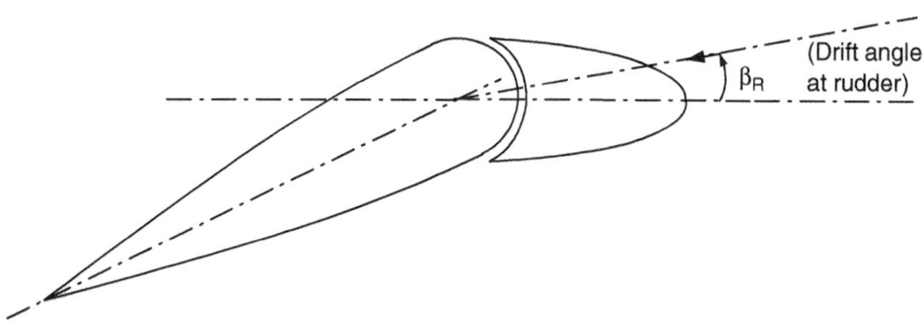

Figure 5.38 Development of negative inflow angle on skeg due to drift

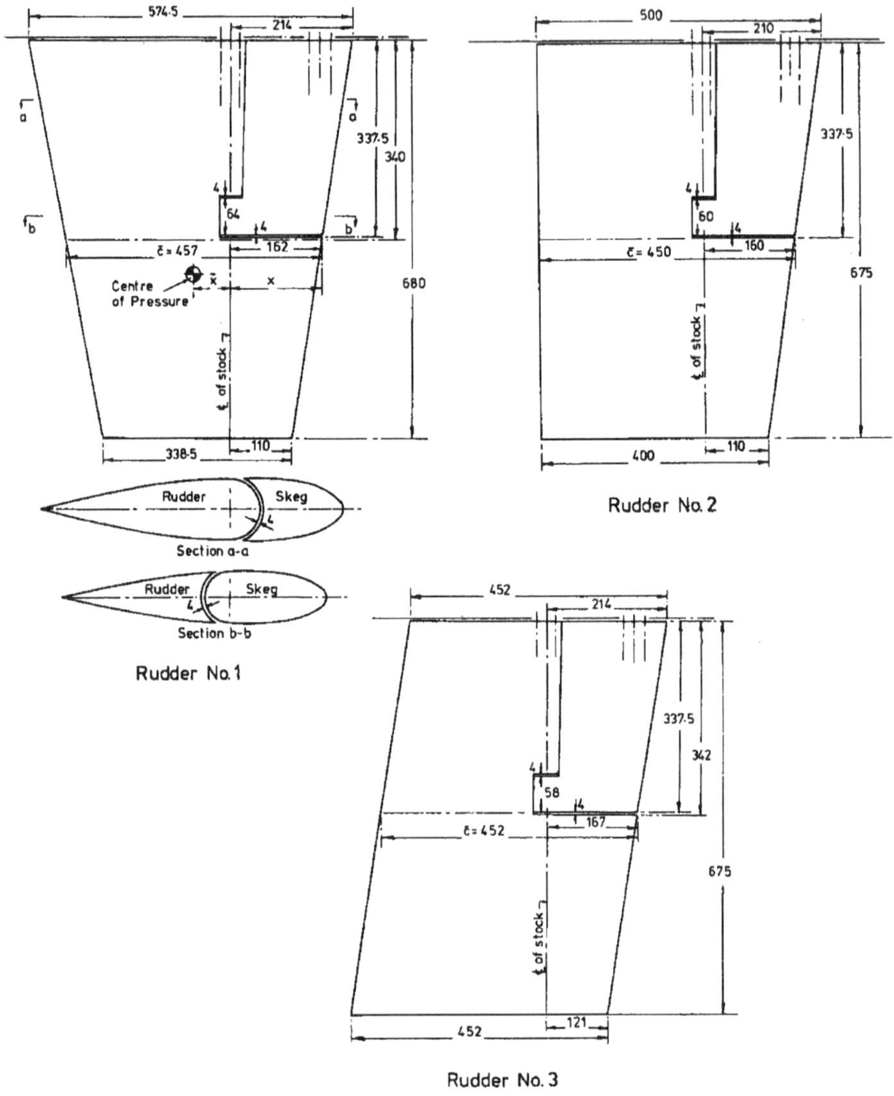

Figure 5.39 Details of model rudders [5.7]

Alternatively, if the rudder is put over whilst the ship is on a turn, a positive angle of attack is developed on the skeg. For test purposes, maximum skeg angles of about ±15° were tested as representing realistic limits on drift angle at the rudder. The basic tests were carried out at a nominal Reynolds number, based on mean chord, of 1.2×10^6 although some tests were limited to a nominal Reynolds number of 0.9×10^6 when loadings on, or oscillations of, the test rig led to a restriction in the wind speed.

Table 5.8 Model rudder dimensions, semi-balanced skeg rudders [5.7]

Rudder Number	*1*	*2*	*3*
Mean chord $\bar{c}$	457	450	452
Span *S* mm	680	675	675
Geometric aspect ratio AR_G	1.49	1.50	1.49
Taper ratio C_T/C_R	0.59	0.80	1.00
Thickness/chord ratio t/c	0.20	0.20	0.20
Section	NACA 0020 root and tip, with square tips		
Skeg depth/span	0.5	0.5	0.5
Skeg area ratio	20.5%	20.5%	21%
Balance area ratio	19%	19%	20%
Horizontal and vertical gaps	4 mm	4 mm	4 mm

In order to provide data for comparison with the skeg rudders, all-movable rudder cases were simulated for each rudder by sealing the gap between rudder and skeg and varying the angles of attack on rudder and skeg simultaneously. The results for the three simulated all-movable rudders in [5.7] showed that there were no significant differences between the lift, drag and centre of pressure characteristics for all three rudders, including no significant effects of taper ratio. The results for Rudder No. 1 were also compared with the free-stream characteristics of all-movable control surfaces published by Jones [5.1] and Whicker and Fehlner [5.2]. The data were corrected to a common geometric aspect ratio of 1.5 and corrections made for sweep, thickness ratio and taper ratio as necessary. The comparisons of the all-movable results are shown in Table 5.9. The lift curve slope for reference [5.7] is a little low but, taking into account the facts that the tests in reference [5.2] were conducted with a groundboard displaced from the tunnel floor, the tests in reference [5.1] were carried out with a complete wing clear of tunnel walls, and that the corrections for variations in the parameters of the various rudders were of necessity approximate, it is seen that there is satisfactory agreement between the different tests.

The notation of angles and coefficients is given in Figure 5.40 and results for skeg rudder No.1 are presented in Figure 5.41(a–f), noting that the data for the movable rudder alone are nondimensionalised using the total rudder plus skeg area. The results in Figure 5.41 show that, with increasing angle of attack, discontinuities occur in the growth of lift and drag together with a large movement of centre of pressure. These discontinuities are evidently due to early separation behind the skeg, which was confirmed by visual flow studies as shown in Figure 5.42. Thus, since separation on the all-movable part of the rudder does not start until fairly high angles, an intermediate situation can exist for the semi-balanced skeg rudder where the flow behind the skeg is fully separated whilst that clear of the skeg is still attached, a phenomenon also noted by Mandel [5.26].

Table 5.9 Characteristics for all-movable rudders; $AR_E = 3.0$, $Re \approx 1 \times 10^6$

	Smooth [5.2]	*Smooth and L.E. roughness* [5.1]	*L.E. roughness* [5.7]
$\left[\frac{dC_L}{d\alpha}\right]_{\alpha=0}$	0.052	0.053 (smooth) 0.051 (rough)	0.048
C_{Lmax}	1.02	0.94 (smooth) 0.77 (rough)	0.95
α_{stall}	22°	18°	21°
C_D at 10°	0.043	–	0.038
C_D at 20°	0.160	–	0.137
CPc at 10°	18%	–	18.5%
CPc at 20°	23%	–	23%
CPs at 10°	45%	–	46%
CPs at 20°	48%	–	48% just pre-stall

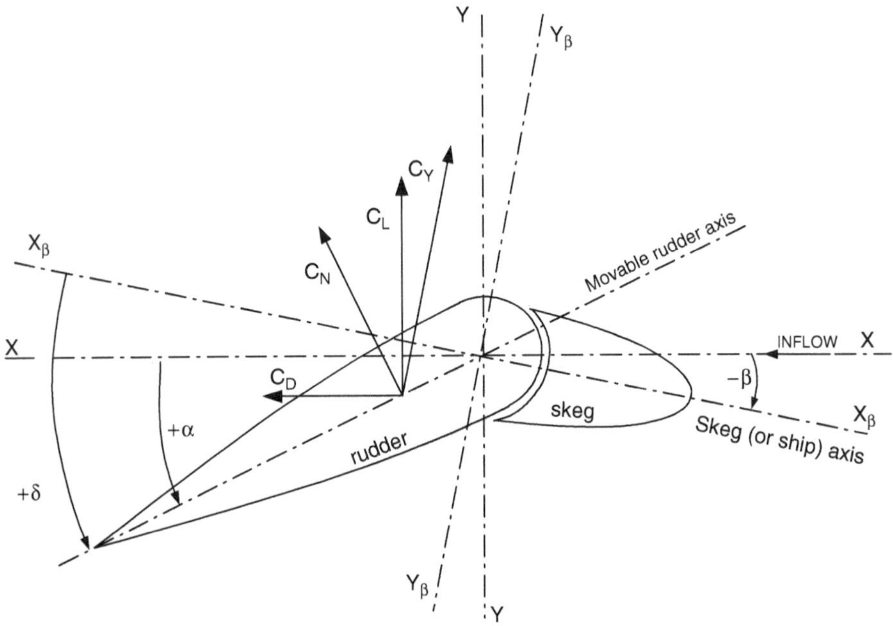

Figure 5.40 Notation of angles and coefficients for skeg rudder

The influence of sealing all the gaps between the skeg and movable part of the rudder, for Rudder No. 1, is shown in Figure 5.43. These results illustrate how the sealed gaps delay separation in the region $\alpha = 10$–$15°$, with consequent increase in lift and decrease in drag. With the vertical gap sealed independently on the high and low-pressure sides the lift is only a little less than with all gaps sealed. Figure 5.44 shows the chordwise pressure distributions with and without the vertical gap sealed at one span position and four angles of attack. It is interesting to note the peaks in

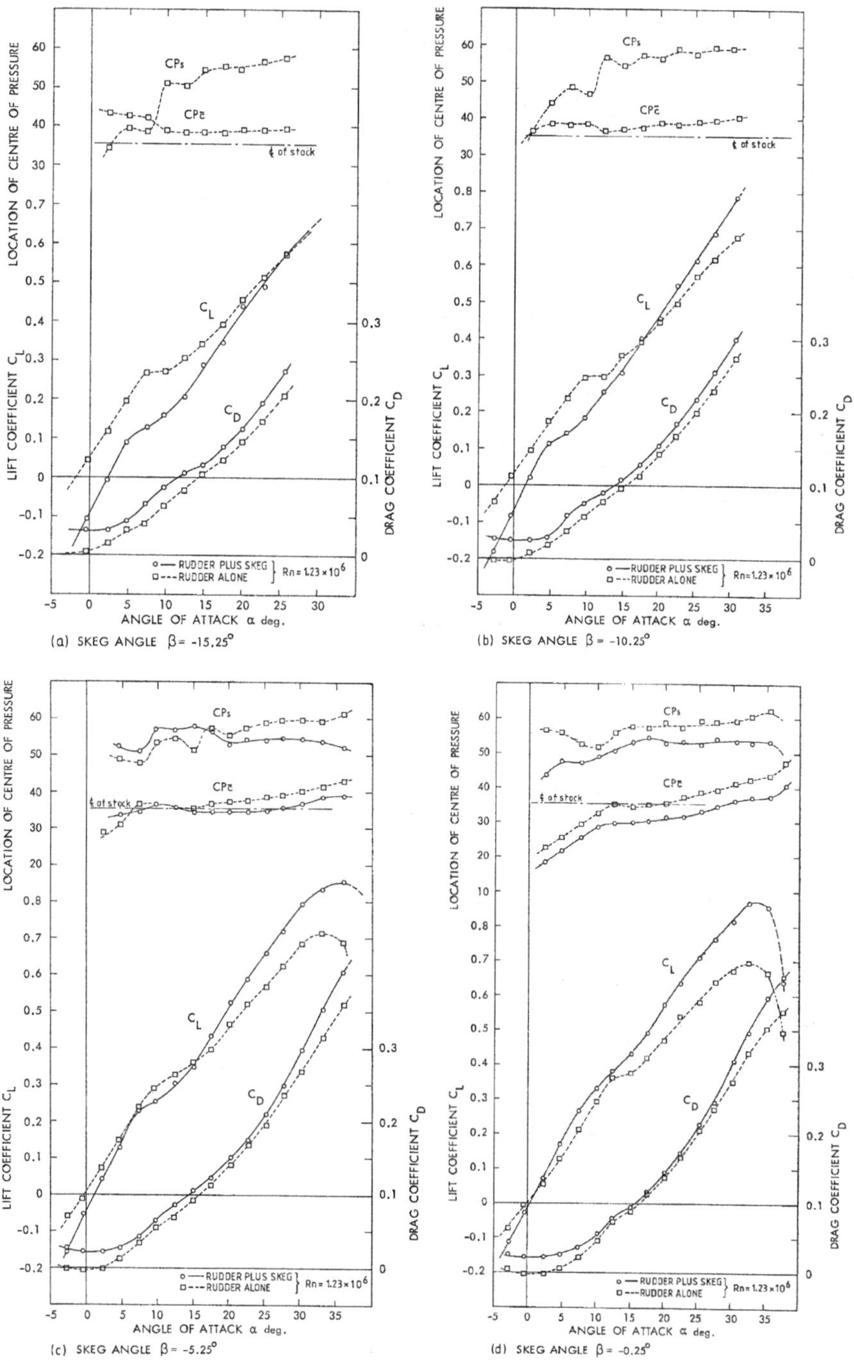

Figure 5.41 Lift, drag and centre of pressure characteristics for Rudder No. 1. (a) Skeg angle $\beta = -15.25°$; (b) Skeg angle $\beta = -10.25°$; (c) Skeg angle $\beta = -5.25°$; (d) Skeg angle $\beta = -0.25°$

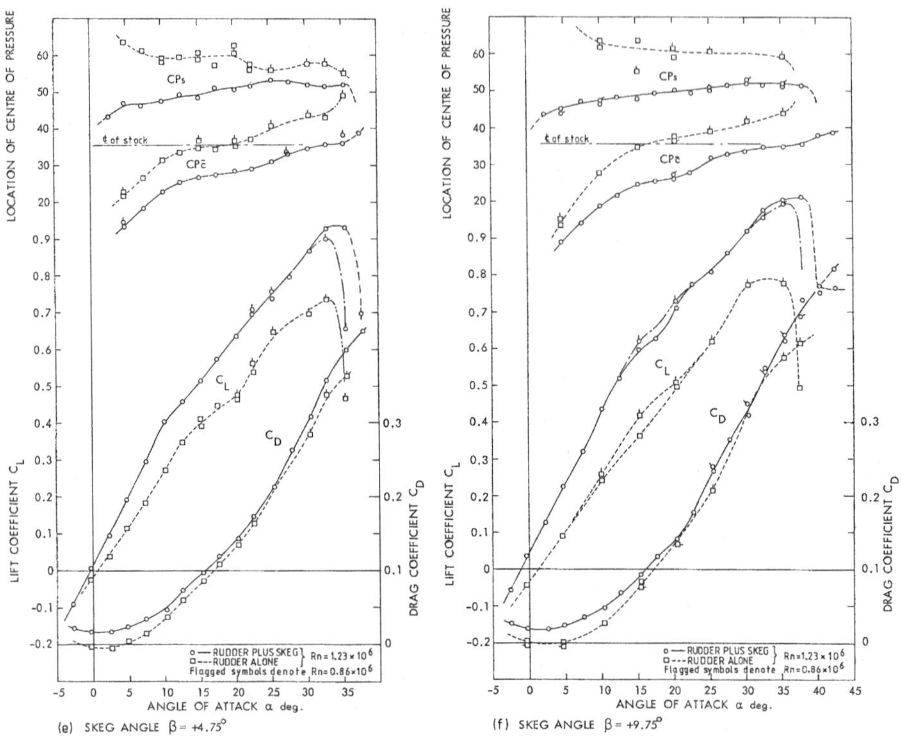

Figure 5.41 (e) Skeg angle $\beta = +4.75°$; (f) Skeg angle $\beta = +9.75°$

the distributions at about 37% chord from the leading edge where the movable part of the rudder starts. These are maintained up to about 10° with the gaps open as well as sealed, whilst at 20° the pressure peak is held only with the gaps sealed, indicating the presence of gap flow when the gaps are open. Tests on Rudder No. 1 with the horizontal gap sealed on the high-pressure side indicated no significant change in performance characteristics compared with the all-gaps-open case. Relevant two-dimensional section data from references [5.40–5.43] for aerofoils with large flaps indicate that discontinuities can occur in the lift curve even when the gap is sealed; these lift characteristics are accompanied by relatively large movements in *CPc*. It is also evident from reference [5.43] that a gap can allow flow leakage with consequent decrease in lift, and that increasing the gap decreases the lift curve slope, a characteristic reported also in reference [5.6].

Figure 5.45 compares for Rudder No. 1 the lift coefficients and lift/drag ratios for the all-movable rudder and skeg rudder (for $\beta = -0.25°$). It is seen that whilst the rate of increase of lift is considerably less for the skeg rudder, its stall angle is delayed by about 12° and the maximum lift coefficient achieved is only about 10% less than the all-movable rudder. However, the lift/drag ratio or measure of efficiency for the skeg rudder is appreciably less than the all-movable rudder. In other words, for the same developed lift at a particular angle, there is a significant increase in drag over the all-movable rudder. These differences in drag increase with increasing negative

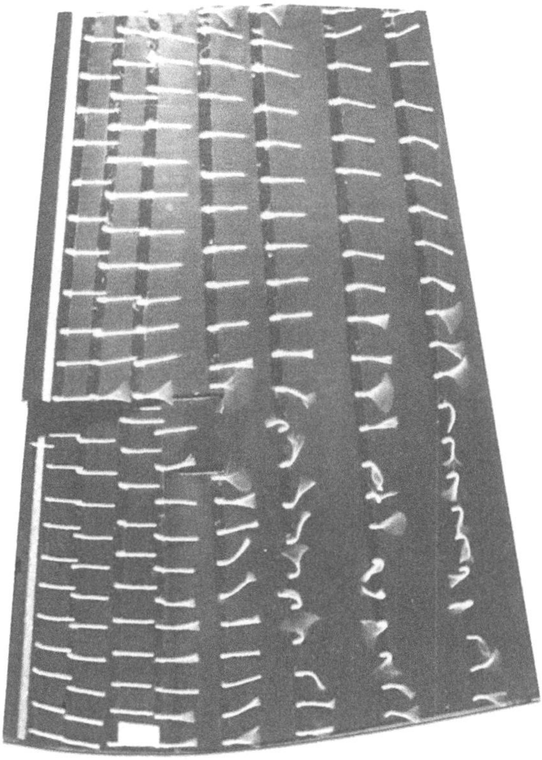

Figure 5.42 Flow pattern over rudder, $\delta = 12.5°$

skeg angle and decrease a little with increasing positive skeg angle. It was also noted that the movement of the centre of pressure for the skeg rudder both chordwise and spanwise with increasing angle of attack is significantly larger than for the equivalent all-movable rudder.

When comparing the overall characteristics of the three semi-balanced skeg rudders in Figure 5.39, it was found that any differences were relatively small. With increase in taper ratio there was some increase in lift at larger angles of attack, although this was found not to be the case with the simulated all-movable rudders.

5.3.6 Rudders behind plane deadwood

A typical layout of a rudder aft of a keel or deadwood is shown in Figure 2.2(f). The rudder–deadwood arrangement can be idealised to the case shown in Figure 5.46. In hydrodynamic terms, the arrangement in Figure 5.46 is equivalent to a very low aspect ratio foil with a trailing edge flap. An approximation to the likely load distribution is shown in Figure 5.47. The rudder *CPc* will be reasonably well forward (say 10–20% chord aft of leading edge) and, as there is no balance area, torques on this type of rudder tend to be high.

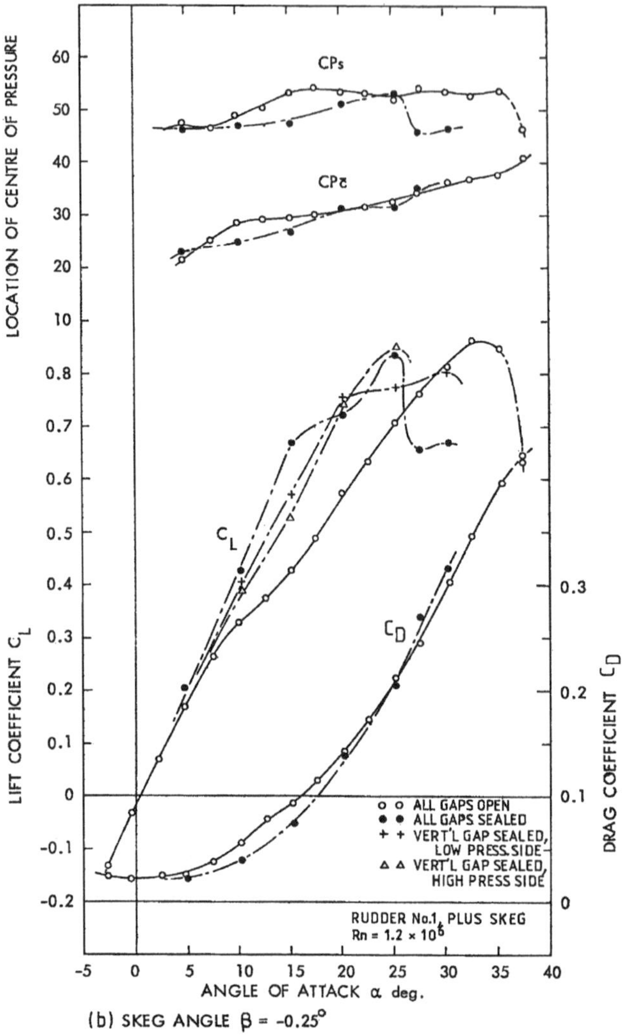

Figure 5.43 Influence of sealed gaps

Such rudder–deadwood arrangements are generally no longer in common use, except for existing older sailing vessels and replicas of older vessels. Their origins lie in the rudders of early sailing ships and they were in effective use until the advent of motor ships and the need to incorporate a propeller in the aft end arrangement. Its applications to twin screw-single rudder layouts was superseded by the semi-balanced rudder, Figure 2.2(e), and the sailing craft using either a spade rudder, Figure 2.2(b) behind a separate fin keel or a transom hung rudder, Figure 2.2(g).

There is little recent published information on the rudder forces in such an arrangement. A review and summary is made of available data and methods that may be useful in assessing the forces and torques for such rudders.

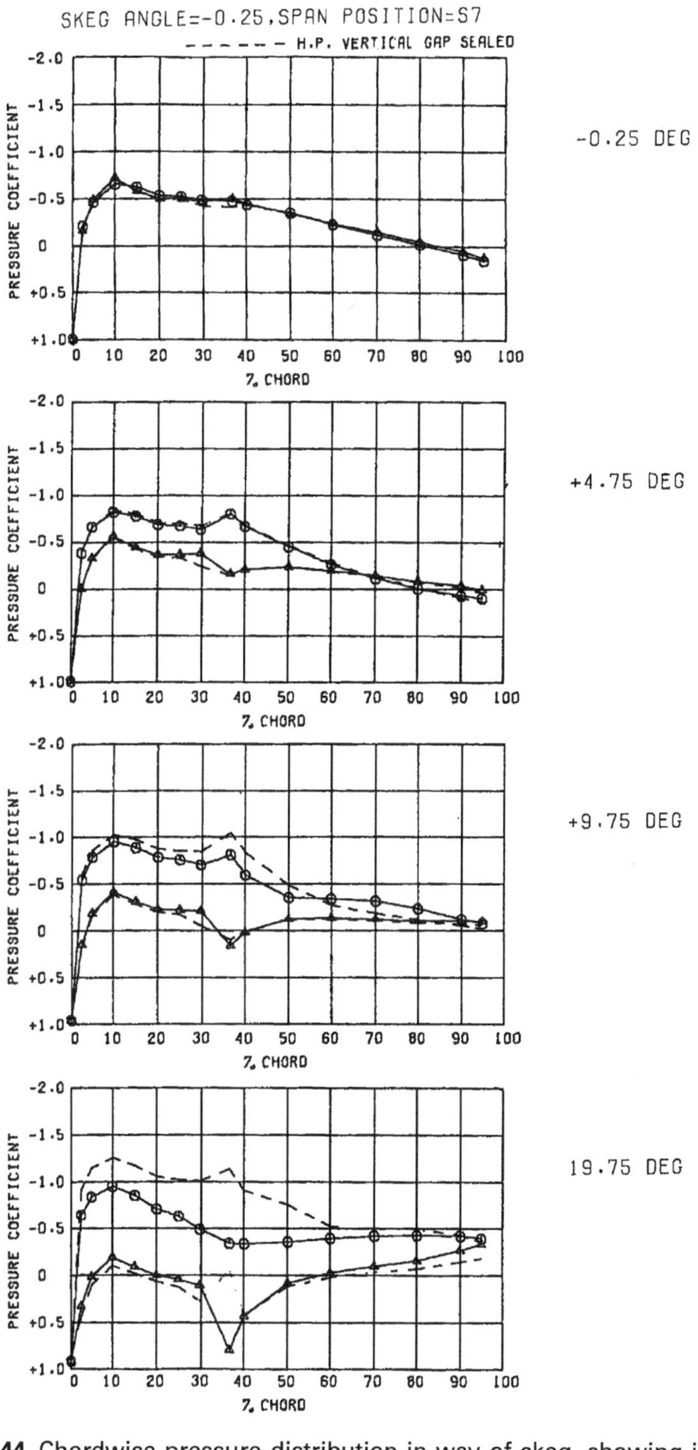

Figure 5.44 Chordwise pressure distribution in way of skeg, showing influence of sealed vertical gap

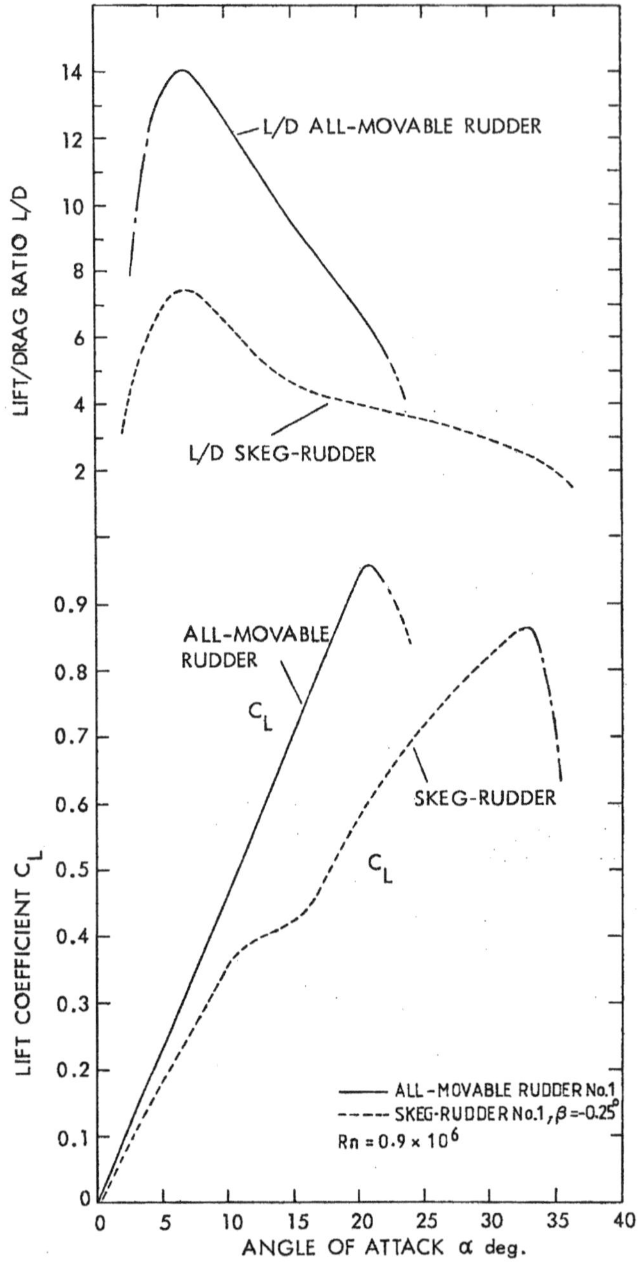

Figure 5.45 Comparison between semi-balanced skeg and all-movable rudders

Abell [5.15] carried out tests on flat plates with aft end rudders, Figures 5.48–5.50. The moments plotted are those on the rudder and deadwood taken together and Abell suggests that the results should be examined qualitatively. In the first set of experiments, Figure 5.48, the effect of endpates and two-dimensional flow was

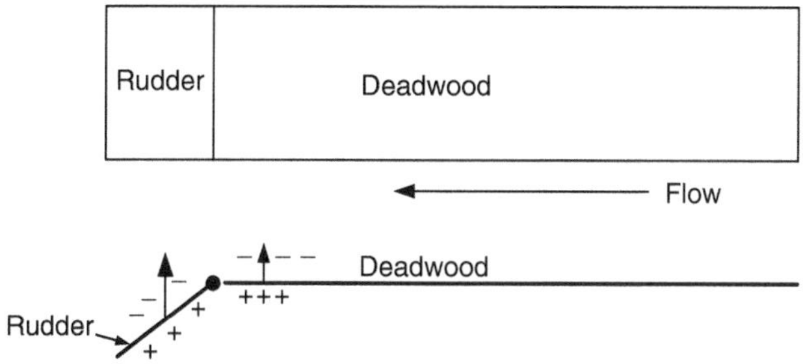

Figure 5.46 Rudder aft of deadwood

Figure 5.47 Load on rudder and deadwood

investigated and, at low angles, there is seen to be a 60% increase in transverse moment. In the second set of experiments, Figure 5.49, a screw aperture was simulated and a drop in transverse moment of about 10% occurs. In the third set of experiments, Figure 5.50, the clearance between the upper edge of the rudder and the plate was varied. A significant decrease in transverse moment is seen to occur with increase in clearance.

Similar tests on plates and flaps were carried out by Cowley *et al.* [5.45] and Munk [5.46]. A summary of their results is shown in Figure 5.51, which has been reproduced, with permission, from reference [5.47]. Rudder alone forces are plotted. Similar trends to the findings of Abell can be observed. Rudders A, B and F have an aspect ratio of 1.5 and rudders M and N an aspect ratio of 5.0. A decrease in rudder lift is observed when going from rudder A behind a fin without gap, to B with a small gap, and to F where the rudder is without a fin. Rudder F has a higher initial lift curve slope but stalls at about 23°, rudder B stalls at about 26° whilst stall is delayed to much bigger angles for case A behind the fin with no gap. It is interesting to note the change in *CPc* from the usual 25–40% chord for rudders N and F alone and 5–20% chord for rudders A and B when behind the deadwood.

The lift curve slopes for rudders N and F in Figure 5.51 are broadly as to be expected for rudders with aspect ratios of 5.0 and 1.5, and their slopes do not change significantly when behind a deadwood. A suggested design approach for this rudder type is therefore to use the aspect ratio for the rudder in question, using appropriate free-stream lift and drag data for that aspect ratio, together with a maximum *CPc* of 20% chord aft of the leading edge.

Bottomley [5.48] carried out tests on the effect of aperture in the after deadwood upstream of the rudder in a twin-screw/single-rudder arrangement. Bottomley

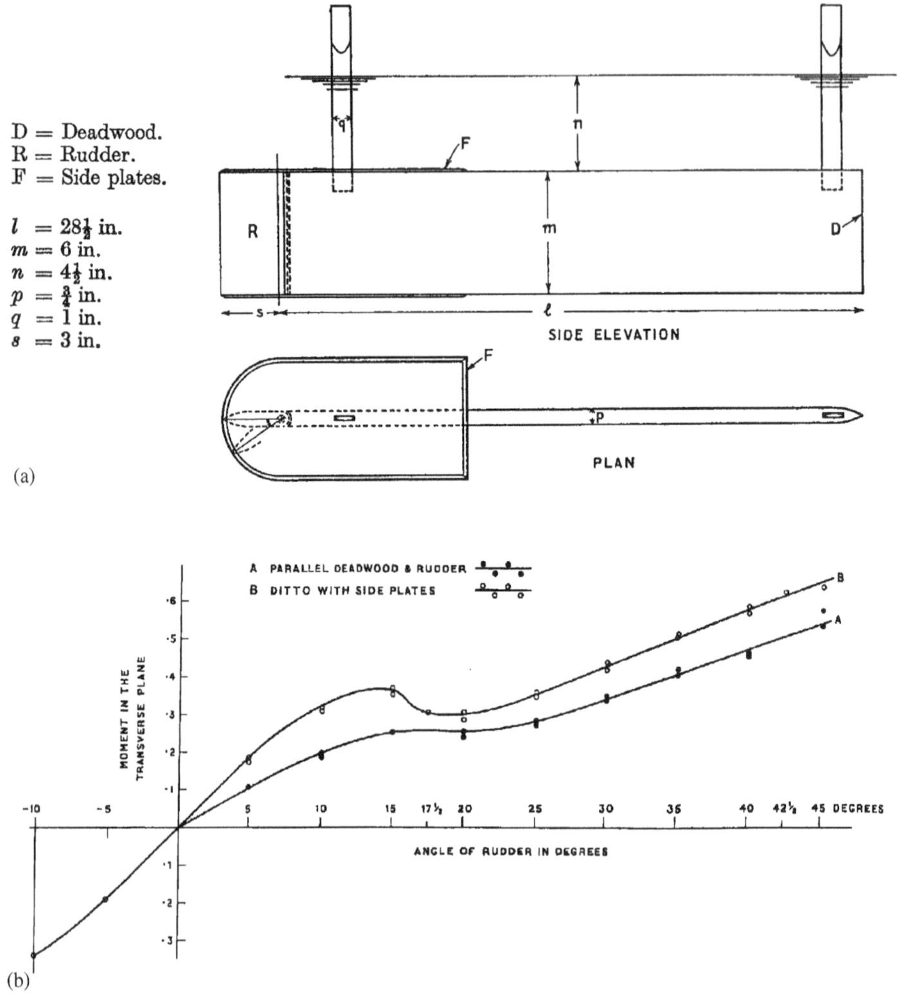

Figure 5.48 Rudder behind deadwood: effect of end plates

concluded that all openings between the rudder and afterbody on twin-screw ships should be avoided or made as small as possible. A thorough review of the effects of various afterbody arrangements and the implications of the alternatives is given in reference [5.49].

An approximate approach to obtaining the forces on a rudder is the use the flap characteristics of flapped rudder data, such as that in reference [5.6]. The main body is taken at zero incidence and the flap treated as the rudder. Like the similar application to full skeg rudders, Section 5.3.4, Figure 5.35 or Equation (5.19) may be used to estimate the force on the flap alone. A shortcoming of this approach is the lack of data for flapped rudders with a low enough aspect ratio.

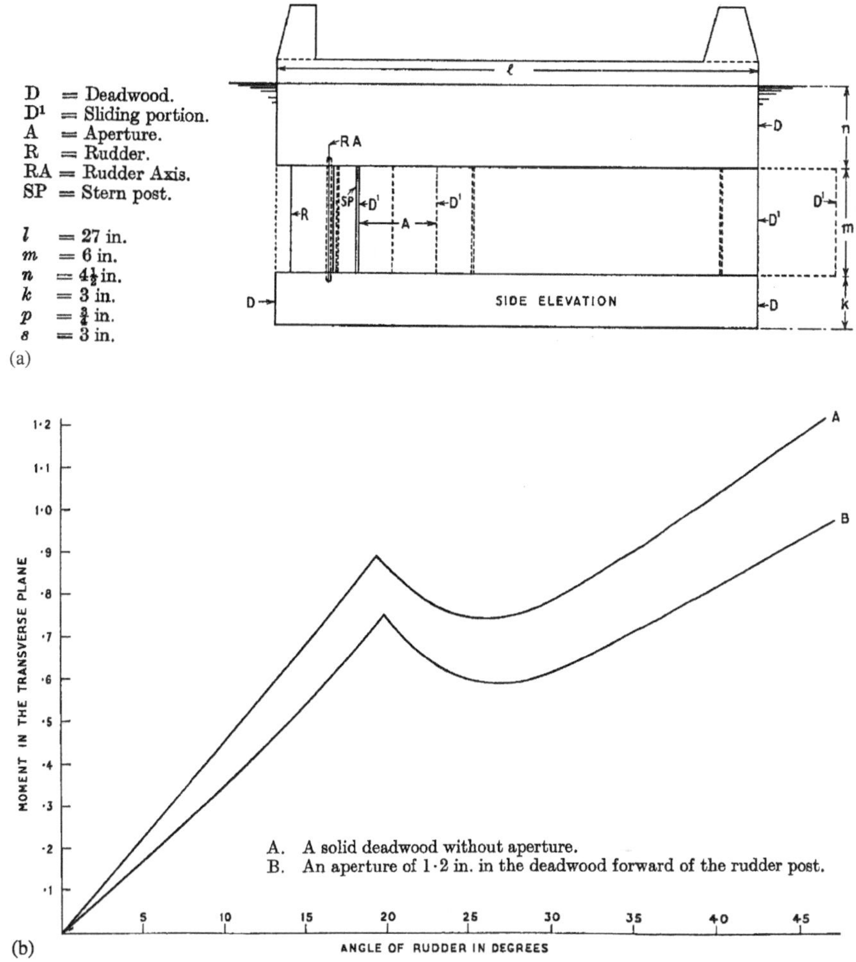

Figure 5.49 Rudder behind deadwood: effect of aperture ahead of rudder

Numerical CFD techniques are now available to estimate the forces on a rudder behind a deadwood using a low aspect ratio for the rudder–deadwood arrangement. Such techniques are discussed in Chapter 6.

5.4 Experimental data for rudder behind propeller

5.4.1 Rudder–propeller interaction data

Much of the early work was carried out by Baker, Bottomley, Cole and Gawn through the 1920–1940s [5.50–5.56]. Lötveit [5.57] carried out one of the earliest detailed investigations into propeller–rudder interaction, using a single rudder–propeller–hull

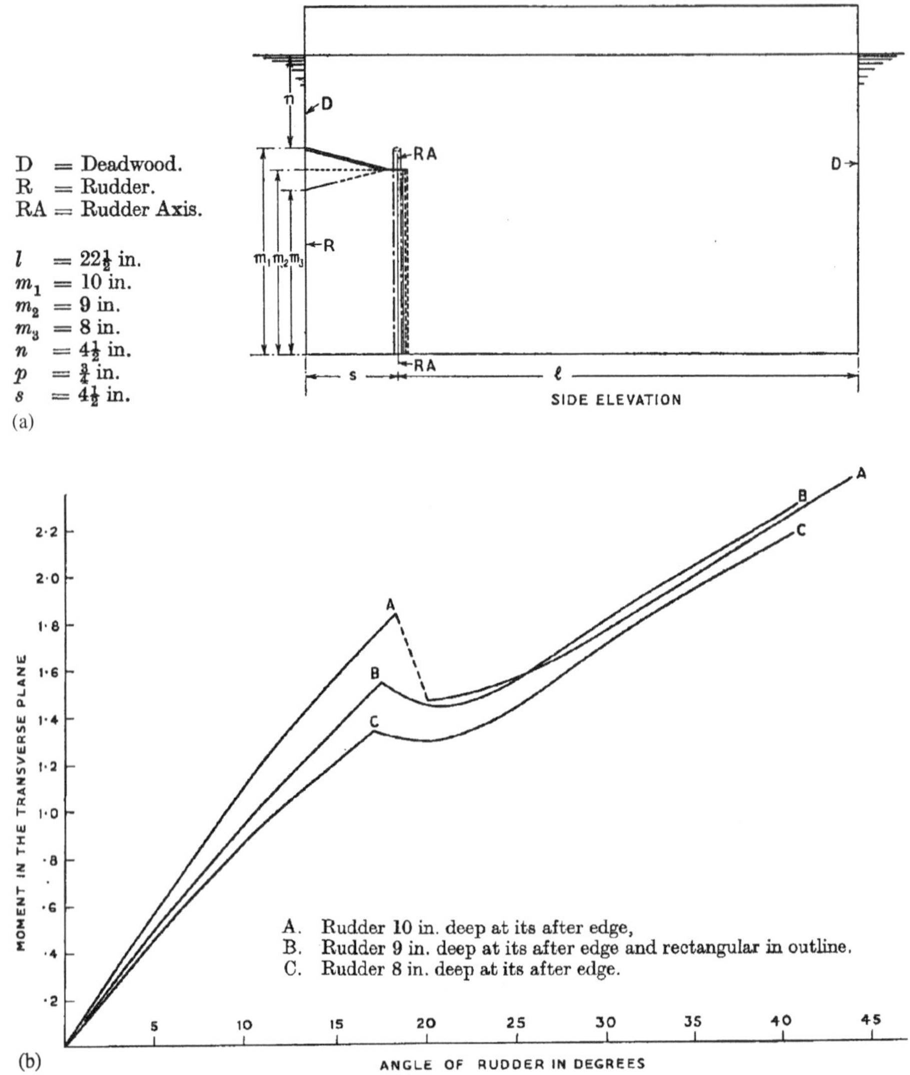

Figure 5.50 Rudder behind deadwood: effect of clearance at top of rudder

geometry and including detailed rudder pressure measurements. Further experimental investigations into propeller–rudder interaction include those of Shiba [5.58], Okada [5.59], English *et al.* [5.60], Landgraf [5.61], Mathis and Gregory [5.62], Kerwin *et al.* [5.63], Van Berlekom [5.64] and Kracht [5.65–5.67]. Extensive experimental investigations into propeller–rudder interaction have been carried out in a wind tunnel by Molland and Turnock [5.23, 5.68–5.73] and Turnock [5.74]. Their investigations provide an extensive database and include all-movable and semi-balanced skeg rudder–propeller combinations at different levels of propeller thrust loading.

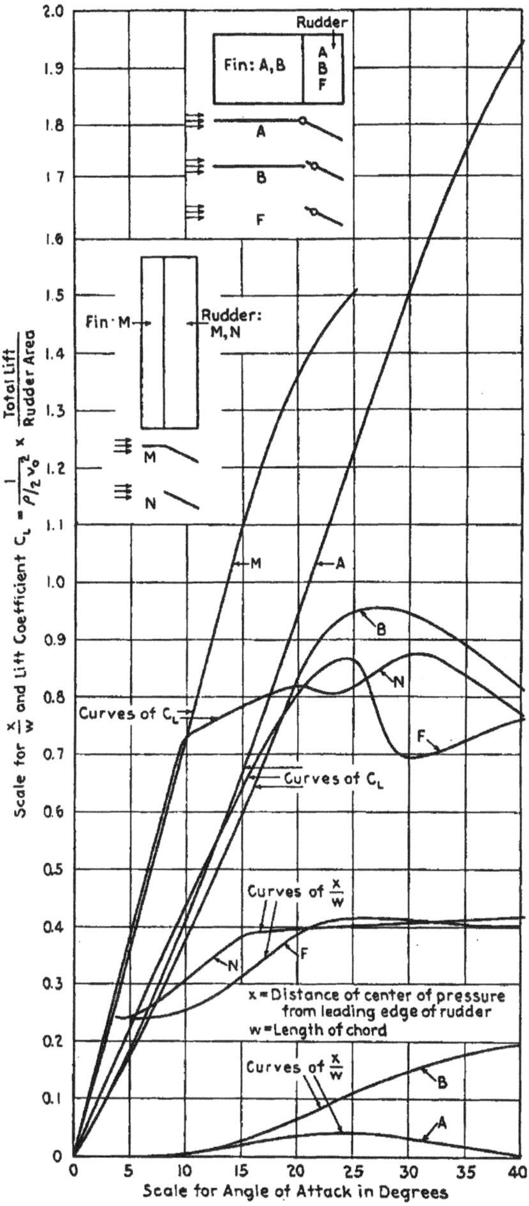

Figure 5.51 Lift coefficient for rudder behind deadwood

5.4.2 Wind tunnel rudder–propeller interaction data

5.4.2.1 *Test rig, models and tests*

The results of the wind tunnel tests on a series of rudders downstream of a propeller carried out by Molland and Turnock are presented in references [5.23, 5.68–5.74].

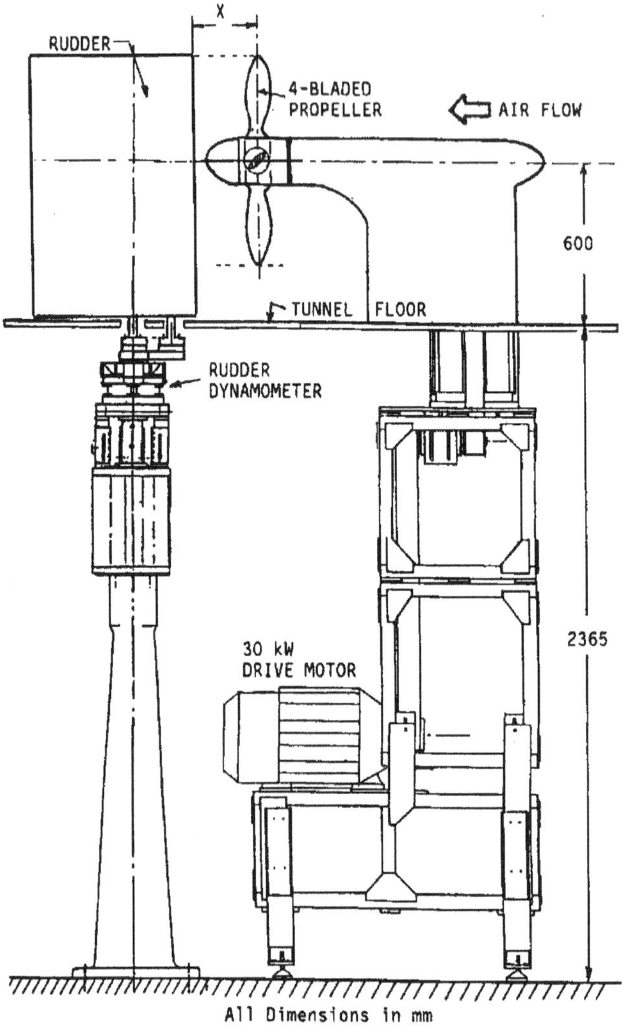

Figure 5.52 Overall layout of wind tunnel rudder–propeller rig

Most of the data have also been published in the form of reports [5.37, 5.38, 5.75–5.80]. Several of the data are included in Figures 5.58–5.74. Examples of tabulated test results are given in Appendix 1, together with the means of access to a complete database of the results.

The tests were carried out in the 3.5 m × 2.5 m low-speed wind tunnel at the University of Southampton. The overall rig for testing the interaction between a ship rudder and propeller is shown in Figure 5.52 and a propeller test set up is shown in Figure 5.53 (See Plate 1 of Colour Plate Section). The rig consists of two independent units, which allow free-stream (open water) tests to be carried out independently on rudders and propellers as well as the investigation of their interaction.

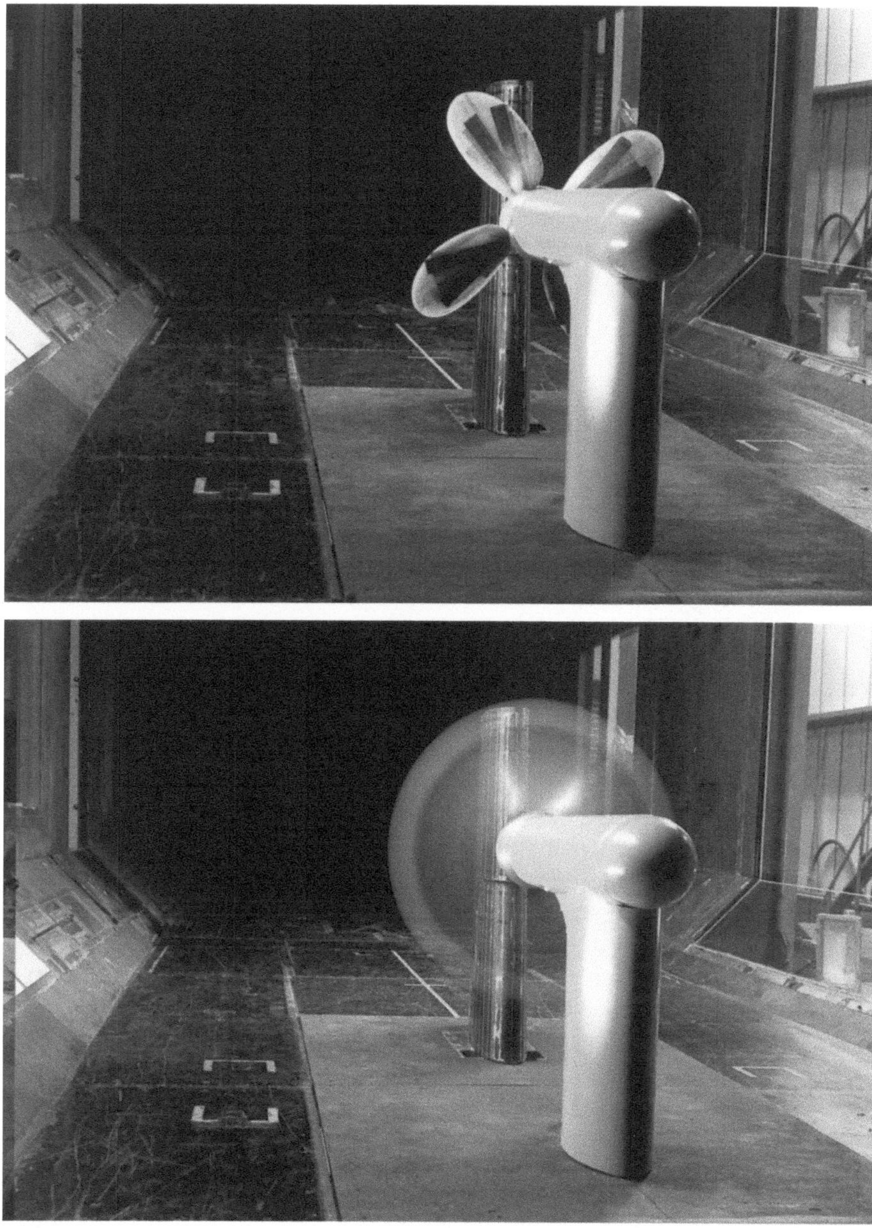

Figure 5.53 Arrangement of rudder–propeller rig in wind tunnel (See Plate 1 of Colour Plate Section)

The rudder is attached to a five-component strain gauge dynamometer supported underneath the tunnel floor. Full details of the dynamometer are given by Molland [5.81] and a summary is given in reference [5.7]. There is a small gap of approximately 2.5 mm (0.004 c) between the rudder root and the floor of the tunnel working section.

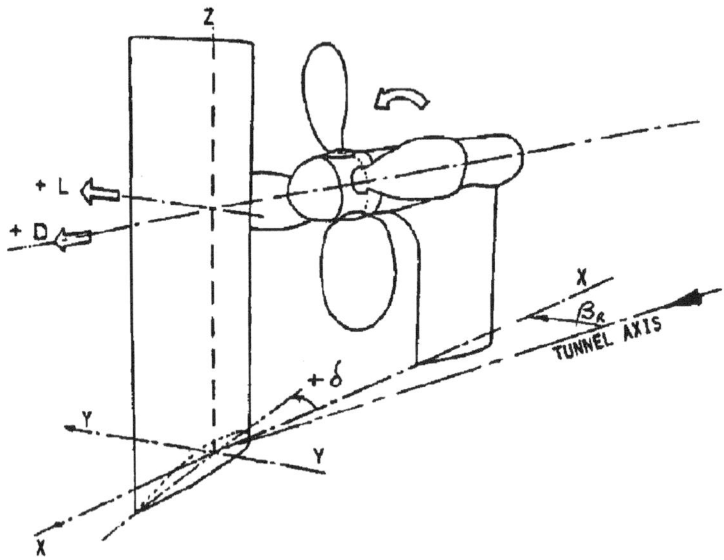

Figure 5.54 Coordinate system of rig

Full details of the propeller rig are given by Turnock [5.82]. The rig is designed in such a way that the propeller can be adjusted vertically, longitudinally and at an angle of drift to the flow, if required. Most tests were carried out with the propeller axis aligned horizontally and at a distance of 600 mm above the wind tunnel floor. The propeller rotates anti-clockwise when viewed from aft (looking upstream). An in-line strain gauge dynamometer mounted on the propeller shaft close to the propeller measures the delivered thrust and torque. Propeller revolutions are measured using an optical shaft encoder. The co-ordinate system of the rig is shown in Figure 5.54.

Seven rudder models were used in the tests. All rudders had a turbulence strip with its leading edge attached at a distance of 5.7% from the leading edge of the chord on both sides of the rudder. The roughness strips consisted of 12 mm wide double-sided tape densely covered with 0.15 mm diameter carborundum grit (no. 100). Dimensions of model Rudders Nos. 0, 1, 2 and 3 are given in Figure 5.55 and the alternative arrangements tested are shown in Figure 5.56 and summarised in Table 5.10. Rudder Nos. 1, 2 and 3 were pressure tapped at spanwise and chordwise positions as shown in Figure 5.55 and Table 5.11.

A four bladed propeller with a diameter of 800 mm and blade area ratio of 0.40 was manufactured for the experiments. The design was modelled on a Wageningen B4.40 series propeller [5.83] with suitable modifications. These modifications consisted of altering the blade root shape to allow an adjustable pitch design with four separate blades and a split hub, removing rake and decreasing blade sweep to reduce centripetal loading moments at the root and increasing the overall boss diameter ratio from 0.167 to 0.25. Details of the model propeller are given in Table 5.12 and its open water characteristics are shown in Figure 5.57. Details of blade outline and

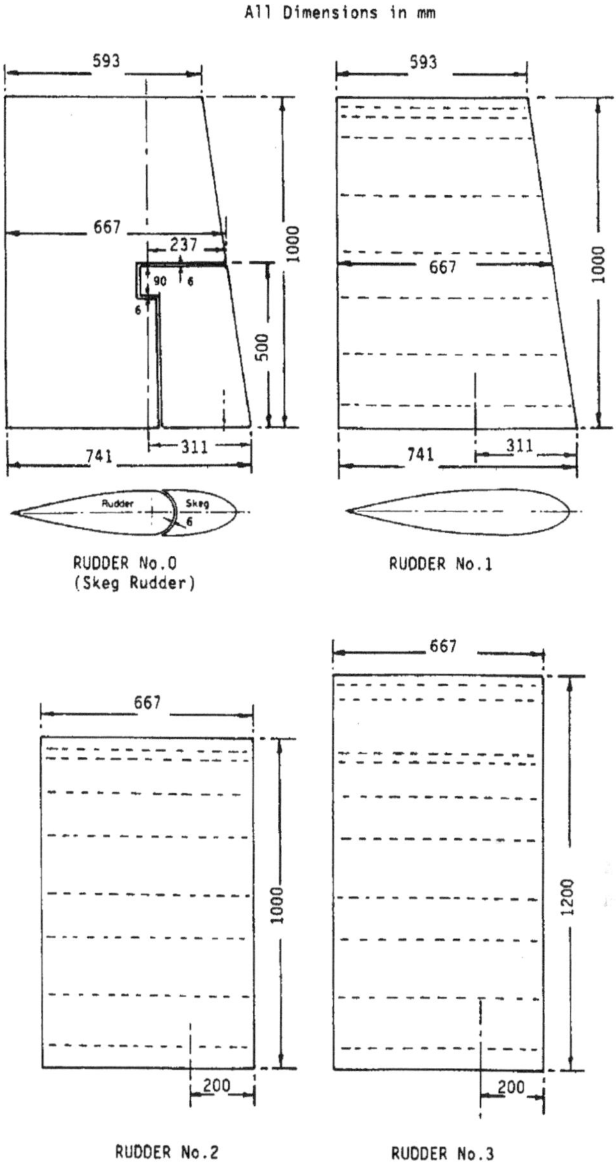

Figure 5.55 Dimensions of model rudders [5.23]

section shapes, together with details of manufacture, tests and test results, are given in reference [5.84].

The test rig does not allow for free surface effects but this restriction is not likely to be significant except where a vessel is in light ballast and/or the rudder or control surface is close to the free surface. Tests in air preclude any investigation of cavitation, which was therefore not considered in this particular test programme. Reynolds

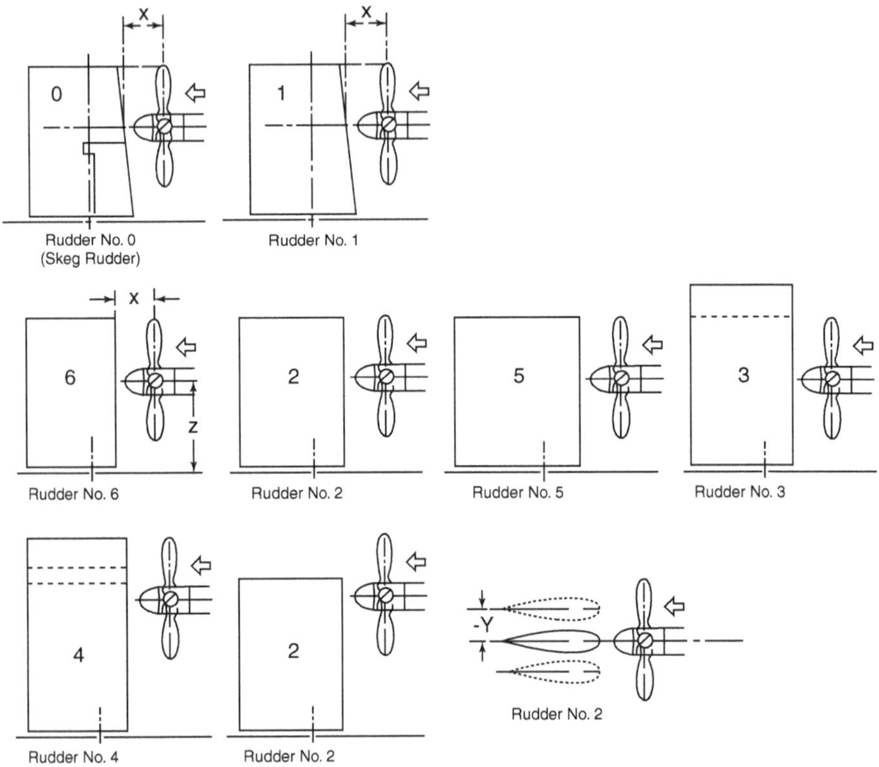

Figure 5.56 Alternative rudder–propeller arrangements tested

number scale effects on propeller drag are similar to those for smaller propellers tested in water. The effects of compressibility are not significant provided limitations on revolutions (hence propeller blade inflow speeds) take this into account. For example, estimates based on the Prandtl–Glauert law (with the correction $\sqrt{1-m^2}$) indicate that compressibility effects lead to increases in K_T and K_Q of up to about 3% at the maximum propeller revolutions of 3000 rpm used in the tests.

A wind speed of 10 m/s was used for the majority of the tests, being a compromise between achieving an adequate rudder Reynolds number and enabling a satisfactory range of J values to be achieved. Based on a free-stream velocity of 10 m/s and rudder chord, the nominal Reynolds number was 0.4×10^6. It should be noted that velocities induced by the propeller at the higher thrust loadings led to effective Reynolds number of up to 1×10^6. Results presented by English *et al.* [5.60] indicate that tests at these conditions should preclude any significant scale effects.

Nominal propeller revolutions of 800, 1470 and 2150 rpm were set for each rudder test condition leading, at a wind speed of 10 m/s, to J values of 0.94, 0.51 and 0.35 and nominal K_T/J^2 values of 0.05, 0.88 and 2.30.

A summary of the programme of work is given in Table 5.13.

Table 5.10 Alternative rudder–propeller arrangements tested [5.68–5.74]

Rudder No.	*0 (skeg)*	*1*	*2*						*3*	*4*	*5*	*6*
S mm	1000	1000	1000						1200	1300	1000	1000
C mm	667	667	667						667	667	800	556
AR	3.0	3.0	3.0						3.6	3.9	2.5	3.6
TR	0.8	0.8	1.0						1.0	1.0	1.0	1.0
X/c	0.355	0.355	0.30	0.30	0.30	0.30	0.30	0.54	0.30	0.30	0.30	0.30
X/D	0.34	0.34	0.39	0.30 0.52	0.39	0.39	0.39	0.39	0.30 0.39 0.52	0.39	0.39	0.39
Z/D	0.75	0.75	0.75	0.75	1.125	0.75	0.75	0.75	0.75	1.125	0.75	0.75
Y/D	0	0	0	0	0	+0.25 −0.25	0	0	0	0	0	0
λ	1.0	1.0	1.0	1.0	0.625	0.866	1.0	1.0	1.0	1.0	1.0	1.0
ξ	0.80	0.80	0.80	0.80	0.500	0.693	0.80	0.80	0.667	0.615	0.80	0.80
P/D	0.95	0.95	0.95	0.95	0.95	0.95	0.69 1.34	0.95	0.95	0.95	0.95	0.95

Rudder thickness/chord ratio *t/c*, in all cases: 0.20

Rudder sections, in all cases: NACA 0020 root and tip, with square tips

Propeller diameter in all cases: 800 mm

Table 5.11(a) Spanwise location of pressure tappings: Rudder Nos. 0, 2 and 3

Spanwise location	*S1*	*S2*	*S3*	*S4*	*S5*	*S6*	*S7*	*S8*	*S9*	*S10*
mm from root Rudders Nos. 2 and 3	70	230	400	530	700	830	940	970	1140	1170
mm from root Rudder No. 0	70	230	400	455	530	700	830	940	970	

Table 5.11(b) Chordwise location of pressure tappings: All rudders

Chordwise location % *c* from LE	0	2.5	5.0	10	20	30	40	50	60	70	80	90	95

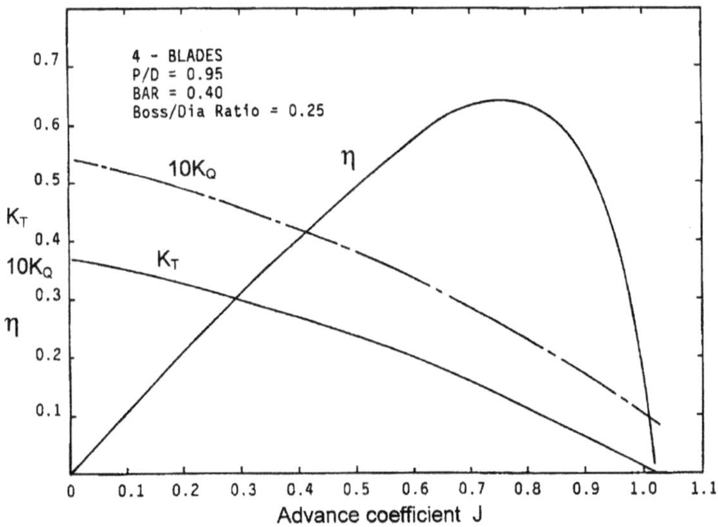

Figure 5.57 Propeller free-stream (open water) characteristics

Table 5.12 Details of model propeller

Number of blades	4
Range of revolutions r.p.m.	0–3000 Positive revolutions are anti-clockwise viewed from aft
Diameter (mm)	800
Blade area ratio	0.40
Boss diameter [max] (mm)	200
Mean pitch ratio *P/D*	0.95 [for main tests] plus 0.69 and 1.34
Rake (deg.)	0.0
Blade root thickness ratio	0.050
Sections shape	Based on Wageningen B Series
Blade outline shape	Based on Wageningen but with reduced skew

Table 5.13 Summary of experimental programme [5.68–5.74]

1	Free-stream tests–rudder alone
2	Free-stream (open water) tests–propeller alone
3	Rudder plus propeller combination–straight flow
4	Rudder plus propeller combination–straight flow–low and zero speed
5	Rudder plus propeller combination–straight flow–four quadrants
6	Rudder plus propeller combination–oblique flow
7	Rudder plus propeller plus centre boards (3 lengths)–straight and oblique flow
8	Rudder plus propeller plus hull–straight and oblique flow

The first phase of the work entailed extensive experimental modeling of the rudder and propeller combination in isolation. This included experimental testing of the seven rudder models at various longitudinal, lateral and vertical separations from the propeller over a range of rudder incidence for various thrust loadings. The tests included measurements of rudder forces and moments and pressure distributions, which provided the distribution of loading over the rudder.

The second phase of the work was to extend the investigation of the rudder plus propeller combination in straight flow to include operation at low and zero ship speed and in four quadrants of operation. In order to provide a basic understanding of flow straightening due to the propeller alone, the next phase of the work entailed experimental testing of the rudder plus combination alone in oblique flow.

The final phase of the work investigated in a systematic manner the influence of an upstream body on flow straightening effects. This firstly entailed tests using two-dimensional centreboards of different lengths upstream of the rudder–propeller combination, which provided basic information on the influence of upstream body length. This was followed by straight and oblique flow tests with a representative hull upstream of the rudder–propeller combination. The hull was pressure tapped which allowed hull-developed sideforce due to the rudder–propeller combination to be derived for change in drift angle, rudder incidence and propeller thrust loading.

The wind tunnel experimental investigations of Molland and Turnock have been supported by theoretical work using lifting line and boundary element methods and RANS codes. These techniques are described and discussed in Chapter 6.

5.4.2.2 *Service speed data*

(A) *All-movable rudders*

(i) *Thrust loading*: Figure 5.58 shows the influence of propeller thrust loading on Rudder No. 2 for the mid-longitudinal separation position ($X/D = 0.39$). It can be seen that, with decrease in J (increase in K_T and increase in K_T/J^2), the lift curve slope $dC_L/d\alpha$ increases. At higher thrust loadings, these increases are found to be less than those using momentum theory to predict the propeller-induced velocities, Section 3.6.

There is a significant delay in stall angle to over 40°, compared with results obtained in a free stream, where stall for this aspect ratio occurs at about 20°, Figure 5.17. It

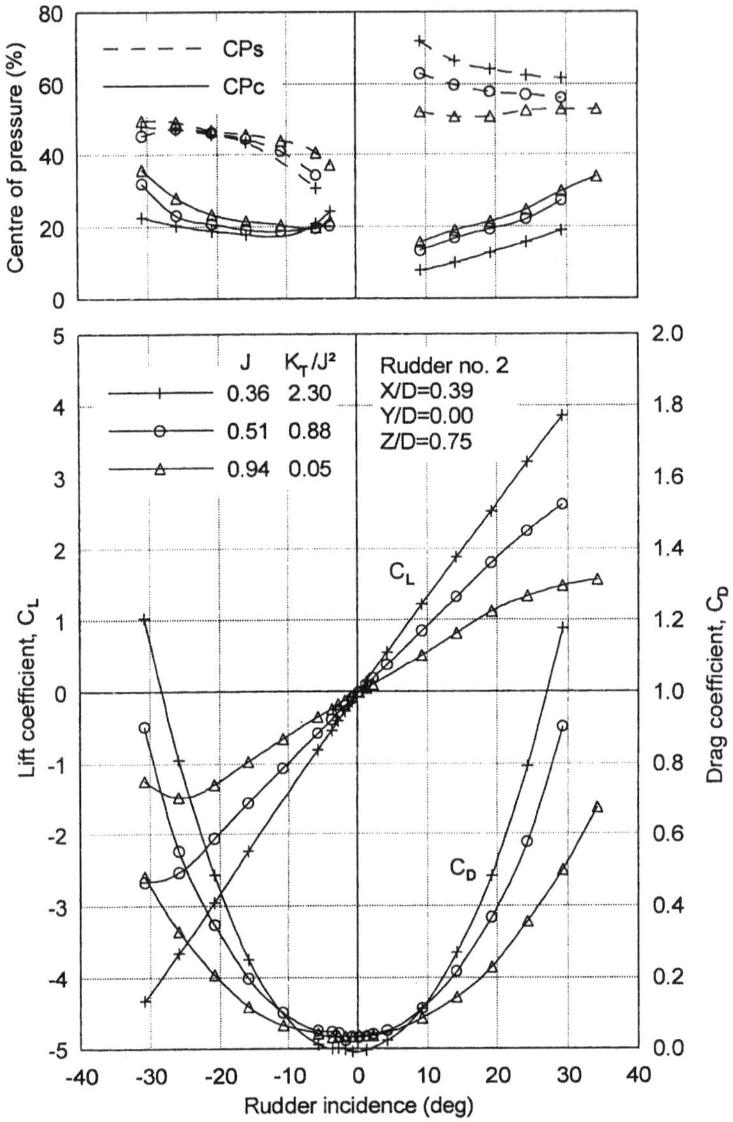

Figure 5.58 Influence of propeller thrust loading: Rudder No. 2

is also noted that stall angle increases with increasing thrust loading. The delay in stall may be attributed to the large-scale turbulence in the propeller slipstream which could affect the boundary layer on the rudder and delay stall, together with the axial pressure gradient in the propeller slipstream which is likely to act favourably in delaying stall. The stall angle is no longer the same at positive and negative incience, stall occurring later for positive incidence. This would appear to be due to the rotational nature of the rudder inflow.

The rudder drag coefficient, C_D, for all three advance ratios is similar between $-10°$ and $+10°$. As rudder incidence is increased, the drag component due to lift increases rapidly for the lower advance ratios (higher thrust loadings).

Centre of pressure chordwise, *CPc*, generally moves forward with increasing thrust loading. *CPc* in the presence of the propeller tends to be at or slightly further forward than the free-stream case, particularly at the highest thrust loadings.

Centre of pressure spanwise, *CPs*, increases with increase in thrust loading at positive rudder angles, whilst for negative rudder angles *CPs* decreases with increase in thrust loading. These characteristics result from the rotational nature of the rudder inflow.

Landgraf [5.61] carried out systematic tests on rudder–propeller interaction. The rudders had an effective aspect ratio 2.4 and three section shapes of type IfS 58 TR15, IfS 63 TR25 and NACA0025. J and X/D were varied. The results for lift, drag, stall angle and *CPc* show similar trends to the findings of Molland and Turnock under discussion.

(ii) *Propeller pitch setting*: For a given advance ratio, changing the propeller pitch ratio (P/D) setting alters the thrust coefficient (K_T) of the propeller. Tests were carried out with three pitch ratio settings of 0.69, 0.95 (datum) and 1.34. In these tests, the advance ratio was varied to maintain the same open-water thrust loading (K_T/J^2) to allow direct comparison of the rudder characteristics between the three pitch ratio settings. Tests were carried out at three K_T/J^2 values of 0.05, 0.88 and 2.30. An example of the results, for the case of $K_T/J^2 = 0.88$, is shown in Figure 5.59.

At the low thrust loading ($K_T/J^2 = 0.05$), there was little difference in lift curve slope for the three pitch settings. Drag coefficient, *CPc* and *CPs* were also not influenced to any great extent. For the mid-thrust loading ($K_T/J^2 = 0.88$), corresponding to an advance ratio $J = 0.51$ for $P/D = 0.95$, differences in the rudder characteristics are more pronounced. The increase in pitch ratio slightly increases the lift curve slope and decreases drag at low rudder incidence. Changes in *CPc* are still small. For positive incidence, *CPs* moves outboard for increasing pitch ratio and inboard for negative incidence. At high thrust loading ($K_T/J^2 = 2.30$), corresponding to an advance ratio $J = 0.35$ at $P/D = 0.95$, the trends are broadly similar. With increase in pitch ratio, the lift curve slope increases and the drag is less. For the pitch setting of 1.34 the rudder generates a net thrust. These results, over a range of pitch ratio settings for each thrust loading value, confirm that it is the propeller thrust loading which controls the performance of the rudder. However, variations were observed in the drag characteristics, which suggest a dependence on the flow structure within the propeller race.

The spanwise distributions of local C_N for the different pitch ratios P/D of 0.69 and 1.34, but with the same propeller thrust loading $K_T/J^2 = 0.88$, are shown in Figures 5.60 and 5.61. It is interesting to note that, even for this significant difference in P/D, the spanwise load distributions remain very similar. More detailed descriptions of the various surface pressure measurements are given in Section 5.4.2.3.

(iii) *Rudder aspect ratio*: The effect of varying aspect ratio for a constant coverage $\xi = 0.80$ was investigated at three thrust loadings K_T/J^2 of 0.05, 0.88 and 2.30. Three rudders were tested, Rudder Numbers 5, 2 and 6, Figure 5.56 and Table 5.10,

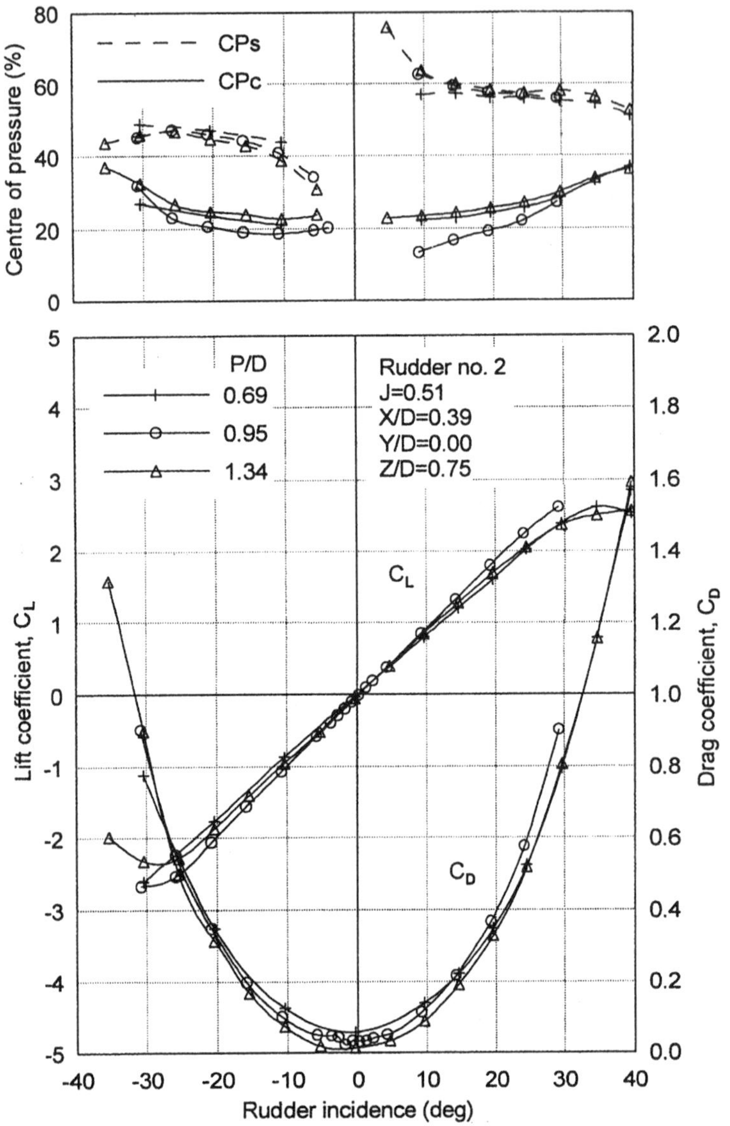

Figure 5.59 Influence of propeller pitch ratio, each case at a nominal open water thrust loading $K_T/J^2 = 0.88$

with (effective) aspect ratios of 2.5, 3.0 and 3.6 respectively. An example of the results, for the high thrust loading (low J) case of $K_T/J^2 = 2.3$, is shown in Figure 5.62. For all three rudders, at the same longitudinal separation, there was an increase in lift curve slope with increase in aspect ratio. This reproduces the effect of aspect ratio observed in the free stream, although at the high thrust loading (Figure 5.62) the effect of aspect ratio is a little greater than that for rudders in a free stream.

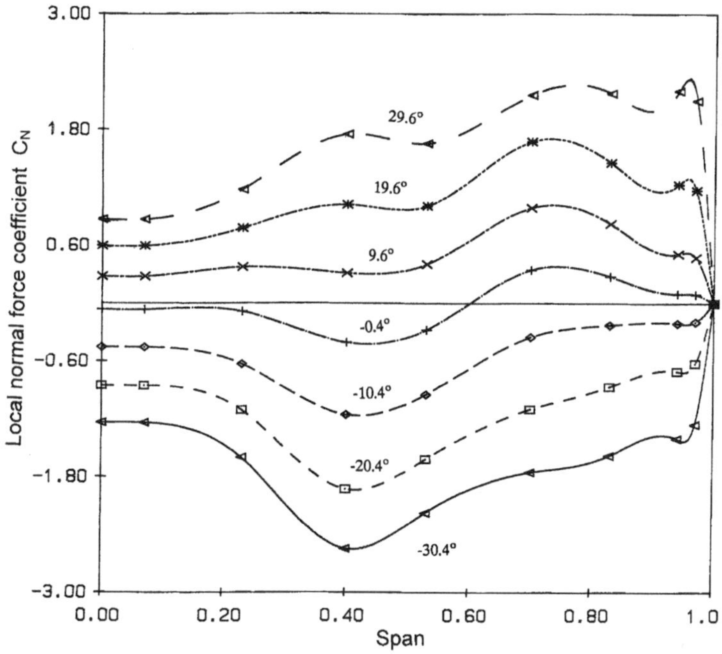

Figure 5.60 Spanwise load distribution; Rudder No. 2, $K_T/J^2 = 0.88$, $P/D = 0.69$ at $J = 0.42$

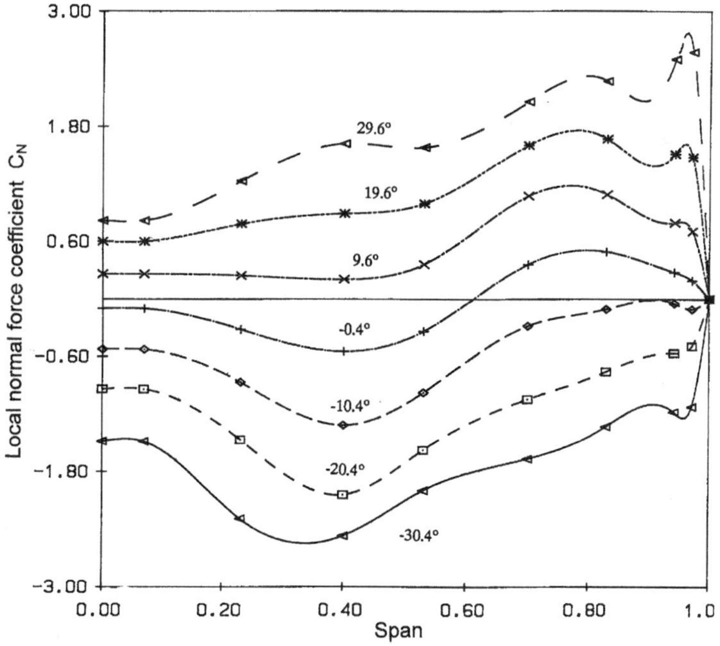

Figure 5.61 Spanwise load distribution; Rudder No. 2, $K_T/J^2 = 0.88$, $P/D = 1.34$ at $J = 0.61$

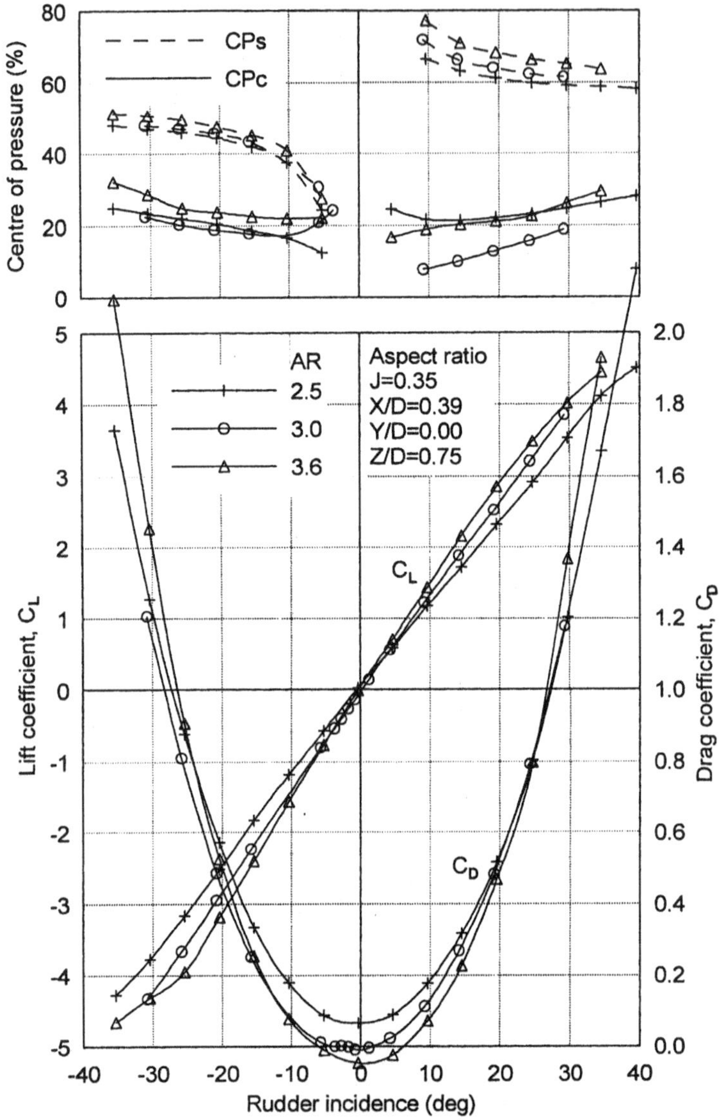

Figure 5.62 Influence of aspect ratio on rudder performance characteristics

Based on a change in lift curve slope with change in aspect ratio described by Equation (5.7b), a satisfactory correction to lift for change in aspect ratio from AR_1 to AR_2 over the range of thrust loadings is given by Equation (5.17).

With increase in aspect ratio, there is little change in *CPc* but there is an increase in *CPs*. At low rudder incidence, the drag appears sensitive to the structure of the propeller race and decreases with increasing aspect ratio; at low *J* a thrust force is produced.

Table 5.14 Range of parameters

X/D	0.30	0.39	0.52
Y/D	−0.25	0.00	+0.25
Z/D	0.625	1.00	–

(iv) *Rudder position relative to propeller*. Part of the work of Molland and Turnock [5.75, 5.77] was to investigate systematic changes in the relative positions of the rudder and propeller, Figure 5.56 and Table 5.10. The geometrical parameters X/D, Y/D and Z/D are defined in Figures 4.8, 4.9 and 4.10 and the range of parameters used in the investigation is shown in Table 5.14.

The propeller pitch was set at 0.95 and open-water thrust loadings (K_T/J^2) of 0.05, 0.88 and 2.3 (J values of 0.94, 0.51 and 0.35) were used. Rudder angles were varied between −40° and +40°.

It was found that changes in the geometrical parameters (X/D, Y/D and Z/D) can have significant influences on the relative magnitudes of the rudder manoeuvring forces and combined rudder–propeller propulsive forces.

(1) *Rudder forces: Manoeuvring*

Longitudinal separation (X/D): The basic influence of X/D on rudder sideforce, for one propeller thrust loading, is illustrated by the experimental results in Figure 5.63(a). Changes in stall angle and maximum C_L were relatively small. X/D effects depend on thrust loading as seen in Figure 5.63(b), which shows the influence on lift curve slope at zero rudder incidence. It is seen that as X/D increases $dC_L/d\alpha$ firstly increases up to an X/D of about 0.4 and then decreases. In a typical ship, operational condition ($J = 0.51$, $K_T/J^2 = 0.88$), a change in X/D from 0.30 to 0.40 leads to a 6.5% increase in $dC_L/d\alpha$ hence the available manoeuvring rudder sideforce; whilst at $J = 0.36$ ($K_T/J^2 = 2.30$) a similar change in X/D leads to a 9% increase in $dC_L/d\alpha$. Above $X/D = 0.4$, $dC_L/d\alpha$ decreases. These changes in lift curve slope are broadly in line with those derived experimentally by Landgraf [5.61] for higher X/D values, but the results of Landgraf do not show a decrease in $dC_L/d\alpha$ at low X/D except at lower propeller thrust loadings.

Lateral separation (Y/D): The main effect of lateral movement of the rudder relative to the propeller, as seen in Figure 5.64(a), is a shift in the lift curve so that zero lift no longer occurs at zero incidence. Corresponding to this is an increase in maximum lift in one direction and a decrease for the other. The lift curve slope decreases with increasing magnitude of Y/D, Figure 5.64(b), the effect being more apparent at higher thrust loading. The zero sideforce offset α_0 increases with increase in propeller thrust loading as shown in Figure 5.64(c) with a typical value of about 3° at $Y/D = 0.20$ and $J = 0.51$.

Vertical position (Z/D): For this particular part of the experimental investigation the propeller axis height was increased from 600 to 900 mm, Figure 5.56, changing Z/D from 1.0 to 0.625 and reducing the fraction of the propeller race in way of the rudder span. The effects on the rudder characteristics, as shown in Figure 5.65(a), are similar to those of a lateral movement. The lift characteristic is shifted so that a positive rudder incidence is required for zero lift. The positive stall angle is

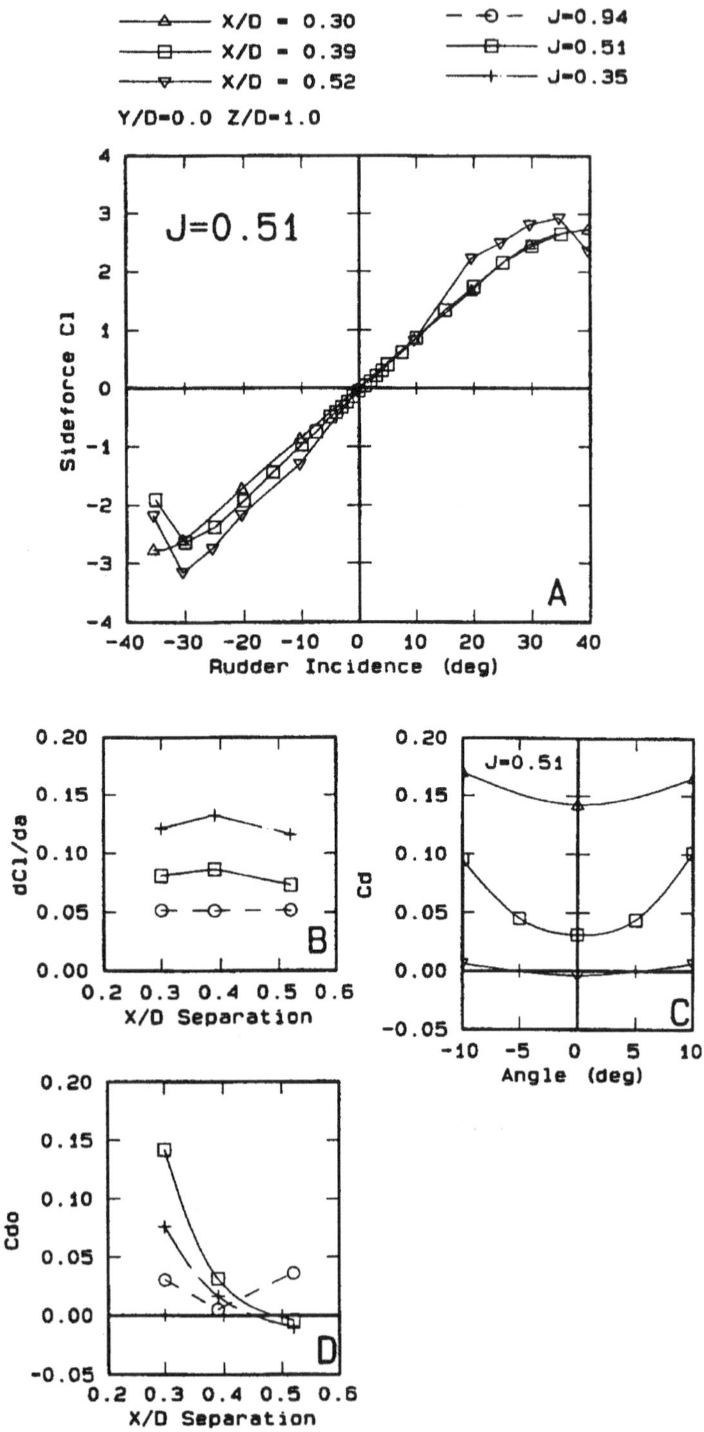

Figure 5.63 Influence of longitudinal separation on rudder performance characteristics

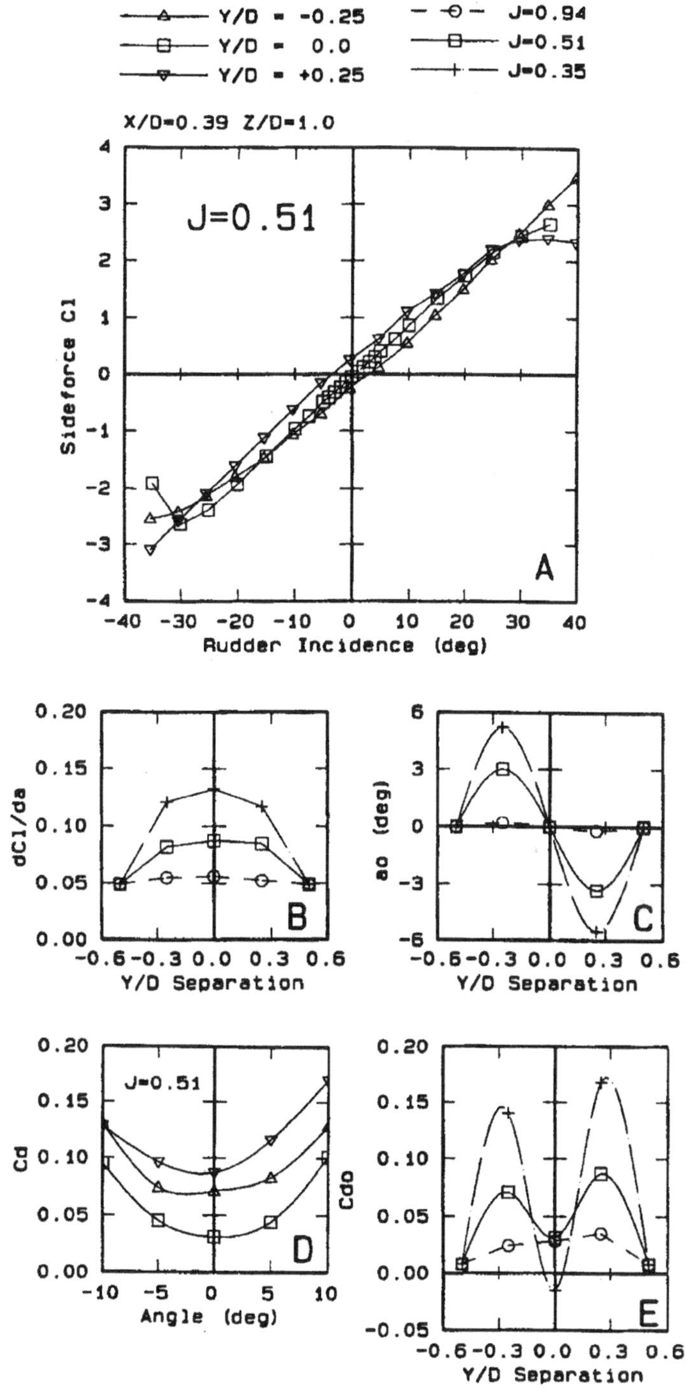

Figure 5.64 Influence of lateral separation on rudder performance characteristics

reduced, as is the negative stall angle. The lift curve slope decreases, Figure 5.65(b), which is to be expected with a reduced amount of the propeller race in way of the rudder. The angle for zero lift increases with propeller thrust loading as shown in Figure 5.65(c).

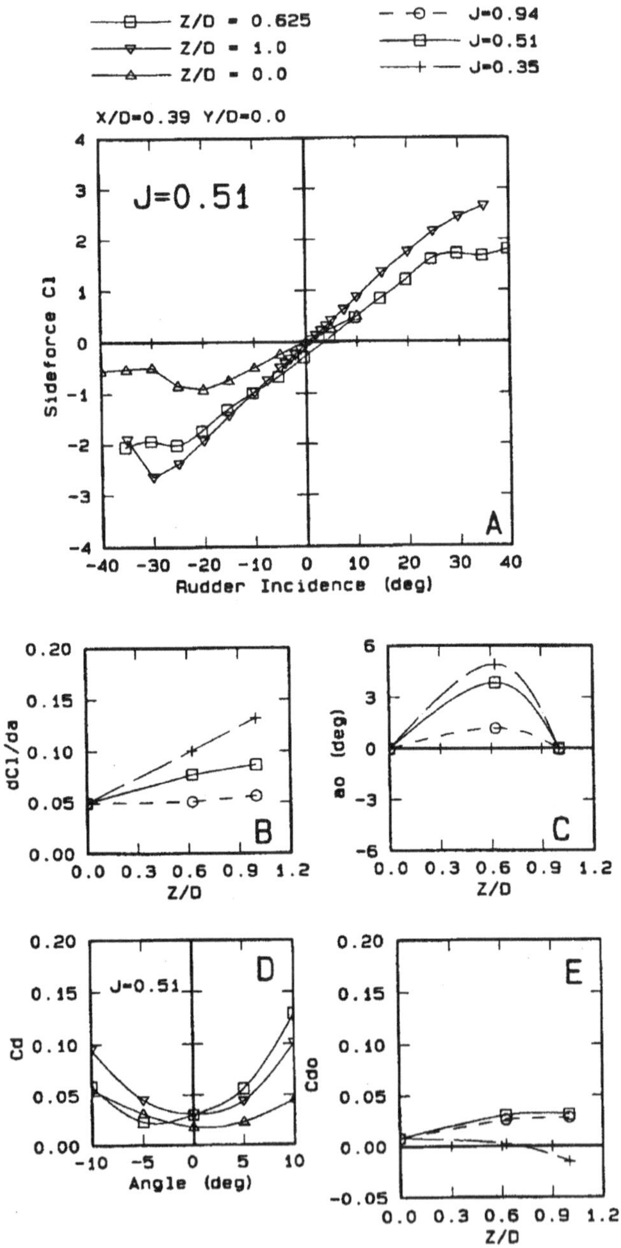

Figure 5.65 Influence of vertical position on rudder performance characteristics

(2) *Rudder forces: Thrust and drag*

Relatively large interactions between the rudder and propeller can occur which affect the net propulsive force of the rudder–propeller combination. Rotational effects in the propeller race lead to significant increases or decreases in net rudder drag, whilst the rudder blockage leads to increases in propeller thrust, both of these depending on the relative positions of the rudder and propeller. Changes in propeller thrust and torque due to rudder blockage are discussed separately in Section 5.9.

Longitudinal separation (X/D): The net rudder drag at zero and low incidence is illustrated by the results in Figures 5.63(c) and 5.63(d). It is seen that high values of C_D are obtained with small X/D whilst increasing X/D leads to significant reductions in C_D. This effect was also observed by Stierman [5.85] and Nakatake [5.86]. Rudder drag is seen to decrease with increase in propeller thrust loading and at high thrust loadings a net rudder thrust is developed. The resolution of the forces in this situation is given in Figure 3.33(b).

Lateral separation (Y/D): The influence of lateral separation on rudder drag at zero and low incidence is shown in Figures 5.64(d) and 5.64(e). This shows some asymmetry due to the differing influences of propeller rotation near the rudder tip. The results indicate that, at a typical Y/D value of 0.20 and operational $J = 0.51$, there is a significant increase in rudder drag of the order of 100%. At higher thrust loading (lower J) this increase in rudder drag is much greater, references [5.75, 5.77].

Vertical position (Z/D): The influence of the vertical rudder–propeller position on rudder drag is shown in Figures 5.65(d) and 5.65(e). With Z/D reduced to 0.625, minimum rudder drag values occur at about $-5°$ rudder incidence, when a significant sideforce is being produced, Figure 5.65(a). At zero rudder incidence, changes in rudder drag with decrease in Z/D are small.

(v) *Coverage* (ξ): Coverage defines the proportion of the area of the rudder that is within the propeller race. The magnitude of the thrust loading imparted to the fluid in the propeller race will control the value of the sideforce generated for a given rudder incidence for that area of the rudder within the propeller race. The coverage, therefore, is a measure of the proportions of the rudder where the force is dictated by the propeller and by the free stream. A larger coverage will be more strongly influenced by changes in propeller thrust loading.

Two rudders with constant chord but spans of 1000 and 1300 mm (Rudders 2 and 4, Figure 5.56, Table 5.10) are used to investigate the effect of coverage. For these tests, the rudder was in way of all the propeller race, so $\lambda = 1.0$ and the coverage for Rudder No. 2 was $\xi = 0.800$ and, for Rudder No. 4, $\xi = 0.615$. The effect of tip flow was the same for both rudders. An example of the results, for the mid-thrust loading case of $K_T/J^2 = 0.88$, is shown in Figure 5.66. From the results for change in aspect ratio, it would be expected that with increase in span the lift curve slope would increase. This is indeed observed at high J, but as J is decreased the lift curve slope for the high aspect ratio rudder becomes progressively less than the other rudder. Also, the stall angle is slightly reduced. Chordwise centre of pressure is similar for both rudders for all J values. However, there is a shift outboard of the spanwise centre of pressure for Rudder Number 4, especially at low J.

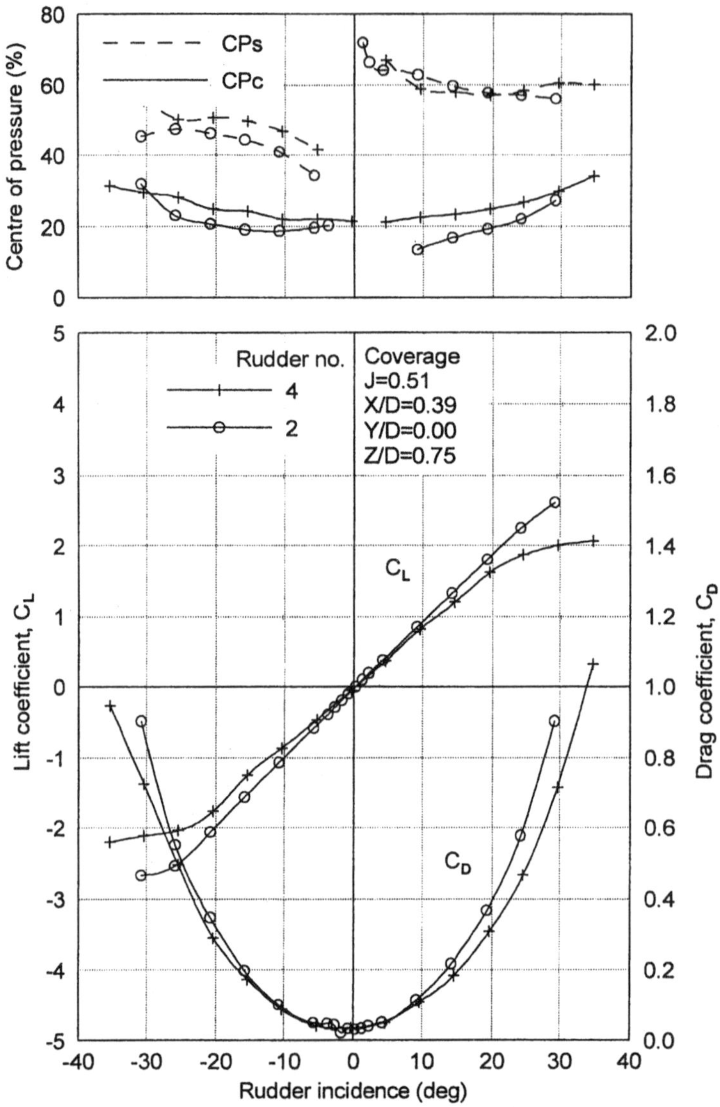

Figure 5.66 Effect of coverage of the rudder span by the propeller race

(B) *Semi-balanced skeg rudders*

Figures 5.67 and 5.68 show the influence of propeller thrust loading on Rudder No.0 for the longitudinal separation position ($X/D = 0.34$). Figure 5.67 is for the movable rudder plus skeg forces, which will be used for the prediction of total forces such as for structural design and manoeuvring purposes, whilst Figure 5.68 is for the movable rudder alone, which will be used for rudder torque prediction purposes. It should be noted that the data for the rudder alone are nondimensionalised using the total rudder plus skeg area. It is seen that the influences of thrust loading on the skeg rudder are similar to the all-movable rudder. It is useful to compare the results in Figure 5.67 with the free-stream semi-balanced skeg rudder results presented in Section 5.3.5. In a free stream the skeg rudder displays a characteristic discontinuity in the growth of lift with increasing rudder incidence, the discontinuity being caused by the early stall of the movable part of the rudder behind the skeg. Figure 5.67 indicates that at the low thrust loading ($J = 0.94$) some discontinuity in the lift curve exists which is similar to the free stream case. There is an indication that a discontinuity also occurs when $J = 0.51$ but at higher rudder incidence. However, for the higher thrust loading of $J = 0.35$, it is quite clear that discontinuities do not occur in the lift curves, the periodic nature of the propeller induced velocities preventing the early development of stall aft of the skeg. These findings are confirmed by flow visualisation tuft studies reported in [5.73].

Kracht [5.87] carried out systematic tests on a semi-balanced skeg rudder in a cavitation tunnel, varying the rudder propeller separation X/D. The results for lift, drag and CPc, which are for the movable rudder alone, show similar trends to the findings of Molland and Turnock under discussion. Kracht monitored the effects on propulsion factors and this is further discussed in Section 5.9. In particular, Kracht studied the cavitation performance of the semi-balanced skeg rudder, and the results are discussed in Section 5.8.

5.4.2.3 *Surface pressure data*

Rudder pressure distributions were measured in the course of the wind tunnel experiments reported in references [5.75–5.80]. Details of the pressure tappings are given in reference [5.89]. The distribution of tappings over the rudders is given in Figure 5.55 and Table 5.11.

Free stream: Rudder No. 2: Examples of chordwise pressure distributions in a free stream for the all-movable Rudder No. 2 at one span position are given in Figure 5.69. The pressures on the face (pressure) and back (suction) sides are shown. The shape of the curves remains relatively constant as incidence is increased up to 20°. It is noted from Figure 5.17 that the rudder has stalled by about 25°, and this is apparent from the back pressure curve in Figure 5.69(c).

Figure 5.70 shows the spanwise distributions for Rudder No. 2 in a free stream. The distributions are shown as the local normal force coefficient C_N to a base of span, where the local total C_N has been derived from the chordwise integration of the sum of the local face pressure and back suction curves, Figure 5.69. It is noted that peaks in the spanwise distribution develop near the rudder tip due to the presence of a tip vortex. These peaks become more prominent as incidence is increased. It can also be noted that integration of the spanwise distribution yields the total C_N at that incidence. The experimental results indicate that integration of the spanwise loadings leads to values close to the total forces measured directly by the force dynamometer.

Rudder with propeller upstream: Rudder Nos. 0, 2 and 3: Figure 5.71 shows the influence of propeller thrust loading, K_T/J^2, on the chordwise pressure distributions for one span position and one angle of attack for Rudder No. 0. It is noted that the shapes of the pressure distributions remain broadly similar as K_T/J^2 is increased.

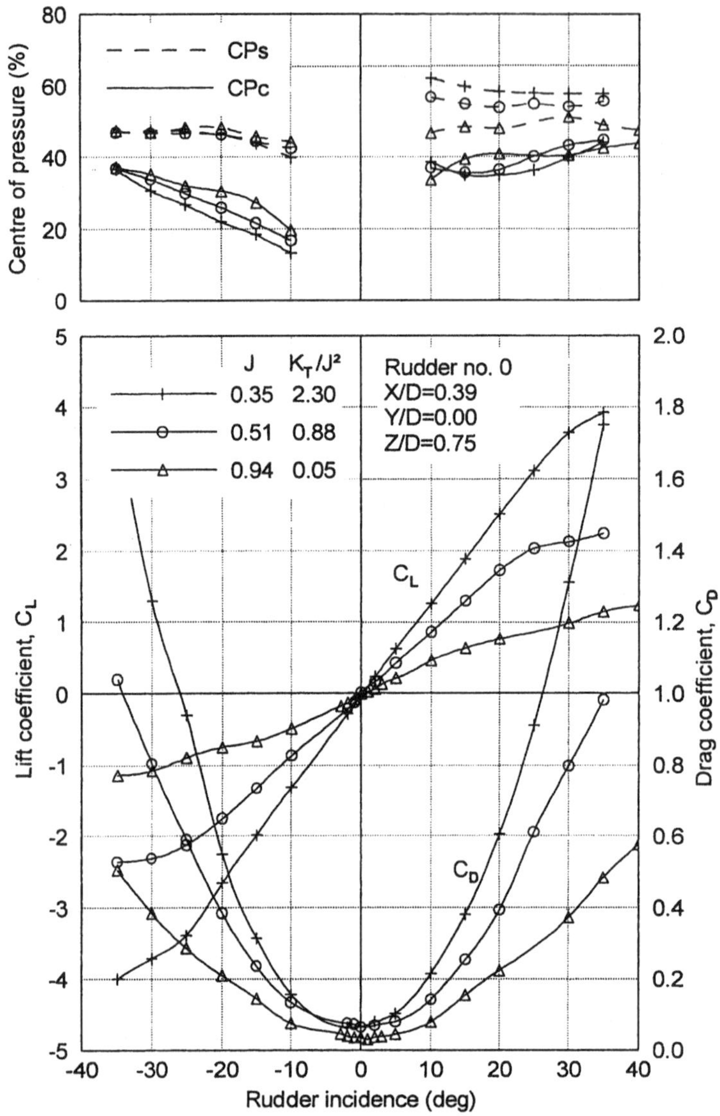

Figure 5.67 Influence of propeller thrust loading: Rudder No. 0, rudder plus skeg

Figure 5.72 shows a three-dimensional plot of the pressure loads (face pressure and back suction) for Rudder No. 2 at a drift angle $\beta = -7.5°$, incidence $\delta = 20°$ and $K_T/J^2 = 0.88$. This presentation shows the overall distribution of normal load C_N over the rudder at a particular incidence and propeller thrust loading.

Spanwise distributions of local load C_N for the case of Figure 5.72 are shown in Figure 5.73. Compared with Figure 5.70, Figure 5.73 shows the asymmetric nature of the load distribution which results from the angular flow change induced by the

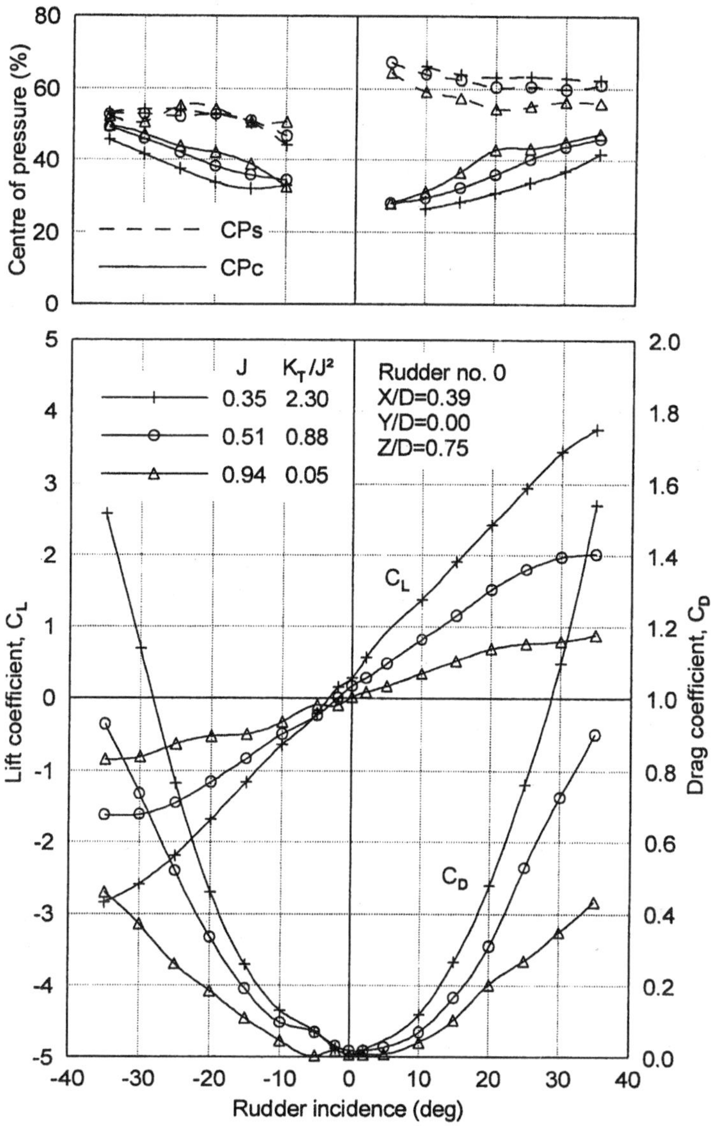

Figure 5.68 Influence of propeller thrust loading: Rudder No. 0, rudder alone

propeller, as described in Section 3.5. The development of a tip vortex is observed, as noted earlier for the free stream results.

Figures 5.74(a–f) show further spanwise load distributions for Rudder No. 2. The distributions are shown for the mid longitudinal separation condition $X/D = 0.39$ and for three propeller thrust loadings. At low propeller thrust loading (high J) (Figure 5.74(a)), the spanwise distributions show only small differences from those expected for the rudder in a free-stream (Figure 5.70). At zero rudder incidence, a flow angle offset is observed between the tunnel floor and the propeller race (<0.2 span).

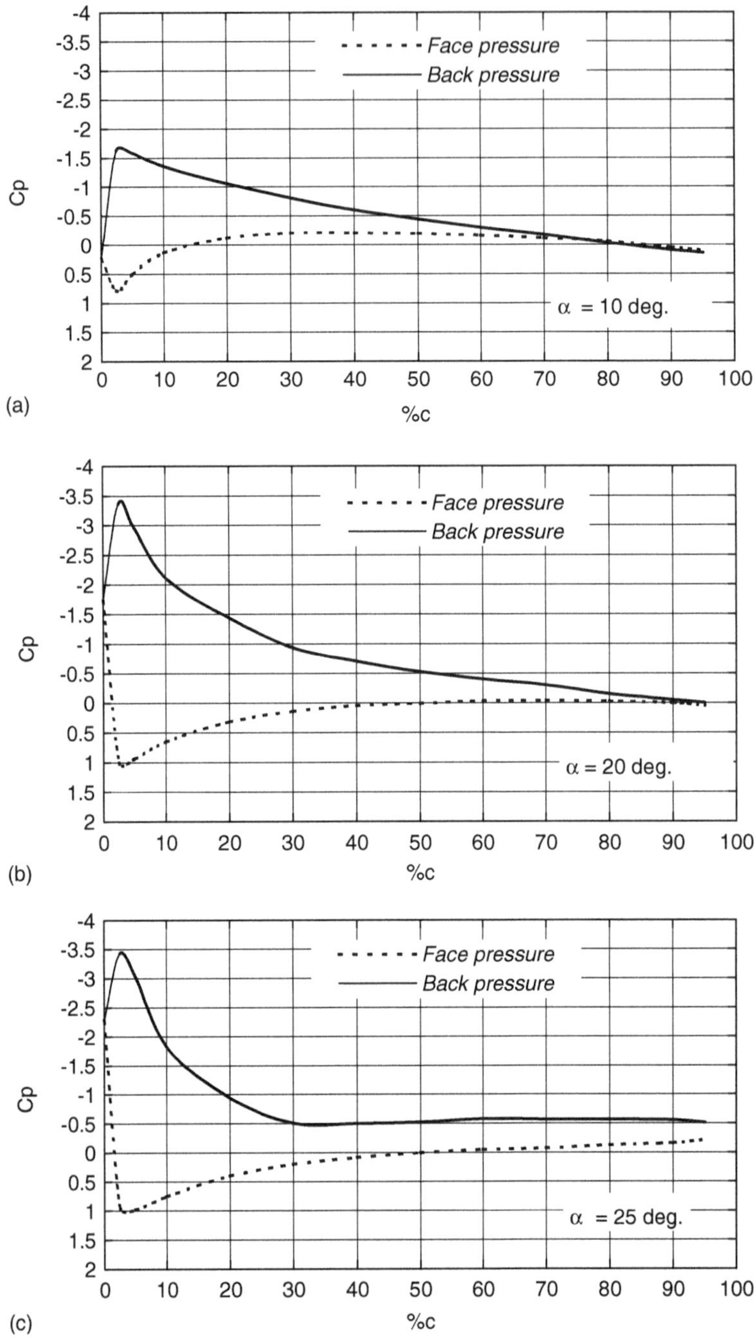

Figure 5.69 Chordwise pressure distributions: free stream; all-movable rudder No. 2, span position 700 mm from root; influence of incidence

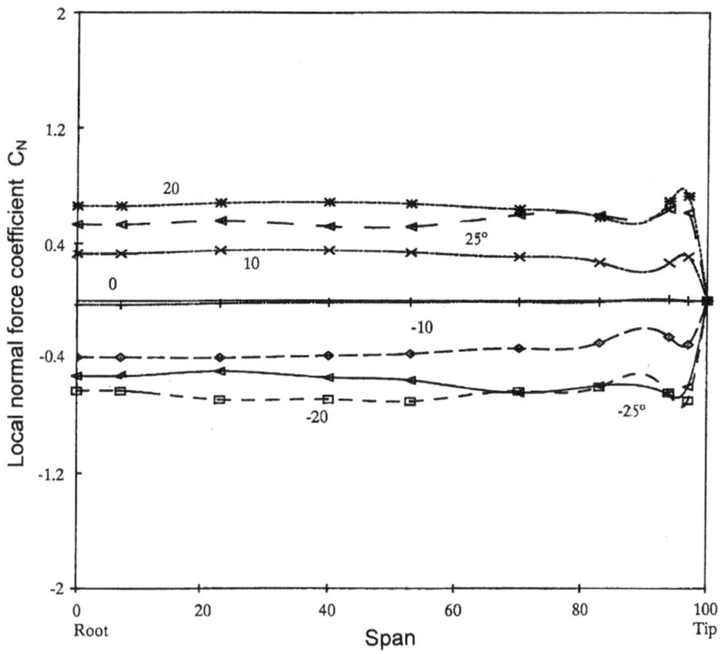

Figure 5.70 Spanwise load distribution: free stream, all-movable Rudder No. 2: influence of incidence

Figures 5.74(b) and 5.74(c) illustrate the effect of increased propeller thrust loading. These figures clearly show the increased asymmetric nature of the load distribution as thrust loading is increased (J is decreased). It is also noted that at lower thrust loadings (high J), the force-induced coefficient outside the propeller slipstream is of the same order as that within the slipstream whereas at high propeller thrust loadings (low J) the slipstream local force coefficients dominate the loading on the rudder.

Similar patterns of results were obtained for Rudder Nos. 1 and 3, although for Rudder No. 3 a free-stream type of behaviour occurs over the outer 20% of the rudder span, as illustrated in Figure 5.74(d) for the mid-separation position.

An example of the spanwise distributions for skeg Rudder No. 0 is shown in Figure 5.74(e). It is noted that, for positive incidence, the loading has migrated towards the tip with a highly asymmetric distribution, leading to a movement in the spanwise centre of pressure and increased bending moments about the root of the rudder.

Figure 5.74(f) illustrates the case for Rudder No. 2 when the propeller axis is 900 mm from the rudder root (Figure 5.56), and the propeller is only partly covering the rudder. This is often the case with smaller twin-screw vessels (Figure 4.10). It is apparent that much of the rudder load is now being carried by the outer half of the rudder.

Four quadrant operation: Examples of pressure load distributions at one propeller thrust loading in four quadrants of operation for Rudder No. 2 are presented in Figure 5.88 of Section 5.4.2.6. The form of the spanwise distributions in the different quadrants are discussed in that section.

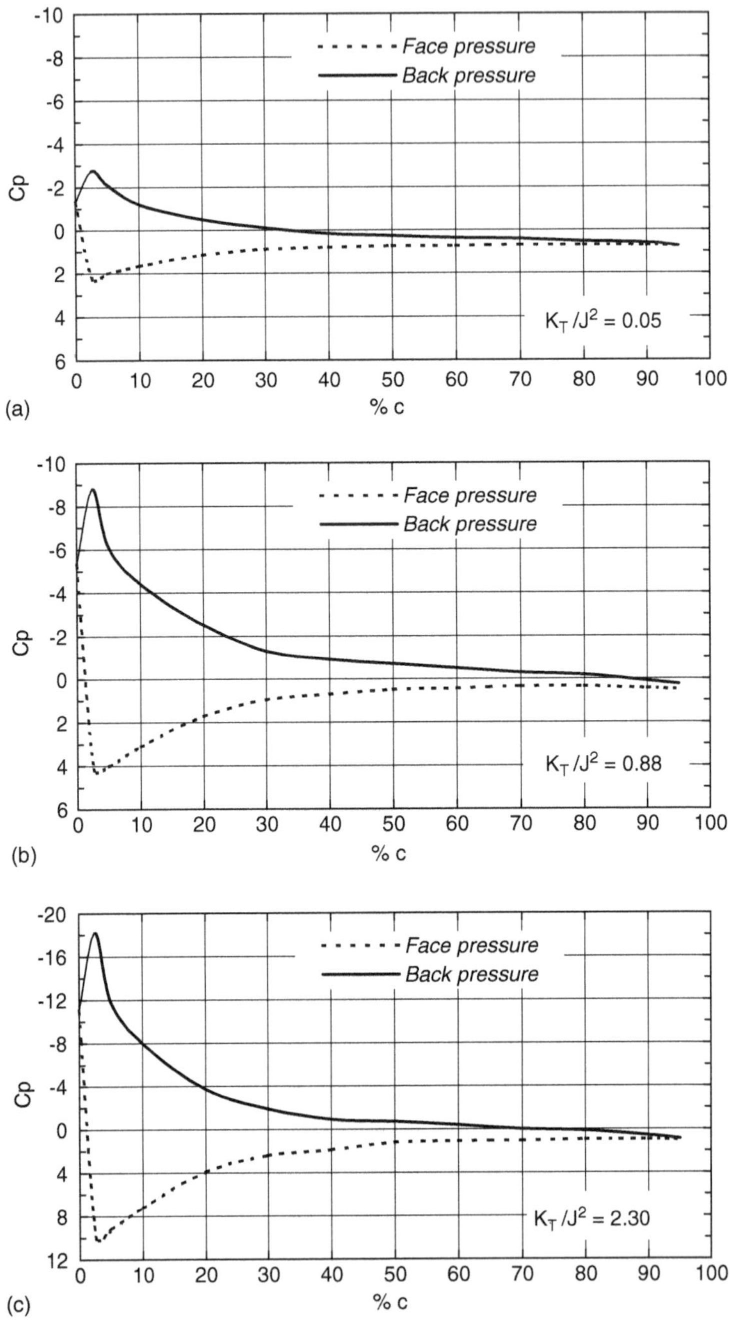

Figure 5.71 Chordwise pressure distributions; Rudder plus propeller, Skeg Rudder No. 0: all-movable section at span position 830 mm from root; $\alpha = 20°$. influence of propeller thrust loading K_T/J^2

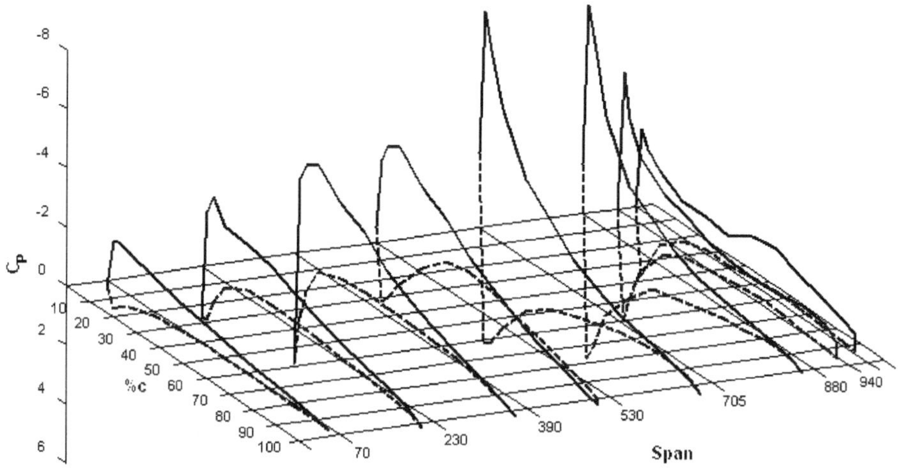

Figure 5.72 Three-dimensional pressure distribution: Rudder No. 2, drift angle $\beta = -7.5°$, $\delta = 20°$, $K_T/J^2 = 0.88$

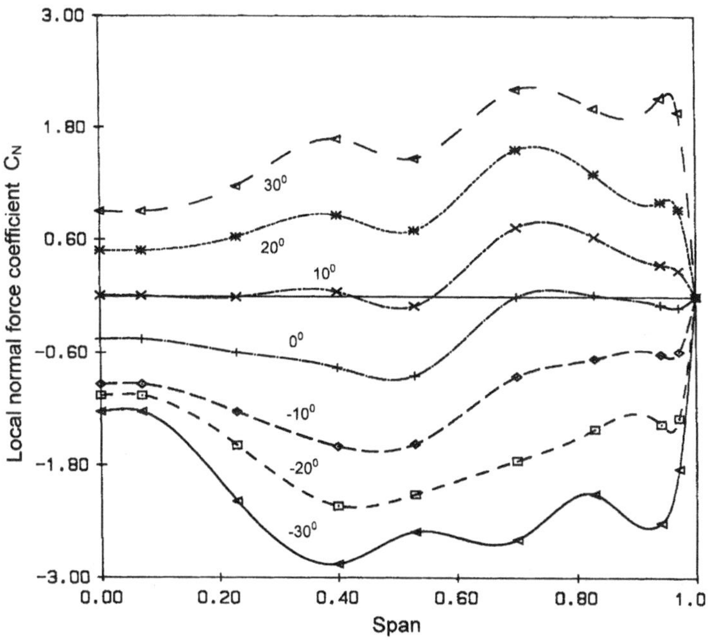

Figure 5.73 Spanwise load distribution: Rudder No. 2, drift angle $\beta = -7.5°$, $K_T/J^2 = 0.88$

3-D pressure/load distributions for structural design: 3-D pressure distributions are available in the experimental database for a limited number of cases, together with a more extensive range of spanwise distributions. A summary of the experimental pressure data tabulated in the database is given in Appendix 1.

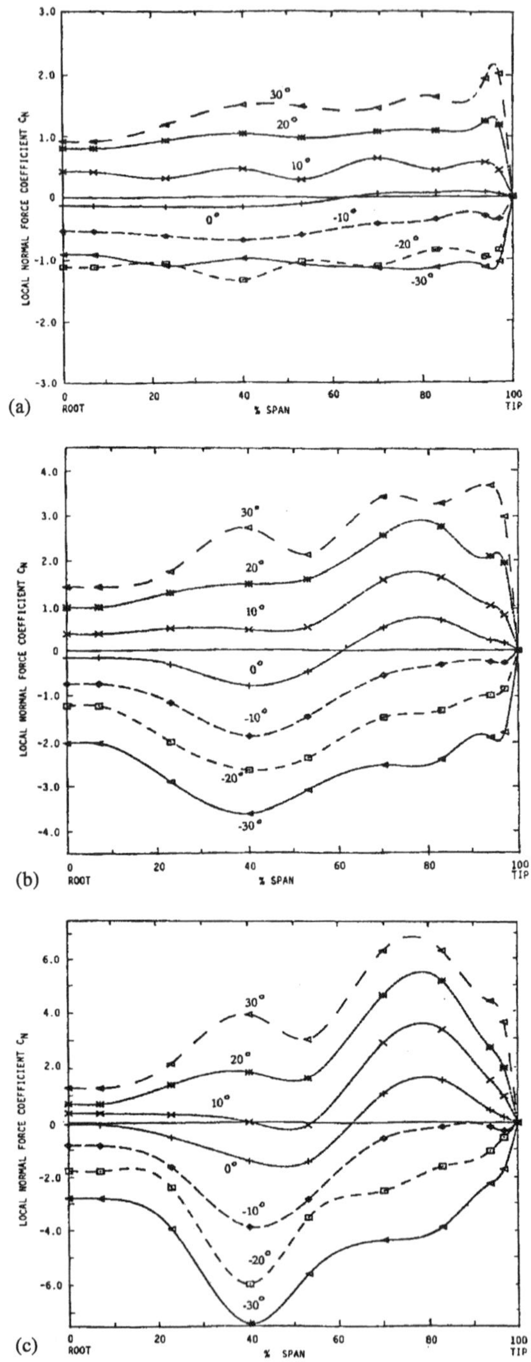

Figure 5.74 (a) Spanwise load distribution: Rudder No. 2, $X/D = 0.39$, $J = 0.94$; (b) Spanwise load distribution: Rudder No. 2, $X/D = 0.39$, $J = 0.51$; (c) Spanwise load distribution: Rudder No. 2, $X/D = 0.39$, $J = 0.35$

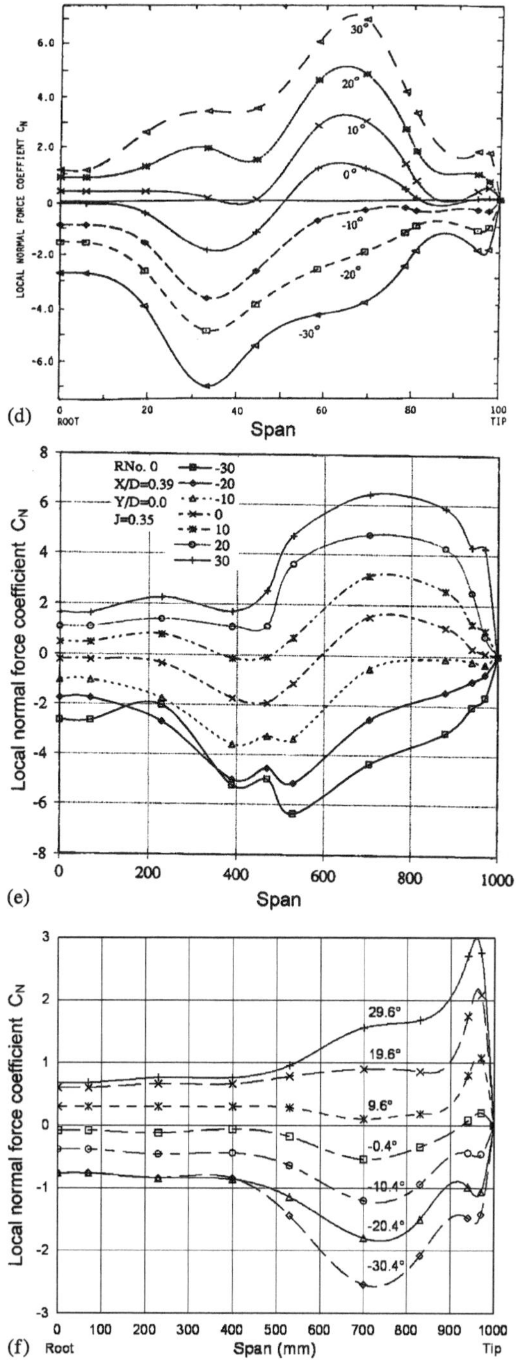

Figure 5.74 (d) Spanwise load distribution: Rudder No. 3, $X/D = 0.39$, $J = 0.35$; (e) Spanwise load distribution: Skeg Rudder No. 0, $X/D = 0.39$, $J = 0.35$; (f) Spanwise load distribution: Rudder No. 2, $J = 0.51$, propeller axis at 900 mm

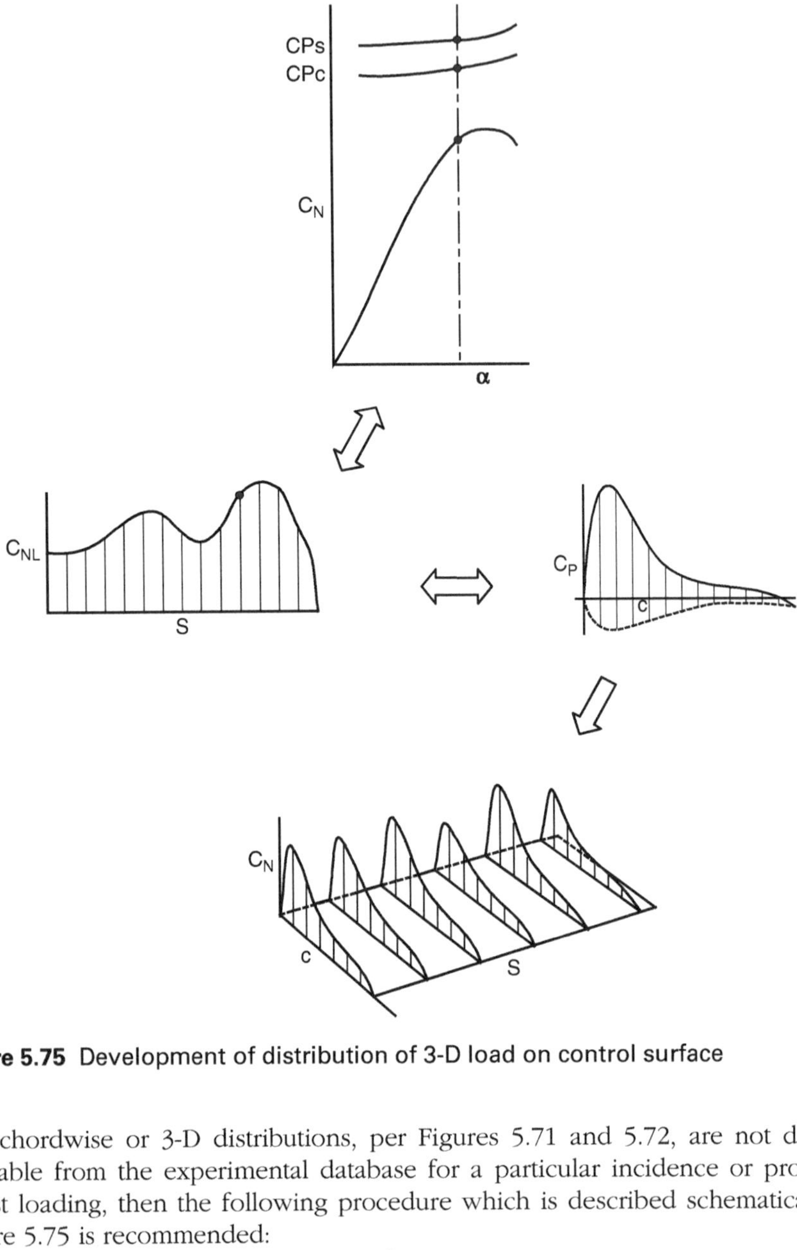

Figure 5.75 Development of distribution of 3-D load on control surface

If chordwise or 3-D distributions, per Figures 5.71 and 5.72, are not directly available from the experimental database for a particular incidence or propeller thrust loading, then the following procedure which is described schematically in Figure 5.75 is recommended:

For a given rudder, given α and K_T/J^2:

(i) Derive total C_N and CP position for the appropriate rudder type from the database, or directly from the data and equations in Sections 5.3 and 5.4.

(ii) Derive spanwise distributions for the closest case from Figure 5.73 or 5.74 or from the database. Adjust the level if necessary whereby integration of the spanwise distribution yields the total required C_N.

(iii) At each span position, choose a chordwise distribution, per Figure 5.71, from the database for the closest case. Integrate to give C_N and scale if necessary to provide the correct local C_N on the spanwise distribution.
(iv) Plot 3-D distribution of load, per Figure 5.72.

The derivation of the three-dimensional distribution of rudder loads by such methods should be adequate for use in structural analyses. The use of such data in an outline structural analysis is described in example application 8, Chapter 11.

Theoretical pressure and load distributions: It should be noted that the spanwise distributions of load can be derived from the experimental database, or via the lifting line theory described in Chapter 6. Similarly, chordwise and spanwise pressure distributions may also be estimated directly using CFD methods such as the Boundary Element or RANS codes described in Chapter 6.

5.4.2.4 *Influence of hull*

The basic influence of the hull is to slow down the flow into the propeller relative to the ship speed, as discussed in Section 3.7. In the course of the wind tunnel experiments described in Section 5.4.2.2, tests were carried out on the rudder–propeller combination in the presence of an upstream hull. This should indicate whether the hull upstream of the rudder–propeller combination would have a significant effect on the rudder characteristics. In particular, this would also relate to the effect on the rudder force and moment characteristics and on the form or shape of the spanwise load distributions over the rudder due to the nonuniform ship wake.

The Mariner hull, Russo and Sullivan [5.90], was used as a representative hull form as it could be used in conjunction with other studies into the manoeuvring performance of this particular hull, such as Sedat and Fuller [5.91]. The hull was truncated to a length of 2.69 m, in a manner similar to some cavitation tunnel hull models, whereby an exact stern shape was used upstream of the propeller for a proportion of the hull length (to station 17), together with a correct thickness for the model (full-scale equivalent) wake using boundary layer stimulation. Section 5.4.2.1 gives further information on the wind tunnel test rig. The rudder–propeller–hull arrangement in the wind tunnel is shown in Figure 5.76 (See Plate 2 of Colour Plate Section) and further detailed information on the models is included in Molland and Turnock [5.78].

The derivation of the wake flow speed was based on the identity of rudder lift at a particular incidence with and without the hull upstream. This process is analogous to obtaining wake speed by applying a thrust (or torque) identity to the propeller open and behind performance, Carlton [5.92]. If subscripts o and b denote open and behind conditions, then:

$$C_{Lo} = L_o/\tfrac{1}{2}\rho A V_o^2 \tag{5.20}$$

$$C_{Lb} = L_b/\tfrac{1}{2}\rho A V_b^2 \tag{5.21}$$

$$C_{Lbo} = L_b/\tfrac{1}{2}\rho A V_o^2 \tag{5.22}$$

where C_{Lbo} is the measured lift, nondimensionalised using the open, or free stream, speed V_o.

Figure 5.76 Arrangement of rudder, propeller and hull in wind tunnel (See Plate 2 of Colour Plate Section)

then from Equations (5.20) and (5.21), $L_b/L_o = V_b^2/V_o^2$
and from Equations (5.20) and (5.22), $L_b/L_o = C_{Lbo}/C_{Lo}$

hence,
$$\frac{V_b}{V_o} = (C_{Lbo}/C_{Lo})^{1/2} \tag{5.23}$$

The basic tests were carried out with skeg Rudder No. 0, Figure 5.55, and are reported in references [5.73, 5.79]. The results for the open, or without hull, condition, together with the behind condition are shown in Figure 5.77. It must be noted that the results in the behind condition have been nondimensionalized using the open or without hull free-stream speed. The open results were obtained from references [5.37 and 5.79]. In Figure 5.77, the lift curve slopes for the behind hull and open conditions are 0.019 and 0.036 respectively, that is the behind hull lift curve slope is 53% of the open condition, or $C_{Lbo} = 0.53\ C_{Lo}$. Noting from Equation (5.23) that, based on a lift identity, lift coefficient varies as V^2 and V as $\sqrt{C_L}$, then $V_b = 0.73\ V_o$. This corresponds to a wake fraction $w_T = 0.27$, which is of the correct order for the Mariner hull form.

Rudder pressure distributions behind the hull were available for Rudder No. 2 in reference [5.79]. These were corrected by multiplying the J values by 0.73 and dividing the pressures by 0.73^2. These corrected curves were then interpolated to provide pressure distributions at J values of 0.94, 0.51 and 0.36. These were then compared with the without hull values, as shown in Figure 5.78, [5.79]. The without hull values are shown in Figures 5.74(b) and 5.74(c) (Section 5.4.2.3). From Figure 5.78 it is seen that for negative rudder angles the hull has a very small effect on the spanwise load distributions. At positive rudder incidence the hull is having some relatively small effect from root to about mid span, where the results without the hull give lower values.

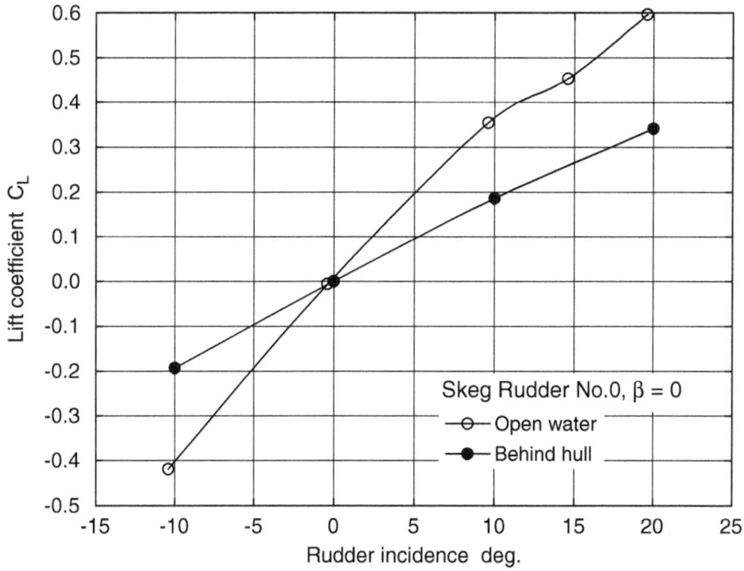

Figure 5.77 Effect of hull on rudder lift

The same correction process can be applied to the force data [5.73], when it is found that the forces also coincide. There was broad agreement for *CPc* and *CPs*, although there are some (relatively small) differences on *CPc* for positive rudder incidence.

The results of the behind hull investigation are very satisfactory in that the hull with its nonuniform distribution of wake tends to have a relatively small effect on the rudder load distributions, total force and centre of pressure results. This might be expected considering the dominating effect that the propeller race has on a rudder. Importantly, it also means that the extensive data for rudder–propeller combinations in the absence of an upstream hull, such as in Section 5.4.2.2, can be used at a corrected J value with acceptable confidence.

5.4.2.5 *Low and zero speed data*

Data for rudders operating at zero and low speeds downstream of a propeller are presented by Molland and Turnock [5.38, 5.76 and 5.80].

Low and zero speed tests were carried out in both a wind tunnel and a laboratory. In the case of the wind tunnel, tests were carried out with the tunnel fan stationary to simulate the bollard pull condition ($J = 0$). The flow induced by the propeller drove the tunnel at a slow but measurable speed, leading to an actual advance ratio of $J = 0.17$ which corresponds to an open water thrust loading (K_T/J^2) of 11.5. The test rig was assembled in a laboratory to simulate the true bollard pull ($J = 0$) condition. The arrangement of the rudder and propeller models and floor boards were identical to that used in the wind tunnel tests. The velocities induced by the propeller at higher revolutions led to effective Reynolds Numbers based on rudder chord of up to 0.75×10^6 over much of the rudder. Results presented by English *et al.* [5.60] indicate that tests at these conditions should preclude any significant scale effect.

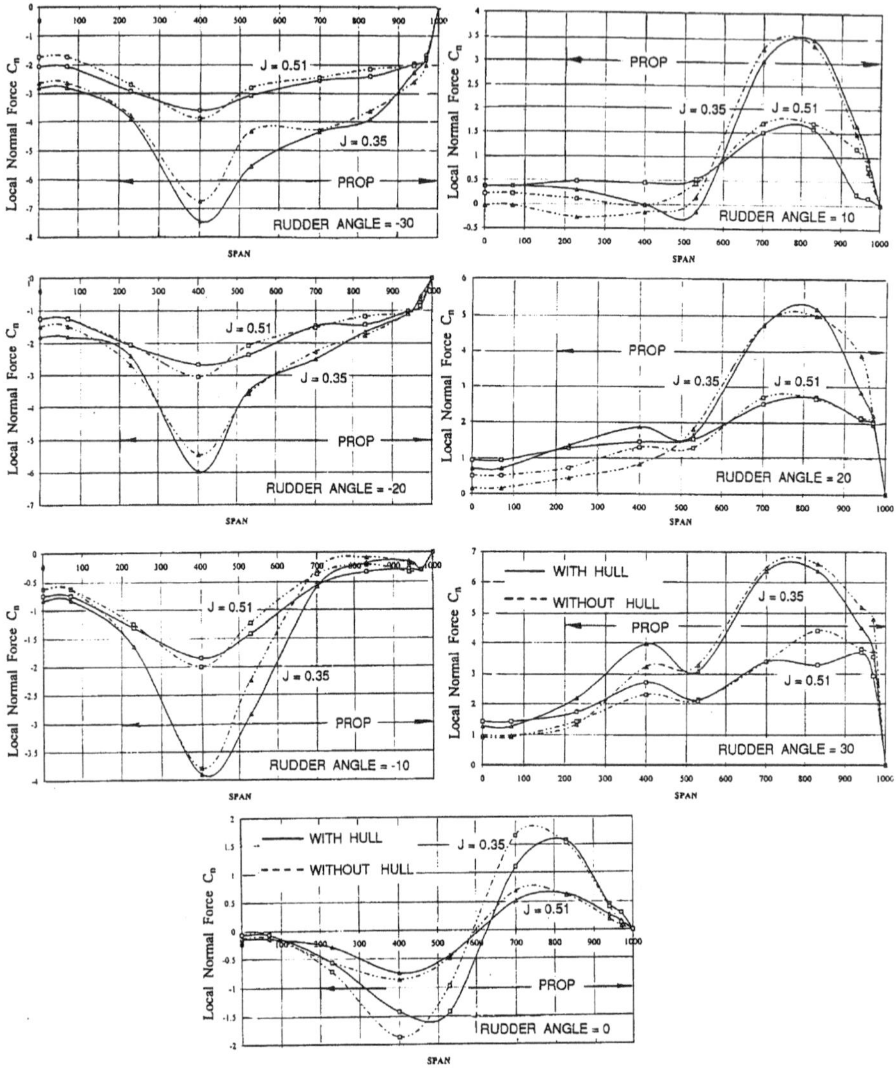

Figure 5.78 Comparison of spanwise load distributions; Rudder No. 2, with and without upstream hull

At low ship speeds in the first quadrant and at zero speed, the rudder forces are nondimensionalised using $K_T n^2 D^2$, a function of the theoretical propeller induced flow speed at zero J (see Equation (3.35a)).

$$C_L^* = \frac{L}{\frac{1}{2}\rho A K_T n^2 D^2} \qquad C_D^* = \frac{d}{\frac{1}{2}\rho A K_T n^2 D^2}$$

Results for low and zero speed are given in Figures 5.79 and 5.80. The values of K_T used in the analysis of the data are those for the test propeller with $P/D = 0.95$

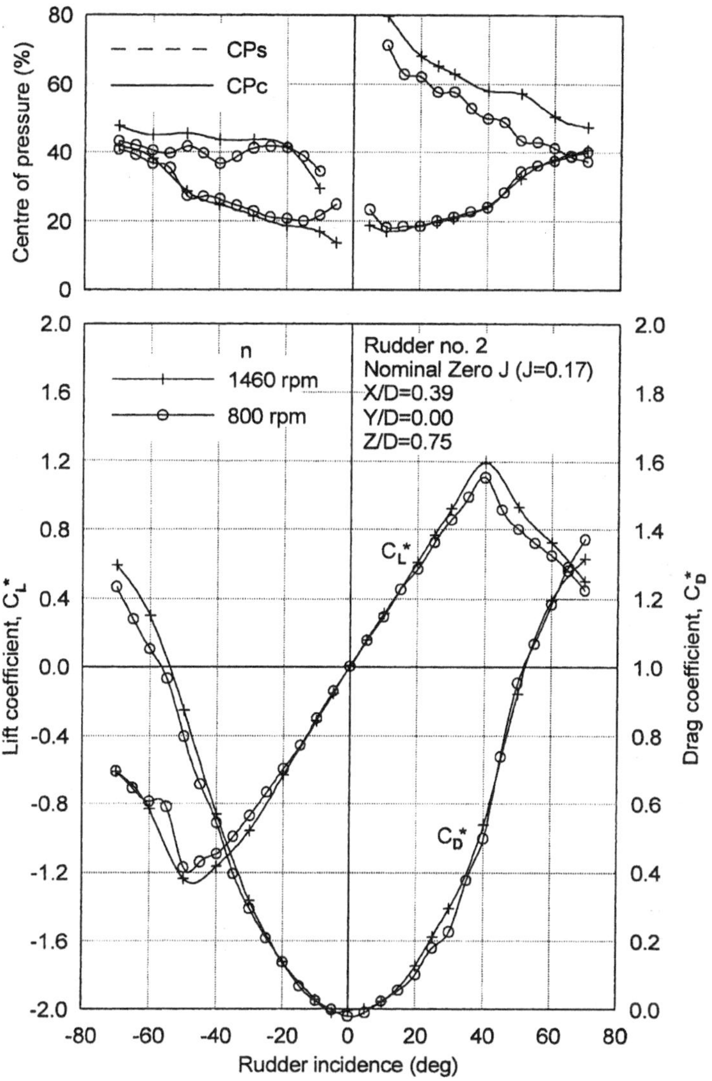

Figure 5.79 Lift, drag and *CPc* characteristics, Rudder No. 2, $J = 0.17$

(Table 5.12 and Figure 5.57) and amounted to $K_T = 0.37$ at $J = 0$ and 0.33 at $J = 0.17$. Results for the low advance ratio of $J = 0.17$ are given in Figure 5.79. It is seen that the results for the two revolutions are broadly similar, with the maximum lift at 1460 rpm being slightly higher. Figure 5.80 shows a similar comparison for the zero speed $J = 0$ case. Overall, the stall angle at $J = 0$ occurs later than when $J = 0.17$ and the lift curve slope at zero incidence $dC_L^*/d\alpha$ is a little lower at $J = 0$. It was noted in Section 5.4.2.2 that the stall angle increases with increase in propeller thrust loading (decrease in J) and this is observed with the further increase in stall angle for $J = 0$ which is at about 45°.

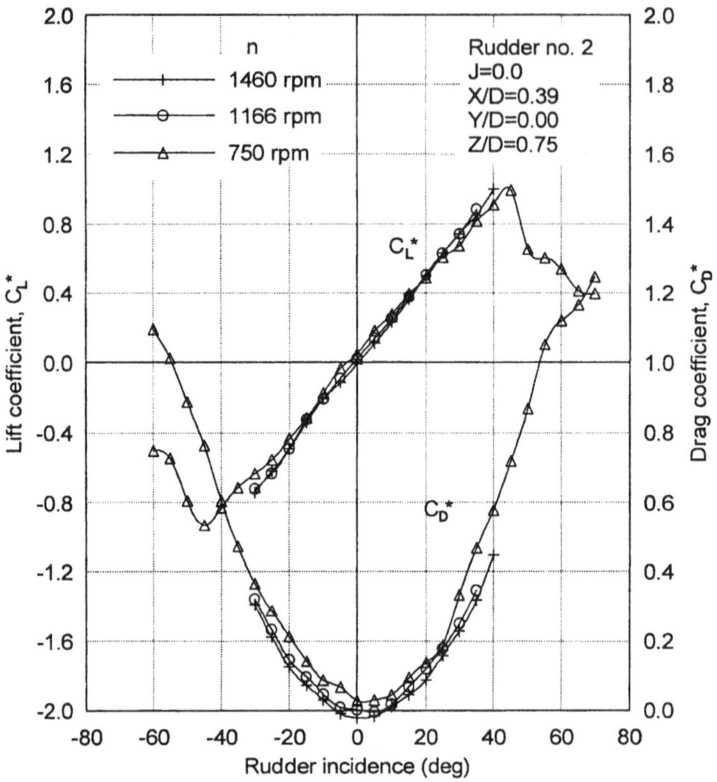

Figure 5.80 Lift and drag characteristics; Rudder No. 2, $J = 0$

Chordwise centre of pressure *CPc* has moved further forward, which follows the trend for increased thrust loading described in Section 5.4.2.2.

Change in lift curve slope over a range of advance ratio J is shown in Figure 5.81 where it is seen that for $J < 0.3$ the change in slope with J is small. This confirms the practice of dividing the presentation of rudder data into distinct domains depending on speed or advance ratio and propeller thrust loading.

Negative propeller revolutions were applied at zero advance ratio, $J = 0$, but the rudder forces developed were very small. It follows that, for zero speed, the use of negative revolutions is unlikely to develop any significant rudder forces.

Longitudinal (X/D) and lateral (Y/D) separations were also investigated at $J = 0$ and the results are shown in Figures 5.82 and 5.83. In Figure 5.82, it is seen that there is very little change in lift curve slope with change in longitudinal separation, as was found for a nonzero advance ratio. Drag reduces with increasing separation and at $X/D = 0.39$ and 0.52 a thrust is produced at zero incidence. Figure 5.83 shows the influence of lateral separation (Y/D). There is a lateral shift in the lift and drag curves and the effects are similar to those for nonzero advance ratio.

Tests with Rudder No. 3, with increased aspect ratio, are shown in Figure 5.84 where a comparison is made with Rudder No. 2. It is found that the difference in lift curve slopes between the two rudders is equal to the difference in rudder area.

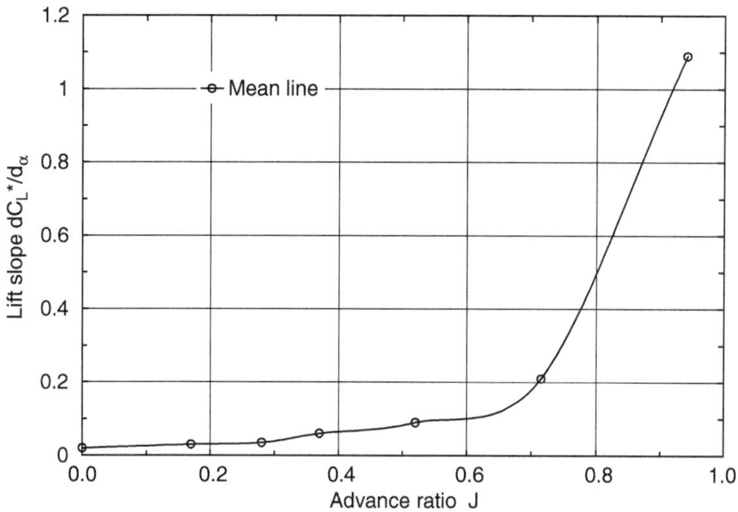

Figure 5.81 Variation of lift curve slope $dC_L^*/d\alpha$ with J

This means that the actual force developed on each rudder is the same and, as expected, confirms that at zero J only the rudder area in way of the propeller race will develop useful lift and generate drag.

Reference [5.80] includes tests at $J = 0$ for the skeg Rudder No. 0. It is found that the performance characteristics for changes in X/D and Y/D are similar to those for Rudder No. 2.

The results and conclusions for the $J = 0$ tests reported by English *et al.* [5.60] are broadly similar to those reported in this section.

It is noted from the data that significant sideforce can be generated in the static $J = 0$ condition. It is an attribute used in the low speed handling of ships. English [5.93] demonstrates, using a measure of merit, that because the stall angle is high, a rudder–propeller combination is an effective manoeuvring arrangement in the static thrust condition.

5.4.2.6 *Four quadrant data*

A set of wind tunnel experiments, which investigated rudder–propeller interaction in four quadrants of operation are reported by Molland and Turnock [5.70]. The tests were carried out in the wind tunnel using Rudder No. 2, Figure 5.56 and Table 5.10. The rudder had pressure tappings at 200 locations (Table 5.11), which gave complete coverage of the rudder surface. Particulars of the propeller are given in Table 5.12, noting that positive revolutions are in the anti-clockwise direction when viewed from aft. The wind tunnel test rig consists of two independent units that allowed the ship astern Quadrants III and IV to be simulated by rotating the whole arrangement through 180°.

The four quadrants of operation of a ship's propeller–rudder system are shown in Figure 5.85 and are defined as: Quadrant 1 (normal state) of ship ahead, propeller

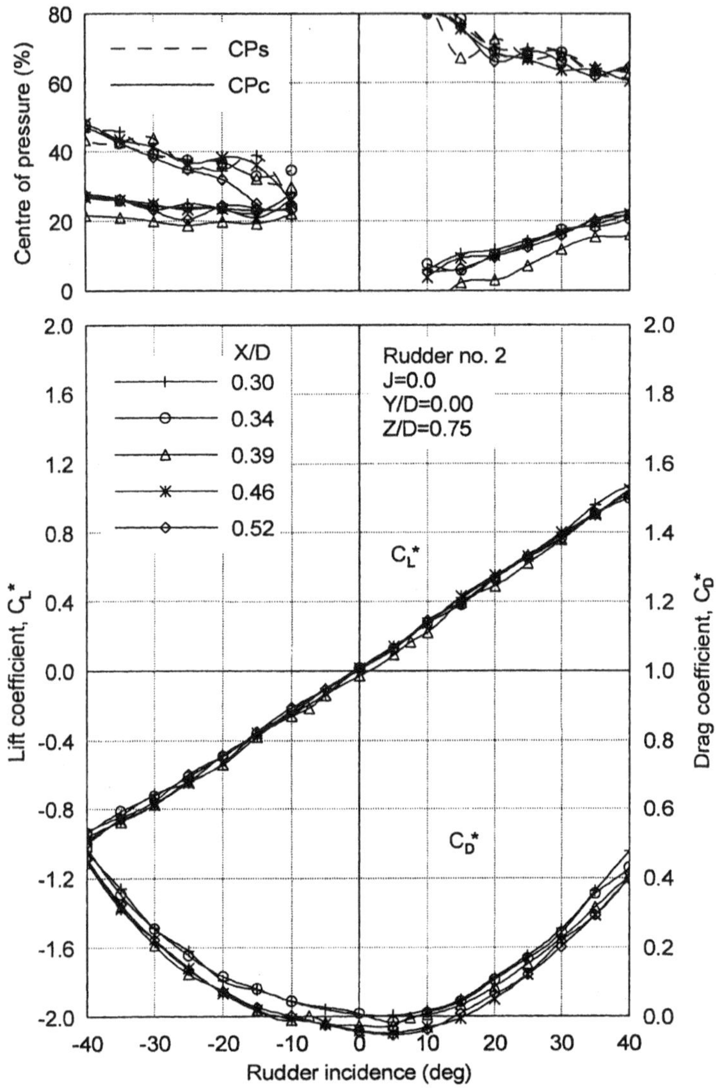

Figure 5.82 Effect of longitudinal separation; Rudder No. 2, $J = 0$

ahead; Quadrant II ship ahead, propeller astern; Quadrant III ship astern, propeller astern and Quadrant IV ship astern, propeller ahead.

The tests in four quadrants of operation were carried out using a mean propeller pitch setting of 0.95, one longitudinal separation of $X/D = 0.39$, a wind speed of 10 m/s, and propeller revolutions up to ±2100 rpm. Based on a free stream velocity of 10 m/s and rudder chord, the nominal Reynolds Number in the ahead condition was 0.4×10^6 whilst velocities induced by the propeller at higher thrust loadings led to effective Reynolds Number of up to 1×10^6 over much of the rudder.

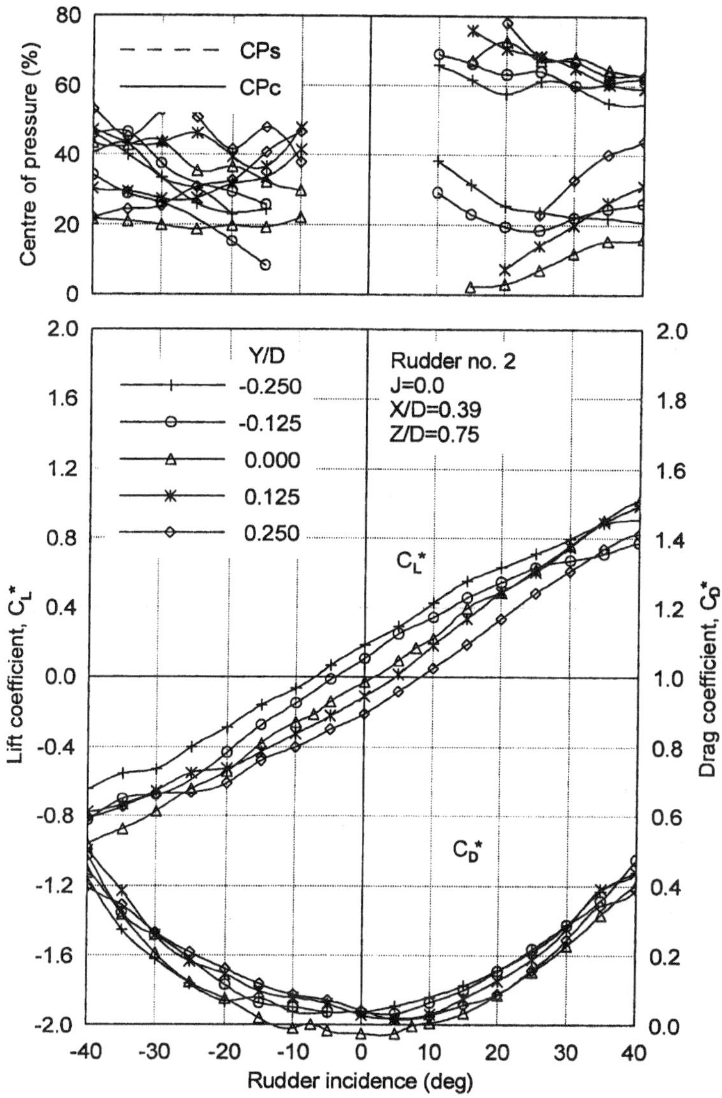

Figure 5.83 Effect of transverse separation; Rudder No. 2, $J = 0$, 1460 rpm

The quadrant of operation is defined in terms of the propeller advance angle ψ, where $\psi = \tan^{-1} J/0.7\pi$. The four quadrant performance of the propeller for a particular pitch ratio is expressed in terms of a propeller thrust and torque coefficients C_T^* and C_Q^*, as used by van Lammeren *et al.* [5.83], where

$$C_T^* = \frac{8T}{\rho\pi D^2[V^2 + (0.7\pi nD)^2]} \qquad C_Q^* = \frac{8Q}{\rho\pi D^3[V^2 + (0.7\pi nD)^2]}$$

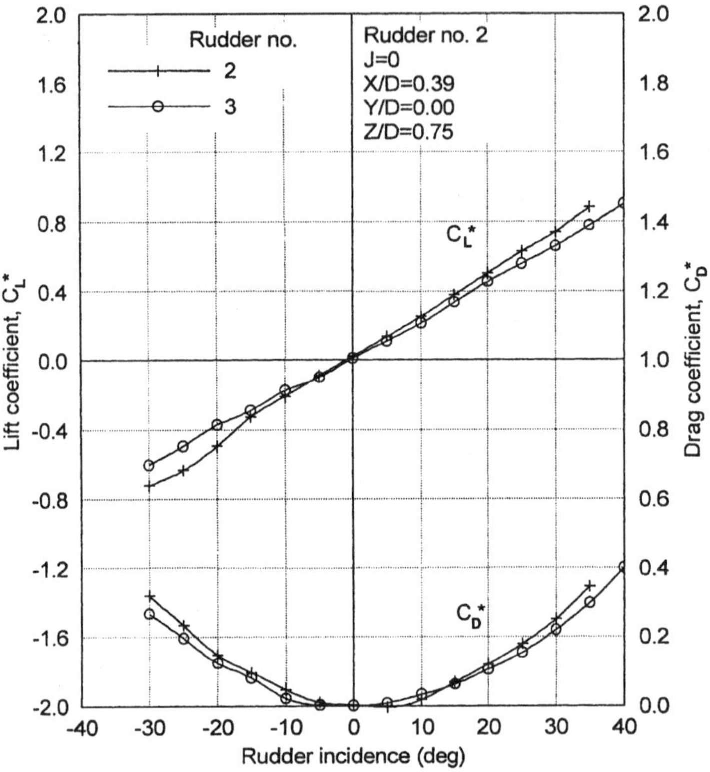

Figure 5.84 Effect of rudder span; Rudder No. 2, $J = 0$, 1166 rpm

QUADRANT I	QUADRANT II	QUADRANT III	QUADRANT IV
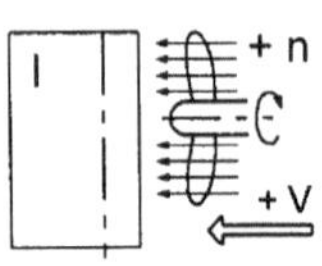	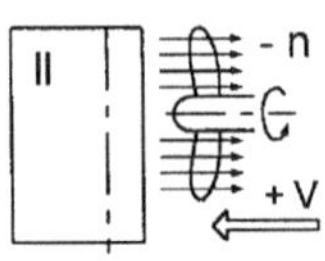	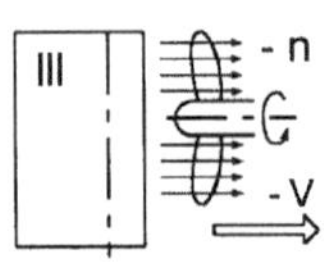	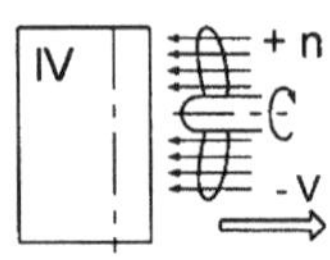
+V, +n	+V, -n	- V, -n	- V, +n
Ship ahead/prop ahead	Ship ahead/prop astern	Ship astern/prop astern	Ship astern/prop ahead

Figure 5.85 Four quadrants of operation

Knowing ψ and C_T^*, K_T/J^2 can be recovered and applied as a thrust loading to the rudder data for a given quadrant. Table 5.15 gives the conditions tested in terms of wind speed, revs. and the measured propeller thrust expressed in the alternative forms of presentation of thrust loading.

Table 5.15 Four-quadrant test cases for rudder–propeller interaction

Quadrant	*Wind speed V m/s*	*Test revs N (rpm)*	*Prop advance angle ψ*	*J*	K_T	K_T/J^2	*Prop* C_T^* *(open)*
	0	750	0	0	0.37	–	0.195
	0	1400	0	0	0.37	–	0.195
I	10	+2780	7.0	0.27	0.31	4.25	0.16
	10	2100	9.2	0.35	0.28	2.30	0.14
	10	1460	13.1	0.51	0.23	0.88	0.114
	10	805	23.0	0.94	0.04	0.05	0.02
	10	(735)	24.9	1.02	0	0	0
	10	400	40.5	1.88	−0.62	−0.18	−0.24
	10	200	59.6	3.75	−3.71	−0.26	−0.46
	10	0	90.0	–	–	–	−0.52
II	10	0	90.0	–	–	–	−0.52
	10	−200	120.4	−3.75	−4.45	−0.32	−0.58
	10	−400	139.5	−1.88	−1.31	−0.37	−0.57
	10	−800	156.9	−0.94	−0.63	−0.71	−0.29
	10	−1453	166.9	−0.51	−0.20	−0.77	−0.12
	0	−n	180.0				−0.12
III	−10	−2105	189.2	0.35	−0.097	−0.79	−0.10
	−10	−1460	193.1	0.51	0	0	−0.07
	−10	−800	203.0	0.94	0.18	0.20	0.02
	−10	−400	220.5	1.88	0.99	0.28	0.24
	−10	0	270.0	–	–	–	0.47
IV	−10	0	270.0	–	–	–	0.47
	−10	+400	319.5	−1.88	1.64	0.47	0.51
	−10	800	336.9	−0.94	0.67	0.76	0.25
	−10	1460	346.9	−0.51	0.40	1.54	0.17
	0	750	360.0	0	0.37	–	0.195
	0	1400	360.0	0	0.37	–	0.195

Four quadrant performance of propeller in free stream: The measured free-stream propeller performance characteristics in four quadrants for a pitch ratio of 0.95 are shown in Figure 5.86. The four quadrant results for the Wageningen B-Series [5.83] are also included. The wind tunnel results are very similar to the Wageningen results except in areas of low thrust loading ($\psi = 90°$ and $\psi = 270°$) where the wind tunnel results are higher. The differences could be due to modifications to the wind tunnel propeller model in way of the hub. These low thrust areas are of lesser importance as far as rudder–propeller interaction is concerned. Similar behaviour was found for the propeller torque, [5.38].

Four quadrant performance of rudder: Figures 5.87(a) and 5.87(b) present the rudder sideforce and drag characteristics for all four quadrants for the range of propeller revs and thrust loadings listed in Table 5.15.

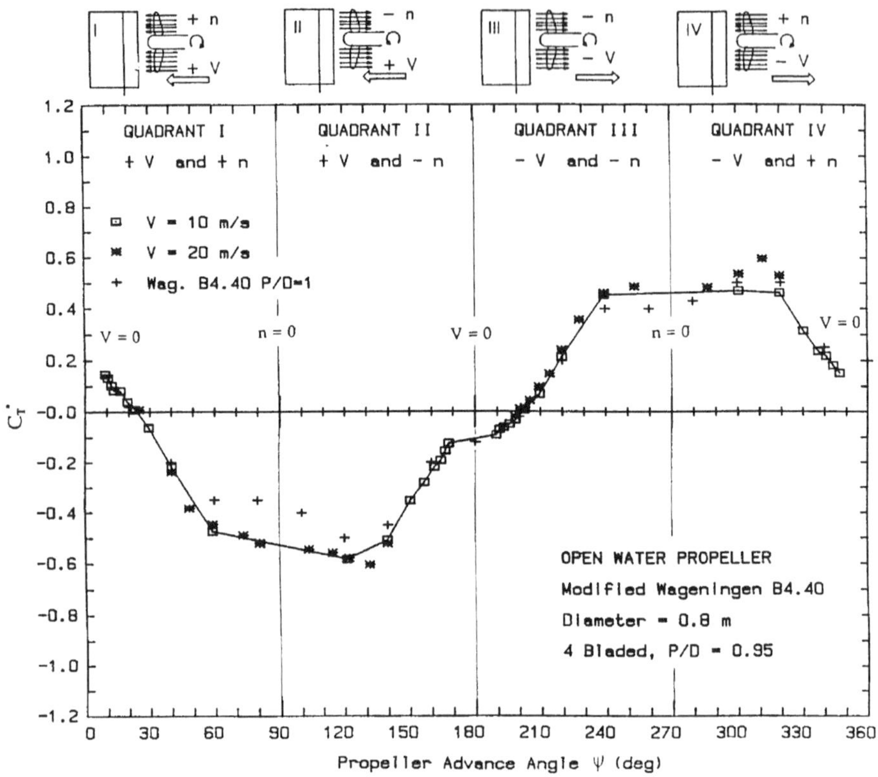

Figure 5.86 Propeller thrust coefficient C_T^*; four-quadrant open-water performance

The results for Quadrant I ($+V$, $+$n) in Figure 5.87(a) show typical increases in lift curve slopes with increase in thrust loading. It is also seen that as revs decrease to less than about 800 rpm, or $K_T/J^2 \to 0$, the sideforce falls below the free stream result and keeps decreasing as revs are further reduced. The propeller is now extracting energy from the fluid and forces developed by the rudder are small.

In Quadrant II ($+V$, $-n$) as revs are increased in a negative sense at about -800 rpm, flow over the rudder and lift force is reversed. With further increase in negative revs to -1460 rpm and negative rudder incidence, there is a further reversal of the flow although it is seen that for positive incidence a breakdown of flow has occurred. Vibration of the test rig precluded further ($-$ve) increase in test revs in this complicated flow situation.

For Quadrant III ($-V$, $-n$) where there is complete reversal of the signs for speed and revs, the rudder is now effectively operating upstream of the propeller. The propeller therefore provides little acceleration and the rudder is effectively working close to its astern free-stream condition. The only major difference is that for higher thrust loadings, stall is delayed.

In Quadrant IV ($-V$, $+n$) it is seen that, as revolutions are increased, rudder forces diminish although, even at relatively high revs, the flow has still not reversed, unlike the converse case in Quadrant II. It is apparent that a further increase in

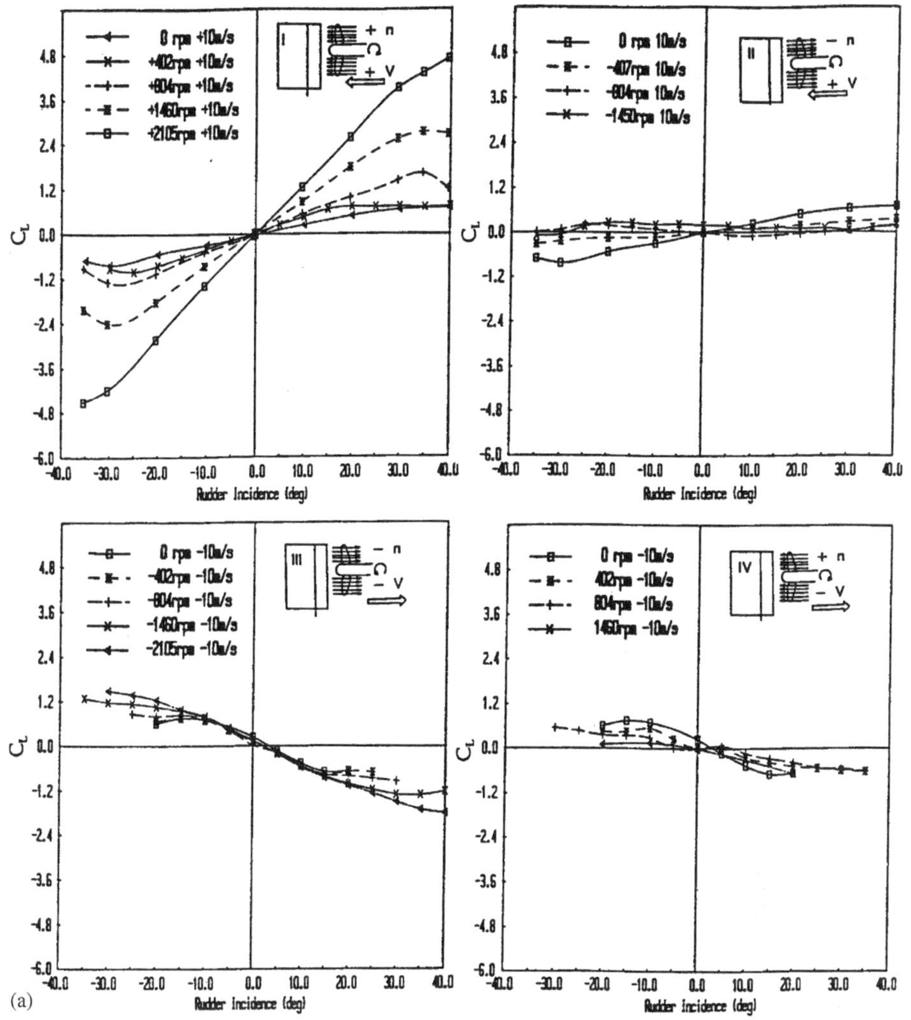

Figure 5.87 (a) Rudder lift (sideforce) characteristics in four quadrants

(+revs) or decrease in (−ve) speed is required before flow reversal will occur. Again, vibration of the test rig precluded further exploration of rudder behaviour.

In Figure 5.87(b) it is seen that, for all four quadrants, the drag behaviour at rudder incidence is generally controlled by the magnitude and sign of the sideforce generated by the rudder. It is seen that some rudder thrust (negative drag) is developed in the second quadrant.

Figure 5.88 gives a detailed picture of the spanwise load distributions over the rudder for particular rpm and wind speed for each of the four quadrants and for a range of rudder incidence. Details of the form of the pressure measurements and derivations of load distributions are given in Section 5.4.2.3. It can be seen that at +400 rpm, +10 m/s (Quadrant I) the propeller is extracting energy from the flow

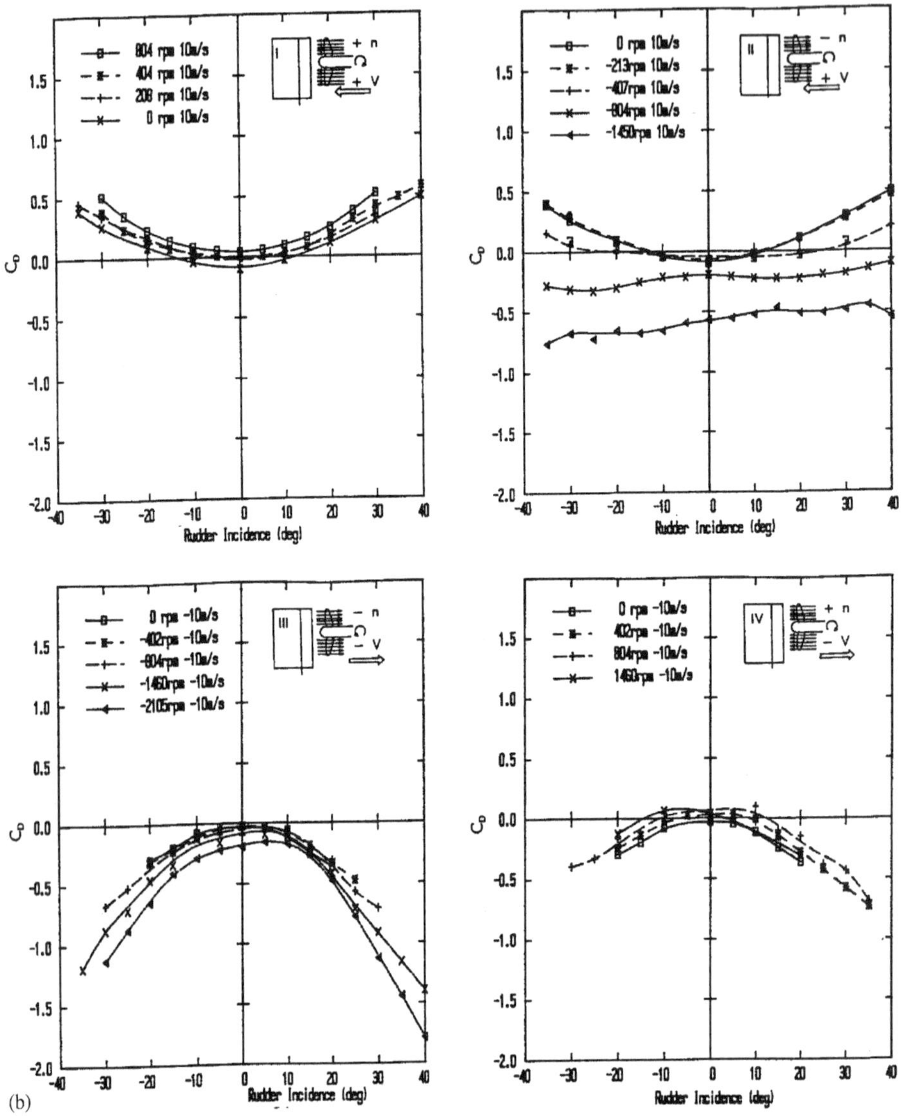

Figure 5.87 (b) Rudder drag characteristics in four quadrants

and this results in a normal force lower in way of the propeller race than in the free stream. When the propeller direction is reversed at −400 rpm (Quadrant II) almost no flow passes over the rudder in way of the propeller race. For astern flow of −10 m/s with −800 rpm (Quadrant III), the rudder is working in the opposite sense to normal, giving negative sideforce for positive rudder incidence. For positive +800 rpm (Quadrant IV), the load distribution remains reversed but the magnitude is lower, implying that the positive rotating propeller is slowing the flow passing over the rudder.

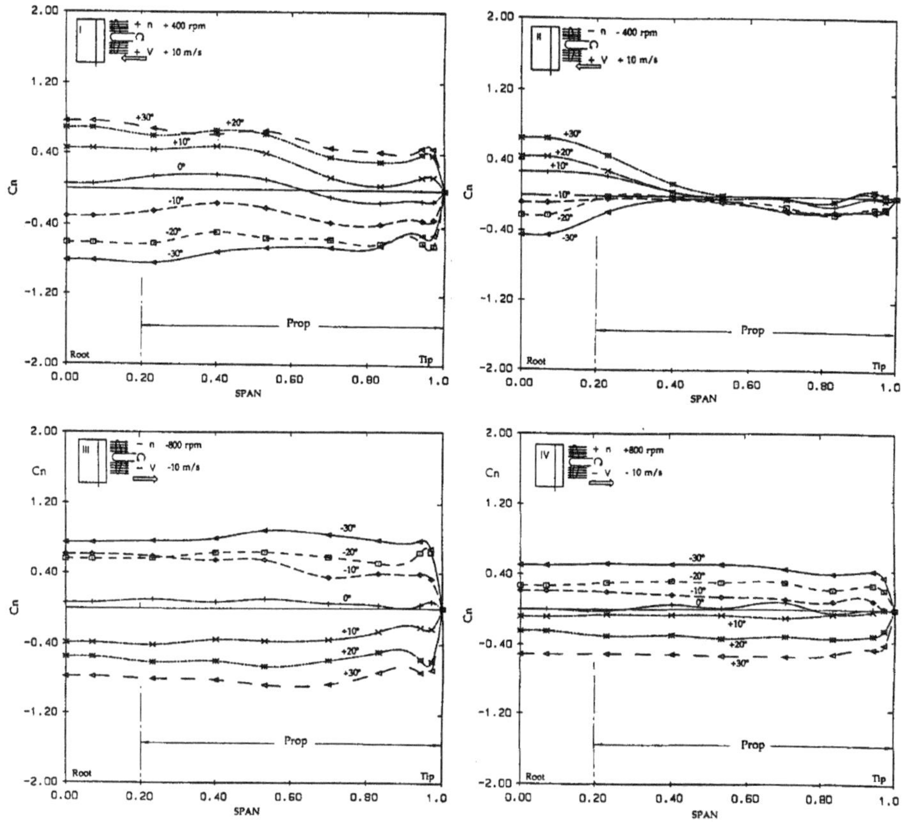

Figure 5.88 Spanwise load distributions: four quadrants; effect of rudder incidence

A useful presentation of the four quadrant rudder data for use in manoeuvring simulations is to present the rudder lift and drag data in terms of the propeller advance angle in a similar manner to that used for propeller four quadrant data, Figure 5.86. Such an approach has been proposed by Molland *et al.* [5.94] and Chislett [5.95].

For the four quadrant case, a presentation of the rudder forces is adopted that allows the cases of $n = 0$ and $V = 0$ to be included. In this case, adopting an approach analogous to a propeller four quadrant presentation [5.83], the rudder forces are nondimensionalised using $[V^2 + K_T n^2 D^2]$:

$$C_L^{**} = \frac{L}{\frac{1}{2}\rho A[V^2 + K_T n^2 D^2]} \qquad C_D^{**} = \frac{d}{\frac{1}{2}\rho A[V^2 + K_T n^2 D^2]}$$

and are presented in terms of the propeller advance angle $\psi = \tan^{-1} J/0.7\pi$ across the four quadrants for different values of rudder incidence δ. Such a presentation, using data from Figure 5.87, is shown in Figure 5.89. To provide stability of data across all quadrants, the K_T value used is that at $J = 0$. Applications of the approach and the data are discussed in Chapter 8.

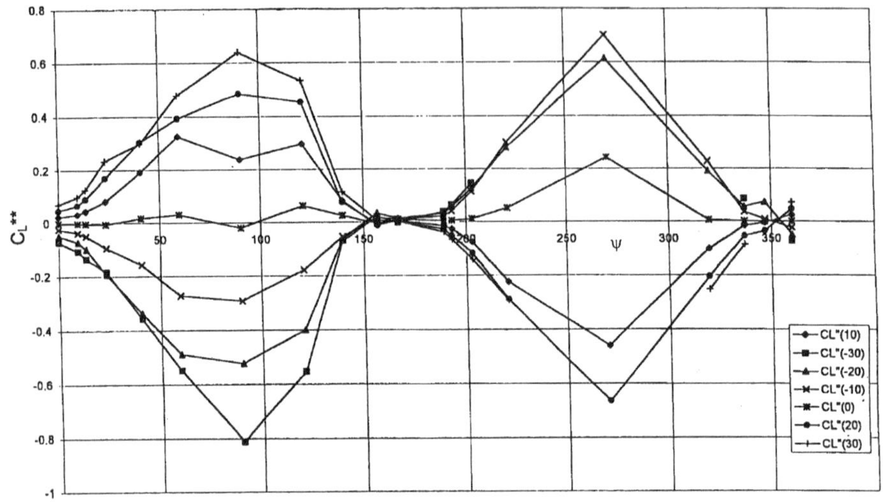

Figure 5.89 Lift coefficient at fixed incidence over four quadrants

5.4.2.7 *Flow straightening effects*

When estimating the forces on a rudder, it is necessary to use the effective angle of attack to the flow. This may be different from the geometric angle of attack of the rudder. When rudder angle is applied, the ship develops a yaw or drift angle β, as discussed in Section 4.1. This leads to a cross flow or drift angle β_R at the aft end of the ship in the vicinity of the rudder, Figure 5.37, which leads to a decrease in the effective inflow angle to the rudder. At the same time, the effects of a propeller and hull upstream of the rudder are to straighten the flow, leading to a recovery, or increase, in the effective inflow angle to the rudder. It is clear that these effects must be incorporated, and a net effective angle of attack used, in any estimate of rudder forces.

A summary of the extensive wind tunnel investigations into rudder–propeller interaction is given in Section 5.4.2.1. The results of experiments with and without a hull upstream of the rudder–propeller combination, together with a comparison of the rudder pressure distributions with and without the hull, are described in Section 5.4.2.4 and Molland *et al.* [5.78]. These investigations showed that, due mainly to the dominating effects of the propeller, the data for the rudder–propeller combination in isolation can be applied successfully downstream of a hull when the appropriate hull wake fraction and hence appropriate inflow velocity is applied to the rudder–propeller combination. In those tests, only straight-line flow was considered with no applied drift angle.

Investigations into the flow straightening effects of the hull and propeller were carried out as part of the wind tunnel investigations summarised in Section 5.4.2.1. This entailed carrying out tests in oblique flow to simulate drift angle. The tests were carried out in three stages.

(i) Using a rudder–propeller combination in isolation at angles of drift between $-15°$ and $+15°$.

(ii) Using centreline boards of relatively small thickness, with three different lengths, situated upstream of the rudder–propeller combination to simulate and investigate the effect of an upstream plate on flow straightening.
(iii) Using an upstream representative hull form to simulate the influence of upstream thickness and curvature.

The results of these tests are contained in Molland and Turnock [5.78]. A reanalysis of much of the data is contained in Molland and Turnock [5.96] and these data are presented, with permission, in the current work.

Models: Rudder Nos. 2 and 3, Figure 5.56, were used in the investigation, together with the propeller whose particulars are given in Table 5.12. The centreboard was designed to be tested at three different lengths, as shown in Figure 5.90. The stern profile was based on the stern profile of the mariner hull form, Russo and Sullivan [5.90]. The rudder–propeller longitudinal separation was fixed at $X/D = 0.39$. The overall thickness of the centreboards was 110 mm, and the tops of all the centreboards were square tipped. The leading edge was circular with a radius of 55 mm and the height of all the boards was 1018 mm. In the case of the hull model, the Mariner hull was used as a representative hull form as described in Section 5.4.2.4.

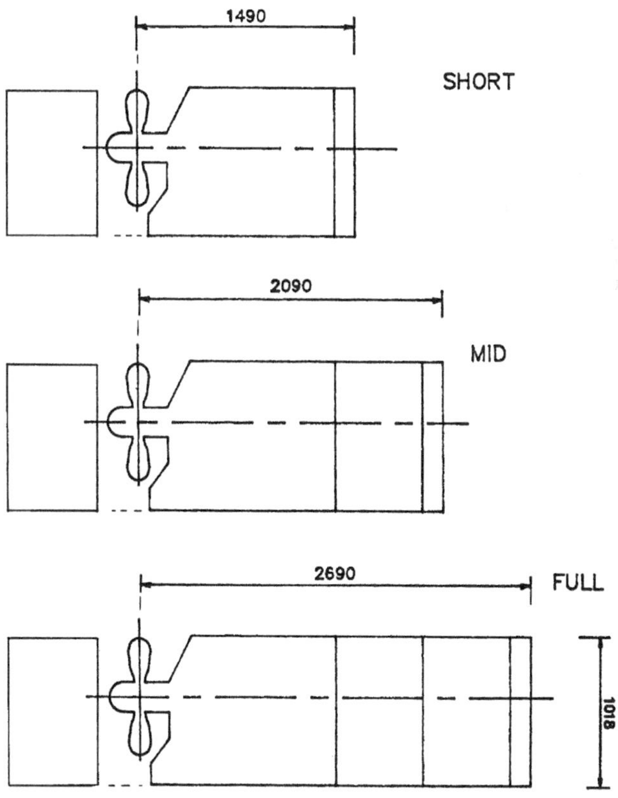

Figure 5.90 Overall dimensions of the three centreboard configurations

Tests: The tests were carried out for a range of rudder incidence of −40° to +40° at a set of *J* vales of 0.94, 0.51 and 0.36, corresponding to propeller thrust loadings K_T/J^2 of 0.05, 0.88 and 2.30. The rudder–propeller combination in isolation was tested at four angles of drift of −15°, −7.5°, +7.5° and +15° for the standard set of flow conditions. The three different lengths of centreboard were tested at drift angles of −7.5° and +15° as well as at 0° for the full length centreboard. The Mariner stern hull form was tested at 0°, −7.5° and −15°. Rudder No. 3 was tested only for the case of rudder plus propeller alone.

Flow straightening terminology: The terminology applied to the flow straightening is shown in Figure 5.91. δ is the rudder angle relative to *ship axis* and β_R is the *geometric* drift angle at the rudder. On a turn, β_R will be larger than the ship drift angle β (at LCG, Figure 5.37), but for oblique model tests in a test tank or wind tunnel β_R will be the same as the ship drift angle β.

With no flow straightening due to propeller and hull:

$$\textit{Geometric}\text{ rudder angle } \alpha = \delta - \beta_R$$

With flow straightening:

$$\textit{Effective}\text{ rudder angle } \alpha_E = \delta - \alpha_0$$
$$= \delta - \gamma\beta_R$$

where

$$\gamma = \alpha_0/\beta_R$$

γ is a flow straightening factor which will depend on drift angle, propeller thrust loading and upstream hull form.

The angle α_0 is the incidence for zero lift at a particular drift angle and may be obtained from the basic lift data at different drift angles, as shown schematically in Figure 5.92.

It is seen from Figure 5.91 that with an increase in flow straightening, for a given drift angle β_R, α_0 decreases, hence γ decreases and there is an increase in the effective angle of attack α_E on the rudder.

Results: An example of the results for the rudder plus propeller in isolation at $J = 0.36$ is shown in Figure 5.93. It is seen that with change in drift angle, there is a change in the no-lift angle. The lift curves appear as a set of near parallel lines, with little change in lift curve slope. It is seen from Figure 5.93 that the rudder angle for minimum drag moves away from zero with change in drift angle and that

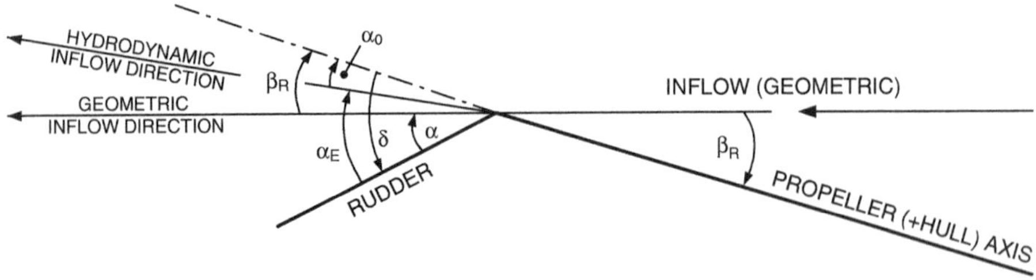

Figure 5.91 Flow straightening terminology

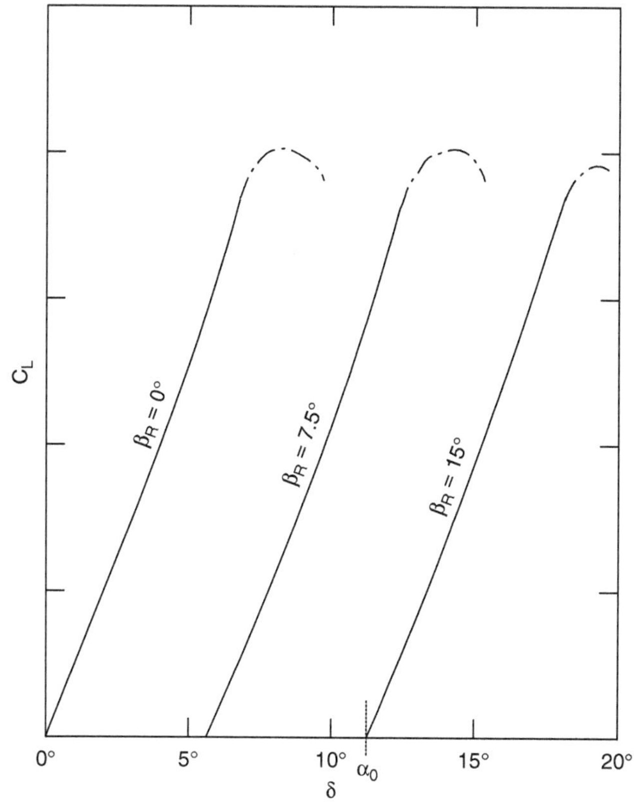

Figure 5.92 Derivation of flow straightening angle (α_0) at zero lift (schematic)

minimum drag is no longer coincident with zero side force. The data for $J = 0.94$ and 0.51, [5.78], showed similar characteristics. The results in Figure 5.93 are important in that, as the slope of the lift curve remains effectively constant with change in drift angle, they confirm the viability of an approach to rudder performance estimates that applies velocity and flow straightening inputs to the basic performance characteristics for a rudder–propeller combination in isolation.

An example of the results for the rudder plus propeller with the full upstream centreboard at $J = 0.36$ is shown in Figure 5.94. Again, the shape and slope of the lift curves remain the same, with the curves offset laterally depending on the drift angle. The data for $J = 0.94$ and 0.51, [5.78], showed similar characteristics.

An example of the results for the rudder plus propeller with upstream hull at $J = 0.36$ is shown in Figure 5.95. The shape and slope of the lift curves are again seen to remain the same, with the curves offset laterally depending on drift angle. The change in drag coefficient due to change in drift is small. The data for $J = 0.94$ and 0.51, [5.78], showed similar characteristics. It should be noted that the results in Figure 5.95 have been nondimensionalised using the upstream (freestream) velocity, not the wake velocity which will be smaller. Hence the effect of the hull is to reduce the lift curve slope compared with that for the rudder–propeller combination in

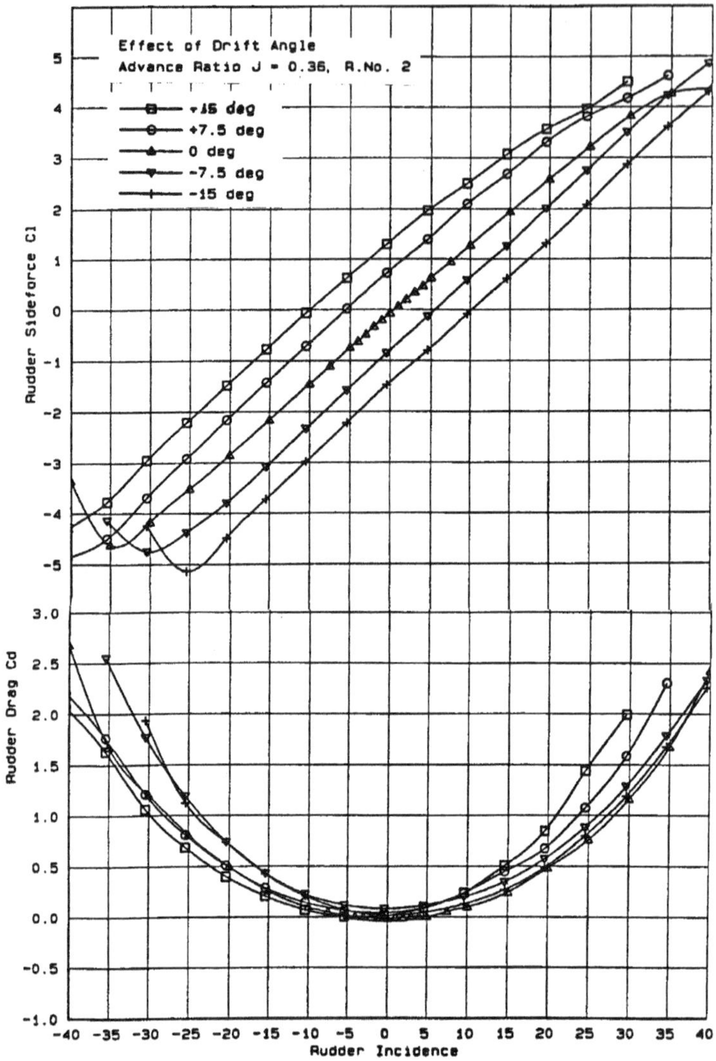

Figure 5.93 Effect of drift angle on the performance of Rudder No. 2 at $J = 0.35$; rudder plus propeller in isolation

isolation. The wake fraction, discussed in Sections 3.6 and 5.4.2.4, can be derived indirectly by comparing the slopes of the rudder–propeller combination alone and behind the hull, noting that J and C_L both need correcting for the wake speed.

Overall effects: The derived no-lift angles (α_0) and flow straightening factors (γ) for the various test combinations are given in Table 5.16. Figure 5.96 illustrates the effects of drift angle and propeller thrust loading J on flow straightening. As the results for positive and negative drift angles were similar for the rudder plus propeller in isolation, the overall data have, for clarity, been averaged and plotted for positive drift angle. It is seen that for the rudder plus propeller alone there is

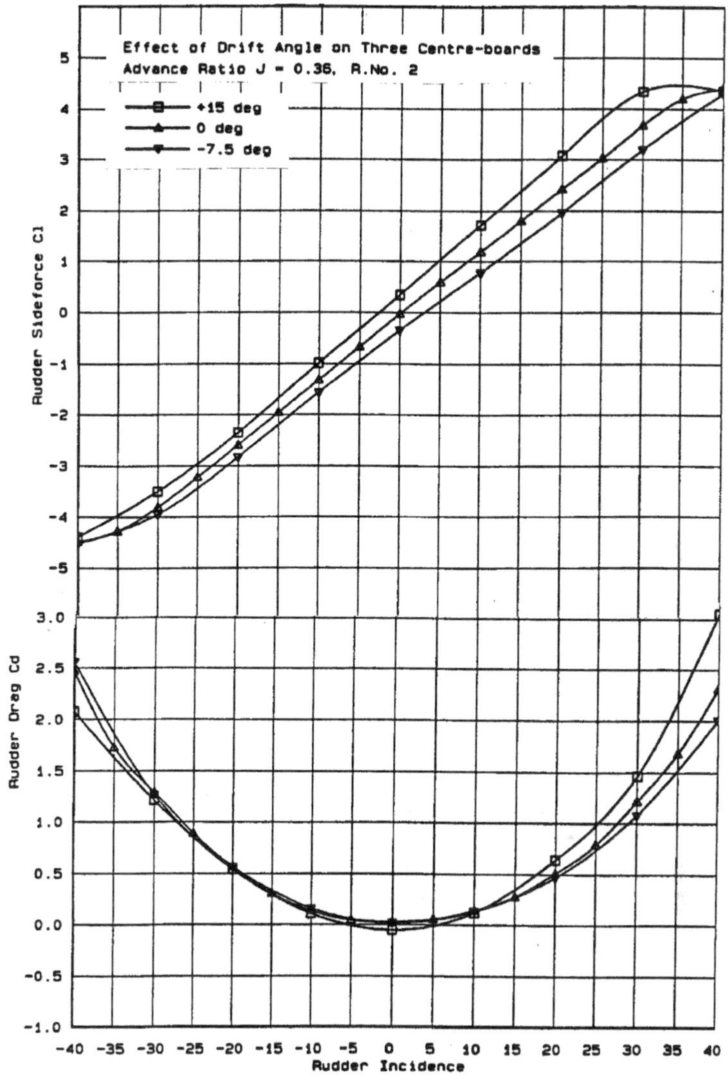

Figure 5.94 Effect of drift angle on the performance of Rudder No. 2, at $J = 0.35$, downstream of full centreboard

an almost linear decrease in straightening (increase in α_0) with increase in drift angle and an increase in flow straightening (decrease in α_0) with decrease in J (increase in propeller thrust loading). As an example, it is noted that for Rudder No. 2 plus propeller at 15° drift and $J = 0.36$, the propeller has straightened the flow (i.e. $\beta_R - \alpha_0$) by almost 5°, leading to a significant increase in sideforce.

When the three centreline boards are upstream of the propeller, it is seen in Figure 5.96 that there is a significant increase in flow straightening (smaller α_0) compared with the propeller alone. At $\beta_R = 15°$ there is a significant increase in

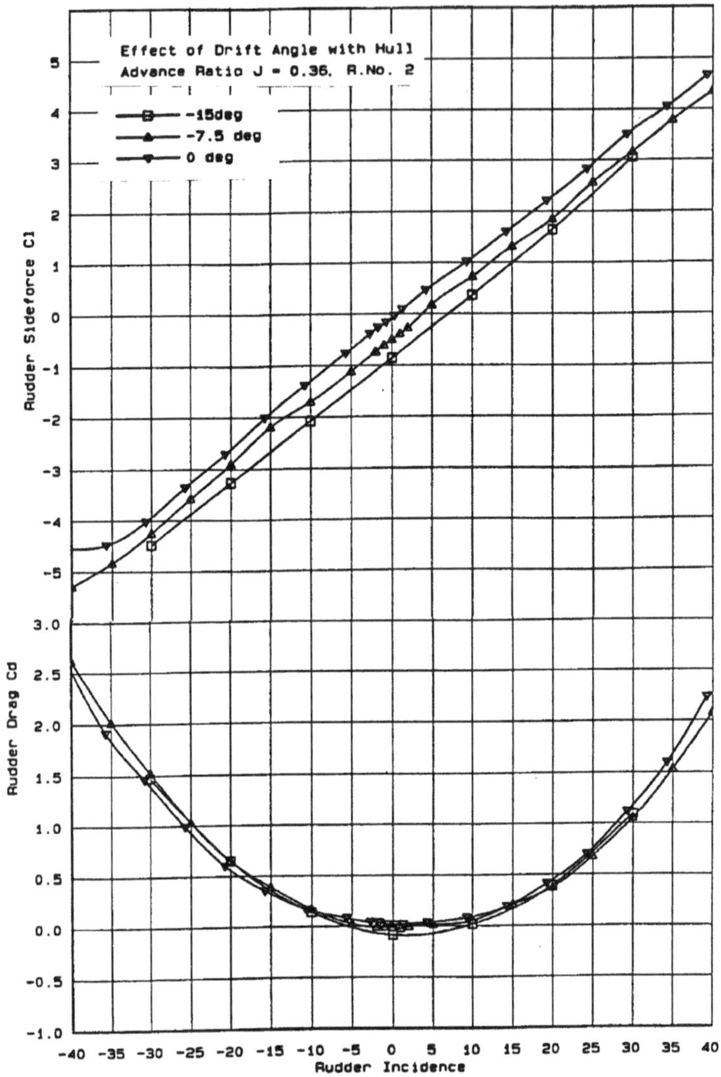

Figure 5.95 Effect of drift angle on the performance of Rudder No. 2, at $J = 0.35$, downstream of Mariner hull form

flow straightening (decrease in α_0) for a decrease in J (increase in propeller loading) whilst at 7.5° drift the effect is reversed.

With the hull upstream there is an increase in flow straightening (decrease in α_0) compared with the propeller alone, but its effects are less than for the three centreline boards. These are important results in that they illustrate the dependence of flow straightening on the angle of the run of the hull at the stern, which will influence the direction of inflow in the stern region. At $\beta_R = 15°$, the flow straightening is seen to increase (decrease in α_0) with increase in thrust loading (decrease in J) whilst at

Table 5.16 No-lift angles ($\alpha°$ deg.) and flow straightening factors (γ), [5.96]

	β_R	−15°		−7.5°		0°		+7.5°		+15°	
Case	J	α_0	γ	α_0	γ	α_0	γ	α_0	γ	α_0	γ
Rudder No. 2											
+Prop	0.94	12.0	0.80	6.3	0.84	–	–	6.0	0.80	11.6	0.77
	0.51	11.5	0.77	6.2	0.83	–	–	5.7	0.76	11.0	0.73
	0.36	10.2	0.68	5.5	0.73	–	–	5.6	0.75	10.0	0.67
+1 Board	0.94	–	–	3.2	0.43	–	–	–	–	6.1	0.41
	0.51	–	–	3.6	0.48	–	–	–	–	4.7	0.31
	0.36	–	–	4.4	0.59	–	–	–	–	4.9	0.33
+2 Boards	0.94	–	–	2.0	0.27	–	–	–	–	5.2	0.35
	0.51	–	–	2.8	0.37	–	–	–	–	4.1	0.27
	0.36	–	–	3.6	0.48	—	–	–	–	3.5	0.23
+3 Boards	0.94	–	–	2.3	0.31	–	–	–	–	4.9	0.33
	0.51	–	–	2.5	0.33	–	–	–	–	3.6	0.24
	0.36	–	–	2.8	0.37	–	–	–	–	2.7	0.18
+Hull	0.94	8.2	0.55	3.8	0.51	–	–	–	–	–	–
	0.51	7.8	0.52	3.2	0.43	–	–	–	–	–	–
	0.36	7.0	0.47	3.8	0.51	–	–	–	–	–	–
Rudder No. 3											
+Prop	0.94	12.8	0.85	6.4	0.85	–	–	–	–	–	–
	0.51	12.6	0.84	6.5	0.87	–	–	–	–	–	–
	0.36	11.3	0.75	6.0	0.80	–	–	–	–	–	–

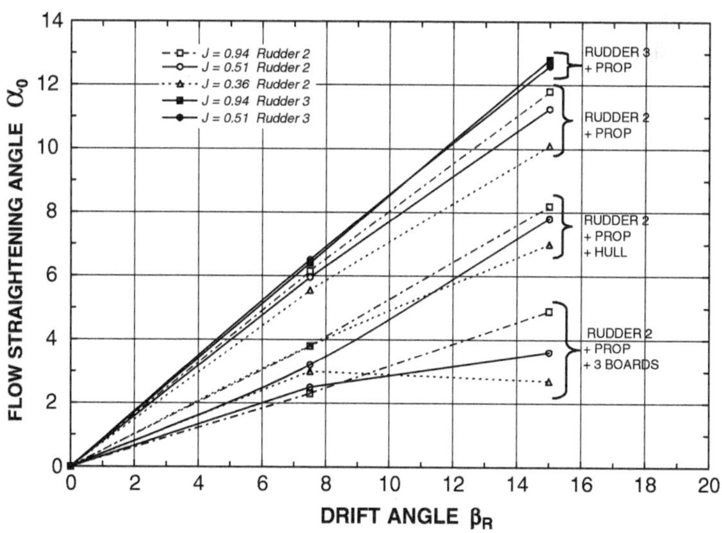

Figure 5.96 Effect of drift angle and propeller thrust loading on flow straightening angle (α_0) [5.96]

7.5° the effects are reversed. This is similar to the case of the three centreline boards and may be due to the fact that the hull and centreline boards were both tested at −7.5° (rather than + 7.5°) which might have led to a reversal of any effects that may be due to the direction of propeller rotation. As the hull was tested at −15° and showed the same trends as for the propeller alone and the centreline boards, both of which were tested at +15°, it is implied that the reversal in effects at 7.5° (i.e. decrease in flow straightening with increase in thrust loading) would be expected for both negative and positive drift angles of 7.5°.

The results for Rudder No. 3 are also included in Figure 5.96. The tip of Rudder No. 3 extends 200 mm beyond the propeller tip, Figure 5.56, and in this case 67% of the rudder span is covered by the propeller diameter compared with 80% for Rudder No. 2. As a result, the flow straightening on Rudder No. 3 is seen to be less than for Rudder No. 2.

An overview of the separate effects on flow straightening of the propeller alone, hull and centreline boards are included in Figure 5.97. Starting from no straightening, it is interesting to note the significant straightening effect (decrease in α_0) of the propeller alone, followed by a further significant increase in straightening with one centreline board and smaller increases in straightening with two and three boards. It is seen that the short single board upstream of the propeller has the most impact. The flow straightening effect of the hull at 7.5° drift is seen to be equivalent to one board upstream, whilst at 15° drift the hull straightening is less than the single board.

An alternative presentation of the data is given in Figure 5.98. This shows the flow straightening factor γ $(=\alpha_0/\beta_R)$, which is a form suitable for practical applications and validation purposes. Limited amounts of published data on flow straightening effects are available as comparators. In Ogawa and Kasai [5.97] the values of γ for the effects of propeller and hull are similar to those in Figure 5.98 with typical values between 0.5 and 0.7 and with similar decreases in γ with increase in propeller thrust loading. Sedat and Fuller [5.91] carried out tests on a Mariner hull and obtained flow straightening factors (for hull without propeller) of

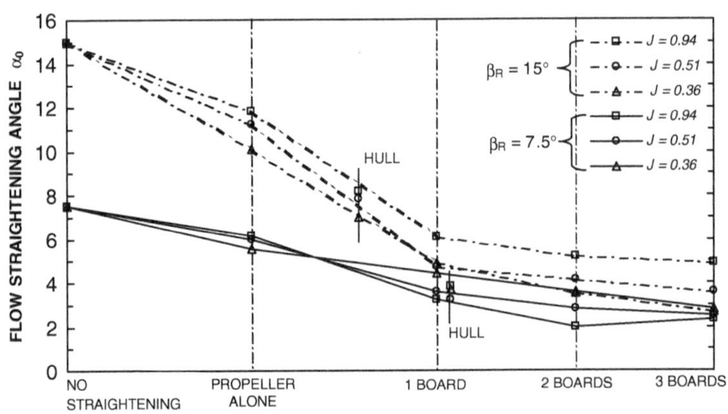

Figure 5.97 Effects of propeller, hull and centreline boards on flow straightening angle (α_0) [5.96]

the order of $\gamma = 0.3$ at $\beta_R = 10°$ to 0.4 at $\beta_R = 20°$, which are similar to the contributions due to the hull alone in Figure 5.98. A commonly used approach to flow straightening in manoeuvring simulations is described by Hirano [5.98] which uses a flow straightening factor $\gamma = CpCs$, where Cp and Cs represent the effects of the propeller and hull respectively. The results of Hirano lead to flow straightening effects due to the propeller alone that are at about the same level as those in Figure 5.98, but with larger changes due to propeller thrust loading. When the hull factor Cs is included, the overall values of γ tend to be a little lower than the values for propeller plus hull shown in Figure 5.98.

From a practical viewpoint, it can be noted that if all the values of α_0 varied linearly with β_R, then γ ($=\alpha_0/\beta_R$) would be independent of drift angle, and this nearly the case for the propeller alone, Figure 5.98. For practical purposes it is likely that relatively small losses in accuracy would be incurred if all the data in Figure 5.96 were linearised with respect to drift angle.

Applications: The flow straightening factors in Figure 5.98 provide data for correcting the inflow angle to the basic rudder–propeller combination in a manoeuvring situation. The influence of the hull might be applied to typical single screw cases. The data for the centreline boards may be used to represent a relatively thin skeg upstream of the propeller, whilst the data for the rudder–propeller combination alone might be applied to the twin screw case, where the rudder–propeller combination is working in relatively clear water.

The application of the flow straightening results to the data for the basic rudder–propeller combination, for use in performance estimates or manoeuvring models, would be carried out as follows [5.96]:

(a) Determine the local speed of advance using a suitable wake fraction at the rudder–propeller location for the given hull form. Calculate the effective propeller thrust loading based on the propeller local advance speed.

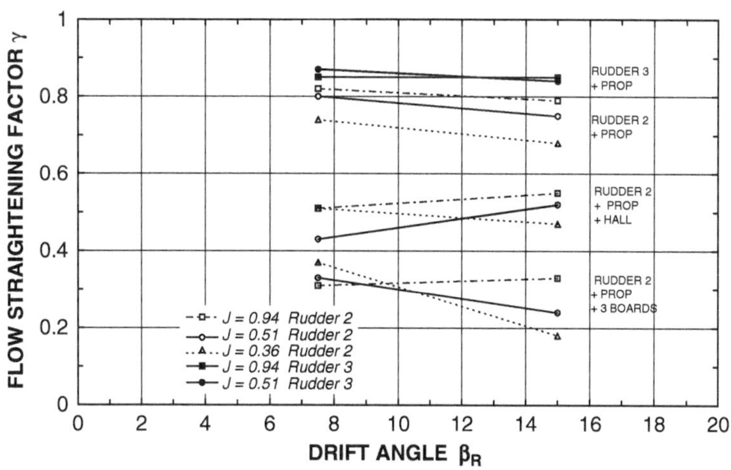

Figure 5.98 Effect drift angle and propeller thrust loading on flow straightening factor (γ) [5.96]

(b) Use the flow straightening factor γ at the local effective drift angle and effective propeller thrust loading to calculate the effective rudder incidence α_E for a given rudder helm.
(c) Apply the effective rudder incidence α_E and appropriate propeller thrust loading (based on local propeller advance speed) to the performance characteristics for the rudder–propeller combination to calculate the rudder forces.

5.4.3 Rudder–propeller combinations in proximity

Some indication of the effect of rudder–propeller combinations working in proximity, and the manoeuvring performance of twin-screw ships may be derived from references [5.99–5.101].

Nakatake [5.99] carried out a theoretical and experimental investigation into interactions among twin rudder–propellers. He investigated the rudder drag and propeller performance for changes in X/D, Y/D and h/D (where h in this case is the separation of the propeller shafts). With results similar to those described in Section 5.9 and Chapter 10, it was found that: the rudder drag affects significantly the propulsive efficiency of the propeller–rudder system; setting the rudder at $Y/D = 0$ behind the propeller axis makes the rudder drag minimum and the propulsive efficiency maximum; the rudder drag decreases with increase in distance between the propeller and rudder and with decrease in rudder thickness. The interactions between twin propellers and twin rudders are almost negligible if the twin propellers are set apart by more than two times the propeller diameter.

Salo and Heikkilä [5.100] carried out static drift angle tests on a twin-screw model and measured hull and rudder forces. The results showed that asymmetry could be found in the rudder forces. The hull-induced force is dependent on propeller load, stern form and the arrangement of the propellers and rudders.

Janke-Zhou [5.101] carries out manoeuvring simulations for twin-screw ships and Brix [5.35] discusses the practical aspects of the manoeuvring of twin-screw vessels, including the effects of the direct steering moments and shaft convergence.

5.5 Effective aspect ratio

5.5.1 General

In order to make satisfactory predictions of rudder or control surface forces and moments, it is necessary to be able to estimate the effective aspect ratio of the rudder, this being the actual aspect ratio on which the hydrodynamic performance will be based.

A rudder of infinite aspect ratio, such as a foil completely spanning a wind tunnel, or a rudder with infinitely large plates top and bottom, has the same flow over all sections and the flow is said to be two-dimensional. For finite aspect ratio, flow occurs

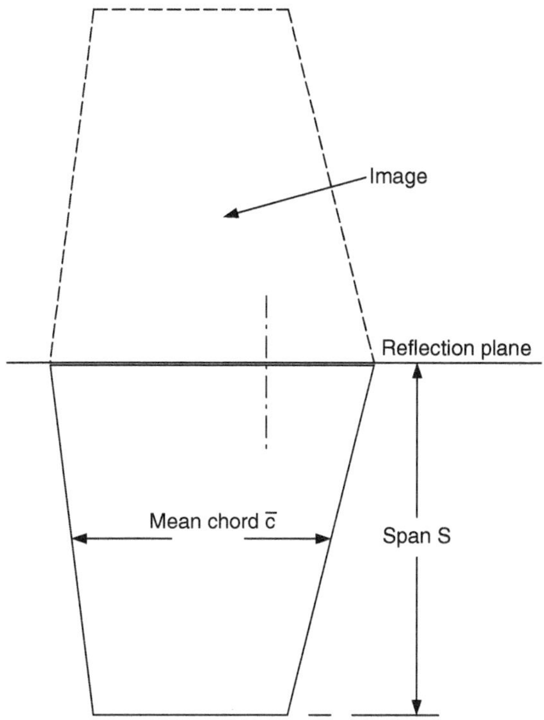

Figure 5.99 Influence of reflection plane on aspect ratio

over the ends from high pressure to low pressure sides, this flow increasing with decreasing span, and the flow is said to be three-dimensional. The implications of three-dimensional flow and resulting induced drag are discussed in Section 3.4. If the root of the rudder is working sufficiently close to a flat surface, say a flat hull, so that there is no cross flow at the root, then the rudder characteristics such as lift and drag become identical to that of a rudder with twice its geometric aspect ratio. This is best illustrated by the concept of a reflection plane and mirror image, Figure 5.99, where the geometric aspect ratio of the actual rudder, AR_G, is $S/\bar{c}$ whereas its hydrodynamic performance will be the same as a rudder with an effective aspect ratio, AR, of $2 \times S/\bar{c}$.

This complete idealised mirror effect cannot normally be achieved in practice since there will be a boundary layer present over the adjacent flat plate. This effect is often minimised in the case of wind tunnel tests by the use of a groundboard displaced from the tunnel wall, for example see Whicker and Fehlner [5.2]. Alternatively, the effective span could be reduced by the boundary layer displacement thickness δ^* (see Section 3.2.13), and the corrected geometric span doubled.

In the case of the actual rudder or control surface design, the boundary layer and wake effect will normally be accounted for in the assumed design velocity. However, in the case of a movable control surface such as the rudder, some practical gap at the root of the rudder is required for its operation and the hull adjacent to the rudder will often not be a flat plane but will be shaped. These features lead to a

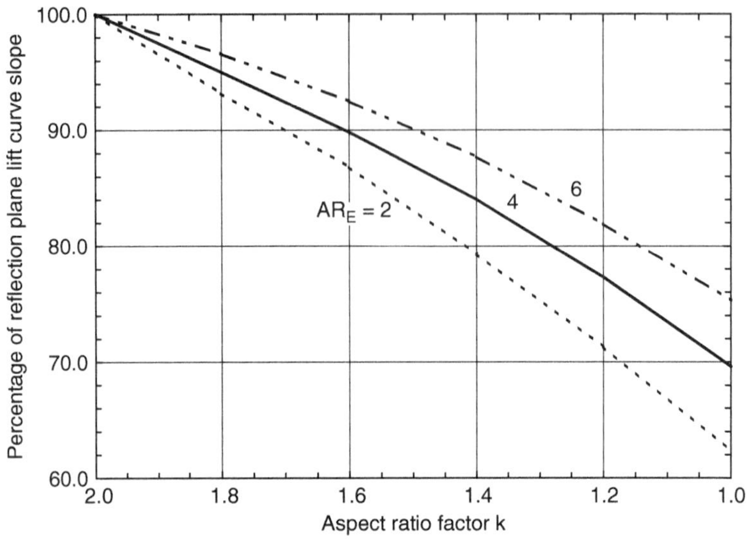

Figure 5.100 Relationship between lift curve slope and aspect ratio factor (based on Equation (5.7b))

decrease in the ideal mirror image effect and the effective aspect ratio, AR, may be expressed as

$$AR = kAR_G$$

where k is the effective aspect ratio factor which will be ≤ 2.

For comparative purposes, the relationship between the loss in lift curve slope relative to a decrease in aspect ratio factor k, derived using Equation (5.7b), has been plotted in Figure 5.100. The effects of rudder–hull gap and hull shape are next discussed.

5.5.2 Rudder root gap

A gap between the root of a rudder and a flat plane will have an effect on the rudder performance. An indication of the influence of root gap on effective aspect ratio factor k is shown in Figure 5.101, which is based on data derived from Kerwin *et al.* [5.6], Molland [5.102, 5.103] and Millward [5.104]. It is seen that in going from zero gap, with $k = 2$, to a gap of $0.005\,\bar{c}$, k drops to 1.9 (e.g. a rudder with AR_G of 2 has AR_E of 3.8) and the decrease in lift curve slope, hence lift, from Equation (5.7b) or Figure 5.100 is approximately 2%. This is not much greater than the order of data scatter in deriving Equation (5.7b), hence it can be assumed for design purposes that for gaps up to about $0.005\,\bar{c}$ (e.g. 2.25 mm for a 450-mm chord rudder) AR_E can be assumed as $2 \times AR_G$. A gap of 5 mm is probably more realistic for a 450-mm chord rudder, or 10 mm for a 900-mm rudder, in which case $G/\bar{c} = 0.011$, $k = 1.8$ and the decrease in lift curve slope, and lift, is approximately 5%.

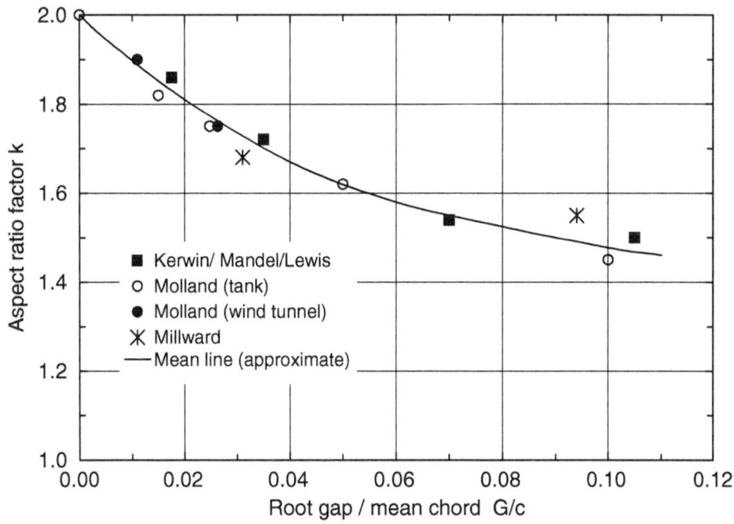

Figure 5.101 Influence of rudder root gap on effective aspect ratio

It is interesting to observe from Figure 5.101 that even for a rudder root gap of $0.1\,\bar{c}$, representing a 45 mm gap for a 450-mm rudder, the effective aspect ratio is still about $1.5 \times AR_G$.

Regarding the influence of gap on maximum lift coefficient, C_{Lmax}, in respect of strength calculations, the data in references [5.6] and [5.102] indicate a relatively insignificant influence on C_{Lmax} for $G/\bar{c}$ values up to about 0.07.

Data relating to the increase in drag for variation in gap size is sparse. However, Figure 5.102, calculated and estimated from data in references [5.6, 5.102], gives some guidance on the influence on C_{Do} and induced drag factor k_i. These increases are relatively small. For example, if an increase in C_{Do} of about 0.001 for a $G/\bar{c}$ of 0.01 is related to the drag coefficients in Figure 5.14(b), it is seen that the increase can be neglected for most practical purposes. Similarly, even for the very large $G/\bar{c}$ of 0.01, the increase in induced drag factor k_i is only about 8%.

Whilst the data indicate that a relatively small and acceptable loss in lift curve slope and increase in drag occurs for a root gap of up to say $0.01\,\bar{c}$, further increases in gap leads to further losses. This may be important in the case of small craft. Although some loss may be acceptable in a workboat or cruising yacht in lieu of a more production friendly design, the data indicate that the gap should be minimised within practical limitations to maximise lift/drag ratios for racing or performance craft.

5.5.3 Shaped hull above rudder

Discussion so far has concerned rudders below flat reflection planes. This may be achieved with rudders below some high-speed chine hull craft with relatively flat bottoms but in many cases the rudder is working under a shaped hull, Figure 5.103.

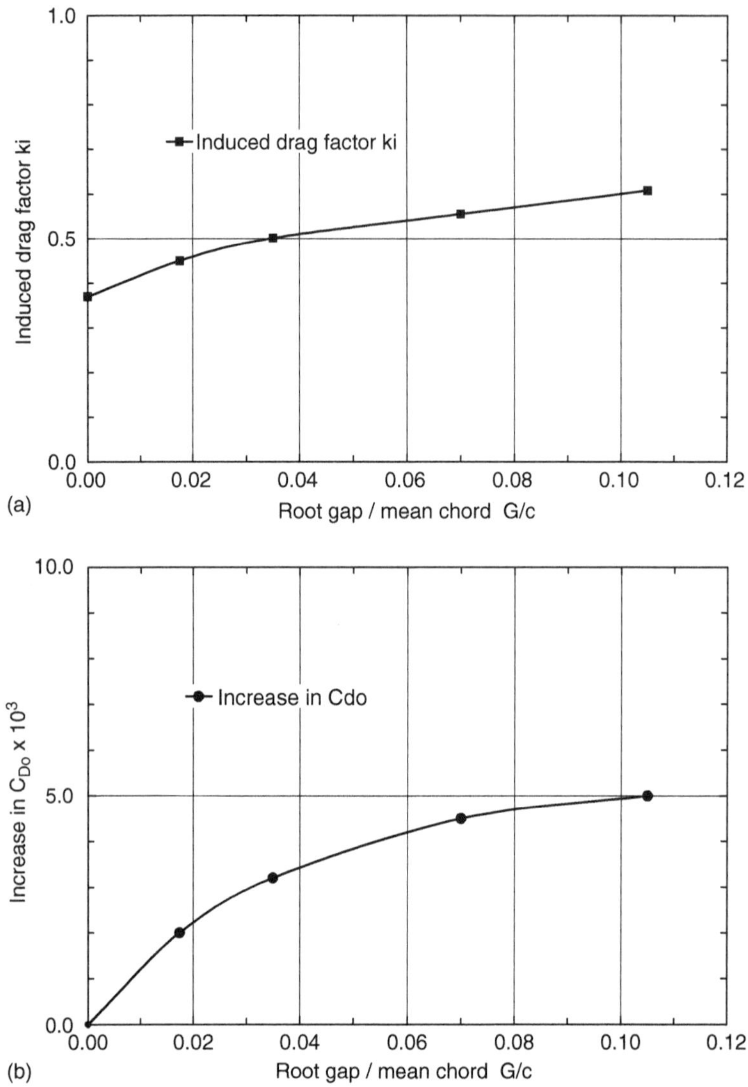

Figure 5.102 (a) Influence of rudder root gap on induced drag factor; (b) Influence of rudder root gap on minimum drag coefficient C_{Do}

With hull shape there is a further loss in the effectiveness of the reflection plane. Figure 5.104 illustrates the approximate magnitude of this effect using data from references [5.103, 5.105, 5.106]. Data from reference [5.103] is for a test-tank model rudder with $AR_G = 1.5$ operating beneath flat plates bevelled at 15° and 30° to the horizontal and sharpened at the leading edge, Figure 5.105. These data show a slightly more rapid decrease in effective aspect ratio factor k than say the data of Harper and Simitses [5.105], which was for a simulated submarine hull. Data in reference [5.103] is for an immersion of 150 mm to the top of the rudder span of 300 mm which should minimise free surface effects (see Section 5.7), and reference [5.105] does not account

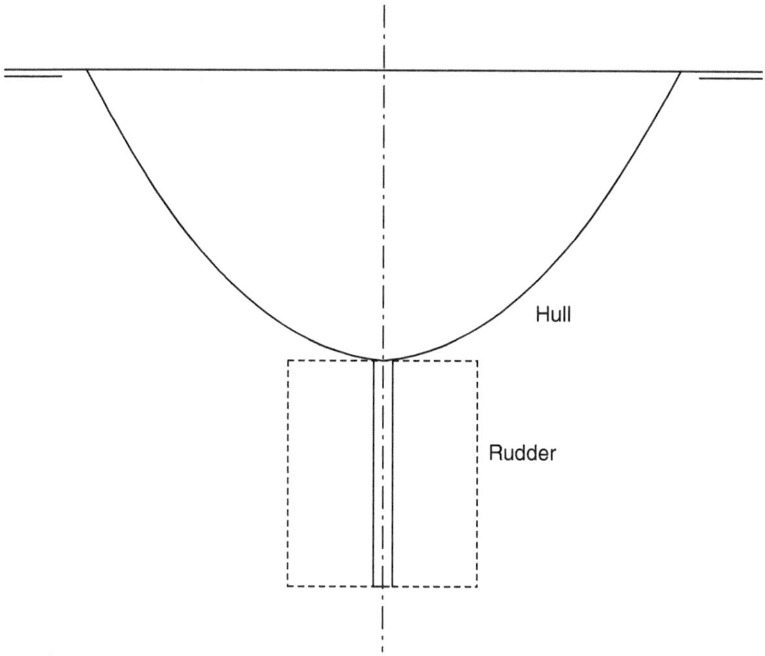

Figure 5.103 Rudder under shaped hull

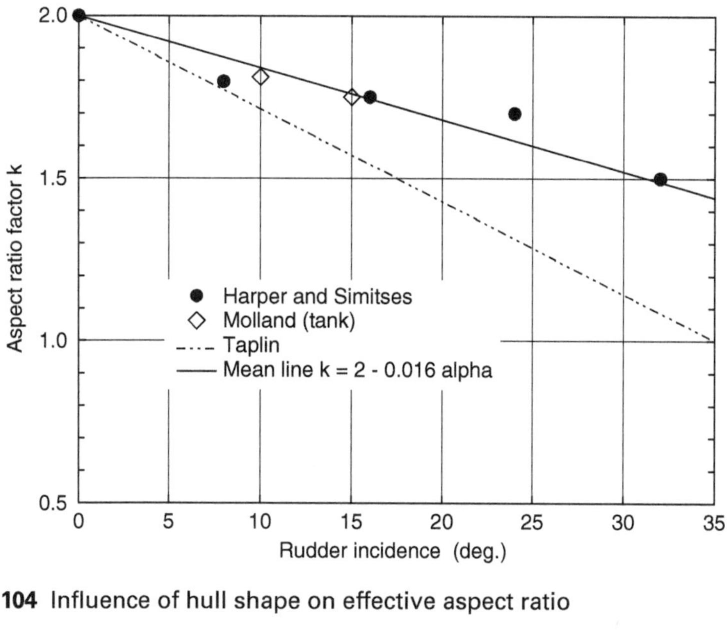

Figure 5.104 Influence of hull shape on effective aspect ratio

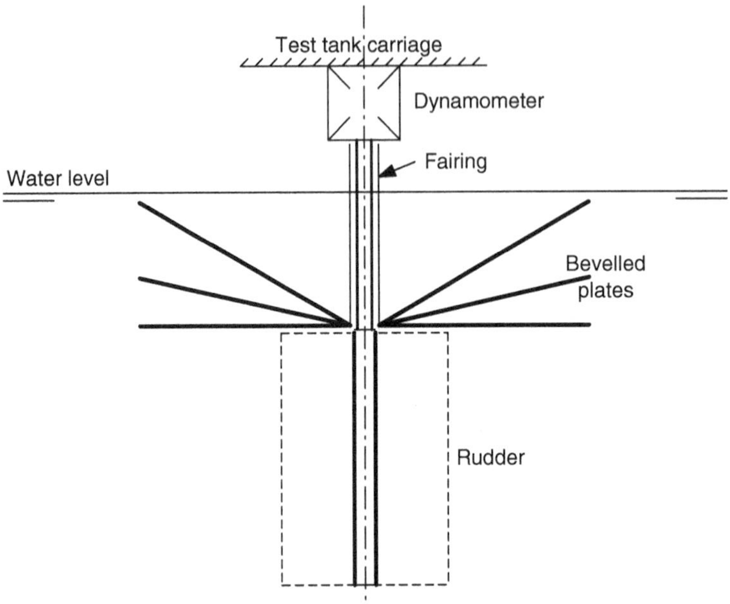

Figure 5.105 Rudder under flat and beveled plates [5.103]

for any influence of free surface. For practical design purposes, Taplin [5.106] proposed the use of a straight line relationship from $k = 2.0$ at zero angle of attack to $k = 1.0$ (i.e. no reflection plane effect) at full rudder angle, that is

$$k = 2 - \frac{\alpha}{\alpha_{max}} \tag{5.24}$$

At large rudder angles, the Taplin line is significantly lower than the other data. This would tend to lead to underestimates of the rudder forces at large rudder angles which will provide some margin in the case of manoeuvring simulations but use forces that are too low in the case of rudder or control surface strength calculations.

The drag results for the reflection plane and bevelled plate cases in reference [5.103] indicated that there was little difference in C_{Do} due to the bevelled plate, simulating hull shape, but that the bevelled case led to an increase in induced drag factor k_i of about 15% compared with the reflection plane; a reduction in C_{Lmax} of about 10% was observed for the plate bevelled up 15° from the horizontal and about 5% for the plate bevelled up 30°.

A mean line is proposed in Figure 5.104, based on the data from references [5.103, 5.105], which gives a satisfactory indication of the influence of hull shape on AR. It should be noted that this is an approximate mean line intending to cover a wide range of effective hull shapes. The line is defined by

$$k = 2 - 0.016\alpha \tag{5.25}$$

It is interesting to note that even as the rudder approaches typical stall angles of 25°–35° the effective aspect ratio is still about $1.5 \times AR_G$. These data can be used to derive the influence of hull shape on aspect ratio which, in conjunction with equation (5.7b), can be used to derive the lift curve slope, hence lift and drag. From equation (5.7b) it can be deduced that the shaped hull has an increasing influence as the design geometric aspect ratio gets smaller.

5.6 Rudder and control surface area

There is no general set rule for the determination of rudder area due to the various steering and manoeuvring requirements for different ships and different operational requirements.

Rudder area for a particular rudder type may be calculated if the required yawing moment to provide a defined steering response is known. This may, for example, be an outcome of mathematical coursekeeping and manoeuvring simulations incorporating the equations of motion. Similarly, rudder area may be calculated in the case of sailing craft if a known contribution to the side force from the rudder is required, or total theoretical balance equations are established as, for example, in references [5.107, 5.108].

For most initial design situations, it is not possible to derive the required rudder force using a fundamental approach and the rudder area is usually estimated from empirical data for similar vessels. In using such data it is important to confirm, where possible, that the base designs had adequate steering and directional stability qualities. It should also be noted that such data normally do not take account of different rudder design parameters such as aspect ratio, which in turn influence hydrodynamic efficiency and hence required rudder area.

The data in Table 5.17 offer guidance for a preliminary estimate of area for a range of coastal and seagoing ships. Vessels such as tugs, trawlers, supply vessels, ferries and warships generally require relatively larger rudder areas than seagoing merchant ships. The rudder area A is described in terms of an approximation to the underwater lateral area ($L \times T$). The area A is the sum of the rudder areas if more than one rudder is fitted.

DNV [5.109] offers the following equation for merchant ships: [SS or TS, with A = total area if more than one rudder];

$$\frac{A}{L \times T} = 0.01\left[1 + 50\ C_B^2 \left(\frac{B}{L}\right)^2\right] \tag{5.26}$$

Table 5.17 Rudder areas

Ship type	$\frac{A}{L \times T}$
Single-screw merchant ships	0.016–0.018
Twin-screw merchant ships	0.016–0.022
Warships	0.024–0.028
Tugs, trawlers etc.	0.030–0.040

for example, if $L/B = 6$ and $C_B = 0.8$, then $A/(L \times T) = 0.019$ and if $L/B = 7$ and $C_B = 0.6$ then $A/(L \times T) = 0.014$.

Shiba [5.58] carried out extensive manoeuvring experiments with 2.5-m models having block coefficients of 0.60, 0.70 and 0.80, representing a high-speed merchant ship, a general cargo ship and a tanker. The experiments included an investigation into rudder area and some results extracted from that work are shown in Figures 5.106, 5.107 and 5.108. These show the influence of rudder area *A/LT* on turning

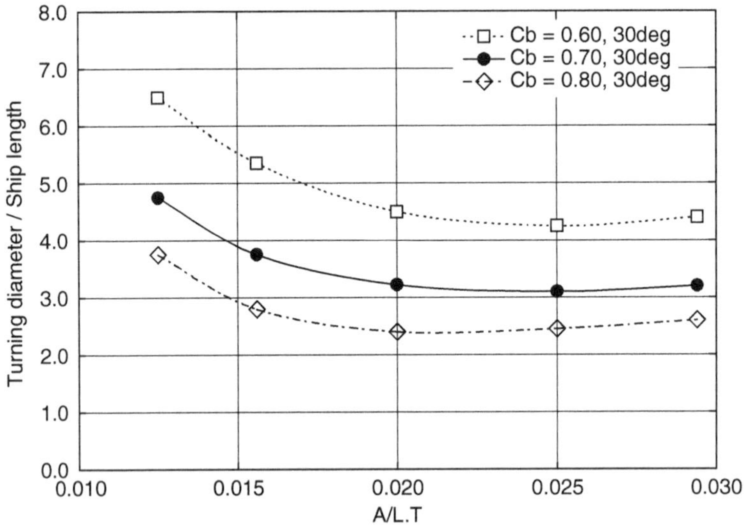

Figure 5.106 Influence of rudder area on turning diameter (derived from Shiba data)

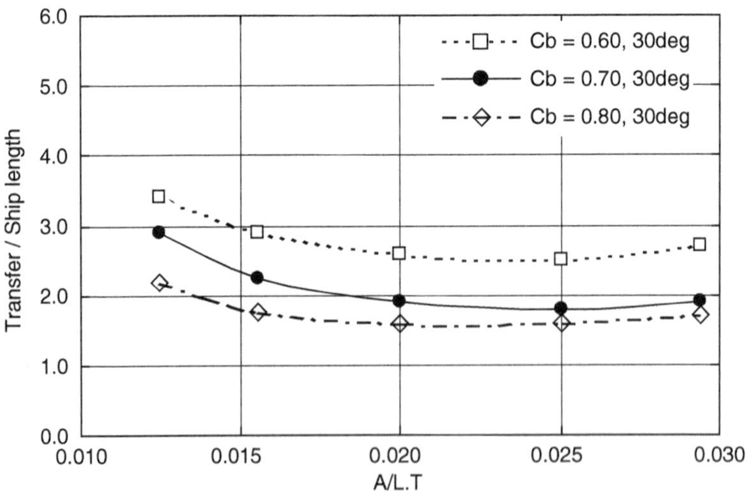

Figure 5.107 Influence of rudder area on transfer (derived from Shiba data)

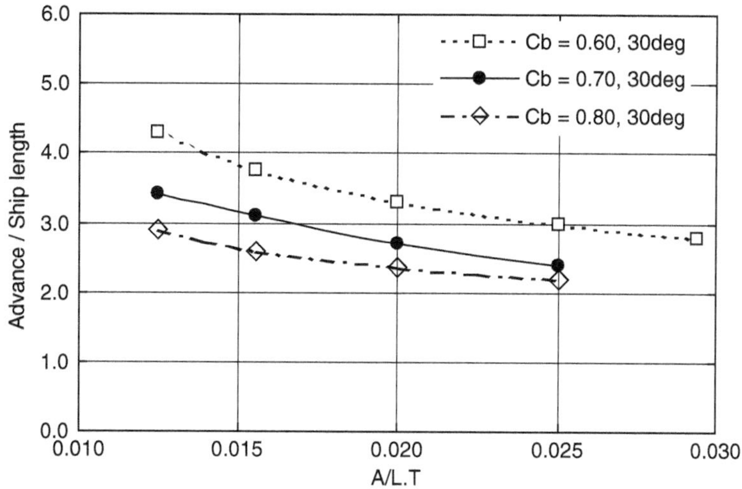

Figure 5.108 Influence of rudder area on advance (derived from Shiba data)

diameter, transfer and advance, see Figure 4.5, for 30° rudder angle and three block coefficients. Tests were also carried out at other rudder angles covering 10°–45°. It is seen that for the turning diameter, Figure 5.106, and transfer, Figure 5.107, there is an improvement in rudder effectiveness (decrease in diameter and transfer) with increase in rudder area up to about $A/LT = 0.025$ (2.5% LT) after which there is a decrease in rudder effectiveness. With advance (Figure 5.108), it is seen that there is a continuous improvement with increase in rudder area, suggesting that even more area may be usefully employed. These results are useful for practical design purposes but caution should be exercised in the detailed interpretation of the data, as the rudder area for the models was increased by adding chord to a fixed rudder span. Consequently, the aspect ratio changed from 2.24 at $A/LT = 0.0125$ down to 0.98 at $A/LT = 0.0286$. The data would indicate that the increase in rudder area has more effect than the decrease in aspect ratio. This is likely to be due to the fact that the propeller diameter is 100 mm and rudder span 98 mm, meaning that the increase in area is fully within the propeller slipstream.

Clarke *et al.* [5.110] propose manoeuvring criteria as a turning index, dynamic stability and manual control and develop a practical mathematical manoeuvring model. The model is based on the linear equations of motion, which limit the technique to small departures from a steady course. Semi-empirical and multiple regression methods were used to deduce the acceleration and velocity derivatives for various hullform parameters. Predictions using the model showed good comparisons with full-scale trials data. The model is used to estimate required rudder area to satisfy a particular turning index for a range of C_B, B/T and L/T. Values of derived A/LT range from about 0.01 to 0.03. As general trends, it was found that as C_B increases, rudder area increases slightly, that as B/T increases rudder area increases significantly, particularly above B/T of about 3.0, and there is a significant increase in area with decrease in L/T, particularly for L/T less than about 25. Further calculations indicated a required increase in rudder area for shallow water conditions.

Gong *et al.* [5.9] investigated in some detail the required rudder area for a ship with a large breadth/draught ratio ($B/T = 3.77$). Model experiments were carried out on three semi-balanced skeg rudders with A/LT ranging from 0.0192 up to 0.0239. The results for the open water tests were found to be in satisfactory agreement with

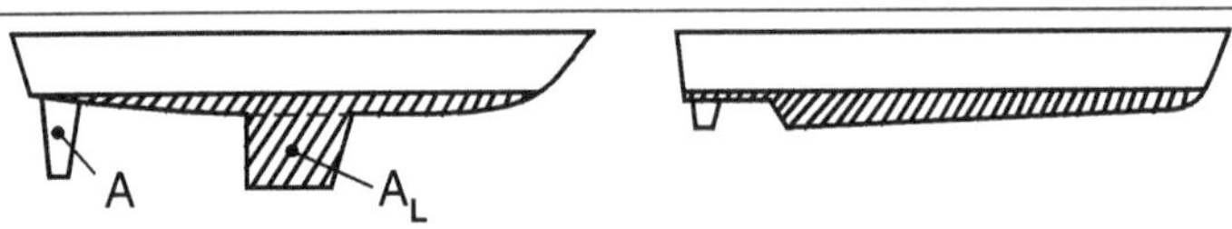

A = Rudder area = sum of rudder areas if more than one rudder fitted
A_L = Immersed lateral area, based on static waterline

Sailing yachts and motor sailers:

$A = 0.07A_L$ to $0.11A_L$
$A = 0.085A_L$ (for $A_L = 40m^2$) to $0.105A_L$ (for $A_L = 5m^2$) Reference [5.111]

Low speed power craft:

$A = 0.03A_L$ to $0.05A_L$

Semi-displacement and planing craft:

$A = 0.03A_L$ to $0.04A_L$ (A_L in static condition)

The following mean curve is based on data from references [5.112, 5.113] for satisfactory existing craft rudder(s) operating in the slipstream of propeller(s).

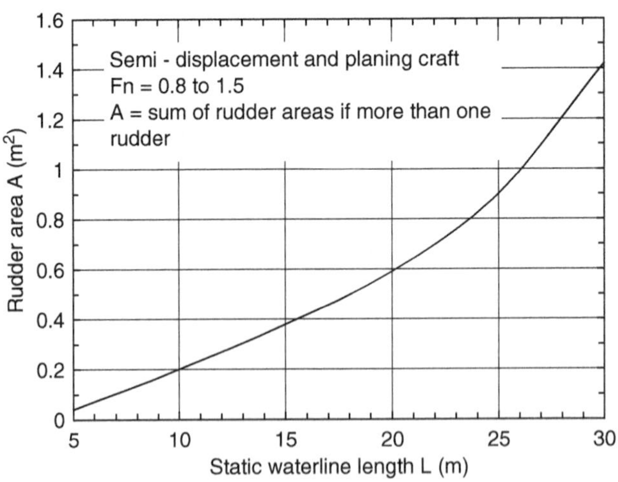

Figure 5.109 Typical rudder areas for small craft

the results for similar skeg rudders in references [5.7, 5.8]. HPMM (Horizontal planar motion mechanism) tests were carried out and the derived hydrodynamic coefficients applied in a manoeuvring simulation that was used to investigate rudder area. It was found that the influence of rudder area on straight-line stability was not significant but increases in rudder area did have significant effects on the advance and tactical diameter of the turning manoeuvre and on the zig–zag manoeuvre. Overall, it was found that the manoeuvring performance was improved with an increase in rudder area and that these improvements seemed to be much more than expected from just the area increment itself.

Rudder areas for small craft are relatively much bigger than for large ships. Figure 5.109 offers some guidance for a preliminary estimate of rudder area for small craft using data obtained from sources such as references [5.111–5.113]. It should be noted that in this case the rudder area is relative to the actual underwater lateral area A_L, rather than simply $(L \times T)$ as for larger vessels. For sailing craft, adequate rudder area might be checked using the modeling techniques and VPPs (velocity prediction programmes) described in references [5.107, 5.108].

Area requirements for other control surfaces, such as stabiliser fins and pitch control foils, are dealt with under each specific topic in Chapter 9.

5.7 Free surface effects

Rudders and other control surfaces operating near or piercing the free water surface will suffer free-surface losses. This also leads to a decrease in the ideal mirror image effect where the water surface is now treated as the reflection plane. For comparative purposes, free-surface effects can be treated as a loss in effective aspect ratio in a manner similar to the effects of a root gap or a shaped hull above the rudder, as described in Section 5.5. Two conditions may be considered, namely when the rudder is fully submerged but relatively close to the surface and when the rudder is piercing the surface.

5.7.1 Rudder submerged condition

A small craft rudder, even whilst in the submerged condition, is working relatively near the surface. Similarly, when large ships are operating in a light load or ballast condition, their rudders are also often working at or relatively close to the surface.

Data showing the influence of surface effects for the submerged rudder case are presented by Millward [5.114] and Ismail [5.115]. The work in these Theses, which is mainly theoretical with some experimental backing, is for foils with a geometric aspect ratio of 4.0. Shiba [5.58] presents results for tests on a ship rudder model with a geometric aspect ratio of 1.68. The data from these three references are presented in Figures 5.111–5.113, which show the influence of depth of immersion on aspect ratio factor k and induced drag factor k_i in Equation (3.25), the induced drag in this case being assumed to also include any drag due to wavemaking. The

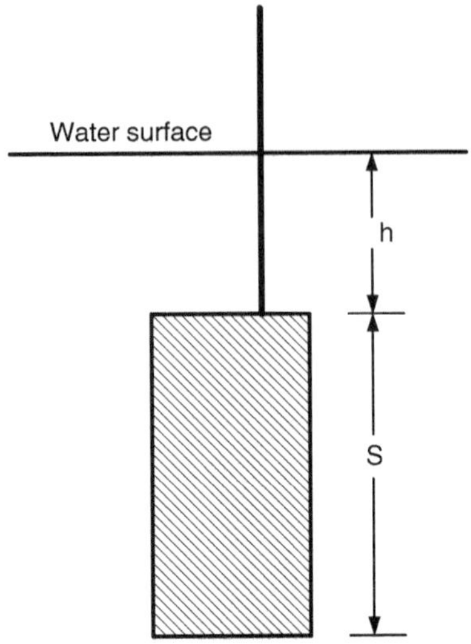

Figure 5.110 Surface effect: submerged rudder

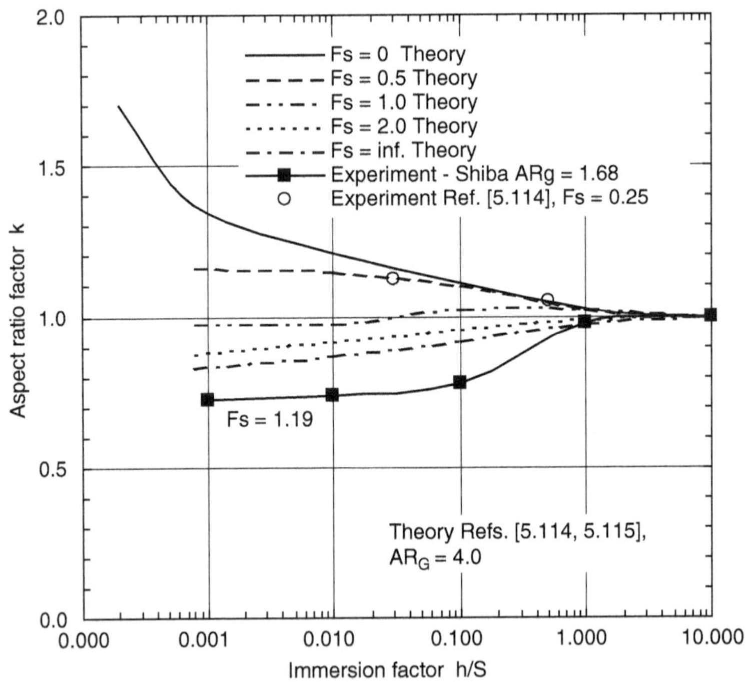

Figure 5.111 Influence of speed and immersion on aspect ratio factor *k*; submerged case

depth of immersion is represented by the nondimensional ratio h/S, Figure 5.110. It is clear from the resulting curves in Figure 5.111 that the influence of immersion is very dependent on Froude number *Fs*. In the submerged case *Fs* is based on rudder span. It is seen that as immersion h/S is decreased, then for small Froude

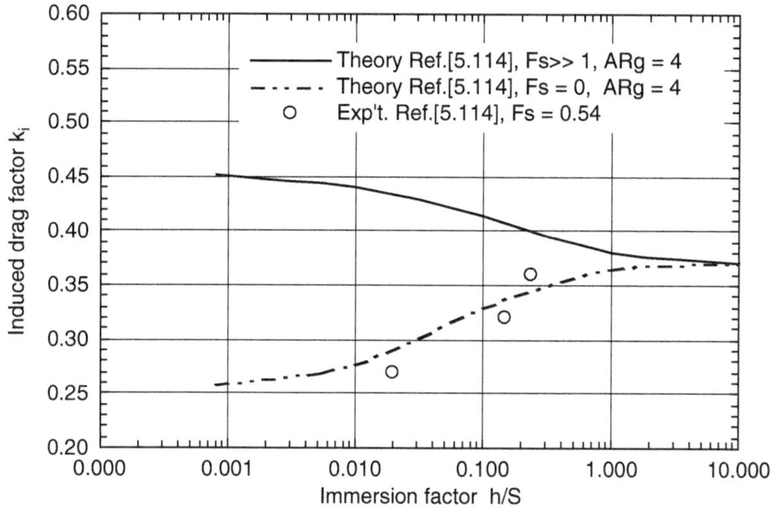

Figure 5.112 Influence of speed and immersion on induced drag factor k_I; submerged case

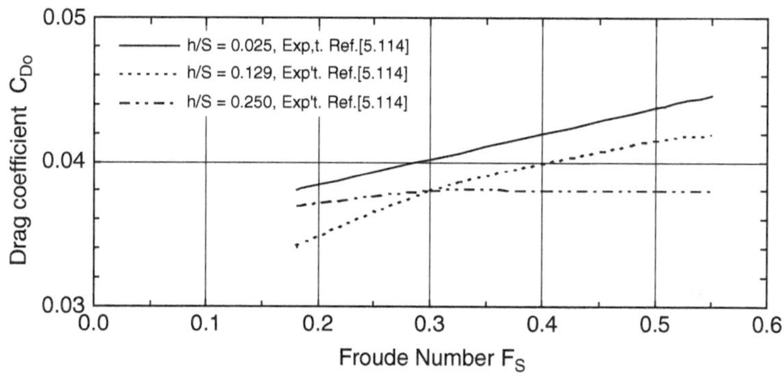

Figure 5.113 Influence of speed and immersion on minimum drag coefficient C_{Do}; submerged case

numbers the value of k increases; in the theoretical limit, as $Fs \rightarrow 0$ and $h/S \rightarrow 0$, then k approaches 2, with the water free surface acting as a reflection plane. At an Fs of about 1, the influence of immersion, from theory, would appear to be negligible whilst for higher Fs there is a decrease in k as h/S is reduced. These trends were confirmed by experiments [5.114, 5.58], the latter being lower but in broad agreement with the conclusions of reference [5.114] in that the surface effect increases with decrease in aspect ratio. References [5.114, 5.115, 5.58] all indicate that free-surface effects tend to negligible amounts, for all speeds, when the immersion of the top of the rudder is greater than about half the span.

Theoretical and experimental data in reference [5.114] indicate a similar trend for the induced drag factor k_i, Figure 5.112. For low Froude number Fs, the value of k_i decreases with increasing immersion h/S whilst at high Fs the value of k_i increases

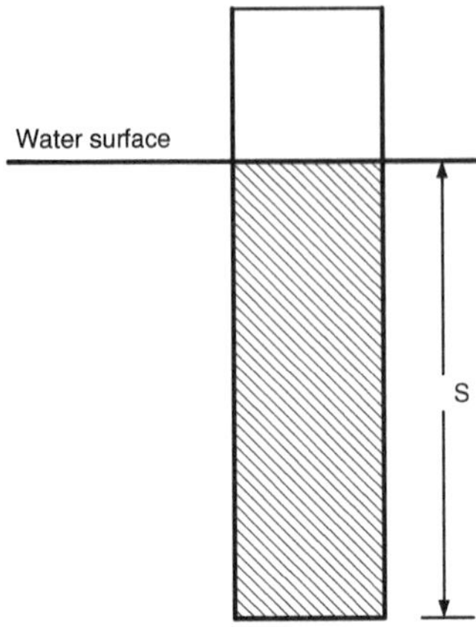

Figure 5.114 Surface effect: surface piercing rudder

with decreasing immersion. Note that in this case the theoretical and experimental values of k_i have been adjusted to give 0.37 at large depth.

The experimental results presented in reference [5.114] show the minimum drag coefficient C_{Do}, Figure 5.113, to be considerably higher than the deeply submerged case, when a value of about 0.01 would be expected; some of this difference will, however, be accounted for by the fact that the experiments were carried out at the relatively low Reynolds number of 0.1×10^6.

5.7.2 Rudder surface piercing condition

Some general indications of the influence of free-surface effects on the surface piercing rudder or control surface are given in references [5.114, 5.115]. In this case, the rudder particulars are defined in Figure 5.114. The data from reference [5.114] is presented in a simplified manner in Figures 5.115–5.117, which show the influence on effective aspect ratio factor k of Froude number Fc, induced drag factor k_i, and minimum drag coefficient C_{Do}. Fc is based on rudder chord for the surface piercing case, and the induced drag for this case is assumed to also include the drag due to wavemaking. It is seen from Figure 5.115 that for very low speeds the water surface forms an effective reflection plane and as $Fc \rightarrow 0$, $k \rightarrow 2$. However, as speed increases, wavemaking losses are incurred and lift, hence effective aspect ratio factor k, is reduced. For example, as $Fc \rightarrow 0.8$, the aspect ratio factor k has fallen to about 1. No surface reflection properties now exist and the rudder has an effective aspect ratio same as its geometric aspect ratio. These data are for the relatively high geometric aspect ratio of 4 but, as discussed in reference [5.114], available experimental data indicate that this effect is likely to increase as aspect ratio decreases.

Figures 5.116 and 5.117, based on the experimental data from reference [5.114], indicate that as Fc is increased, appreciable increases in induced drag factor k_i and minimum drag coefficient C_{Do} are likely.

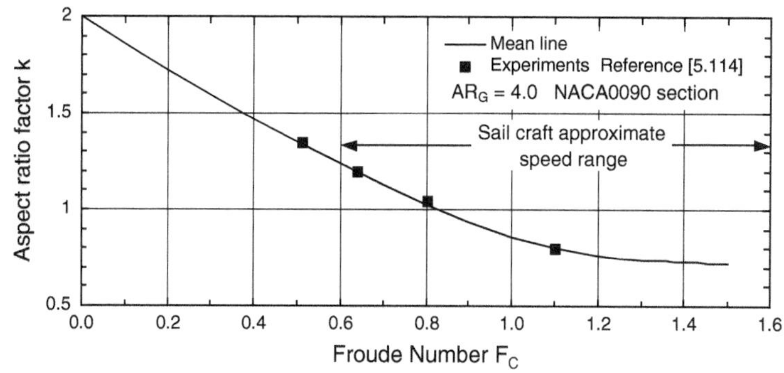

Figure 5.115 Influence of speed on aspect ratio factor *k*; surface piercing case

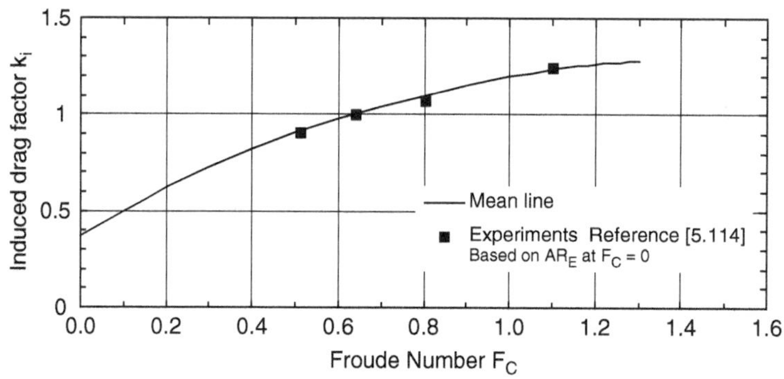

Figure 5.116 Influence of speed on induced drag factor k_i; surface piercing case

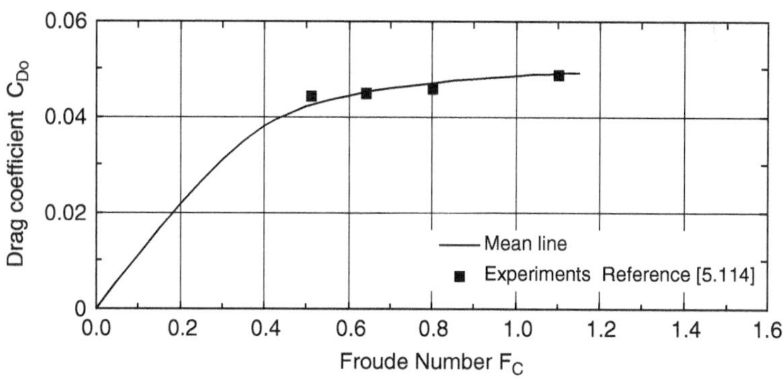

Figure 5.117 Influence of speed on minimum drag coefficient C_{Do}; surface piercing case

With a transom hung, or surface piercing rudder, there is the possibility of ventilation occurring on the rudder at higher angles of attack, when the drop in pressure on the back of the rudder is sufficient to draw air down from the free surface. With ventilation, there is a loss of low pressure on the back of the rudder and a consequent loss in lift leading to premature stall. Similar conditions can arise with high performance yachts whose rudders may approach the free surface at large heel angles and with large tankers and bulk carriers in ballast where the top of the rudder may be at or near the free surface. To decrease the possibility of ventilation, fences made up of thin plate may be used, Figure 5.118, which are mounted on the rudder near the water surface. These create a barrier on the rudder surface making it difficult for the air to be drawn down from the free surface. Fences need to be aligned to the local flow direction, otherwise significant drag penalties can occur. At the same time, the flow direction changes with speed, heel and trim and it is generally not possible to achieve a satisfactory alignment of the fence without some increase in drag. Fences have been used successfully on fixed support struts and inclined foils, such as on hydrofoil craft, where the speed and flow conditions are relatively constant.

A more detailed analysis of ventilation cavities, including experimental results, is given by Swales *et al.* [5.116, 5.117].

It is difficult to draw up general conclusions from the surface effect data. In the case of large commercial ships with submerged rudders, it can be assumed that free-surface effects will be small if the immersion of the top of the rudder is about half the span. If this criterion is decreased, such as in a light draught or ballast condition, then Figures 5.111–5.113 provide information for assessing the free-surface effects on rudder performance.

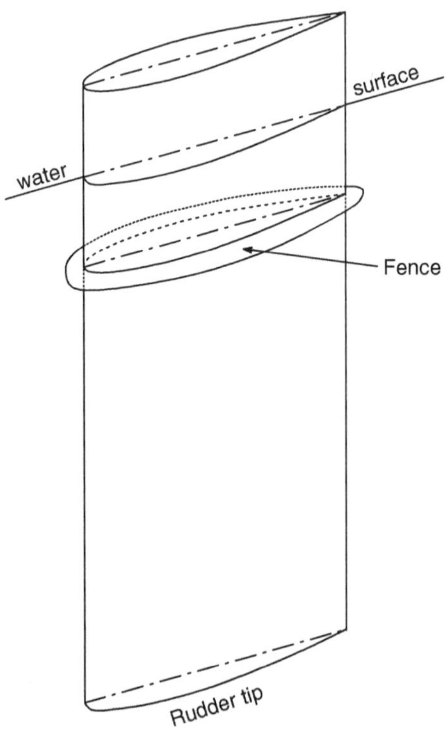

Figure 5.118 Fence near free surface

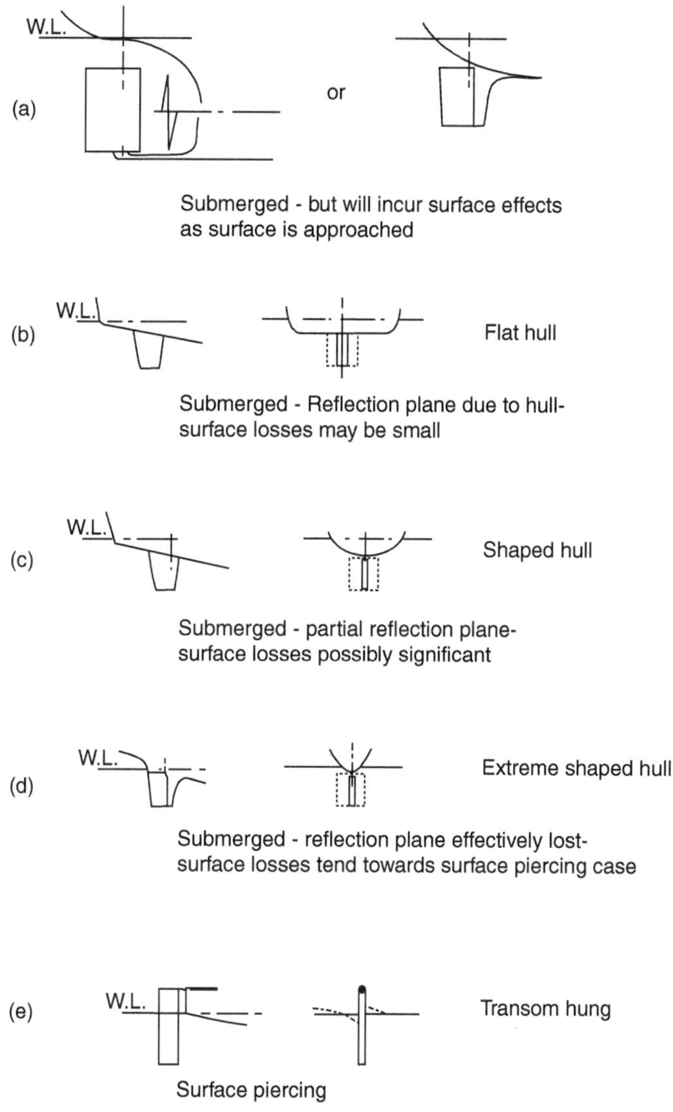

Figure 5.119 Free surface effects

Some conclusions with respect to small craft can also be made. Figure 5.119 illustrates some of the free-surface effects and comments are included in the following paragraphs.

The surface piercing case, for example the transom rudder, is suitable for low speed, hence small sailing craft operation. The data do, however, illustrate the significant surface losses likely to be incurred in the case of the transom rudder as speed is increased; these losses are likely to exceed the effects of the increased lever (from the turning centre) enjoyed by this rudder type.

Most sailing craft operate in an *Fs* range of about 1.1–1.3. Hence for those rudder types that admit flow over the top, such as Figure 5.119(a) the data would suggest a relatively small change due to a lack of immersion; any possible gain suggested by Figure 5.111 at low *Fs* is likely to be offset by the lower aspect ratio (i.e. $AR_G < 4$) normally employed in small craft. Further, as the submerged rudder is brought nearer to the surface, and if the hull effect above the rudder is small, then the submerged rudder effectively becomes surface piercing, as per Figure 5.119(d) and will therefore incur losses as for the surface piercing case.

The deeply submerged rudder for the sailing craft is likely to suffer surface effects in the inclined condition when the low pressure side may effectively become surface piercing. Adequate hull surface above the rudder would partly overcome this effect.

Attempts should be made on workboats and the like to keep the top of the rudder immersed greater than about half its span, when surface effects become negligible. Powercraft with reasonable hull width over the rudder, Figure 5.119(b) are unlikely to suffer significant free surface effects.

The flow conditions near the top of small craft rudders mounted well aft, or transom hung, can be confused and detailed conclusions are not possible. However, the data presented in Figures 5.110–5.117 provide useful guidance in assessing the location, longitudinally and vertically, of a new rudder proposal for particular design speed conditions.

5.8 Cavitation on control surfaces

5.8.1 General

Cavities filled with vapour are formed in places where the pressure falls to the vapour pressure of the fluid. Cavitation occurs where peaks of low pressure arise, such as on the suction side of lifting sections. Cavitation can occur, in particular, on marine propellers [5.92]. It can also occur on support struts, rudders, stabiliser fins, pitch control foils and other control surfaces. Sheet cavitation tends to occur near the nose of a lifting section and bubble cavitation on the back. The magnitude and peaks of the pressure distribution depend on the lift coefficient and on section shape and thickness.

The effects of cavitation are to disturb the flow along the surface, causing changes in the effective profile properties and a reduction in lift. It also leads to possible surface erosion attributed to the collapse of cavitation bubbles as they move downstream into regions of higher pressure, noise as the cavities collapse and possible excitation of vibration.

It is thus desirable to size the area of the control surface, and possibly to limit the incidence, whereby the loadings and hence magnitude of pressure peaks are limited so as to avoid cavitation. Careful choice of section shape may be necessary to lower and smooth out the pressure peaks.

5.8.2 Cavitation criterion

Cavitation inception is assumed to occur on the section when the local pressure P_L on the section falls to, or below, the vapour pressure P_V of the fluid, and can be predicted from the pressure distribution.

A cavitation number σ is defined as

$$\sigma = \frac{(P_0 - P_V)}{0.5\rho V^2} = \frac{(P_{AT} + \rho gh - P_V)}{0.5\rho V^2} \tag{5.27}$$

and the pressure coefficient C_P as

$$C_P = \frac{(P_L - P_0)}{0.5\rho V^2} \tag{5.28}$$

Cavitation inception can be predicted from the pressure distribution since cavitation will occur when $P_L = P_V$, or the minimum negative pressure coefficient, C_P, is equal to σ. A simple representation of the point of inception for a lifting section is shown in Figure 5.120. The cavitation characteristics for a particular section can be described by a minimum pressure envelope, or cavitation-free bucket, as a function of the section cavitation number. Since the section lift coefficient C_L is a function of the pressure distribution then, for a particular section, the cavitation-free bucket can be represented as a limiting C_L envelope to a base of σ. A schematic outline of a cavitation-free bucket for a cambered section typically used on propeller blades, together with the likely types of cavitation, is shown in Figure 5.121. The (vertical)

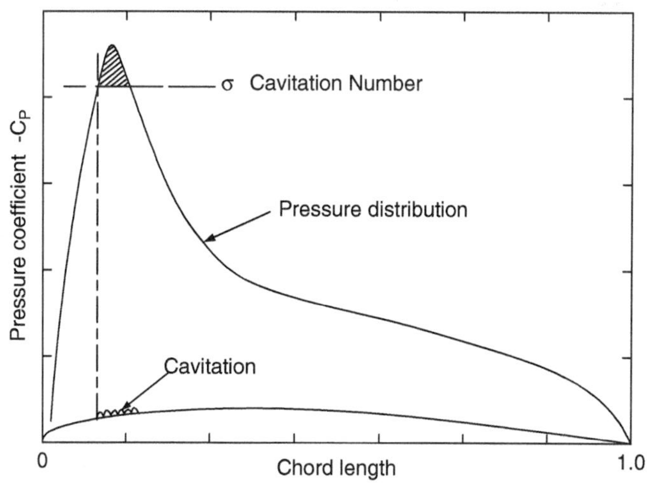

Figure 5.120 Cavitation inception (idealised)

width of the bucket is a measure of the tolerance of the section to cavitation-free operation, i.e. with a wider bucket, the section will be able to tolerate a much wider variation in angle of attack without cavitating. The width and shape of the bucket will depend on section characteristics such as thickness, camber, overall shape and nose shape. For example, an increase in section thickness tends to widen the bucket whilst an increase in section camber tends to move the same bucket width and shape vertically to higher values of C_L. A symmetrical, noncambered section, typically used for movable control surfaces that are generally bi-directional, has a symmetrical cavitation inception envelope as shown schematically in Figure 5.122.

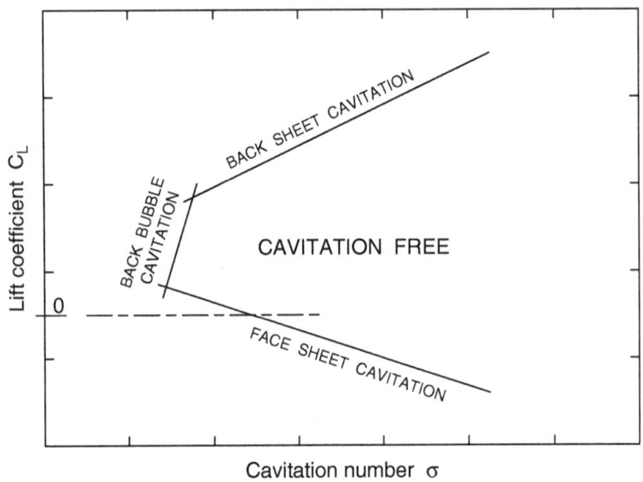

Figure 5.121 Cavitation inception envelope: cambered sections

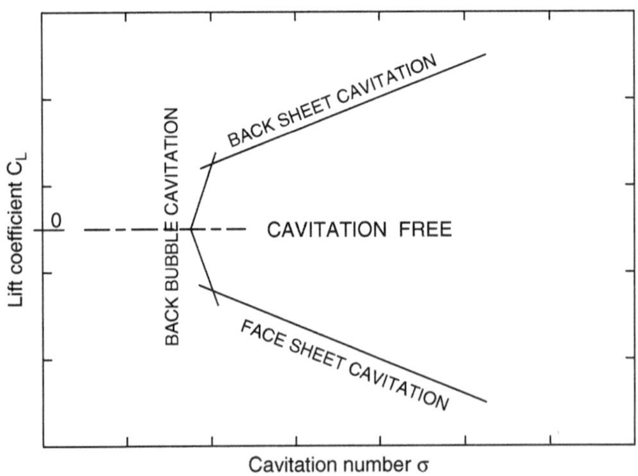

Figure 5.122 Cavitation inception envelope: symmetrical sections

Plate 1 Arrangement of rudder–propeller rig in wind tunnel (See also page 129 of this book)

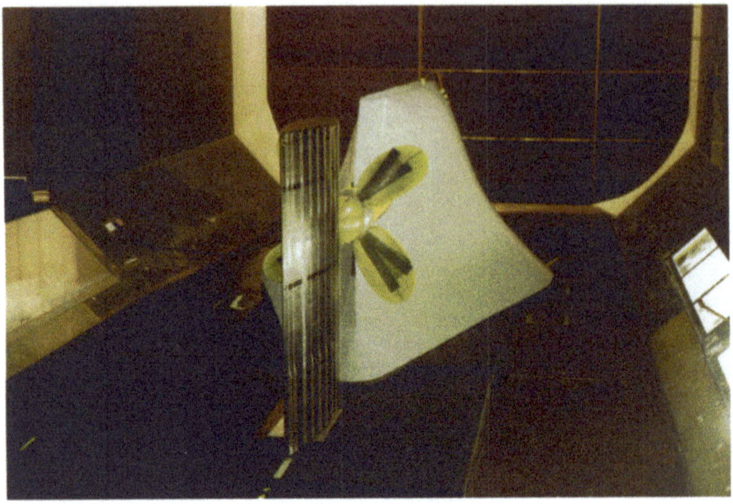

Plate 2 Arrangement of rudder, propeller and hull in wind tunnel (See also page 158 of this book)

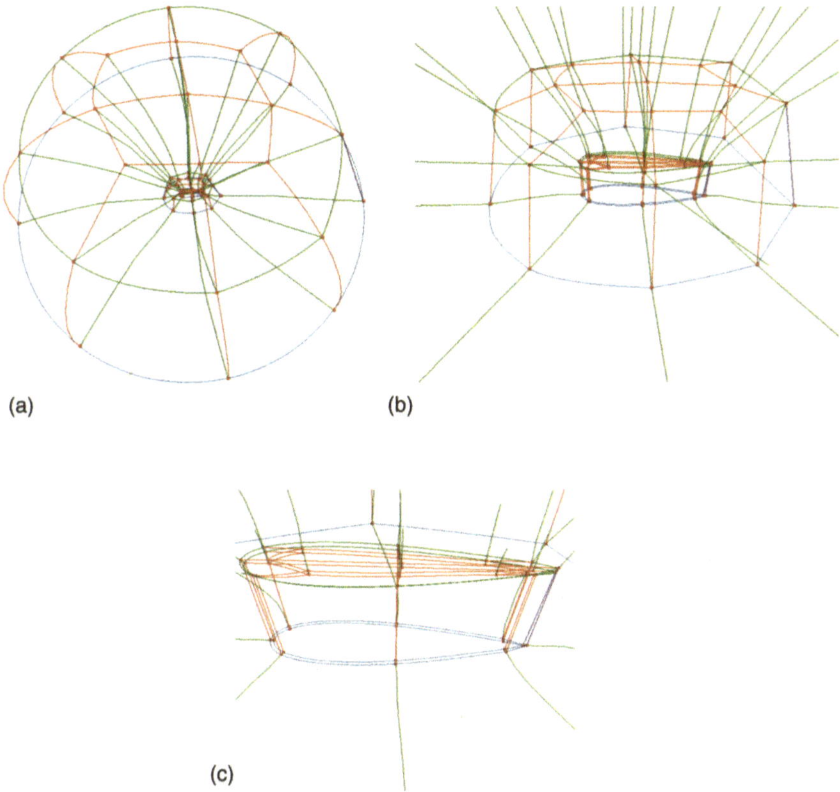

Plate 3 Three-layer multiblock mesh topology for control surface CFD analysis, showing block edges. (a) Outer blocks – showing rudder within a hemispherical domain. (b) Middle blocks – showing transition from boundary layer to outer domain. (c) Inner blocks – showing small volume required to capture boundary layer (See also page 255 of this book)

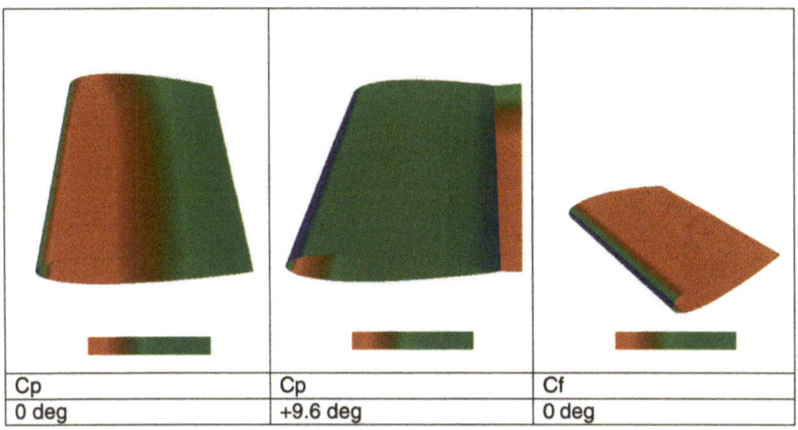

Plate 4 Visualisation of surface pressure and skin friction for coupled boundary layer calculation on Rudder No. 2 (See also page 276 of this book)

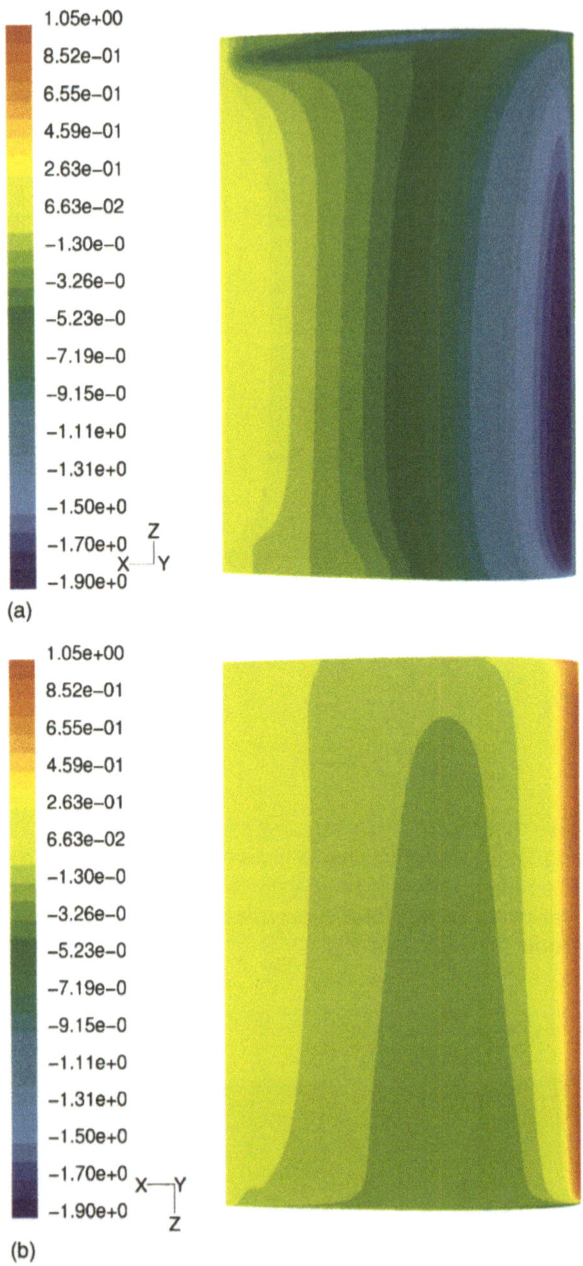

Plate 5 Pressure distribution on rudder surface; NACA0020, $Re = 8 \times 10^5$, $\alpha = 10°$, $k-\varepsilon$ RNG. (a) Pressure side, (b) suction side (See also page 284 of this book)

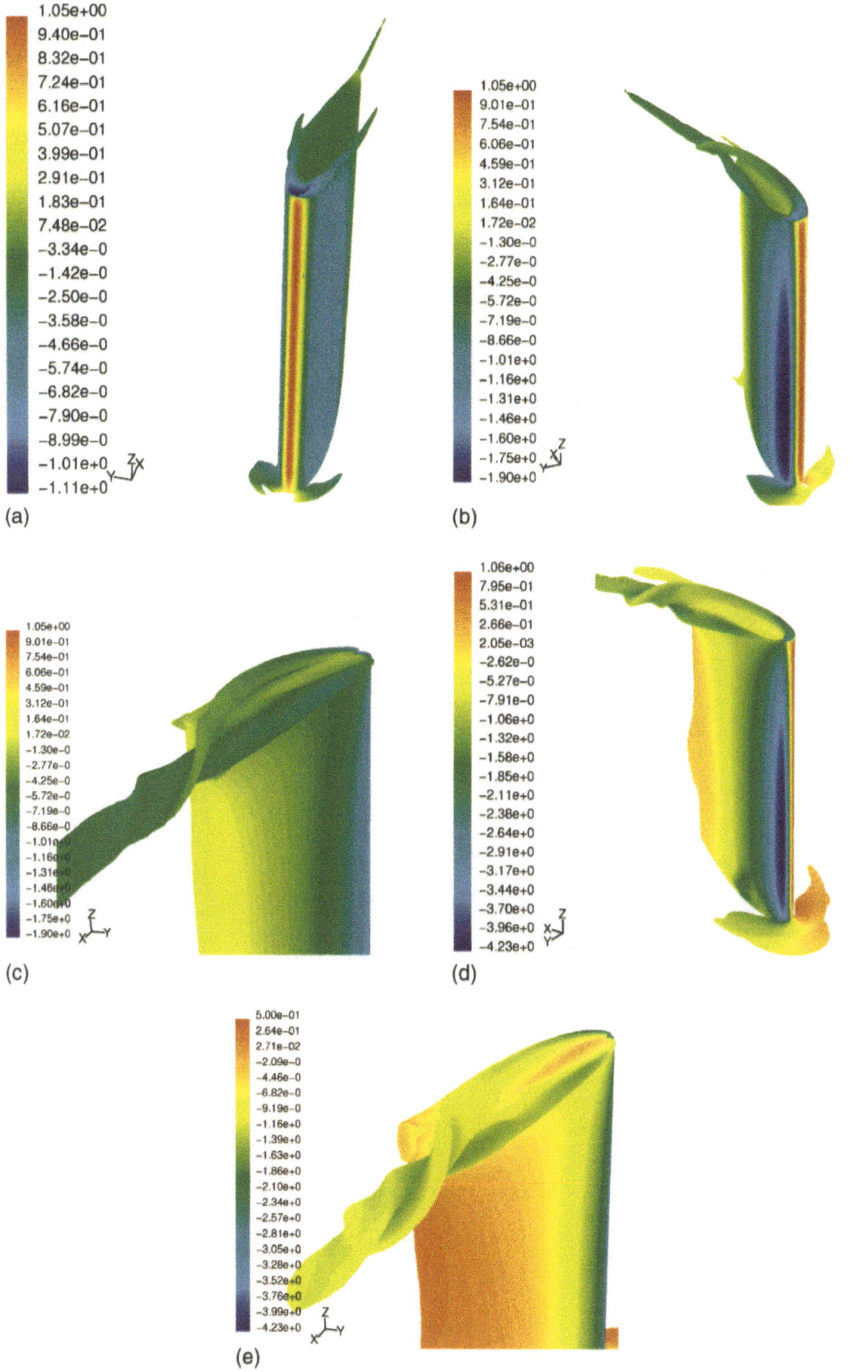

Plate 6 Tip vortex development captured using a constant vorticity surface (500), NACA0020, *Re* = 8×10^5, $\alpha = 10°$, $k-\varepsilon$, RNG. (a) $\alpha = 0°$, (b) front view, $\alpha = 10°$, (c) rear view, $\alpha = 10°$, (d) front view, $\alpha = 20°$, (e) rear view, $\alpha = 20°$ (See also page 285 of this book)

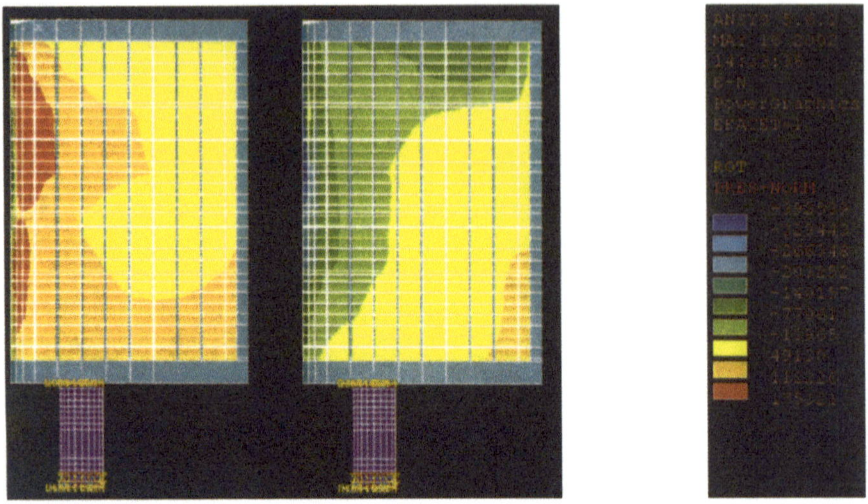

Plate 7 Total pressure distribution (See also page 379 of this book)

Plate 8 Deflection in *x* direction (See also page 380 of this book)

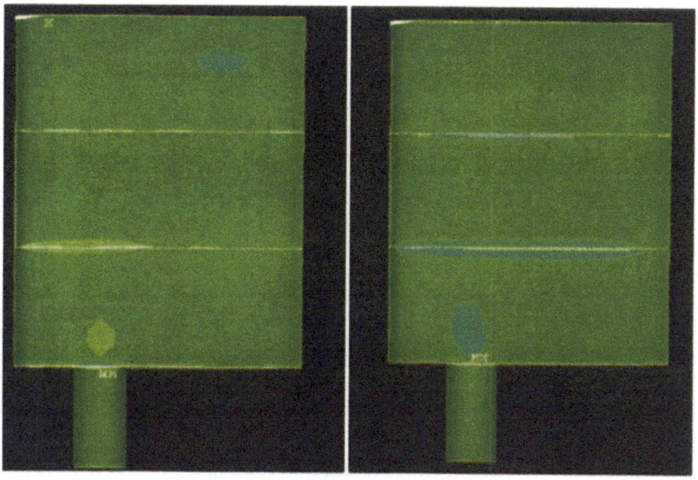

Plate 9 *Z*-component stresses (See also page 381 of this book)

Plate 10 Bending stresses (See also page 382 of this book)

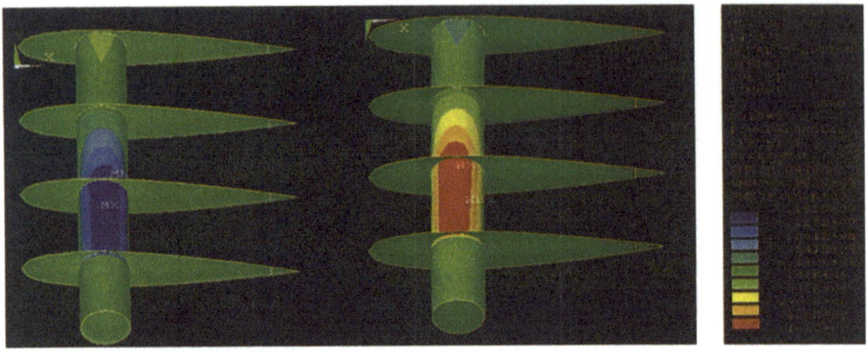

Plate 11 *Z*-component stresses in the stock (See also page 382 of this book)

5.8.3 Cavitation on marine control surfaces

5.8.3.1 *Rudders*

Cavitation can occur on rudders on small high-speed craft such as lifeboats and patrol boats and on large high-speed ships such as ferries, warships and container ships. It arises from the high ship speeds and relatively small rudder immersion, leading to low cavitation numbers and the potential onset of cavitation. If the rudder is operating downstream of a propeller, these speeds are increased further and the tip and root core vortices from the propeller also impinge on the rudder, Figures 5.123 and 5.124. The centres of the propeller vortices are at low pressure, with the potential for exciting cavitation when they impinge on the rudder. Cavitation on the rudder can

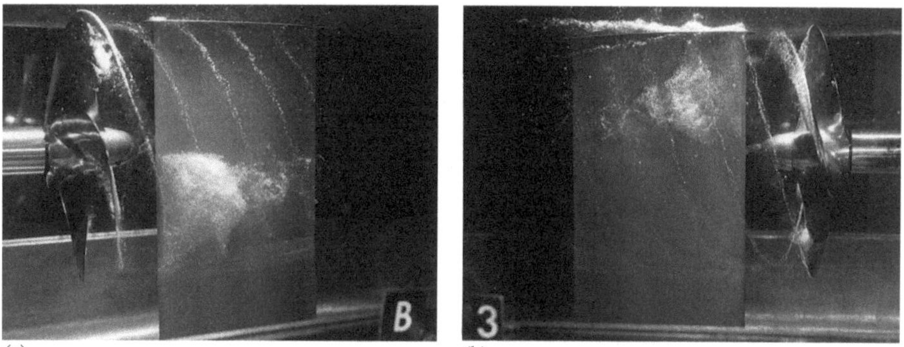

(a) (b)

Figure 5.123 Examples of rudder cavitation at two different angles. Photographs courtesy of VWS Berlin

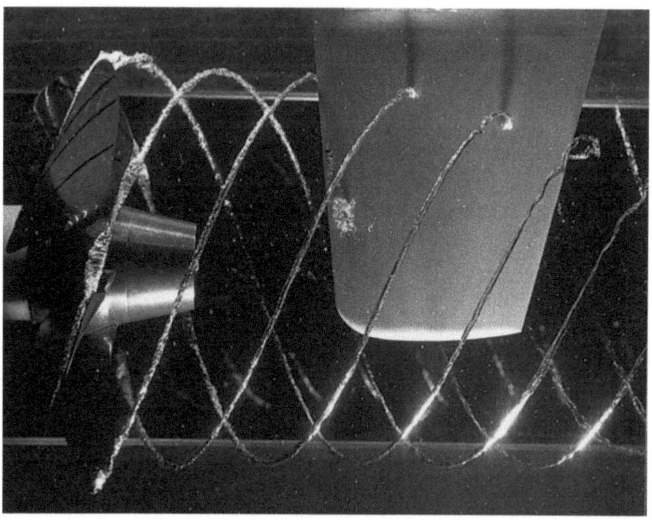

Figure 5.124 Propeller tip vortices impinging on rudder. Photograph courtesy of VWS Berlin

lead to erosion of the rudder surface, increased noise levels and possible decreases in lift. Thus erosion and noise may be caused either by self-induced cavitation on the rudder or by the cavitating propeller tip and hub vortices. Such phenomena have, for example, been reported by Gregory and Dobay [5.118], McGeough and Millward [5.119], Kracht [5.34, 5.65–5.67], Shen *et al.* [5.120], Friesch [5.121], Mevis and Klug [5.122] and ITTC 2005 [5.123].

Due to high ship speeds and large induced velocities from heavily loaded propellers, there is a high risk of cavitation, even at small rudder angles. These small rudder angles, up to about ±5°, occur in service due to coursekeeping. Further, significant rudder inflow angles are induced by the propeller, even at zero rudder incidence, due to the rotational nature of the propeller slipstream, Figure 3.32, Section 3.5. Spanwise load (C_N) distributions for an all-movable (spade) rudder at zero incidence downstream of a propeller are shown in Figure 5.125. These have been derived from Figures 5.74(b and c) in Section 5.4.2.3. The variation in local normal force coefficient, C_N, closely follows the change in effective onset incidence. Even at zero rudder incidence, there is seen to be a significant variation in C_N over the rudder span due to the propeller induced inflow angles. There is also a significant increase when J is reduced from 0.51 to 0.35, that is an increase in propeller thrust loading. At zero incidence and $J = 0.51$, which might represent a typical service condition, C_N values of up to about ±0.8 are seen to occur across the span. Lift curve slopes from Figure 5.13 have been used to calculate the distributions of effective incidence across the span, as shown in Figure 5.126. These would suggest

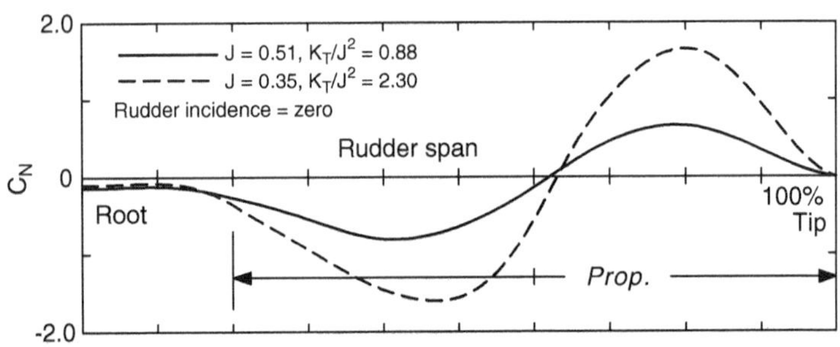

Figure 5.125 Distribution of normal force across span at zero rudder angle (derived from Figures 5.74b and 5.74c)

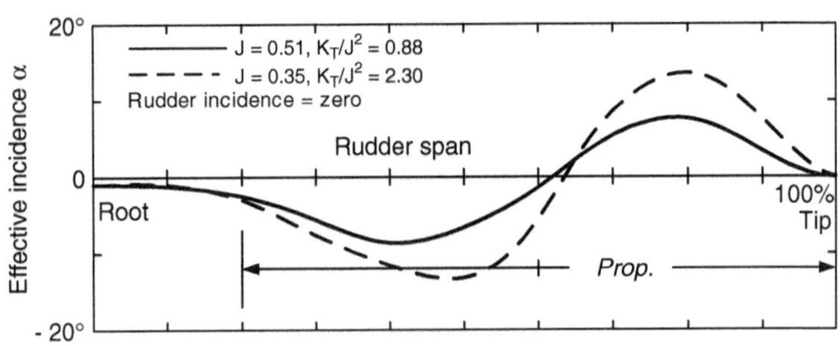

Figure 5.126 Distribution of effective incidence across span at zero rudder angle

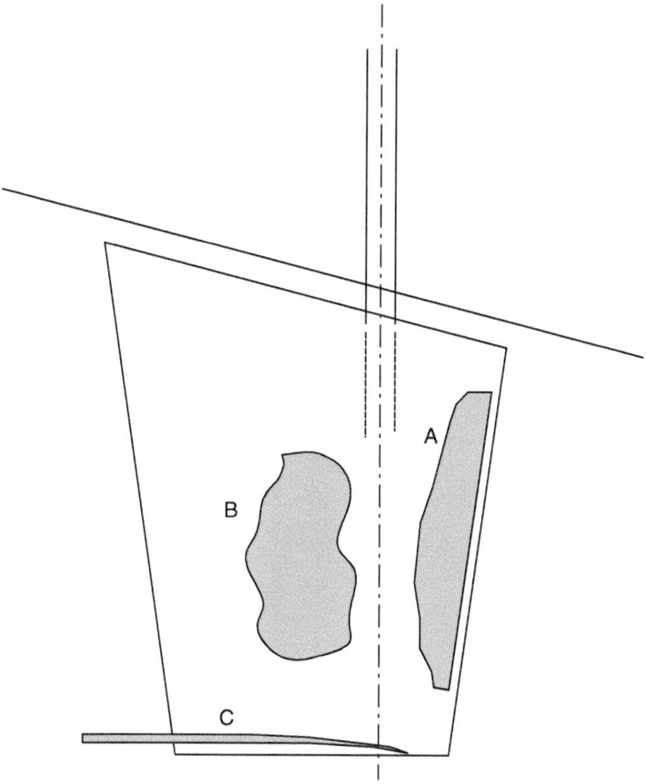

Figure 5.127 Potential areas of cavitation on spade rudder

a variation in effective local incidence of up to about ±9° for $J = 0.51$ and ±13° for $J = 0.35$. These distributions of effective incidence would also broadly indicate the level of twist required across the span to reduce the effective incidence and load to zero across the span.

Typical areas of cavitation and potential erosion on all-movable spade rudders reported in references [5.34, 5.118, 5.120] are shown in Figure 5.127. Area A is back sheet cavitation, area B is back bubble cavitation and C is vortex cavitation at the bottom edge. Erosion damage due to cavitation has occurred on the semi-balanced skeg rudders fitted to large high-speed container ships, such as that reported in references [5.121, 5.122, 5.123]. There is a particular risk for the skeg rudder, which has vertical and horizontal gaps and typical reported areas of cavitation erosion damage are indicated in Figure 5.128. Following the description in [5.123], area A is due to the collapse of vortex cavitation generated from the propeller tip, B and C are due to the surface discontinuity and flow through the gap between skeg and rudder, with the potential for bad erosion damage in these areas, D is due to the rotational flow induced by the propeller, and E is vortex cavitation at the bottom edge.

The presence of gap flow on semi-balanced skeg rudders was reported by Molland [5.8, 5.89] and Goodrich and Molland [5.7], as mentioned in Section 5.3.5.

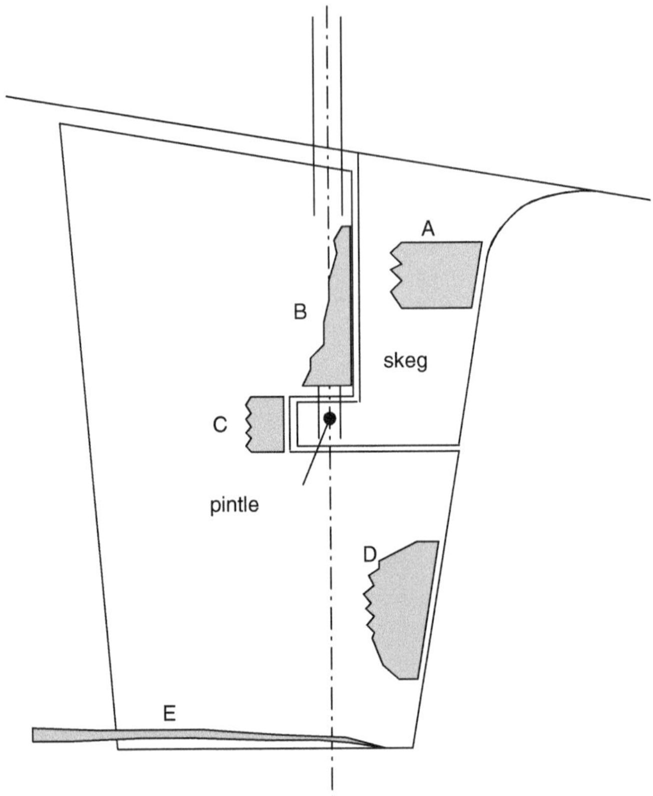

Figure 5.128 Potential areas of cavitation on semi-balanced skeg rudder

Due to the low-pressure peak in way of the rudder–skeg junction, Figure 5.129, a large difference in pressure is developed across the gap. This leads to a gap flow with a relatively high velocity and low pressure which, together with the discontinuity at the rudder–skeg junction, leads to potential conditions for the inception of cavitation. Figures 5.130 and 5.131 show the location of the pressure tappings (for Rudder No. 1) and Figure 5.132 shows the pressures in the gap for changes in rudder incidence. The end of the gap is approximately at tapping No. 33 on the high-pressure side and No. 20 on the low-pressure side. The pressure is seen to be reasonably constant across the gap between tappings No. 18 and No. 31. However, there are pressure peaks at the entry and exit to the gap, particularly at span station No. 6, Figure 5.130, which is located close to the horizontal skeg gap. The low pressure recorded at tapping No. 32 (which is located just inside the gap on the high pressure side) indicates the presence of gap flow, this tapping being situated approximately where the flow accelerates into the gap on the high-pressure side. This gap flow was also confirmed by flow visualisation studies. The pressures measured near the entry to/exit from the gap indicate the likelihood of the occurrence of cavitation in and around the gap on the rudders of high-speed vessels.

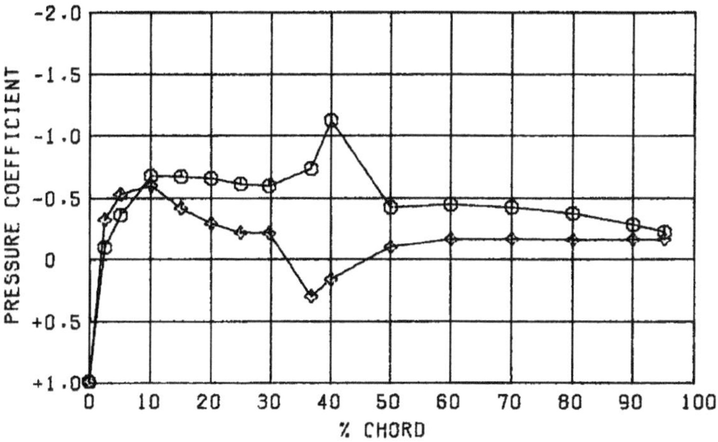

Figure 5.129 Low-pressure peak in way of rudder–skeg junction (span position S6, Figures 5.130, 5.131)

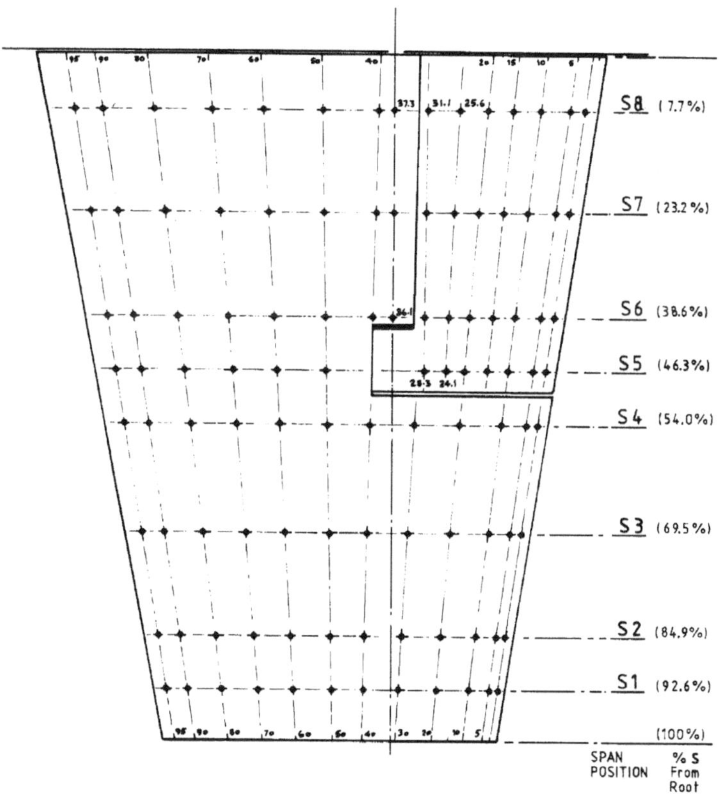

Figure 5.130 Distribution of pressure tappings: Rudder No. 1 [5.8]

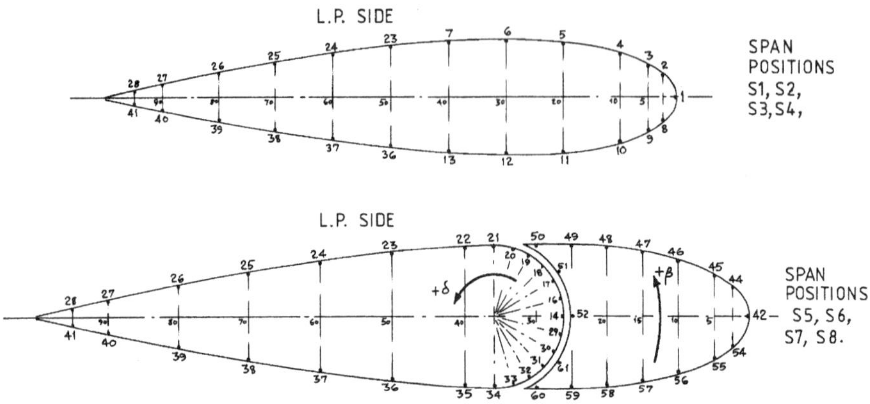

Figure 5.131 Chordwise distribution of pressure tappings; Rudder No. 1 [5.8]

5.8.3.2 *Other control surfaces*

In a manner similar to a rudder, cavitation can occur on other control surfaces such as stabiliser fins, hydroplanes and pitch damping foils. Cavitation may occur when vessel speeds are typically greater than about 20 knots. These control surfaces can, at times, be working relatively close to the water surface, hence reducing the cavitation number. Cavitation is generally avoided by careful section design and restricting operational angles of incidence. The methods of assessing cavitation inception for these control surfaces are broadly the same as for a rudder and are discussed further in Chapter 9.

5.8.4 Design to avoid cavitation and erosion

5.8.4.1 *Cavitation data*

Published cavitation design data for *symmetrical* sections is relatively limited. Relevant cavitation investigations on rudders include those carried out by Gregory and Dobay [5.118], Kracht [5.34, 5.65–5.67] and Shen *et al.* [5.120] and these are discussed further.

Gregory and Dobay [5.118] present experimental results for six high-speed craft rudders operating under cavitating conditions. Section shapes tested were a NACA0015, four wedge sections and a flat plate. The tests were carried out in a cavitation tunnel at a Reynolds Number *Re* of 1×10^6 and a cavitation number σ varying from 4.0 down to 0.5. Details of two of the rudders, the NACA00 section Rudder No. 1 and a wedge section, Rudder No. 3, are shown in Figures 5.133 and 5.134. Results are shown in Figure 5.135 for the NACA00 section and Figure 5.136 for the wedge section. It is seen that there is a sudden loss in the increase in lift together with a decrease in drag as the cavitation number is reduced down to about 0.5. These results show reasonable correlation with the cavitation inception limits presented later in Figure 5.139. Examples of the cavitation patterns for different cavitation numbers and incidence for Rudder No. 1 are shown in Figure 5.137.

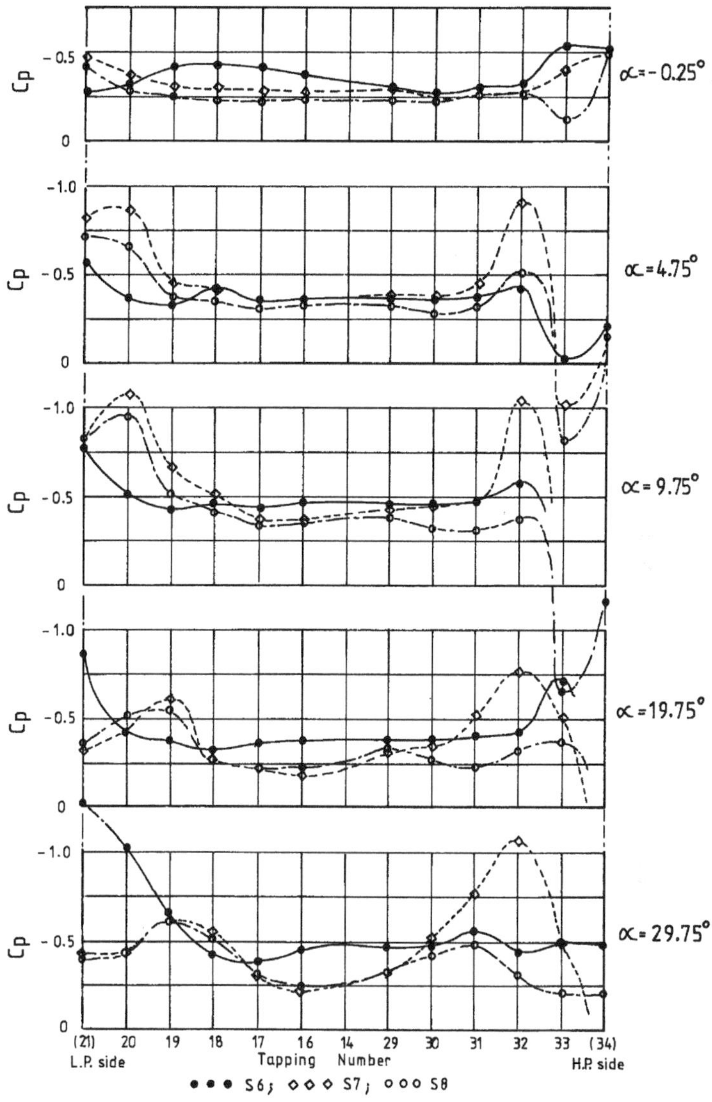

Figure 5.132 Skeg rudder gap pressures; Rudder No. 1 [5.8]

Overall conclusions from the work were that section shape had little effect on the lift produced by the rudder, although it had a substantial effect on the drag, the moment coefficient, and the cavitation inception point. The NACA section was found to be the best section for speeds below the equivalent of about 30 knots, whilst above 30 knots a parabolic or wedge section is recommended. Rudders 1–5 were also tested downstream of a propeller by Mathis and Gregory [5.62].

A comprehensive review of rudder cavitation including the influence of an upstream propeller is presented by Kracht [5.67]. He points out the advantages of

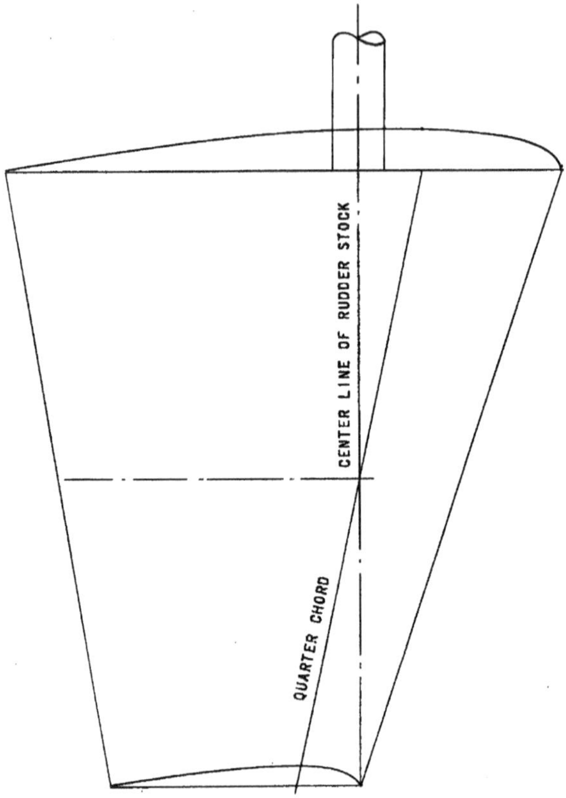

Figure 5.133 Details of Rudder No. 1 [5.118]

spade rudders without gaps, the problems of cavitation damage on semi-balanced skeg rudders, rudder tip cavitation, rudder gap cavitation and the need to use sections with low-grade cavitation such as the HSVA-MP-Profiles. Kracht [5.34] indicates that the NACA00 thickness distribution is not recommended for rudders vulnerable to cavitation, and presents test results for all-movable spade rudders that demonstrate the favourable qualities of the HSVA sections in respect of cavitation. Kracht [5.65, 5.66] presents further results of propeller–rudder interaction, including results for a semi-balanced skeg rudder. A comprehensive presentation of many of Kracht's data is given in reference [5.87].

Shen *et al.* [5.120] present the results of an investigation into the cavitation on the model of a rudder for a Naval vessel. The propeller axis was approximately at the rudder tip with about 60% of the rudder span in the propeller slipstream. The objectives were to identify rudder cavitation problems, investigate propeller effects on rudder cavitation, and develop methods to incorporate propeller–rudder interaction into future rudder design for cavitation improvement. LDV measurements were carried out to identify propeller-induced velocities and angles upstream of the rudder, together with pressure measurements at one chordline. Inflow angles of up

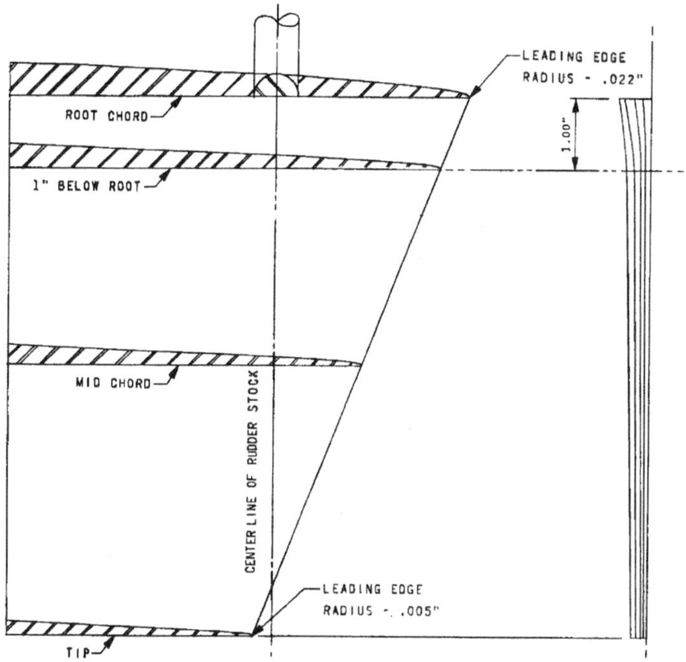

Figure 5.134 Details of Rudder No. 3 [5.118]

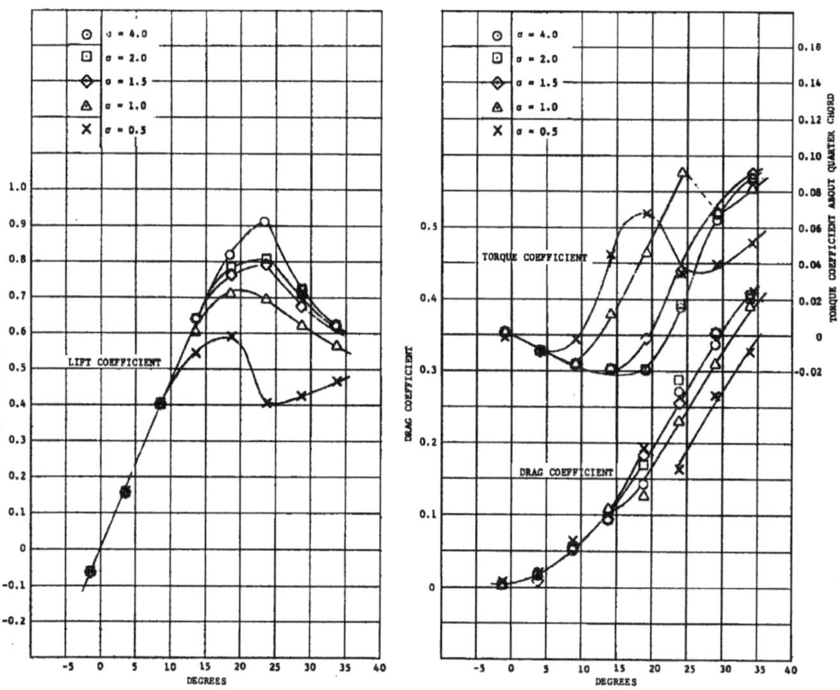

Figure 5.135 Results for Rudder No. 1 [5.118]

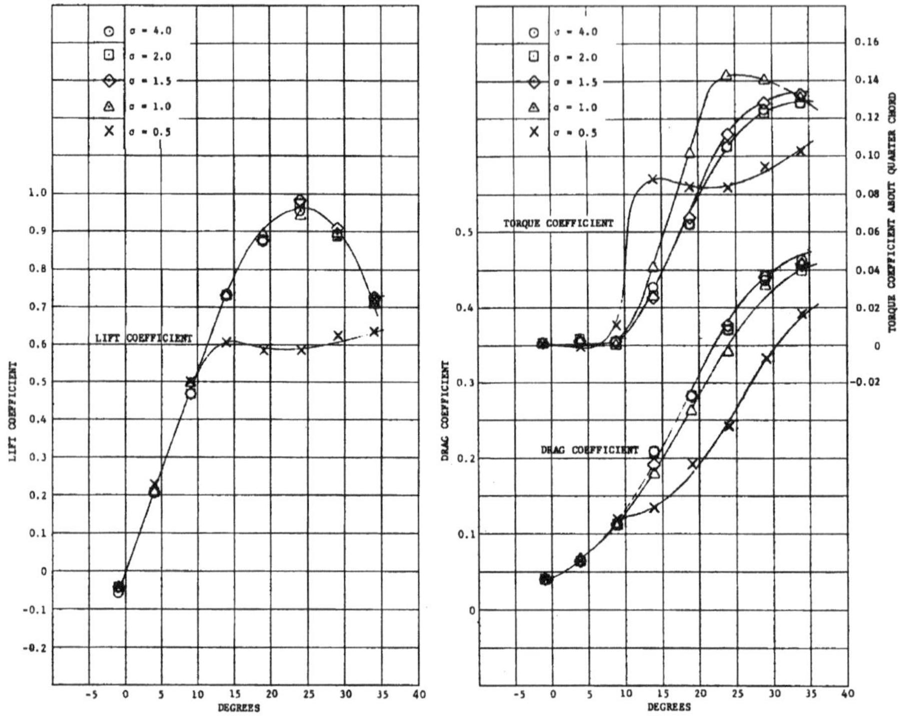

Figure 5.136 Results for Rudder No. 3 [5.118]

to about 8° were measured at about 50–70% of the rudder span. Surface cavitation was observed and an inception envelope developed for full-scale cavitation prediction. Calculation methods were developed that showed reasonable agreement with the experimental results.

Eppler and Shen [5.124] describe the development of symmetrical sections suitable for hydrofoil applications using theoretical methods. Whilst the sections are not directly suitable for movable control surfaces, the paper provides guidance on the likely form of the cavitation inception envelopes for symmetrical sections.

In order to provide suitable design data, pressure distributions and cavitation inception limits for symmetrical sections of NACA00 form have been generated using the 2-D panel code in [5.125]. The technique had been applied successfully to the prediction of cavitation inception on sections suitable for marine current turbines [5.126]. Satisfactory correlation was obtained between the 2-D panel code and cavitation tunnel experimental results, Figure 5.138, although the panel code was a little conservative on the face. The 63–215 section in Figure 5.138 has only 2% camber and the results closely resemble those expected for a symmetrical section. Figure 5.139 shows inception envelopes generated by the Xfoil panel code for NACA00 sections with thickness/chord ratios from 10% to 20%. As expected, these show reasonable agreement with the envelopes shown in Figure 5.138. They

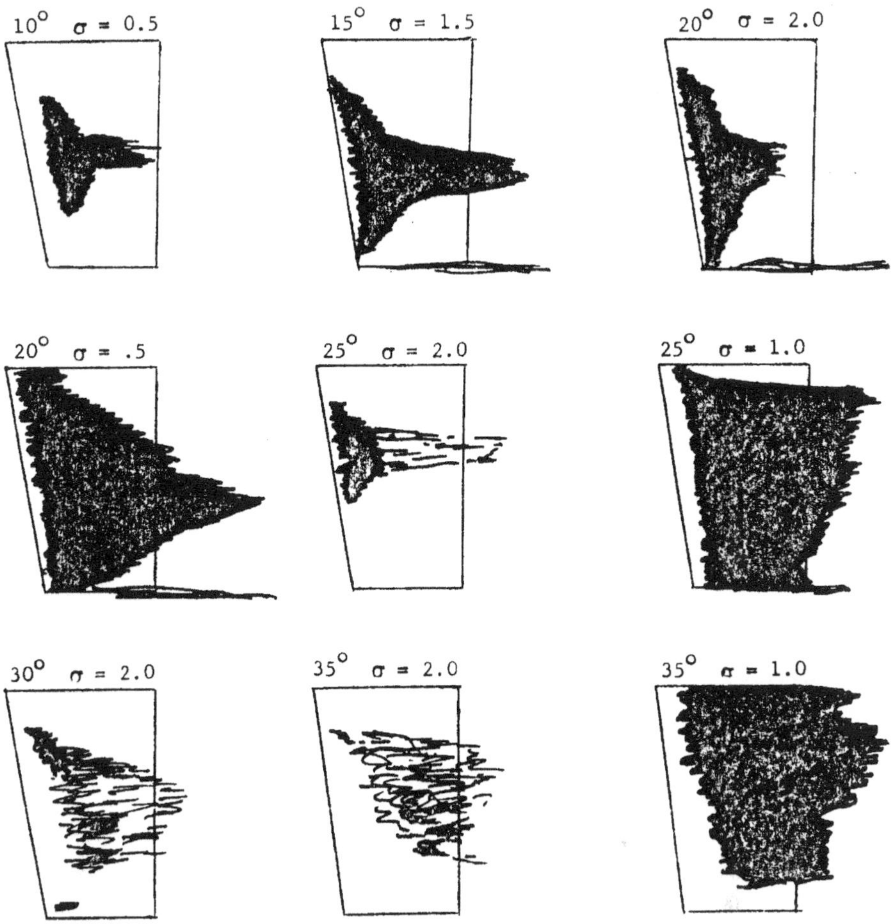

Figure 5.137 Cavitation patterns, Rudder No. 1 [5.118]

are also very similar to the predicted and experimental results for a spade rudder with a mean NACA0018 section reported by Shen *et al.* [5.120].

The inception envelopes shown in Figure 5.139 are suitable for preliminary design checks on the likely occurrence of cavitation on a movable control surface with a symmetrical section. As an example, the results from example application 7 in Chapter 11 can be used which considers cavitation on the rudder of a 7000 TEU container ship with a service speed of 25 knots. The resulting cavitation number in a ballast condition is about 1.23 and, using Figure 5.139, this would suggest limits on C_L of about ±0.35, which would suggest limits on rudder incidence of up to about ±6.5°. Such rudder angles may arise from the propeller induced inflow angles and rudder angle changes needed for coursekeeping. The detailed calculations for this cavitation example are included in Chapter 11.

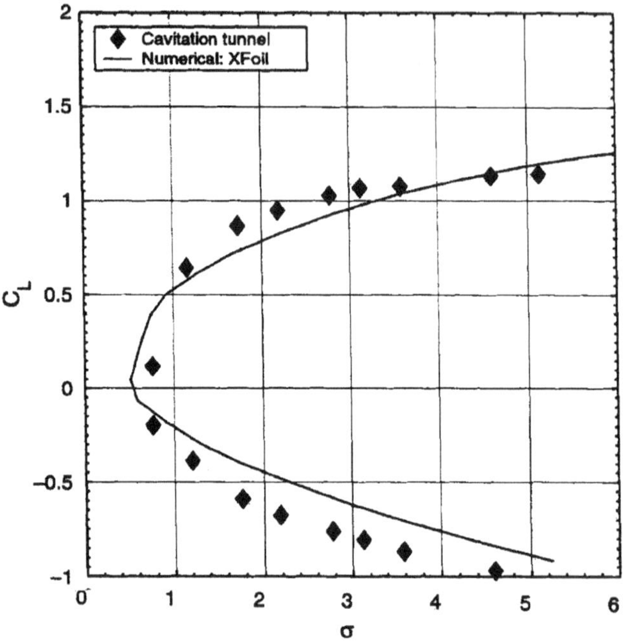

Figure 5.138 Cavitation inception envelopes for NACA 63-215 section; comparison of numerical and experimental results [5.126]

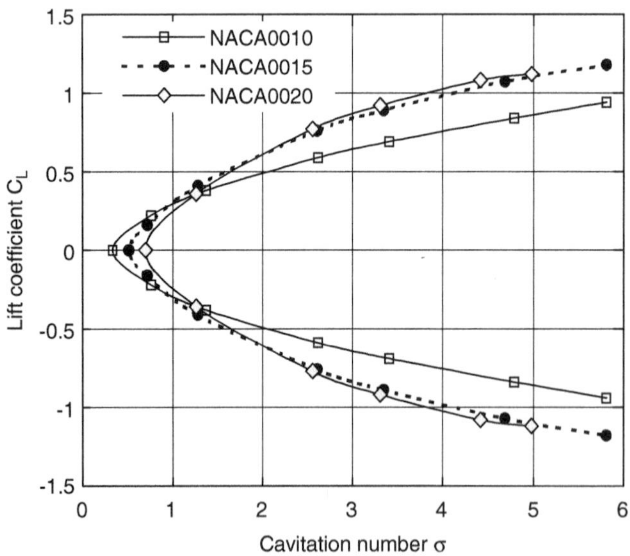

Figure 5.139 Cavitation inception envelopes: numerical data for symmetrical sections

5.8.4.2 *Methodology to avoid cavitation and erosion*

Recommended methods of avoiding or minimising cavitation on rudders are given by Brix [5.35], Kracht [5.67], Friesch [5.121] and ITTC 2005 [5.123]. These may be summarised as

- Use of cavitation-resistant section shapes with smooth pressure distributions and sufficiently small absolute C_P at moderate angles of attack ($<5°$) should be considered. The NACA00 series sections have proved very popular and successful for most movable control surfaces in the marine field. However, in high speed applications where the marine control surface becomes vulnerable to cavitation, then sections with flatter pressure distributions such as the NACA64xx series or HSVA-MP-71-xx series can be usefully employed. Typical pressure distributions for these sections, using data from Abbot and von Doenhoff [5.25] and Brix [5.35], are shown in Figure 5.140 and the wider cavitation inception envelopes for the NACA64 and HSVA sections are shown in Figure 5.141. For higher speeds, say >30 knots, a wedge shape section may be suitable, following the earlier discussion of the results of Gregory and Dobay [5.118].
- 2-D CFD analysis can be usefully employed to produce sections with suitable shape and thickness, noting that sharp section noses with high negative pressure peaks should be avoided.
- Thick section profiles have higher drag and increased susceptibility to mid-chord bubble cavitation, although an increased leading edge radius reduces the magnitude of the leading edge suction peaks. If possible, section thickness should be reduced from a maximum at the root to a lesser thickness at the tip whilst still meeting strength requirements.
- For rudders with gaps and lower pintle, such as semi-balanced skeg rudders: minimise the size of the gaps within practical limits, rounding the edges around the gaps; design to keep gaps away from the slipstream of propeller hub and tip vortices; possible use of horizontal scissor and deflection plates near gap regions and guide and deflection plates at positions where cavitation may occur; possible addition of a strip at the centre of the gap to prevent gap flow. For gap cavitation in the lower pintle area, use CFD to check for gap pressure and gap flow velocity, separated flow and to optimise gap size and inlet geometry.
- Consider the use of twisted sections across the span, that is upper and lower sections twisted in opposite directions.
- Sharp edged tips or end plates can increase rudder effectiveness, but also increase the risk of cavitation. Rudder sole (tip) erosion can be reduced by rounding the lower front edge of the rudder. Research was carried out at HSVA by Brix *et al.* [5.127] on cavitation on the sole at the tip of a rudder. From the results of the cavitation tunnel tests carried out on alternative shapes for the tip profile, it was concluded that a well rounded fore end to the sole, Figure 5.142, would make the rudder less susceptible to sole cavitation.
- Avoid welds where cavitation may occur and round off all edges. Grind and polish all welds.
- Prevent or delay plate erosion by the use of steel overlay or special plastic coatings in appropriate areas.

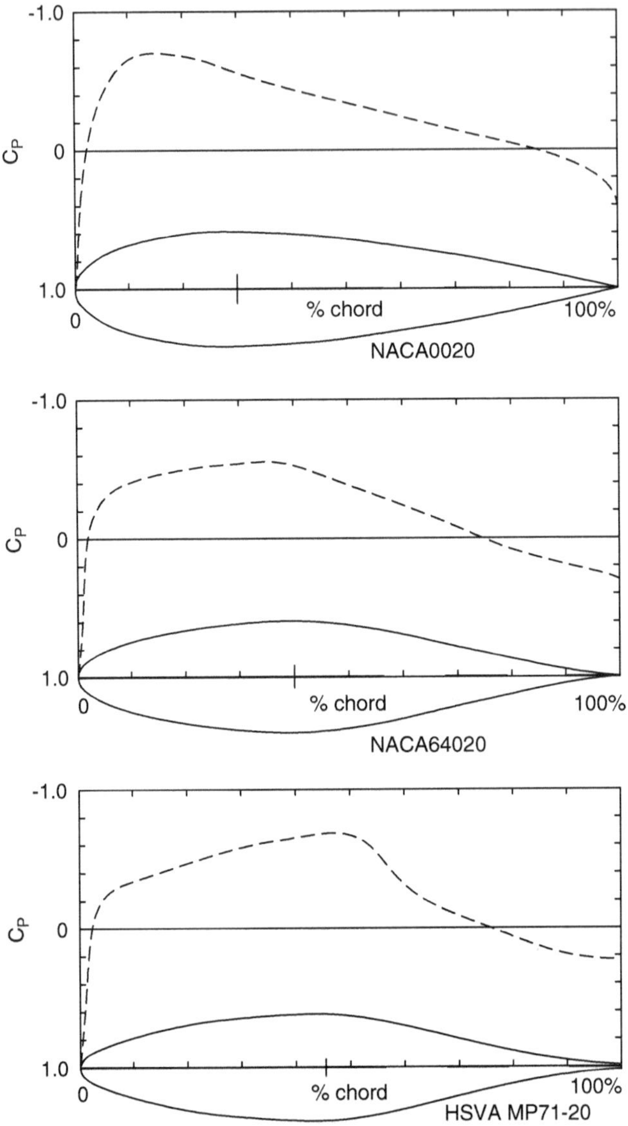

Figure 5.140 Pressure distributions for different section types

- Pay attention to the size and shape of the cathodic protection mounted on a rudder in a propeller race.
- It is generally accepted that model experiments are, at present, probably still the only reliable way to make predictions of cavitation and erosion.

5.8.4.3 *Theoretical investigations into gap flow and skeg rudder cavitation*

Specific investigations into the skeg rudder gap problem have been carried out by Boo *et al.* [5.128]. A RANS code and cavitation tunnel tests were used to investigate

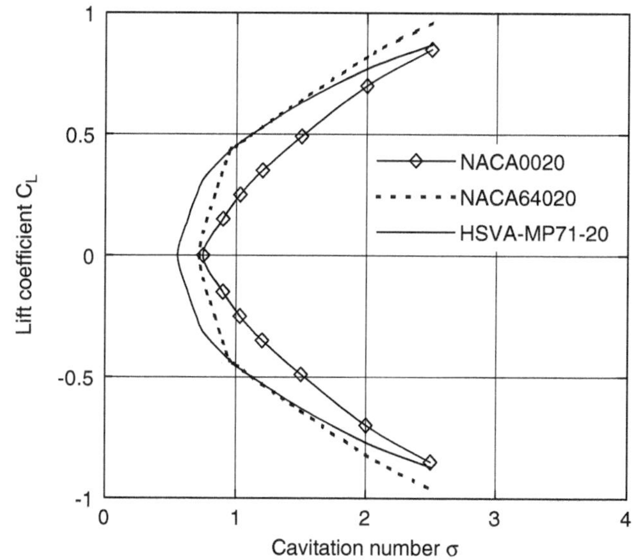

Figure 5.141 Comparison of cavitation inception envelopes for different section types

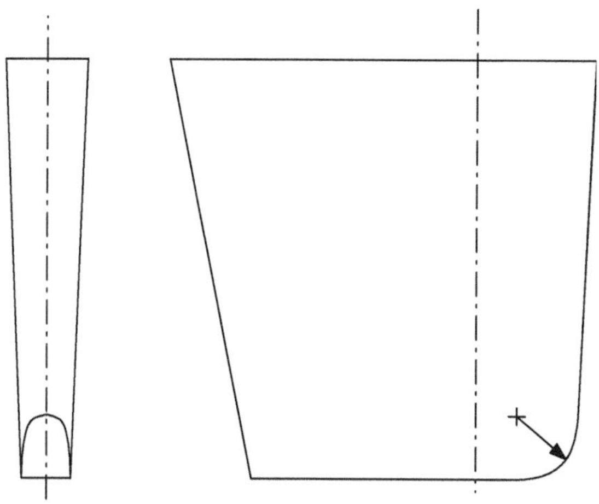

Figure 5.142 Rounded tip at fore end to minimise cavitation on sole of rudder

gap flow, different gap entry/exit shapes and the effects of Reynolds number on gap flow. It was found that the gap flow was sensitive to the gap entry/exit curvature, that the gap flow could be reduced by installing a round bar inside the gap and that pressure variation would increase with increasing Reynolds number, so that cavitation would be expected on the full scale ship.

Han *et al.* [5.129] developed a numerical model for a horn-type rudder (semi-balanced skeg rudder) working behind a propeller. The propeller is modelled using a classical vortex lattice method and the rudder by a surface panel method. The effect of the gap is modelled, with gap flow modelled as Couette flow. Results for the free-stream case are compared with the experimental data for rudder No. 1 in Molland [5.8], and the correlation is reasonably good. With the rudder in the presence of a propeller, the numerical method shows good agreement with measured results and that the method is applicable to check the likelihood of cavitation.

5.8.4.4 *Use of twisted rudders*

There have been a number of proposals whereby the rudder is twisted across its span to take account of the propeller induced twist in the flow. Brix and Baumgate [5.130] carried out model tests on the rudder for a container ship that had the upper half turned 7.5° to starboard and the lower half 7.5° to port. This was found to reduce cavitation by equalising the pressure distribution on the rudder in the straight-ahead position, and to reduce vibration and rudder drag. There was no impairment of steering performance or propulsive efficiency.

Shen *et al.* [5.131] developed and tested a model of a twisted rudder. In this case, there was a distribution of twist varying from zero at the root, to about 5° at 65% span and zero at the tip. The results were compared with the equivalent non-twisted rudder reported by Shen *et al.* [5.120]. The results indicated that the effectiveness of a twisted rudder would be equal or slightly better than a non-twisted rudder, that the drag was lower for the twisted rudder and that the twisted rudder would not impair the steering performance. It was found that the use of twisted rudders should allow a ship to make greater rudder deflections without cavitation. The cavitation inception envelope of the twisted rudder is significantly wider and deeper than that of the non-twisted rudder. It was also noted that the need to twist the rudder is even more demanding at high speeds when the propeller is more heavily loaded.

Twisted rudders are now manufactured and marketed by commercial companies for large fast vessels such as container ships [5.132]. In this case the rudder is also of the spade type, which avoids the gaps associated with skeg rudders and their cavitation problems discussed in Section 5.8.3. Twisted rudders manufactured in composite materials are under construction and evaluation for warships [5.133].

Theoretical methods, Chapter 6, can be used to estimate likely rudder inflow angles induced by the propeller for zero rudder incidence. Such an approach is described by Jurgens [5.134]. Local lift can be estimated and the likelihood of cavitation checked using Figure 5.139. Such a technique can also be used to estimate suitable levels of twist to be applied to a twisted rudder for a particular propeller thrust loading. Similarly, experimental pressure measurements at zero rudder incidence may be used to determine the likely distribution of local effective incidence and hence twist required, as described earlier in Section 5.8.3 and Figure 5.126.

5.9 Propulsive effects

When a rudder is situated downstream of a propeller, the mutual interaction of the rudder and propeller determines both the magnitude of the forces generated by the rudder and the net propulsive efficiency of the rudder–propeller combination as a whole. A propeller upstream of the rudder accelerates and rotates the inflow onto the rudder, as described in Section 3.5. At the same time, the rudder blocks and diverts the flow through the propeller, Figures 3.30 and 3.31, which affects the thrust produced and the torque developed by the propeller. The practical layout of the hull, propeller and rudder is driven by a number of factors, as discussed in Section 4.2. It is, however, found that the rudder position relative to the propeller can have a significant influence on the rudder and propeller characteristics and this should be borne in mind when formulating rudder–propeller arrangements.

The influence of the propeller on the rudder forces for changes in the geometrical properties X/D, Y/D and Z/D in Figures 4.8–4.10, were discussed in Section 5.4. It was seen that significant interactions could take place with changes in X/D Y/D and Z/D.

The influences of a rudder on the propeller characteristics and propulsive efficiency have been studied by a number of investigators including Stiermann [5.85], Suhrbier [5.135], Kracht [5.66, 5.87], Nakatake [5.86] and Molland and Turnock [5.23, 5.38, 5.71, 5.79, 5.80].

Stiermann [5.85] carried out systematic tests in a tank on various propeller–rudder combinations. Three propellers with diameter 240 mm and pitch ratios, $P/D = 0.6$, 1.0 and 1.4 were used. Two rudders, with NACA00 sections and thickness ratios $t/c = 0.12$ and 0.18, were positioned at three longitudinal locations with $X/D = 0.10$, 0.30 and 0.50. The rudder was placed on the centreline of the propeller and was at zero incidence for all the tests. The tests were carried over a range of advance ratios, J, and the changes in propeller K_T, K_Q and rudder thrust/drag K_R were measured. The rudder thrust K_R is derived as the difference between the test results with and without the propeller, the rudder thrust K_R then being the result of propeller induction only. In this case rudder thrust is nondimensionalised in the same way as the propeller coefficients as $K_R = d/\rho n^2 D^4$. Regression equations were fitted to the experimental results enabling interpolation and wider use of the data to be made.

Figures 5.143, 5.144 and 5.145 are cross plots of some of the data from Stiermann [5.85], illustrating the influences of X/D, t/c and J (K_T/J^2). J values were chosen to give approximately the same K_T/J^2 values as those used in references [5.38, 5.71, 5.79, 5.80], namely J values of 0.38, 0.54 and 0.96 which, for the propeller with $P/D = 1.0$, correspond to K_T/J^2 values of 2.35, 0.89 and 0.05. With increase in X/D it is seen from Figure 5.143 that rudder thrust increases. This is broadly in line with the changes in rudder drag/thrust reported in references [5.23, 5.71] and discussed in Section 5.4.2.2, Figure 5.63(d). It is noted that, like the work of Molland and Turnock who used a NACA00 section, the Stiermann data in Figure 5.143 for the thicker NACA0018 section show that a small amount of net rudder thrust is produced only at large X/D and low J (high thrust loading K_T/J^2). It is seen from

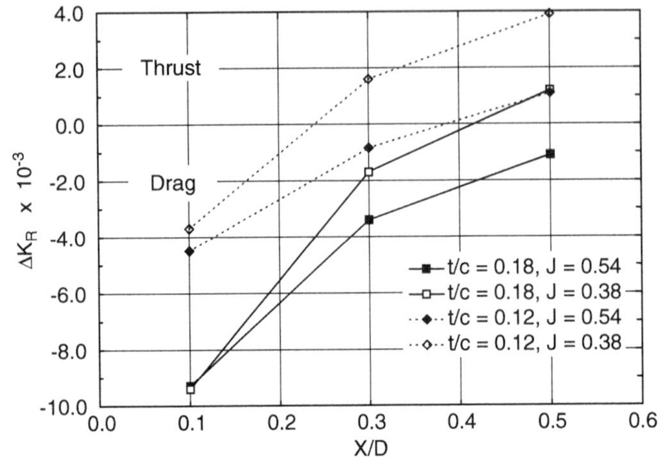

Figure 5.143 Changes in rudder thrust

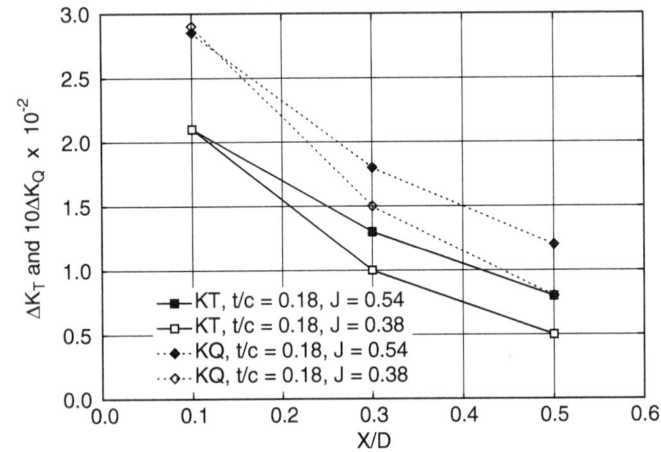

Figure 5.144 Changes in propeller thrust and torque, $t/c = 0.18$

Figure 5.144 that, as X/D increases, ΔK_T and ΔK_Q both decrease. With increase in rudder thickness ratio t/c, the rudder thrust decreases, Figure 5.143, and ΔK_T and ΔK_Q increase by small amounts, Figures 5.144 and 5.145. It was found that, with increase in propeller pitch ratio P/D, the rudder thrust increased at $J = 0$ and ΔK_T and ΔK_Q also increased. Overall it was found that, as the changes in K_Q with change in X/D showed similar trends to the changes in K_T, the net effects on propulsive efficiency are small.

Molland and Turnock carried out systematic variations in the parameters X/D, Y/D and Z/D in the wind tunnel tests described in Section 5.4.2. The results of these tests for changes in K_T are summarised in Figures 5.146–5.148.

Longitudinal separation (X/D): The change in K_T with change in X/D is shown in Figure 5.146, which indicates that ΔK_T increases as X/D is reduced. This tends

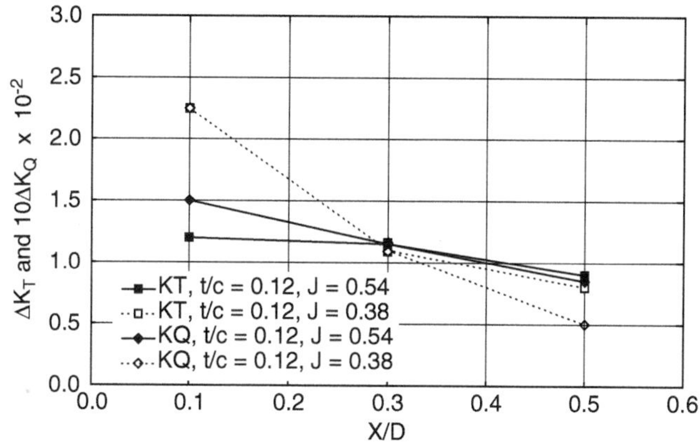

Figure 5.145 Changes in propeller thrust and torque, $t/c = 0.12$

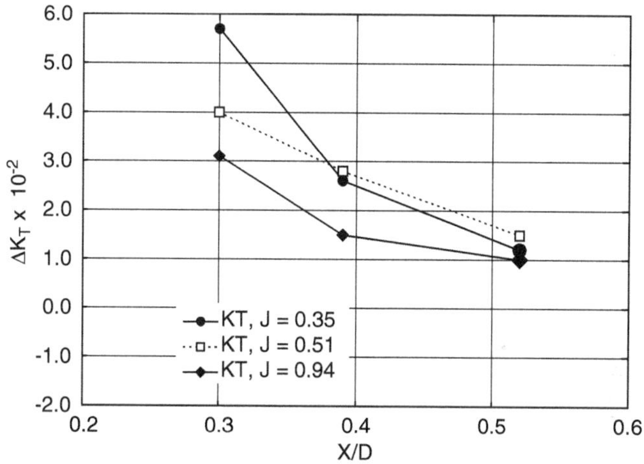

Figure 5.146 Changes in propeller thrust with change in X/D

to be the converse of the effect of the propeller on the rudder, Section 5.4.2.2. Similar conclusions were reached by Stierman [5.85], discussed in the previous section, Nakatake [5.86] and Kracht [5.66]. The results of Molland and Turnock [5.38, 5.79] indicate that these changes in ΔK_T occur over a wide range of rudder incidence. In general, the net effect of ΔK_T and rudder C_{D0} is an increase in overall effective thrust. This can also be concluded from inspection of the Stierman data in Figures 5.143 and 5.144.

Lateral separation (Y/D): Changes in K_T as a result of changes in Y/D, for fixed X/D, are shown in Figure 5.147. These indicate that there is a significant decrease in ΔK_T as Y/D is increased, with a marked asymmetry between positive and negative Y/D.

Vertical position (Z/D): Changes in K_T as a result of changes in Z/D (for fixed $X/D = 0.39$ and $Y/D = 0$) are shown in Figure 5.148. This indicates that as Z/D is decreased, with a reduced amount of the propeller race impinging on the rudder,

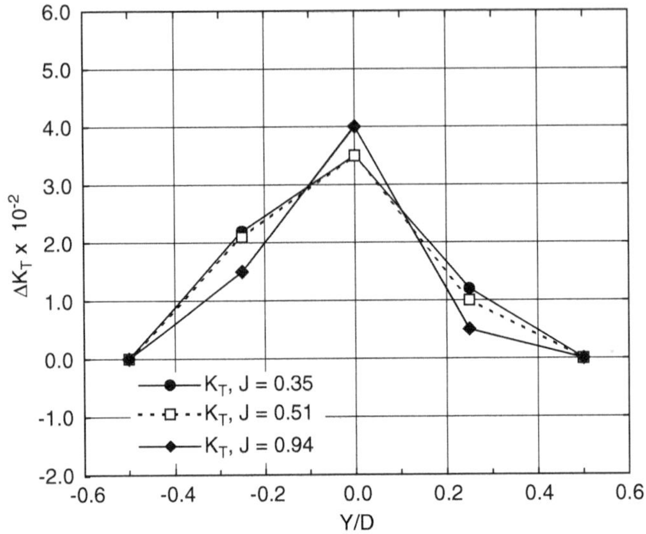

Figure 5.147 Changes in propeller thrust with change in *Y/D*

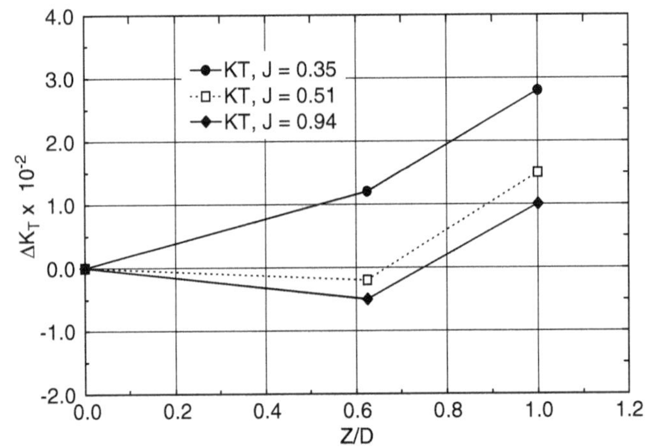

Figure 5.148 Changes in propeller thrust with change in *Z/D*

the influence of rudder blockage decreases and ΔK_T is reduced significantly. These results are broadly in line with those derived by Surhbier [5.135] who tested rudders not fully within the propeller race.

The relative effect of the rudder and propeller is a complicated issue. The results of the Molland and Turnock tests [5.23, 5.71], like the results of others such as Stierman [5.85] and English [5.136], indicate that, as far as propulsion is concerned, the rudder and propeller should be treated as a single propulsive unit.

A discussion of the interrelationship between the rudder and propeller, and the influences on performance and design, is included in example application 9 in Chapter 11.

5.10 Hull pressures

Section 5.4.2 describes wind tunnel tests on a representative hull form upstream of a rudder–propeller combination. Section 5.4.2.4 describes the influence of the hull on the flow speed and Section 5.4.2.7 describes its influence on flow straightening at various drift angles. The hull was also pressure tapped to monitor changes in hull pressures and forces with change in rudder angle, propeller thrust loading, longitudinal separation X/D and drift angle β. Details of the tests are reported in references [5.78,5.79].

Examples of hull pressures along waterlines at a yaw angle of $-7.5°$ are shown in Figure 5.149. Figure 5.150 shows the hull lift coefficient derived from an integration

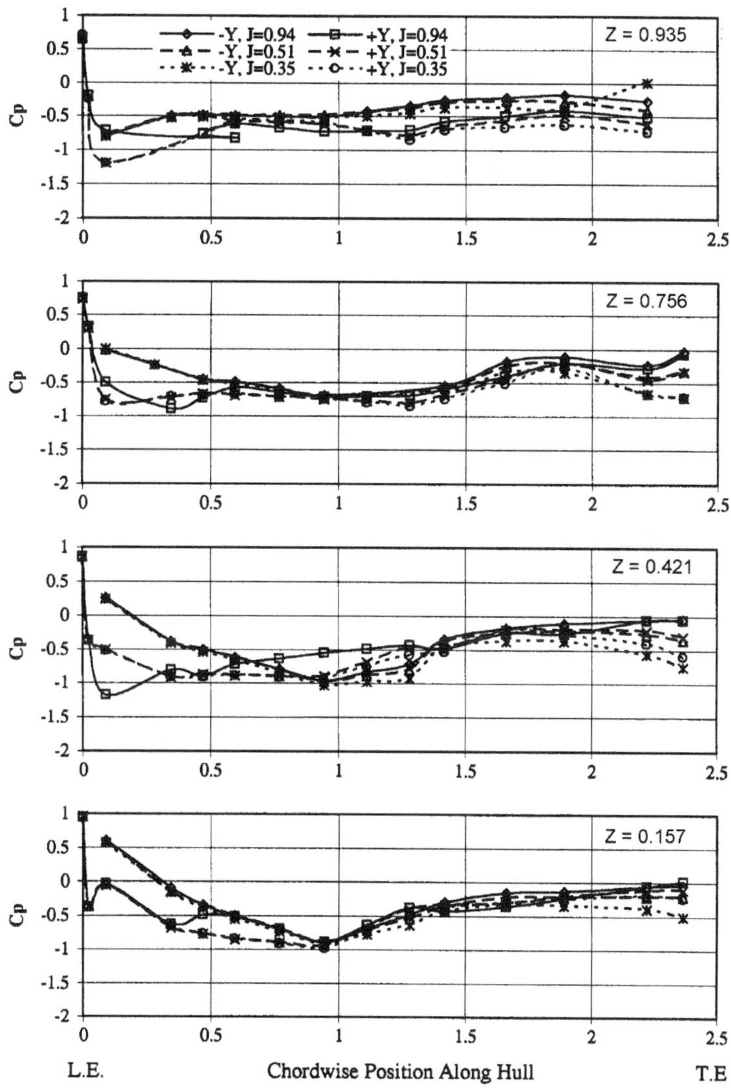

Figure 5.149 Hull chordwise pressure distributions; $X/D = 0.39$, $\beta = -7.5°$, $\delta = 20°$

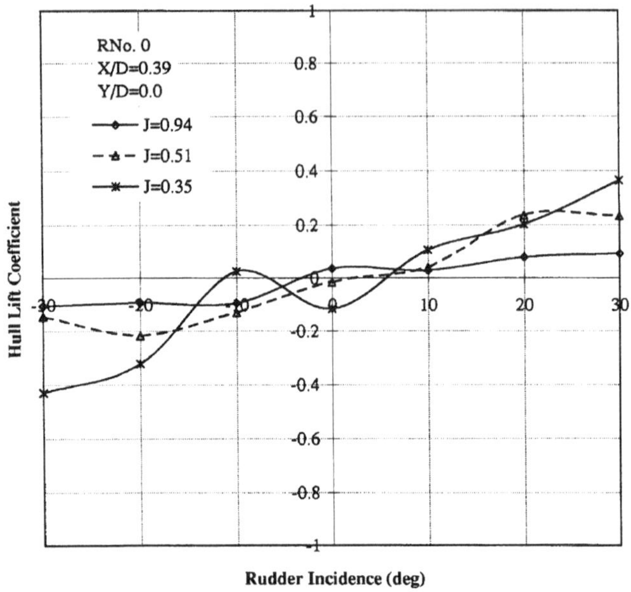

Figure 5.150 Hull lift; $X/D = 0.39$, $\beta = 0°$

of the hull pressures and, for comparative purposes, has been nondimensionalised using rudder area. The hull sideforce was found typically to be of the order of 10–15% of rudder lift for a given rudder incidence.

Further details of the hull pressure measurements and access to the tabulated hull pressure data are given in Table A1.2, Appendix 1.

References

5.1 Jones, G.W. Jr. Aerodynamic characteristics of three low aspect ratio symmetrical wings with rectangular planforms at Reynolds numbers between 0.4×10^6 and 3.0×10^6. *NACA R.M.* L52G18, 1952.

5.2 Whicker, L.F. and Fehlner, L.F. Free stream characteristics of a family of low aspect ratio control surfaces for application to ship design. *DTMB Report* 933, December 1958.

5.3 Thieme, H. Design of ship rudders. *Jarbuch der Schiffbautechnischen Gesellschaft*, Vol 56, 1962 or DTMB Translation No. 321, 1965.

5.4 Hagen, G.R. A contribution to the hydrodynamic design of rudders. Ministry of Defence, *Third Ship Control Systems Symposium*, Bath, September 1972.

5.5 Kato, H. and Motora, S. Studies on rudders with a flap. First report results of open water tests. *Journal of Society of Naval Architects, Japan*, Vol. 124, 1968, pp. 93–104.

5.6 Kerwin, J.E., Mandel, P. and Lewis, S.D. An experimental study of a series of flapped rudders. *Journal of Ship Research*, The Society of Naval Architects and Marine Engineers, December 1972, pp. 221–239.
5.7 Goodrich, G.J. and Molland, A.F. Wind tunnel investigation of semi balanced ship skeg-rudders. *Transactions of the Royal Institution of Naval Architects,* Vol. 121, 1979, pp. 285–307.
5.8 Molland, A.F. *The Free-Stream Characteristics of Ship Skeg-Rudders*, Ph.D. Thesis, University of Southampton, 1982.
5.9 Gong, I-Y., Kang, G-G., Kim, Y-G., You, Y-B., Huh, D. and Kim, Y-H. The influence of rudder area on the manoeuvrability of a ship with large beam-to-draught ratio. Proceedings of the *Sixth International Symposium on Practical Design of Ships and Mobile Units,* PRADS '95, The Society of Naval Architects of Korea, Soeul, 1995.
5.10 Kang, G-G., Gong, I-Y., Kim, S-Y., Yum, D-J., Kim, H-C., Ha, M-K., Son, B-R. and Ahn, S-P. The manoeuvrability of full form ships with low speed. Proceedings of *International Symposium on Forces Acting on a Manoeuvring Vessel,* MAN '98, Val de Reuil, France, 1998.
5.11 Joessel, Rapport sur des experiences relatives aux gouvernails. *Memorial du Genie Maritime,* Rapport 9, 1873.
5.12 Johns, A.W. Normal pressure on thin moving plates. *Transactions of the Royal Institution of Naval Architects,* Vol. 45, 1904.
5.13 Denny, M.E. The design of balanced rudders of the spade type. *Transactions of the Royal Institution of Naval Architects,* Vol. 63, 1921, pp. 117–130.
5.14 Darnell, R.C. Hydrodynamic characteristics of twelve symmetrical hydrofoils. *United States Model Basin,* Report No. 341.
5.15 Abell, T.B. Some model experiments on rudders placed behind a plane deadwood. *Transactions of the Royal Institution of Naval Architects,* Vol. 78, 1936, pp. 135–144.
5.16 Voepel, H. Tests on wings of small aspect ratio. *Royal Aircraft Establishment.* Translation No. 276, 1948.
5.17 Flax, A.H. and Laurence, H.R. The aerodynamics of low-aspect-ratio wings and wing-body combinations. *Cornell Aeronautical Laboratory Report* No. 37, 1951.
5.18 Bartlett, G.E. and Vidal, R.J. Experimental investigation on influence of edge shape on the aerodynamic characteristics of low aspect ratio wings at low speeds. *Journal of Aeronautical Science,* No. 8, 1955.
5.19 Windsor, R.I. Wind tunnel tests of a DTMB NACA0015 aspect ratio 2, flapped surface. *University of Maryland Report* No. 268, 1959.
5.20 Harrington, R.L. Rudder torque prediction. *Transactions of the Society of Naval Architects and Marine Engineers,* Vol. 89, 1981, pp. 23–90.
5.21 Hembold, H.B. Der unverwundene ellipsenflügel als trägendeFläche. *Jahrbuch der Deutschen Luftfahrt-forschung,* 1942, Sect 1, pp. 111–113.
5.22 Polhamus, E.C. A simple method of estimating the subsonic lift and damping in roll of sweptback wings. *NACA, TN* 1862, 1949.
5.23 Molland, A.F. and Turnock, S.R. Wind tunnel investigation of the influence of propeller loading on ship rudder performance. *Transactions of the Royal Institution of Naval Architects,* Vol. 135, 1993, pp. 105–120.

5.24 Molland, A.F. Rudder design data for small craft. *University of Southampton, Ship Science Report* No. 1/78, 1978.
5.25 Abbot, I.H. and Von Doenhoff, A.E. *Theory of Wing Sections.* Dover Publications, New York, 1958.
5.26 Mandel, P. Some hydrodynamic aspects of appendage design. *Transactions of the Society of Naval Architects and Marine Engineers,* Vol. 61, 1953, pp. 464–515.
5.27 Burcher, R.K. Developments in ship manoeuvrability. *Transactions of The Royal Institution of Naval Architects,* Vol. 114, 1972, pp. 1–32.
5.28 Hoerner, S.F. *Fluid Dynamic Drag.* Published by the author, New York, 1965.
5.29 Thiemann Von H. Windkanaluntersuchen von rechteckrudern mit staukeilen und profilkurzungen (in German). *Schiff und Hafen,* Heft 1, 1962.
5.30 Kwik, K.H. Windkanalunterschungen an einem modell eines schiffruders mit dem profil IfS 51 TR 25. *HANSA – Schiffahrt-Schiffbau Hafen,* 107, Jahrgang, Nr. 9, 1970.
5.31 Kwik, K.H. Sytematische windkanalversuche mit schiffsrudern. *Schiffstechnik* Bd.18, Heft 92, 1971.
5.32 Kwik, K.H. Comparative wind tunnel investigation of a rudder with fixed structure and an all-movable balance rudder. (in German). *Institut für Schiffbau, Hamburg University,* Report 285, March 1972.
5.33 Landgraf, J. Increasing rudder effect by means of the propeller stream (in German). *Hansa* 110 (Special No. STG) November 1973.
5.34 Kracht, A.M. Rudder in the slipstream of a propeller. Proceedings of *International Symposium on Ship Resistance and Powering performance.* Shanghai, China, 1989.
5.35 Brix, J. (ed.). *Manoeuvring Technical Manual,* Seehafen Verlag, Hamburg 1993.
5.36 Sears, R.I. Wind tunnel investigation of control surface characteristics. I – Effect of gap on the aerodynamic characteristics of a NACA 0009 Airfoil with a 30-percent-chord plain flap. Report L-377, NACA, June 1941.
5.37 Molland, A.F. and Turnock, S.R. Wind tunnel investigation of the influence of propeller loading on a semi-balanced skeg rudder. *University of Southampton, Ship Science Report* No. 48, 1993.
5.38 Molland, A.F. and Turnock, S.R. Wind tunnel tests on the influence of zero speed operation. *University of Southampton, Ship Science Report* No. 64, 1993.
5.39 Lumley, H. On the steering of ships. *Transactions of The Royal Institution of Naval Architects,* Vol. 5, 1864, pp. 128–134.
5.40 Ames, M.B. Jr. and Sears, R.I. Pressure-distribution investigation of an NACA 0009 airfoil with a 30-percent chord plain flap and three tabs. T.N. No. 759, *NACA,* 1940.
5.41 Street, W.G. and Ames, M.B. Jr. Pressure-distribution investigation of an NACA 0009 airfoil with a 50-percent chord plain flap and three tabs. T.N. No. 734, *NACA* 1939.
5.42 Ames, M.B. Jr. and Sears, R.I. Pressure-distribution investigation of an NACA 0009 airfoil with a 80-percent chord plain flap and three tabs. T.N. No. 761, *NACA,* 1940.

5.43 Comstock, J.P. (ed.) *Principles of Naval Architecture.* The Society of Naval Architects and Marine Engineers, New York, 1967.
5.44 Jaeger, H.E. Approximate calculation of rudder torque and rudder pressures. *International Shipbuilding Progress,* Vol. 2, No. 10, 1955, pp. 243–257.
5.45 Cowley, W.L., Simmons, L.F.G. and Coales, J.D. The effect of balancing a rudder, by placing the rudder axis behind the leading edge, upon the controlling moment of the motion. *Technical Reports of the Advisory Committee for Aeronautics,* 1916–1917, p. 154.
5.46 Munk, M. Systematische Versuche an Leitwerkmodellen. *Technische Berichte der Flugzuegmeisterei,* 1917, p. 168.
5.47 Rossell, H.E. and Chapman, L.B. (eds.) *Principles of Naval Architecture.* The Society of Naval Architects and Marine Engineers, New York, 1939.
5.48 Bottomley, G.H. Manoeuvring of ships – Semi-balanced rudders of twin screw ships. *Transactions of the North East Coast Institution of Engineers and Shipbuilders,* Vol. 48, 1931/32, pp. 97–114.
5.49 Van Lammeren, W.P.A., Troost, L. and Koning, J.G. *Resistance, Propulsion and steering of Ships.* The Technical Publishing Co., H. Stam–Haarlem, Holland, 1948.
5.50 Baker, G.S. and Bottomley, G.H. Manoeuvring of ships. Part I – Unbalanced rudders of single screw ships. *Transactions of the Institute of Engineers and Shipbuilders in Scotland,* Vol. 65, 1921/22, pp. 522–583.
5.51 Bottomley, G.H. Manoeuvring of ships. Part II – Unbalanced rudders of twin screw ships. *Transactions of the Institute of Engineers and Shipbuilders in Scotland,* Vol. 67, 1923/24, pp. 509–559.
5.52 Bottomley, G.H. Manoeuvring of ships. Part III – Unbalanced rudders behind single screw ships, effect of varying fullness of form. *Transactions of the Institute of Engineers and Shipbuilders in Scotland,* Vol. 70, 1926/27, pp. 463–500.
5.53 Bottomley, G.H. Manoeuvring of ships. Part IV – Unbalanced rudders behind twin screw ships, effect of varying fullness of form. *Transactions of the Institute of Engineers and Shipbuilders in Scotland,* Vol. 74, 1930/31, pp. 94–123.
5.54 Bottomley, G.H. Manoeuvring of single screw ships – The effect of rudder proportions on manoeuvring and propulsive efficiency. *Institution of Civil Engineers,* Selected Engineering Papers, No. 175, 1935, pp. 1–18.
5.55 Cole, A.P. Destroyer turning circles. *Transactions of The Royal Institution of Naval Architects,* Vol. 80, 1938, pp. 32–57.
5.56 Gawn, R.W.L. Steering experiments, Part I. *Transactions of the Royal Institution of Naval Architects.* Vol. 85, 1943, pp. 35–73.
5.57 Lötveit, M. A study of rudder action with special reference to single screw ships. *Transactions of the NECIES,* Vol. 75, 1958/59, pp. 87–124.
5.58 Shiba, H. Model experiments about the manoeuvrability and turning of ships. *First Symposium on Ship Manoeuvrability,* DTMB Report 1461, October 1960.
5.59 Okada, S. On the performance of rudders and their designs. *The Society of Naval Architects of Japan,* 60th Anniversary Series, Vol. 11, 1966, pp. 61–137.
5.60 English, J.W., Rowe, S. and Bain, D. Some manoeuvring devices for use at zero and low ship speed. *Transactions of the NECIES,* Vol. 88, 1972, pp. 31–50.

5.61 Landgraf, J. Increasing rudder effect by means of the propeller stream (in German). *Hansa* 110 (Special No. STG) November 1973.

5.62 Mathis, P.B. and Gregory, D.L. Propeller slipstream performance of four high-speed rudders under cavitating conditions. *US Naval Ship Research and Development Center*, Report 4361, May 1974.

5.63 Kerwin, J.E., Lewis, S.D. and Oppenheim, B.W. Experiments on rudders with small flaps in a free-stream and behind a propeller. Report No. 74-16, *Department of Ocean Engineering, Massachusetts Institute of Technology*, 1974.

5.64 Berlekom, W.B. Van. Effects of propeller loading on rudder efficiency. Royal Netherlands College, Proceedings of *Fourth Ship Control Systems Symposium*, The Hague, October 1975.

5.65 Kracht, A.M. On propeller–rudder interaction. Proceedings of *International Symposium on Propulsors and Cavitation*, Hamburg, 1992.

5.66 Kracht, A.M. Ship-propeller–rudder interaction. Proceedings of *Second International Symposium on Propulsors and Cavitation*, Hangzhou, China, 1992.

5.67 Kracht, A.M. Cavitation on rudders. Proceedings of *International Conference on Propeller Cavitation, PROPCAV '95*, Newcastle upon Tyne, 1995.

5.68 Molland, A.F. and Turnock, S.R. The prediction of ship rudder performance characteristics in the presence of a propeller. Proceedings of *Conference, Manoeuvring and Control of Marine Craft, MCMC '92,* Computational Mechanics Publications, 1992, pp. 475–492.

5.69 Molland, A.F. and Turnock, S.R. Developments in modelling ship rudder–propeller interaction. Proceedings of *Conference CADMO '94,* Southampton, Computational Mechanics Publications, 1994, pp. 255–263.

5.70 Molland, A.F. and Turnock, S.R. Prediction of ship rudder–propeller interaction at low speeds and in four quadrants of operation. Proceedings of *Conference, Manoeuvring and Control of Marine Craft, MCMC '94,* Maritime Research Institute, Southampton Institute, 1994, pp. 319–333.

5.71 Molland, A.F. and Turnock, S.R. Some effects of rudder–propeller–hull arrangements on manoeuvring and propulsion. Proceedings of *International Symposium on Practical Design of Ships and Mobile Units, PRADS '95*, Society of Naval Architects of Korea, Korea, September 1995, pp. 1.333–1.345.

5.72 Molland, A.F., Turnock, S.R. and Smithwick, J.E.T. Design studies of the manoeuvring performance of rudder–propeller systems. Proceedings of *7th International Symposium on Practical Design of Ships and Mobile Units, PRADS '98,* The Hague, The Netherlands, September 1998, pp. 807–816.

5.73 Molland, A.F., Turnock, S.R. and Smithwick, J.E.T. Design and performance characteristics of semi-balanced ship skeg rudders. *Transactions of the Royal Institution of Naval Architects*. Vol. 142, 2000, pp. 230–247.

5.74 Turnock, S.R. *Prediction of ship rudder-propeller interaction using parallel computations and wind tunnel measurements.* Ph.D. Thesis, University of Southampton, 1993.

5.75 Molland, A.F. and Turnock, S.R. Wind tunnel investigation of the influence of propeller loading on ship rudder performance. *University of Southampton, Ship Science Report* No. 46, 1991.

5.76 Molland, A.F. and Turnock, S.R. Preliminary wind tunnel investigation of the influence of propeller loading on a ship rudder in the bollard ($J = 0$) condition. *University of Southampton*, Ship Science *Report* No. 49, 1993.
5.77 Molland, A.F. and Turnock, S.R. Further wind tunnel tests on the influence of propeller loading on ship rudder performance. *University of Southampton, Ship Science Report* No. 52, 1992.
5.78 Molland, A.F. and Turnock, S.R. Wind tunnel tests on the effect of a ship hull on rudder–propeller performance at different angles of drift. *University of Southampton, Ship Science Report* No. 76, 1994.
5.79 Molland, A.F., Turnock, S.R. and Smithwick, J.E.T. Wind tunnel tests on the influence of propeller loading and the effect of a ship hull on skeg–rudder performance. *University of Southampton, Ship Science Report* No. 90, 1995.
5.80 Molland, A.F., Turnock, S.R. and Smithwick, J.E.T. Investigation of the influence of propeller loading on a ship rudder in the bollard ($J = 0$) condition. *University of Southampton, Ship Science Report* No. 94, 1996.
5.81 Molland, A.F. The design, construction and calibration of a five-component strain gauge wind tunnel dynamometer. *University of Southampton, Ship Science Report* No. 1/77, 1976.
5.82 Turnock, S.R. A test rig for the investigation of ship propeller/rudder interactions. *University of Southampton, Ship Science Report* No. 45, 1990.
5.83 Van Lammeren, W.P.A., Van Manen, J.D. and Oosterveld, M.W.C. The Wageningen B-Screw Series. *Transactions of the Society of Naval Architects and Marine Engineers*, Vol. 77, 1969.
5.84 Molland, A.F. and Turnock, S.R. Wind tunnel test results for a model ship propeller based on a modified Wageningen B4.40. *University of Southampton, Ship Science Report* No. 43, 1990.
5.85 Stierman, E.J. The influence of the rudder on the propulsive performance of ships. Parts I and II, *International Shipbuilding Progress*, Vol. 36, No. 407, pp. 303–334 and 408 pp. 405–435, 1989.
5.86 Nakatake, K., Ando, J., Tamashima, M. and Tashiro, K. Interaction among twin screw propellers and twin rudders in uniform flow. Proceedings of *2nd International Symposium on Propellers and cavitation*, Hangzhou, China, 1992.
5.87 Kracht, A.M. Ruder im Schraubenstrahl (Rudder in the slipstream). VWS Report No. 1178/90, *Versuchsanstalt für Wasserbau und Schiffbau*, Berlin. 1990. (in German).
5.88 Kracht, A.M. Ship–propeller–rudder interaction. Proceedings of *Second International Symposium on Propulsors and Cavitation*, Hangzhou, China, 1992.
5.89 Molland, A.F. Pressure–distribution investigation of a semi-balanced ship skeg-rudder. *University of Southampton, Ship Science* Report No. 5/81, 1980.
5.90 Russo, V.L. and Sullivan, E.K. Design of the mariner-type ship. *Transactions of The Society of Naval Architects and Marine Engineers*, Vol. 61, 1953, pp. 98–151.
5.91 Sedat, R. and Fuller, N. U.S. coast guard modular maneuvering tests. Proceedings of the *24th American Towing Tank Conference,* Offshore Technology Research Centre, Texas A&M University, USA, November 1995.

5.92 Carlton, J.S. *Marine Propellers and Propulsion.* Butterworth Heinemann, Oxford, 1994.

5.93 English, J.W. Rudder performance at static thrust. *The Naval Architect,* The Royal Institution of Naval Architects, London, January, 2000, pp. 31–32.

5.94 Molland, A.F., Turnock, S.R. and Wilson, P.A. Performance of an enhanced rudder force prediction model in a ship manoeuvring simulator. Proceedings of *International Conference on Marine Simulation and Ship Manoeuvrability, MARSIM '96,* Copenhagen, Denmark, 1996, pp. 425–434.

5.95 Chislett, M.S. A generalised math model for manoeuvring. Proceedings. of *International Conference on Marine Simulation and Ship Manoeuvrability, MARSIM '96,* Copenhagen, Denmark, 1996, pp. 593–606.

5.96 Molland, A.F. and Turnock, S.R. Flow straightening effects on a ship rudder due to upstream propeller and hull. *International Shipbuilding Progress, 49,* No. 3, 2002, pp. 195–214.

5.97 Ogawa, A. and Kasai, H. On the mathematical model of the maneuvering of ships. *International Shipbuilding Progress,* Vol. 25, December 1978, pp. 306–319.

5.98 Hirano, M. A practical calculation method of ship maneuvering motion at initial design stage. *Journal of The Society of Naval Architects, Japan,* Vol. 147, June 1980, pp. 68–80.

5.99 Nakatake, K., Ando, J., Tamashima, M. and Tashiro, K. Interaction among twin screw propellers and twin rudders in uniform flow. Proceedings of *Second International Symposium on Propellers and cavitation,* Hangzhou, China, 1992.

5.100 Salo, M. and Heikkilä, M. On the modelling of hull–propeller–rudder interactions in manoeuvring of twin screw ships. Proceedings of *MARSIM and ICSM,* Tokyo, Japan, 1990.

5.101 Janke-Zhou, Y. Manoeuvring motion simulation of twin-screw ships. *Ship Technology Research,* Vol. 41, 1994, pp. 3–16.

5.102 Molland, A.F. The free stream characteristics of a semi-balanced ship skeg rudder. *University of Southampton, Ship Science Report* No. 3/77, 1977.

5.103 Molland, A.F. Free stream tank test data for a spade rudder. 1977 (Report unpublished).

5.104 Millward, A. The design of spade rudders for yachts. *Southampton University Yacht Research Report,* SUYR No. 28, 1969.

5.105 Harper, J.J. and Simitses, G.J. Effect of a simulated submarine hull on the characteristics of all-movable control surfaces. *Georgia Institute of Technology,* Report 439, August 1959.

5.106 Taplin, A. Notes on rudder design practice. *First Symposium on Ship Manoeuvrability,* DTMB. Report No. 1461, October, 1960.

5.107 Myers, P.A. Theory of sailing applied to ocean racing yachts. *Marine Technology, Society of the Naval Architects and Marine Engineers,* Vol. 12, No. 3, 1975.

5.108 Claughton, A., Wellicome, J.F. and Shenoi, R.A. (eds.) *Sailing Yacht Design: Theory.* Published by the University of Southampton, Southampton 2006.

5.109 DNV rules for classification of ships, Part 3, Ch. 3, *Hull Equipment and Safety,* Det Norske Veritas, Oslo, January 2006.

5.110 Clarke, D., Gedling, P. and Hine, G. The application of manoeuvring criteria in hull design using linear theory. *Transactions of the Royal Institution of Naval Architects*, Vol. 125, 1983, pp. 45–68.
5.111 Barnaby, K. *Basic Naval Architecture.* Huchinson, London, 1963.
5.112 Du Cane, P. *High Speed Craft.* Temple Press Ltd., London, 1951.
5.113 Performance Prediction of Small Craft. *Royal Institution of Naval Architects, Occasional Publication* No. 1, 1974.
5.114 Millward, A. *The induced drag of a vertical hydrofoil.* Ph.D. Thesis, University of Southampton, 1967.
5.115 Ismail, K.A.R. *The hydrodynamic theory of vertical swept hydrofoils near a free surface and comparison with measurement.* Ph.D. Thesis, University of Southampton, 1973.
5.116 Swales, P.D., Cole, B.N. and Smith, G.L. Some aspects of the ventilation of surface piercing struts and foils. *Transactions of The Royal Institution of Naval Architects*, Vol. 113, 1971, pp. 365–374.
5.117 Swales, P.D., McGregor, R.C. and Wright, A.J. Correlation of force measurements and separated flow regions on surface piercing struts. Engineering Notes, *Journal of Hydronautics*, Vol. 8, No. 2, 1974. pp 72–73.
5.118 Gregory, D.L. and Dobay, G.F. The performance of high-speed rudders in a cavitating environment. Paper Q, *Spring Meeting of Society of Naval Architects and Marine Engineers*, 1973.
5.119 McGeough, F.G. and Millward, A. The effect of cavitation on the rotating cylinder rudder. *International Shipbuilding Progress*, Vol. 28, No. 317, 1981, pp. 2–9.
5.120 Shen, Y.T., Remmers, K.D. and Jiang, C.W. Effects of ship hull and propeller on rudder cavitation. *Journal of Ship Research*, Vol. 41, No. 3, September 1997, pp. 172–180.
5.121 Friesch, J. Erosion problems on fast high powered ships. Proceedings of *Seventh International Conference on Fast Sea Transportation, FAST'2003*, Ischia, Italy, October, 2003.
5.122 Mevis, F. and Klug, H. The challenge of very large container ships. Proceedings of *Ninth International Symposium on Practical Design of Ships and other Floating Structures, PRADS'2004*, Lübeck-Travemünde, September 2004.
5.123 ITTC 2005. Report of the specialist committee on cavitation erosion on propellers and appendages on high powered/high speed ships. Proceedings of *24th ITTC*, Vol. II, Edinburgh, 2005.
5.124 Eppler, R. and Shen, Y.T. Wing sections for hydrofoils- Part I: Symmetrical profiles. *Journal of Ship Research*, Vol. 23, No. 3, Sept. 1979, pp. 209–217.
5.125 Drela, M. Xfoil: an analysis and design system for low Reynolds number aerofoils. *Conference on Low Reynolds Number Airfoil Aerodynamics*, University of Notre Dame, Indiana, 1989.
5.126 Molland, A.F., Bahaj, A.S., Chaplin, J.R. and Batten, W.M.J. Measurements and predictions of forces, pressures and cavitation on 2-D sections suitable for marine current turbines. Proceedings of *Institute of Mechanical Engineers*, Vol. 218, Part M, 2004, pp. 127–138.

5.127 Brix, J., Nolte, A. and Heinzel, S. Konstruktive massnahmen zur verhinderung der kavitation an der rudersohle. *HANSA*, Nr. 16, 1971.

5.128 Boo, K-T., Song, I-H. and Shin, S-C. Numerical simulation for the rudder in order to control the cavitation phenomena. *Journal of Ship and Ocean Technology*, Vol. 8, No. 1, 2004, pp. 42–50.

5.129 Han, J-M., Kong, D-S., Song, I-H., and Lee, C-S. Analysis of the cavitating flow around the horn-type rudder in the race of a propeller. 4th International Conference on Cavitation CAV2001, Pasadena, 2001.

5.130 Brix, J. and Baumgarten, B. Development of a rudder having low cavitation and vibration, low drag and a Quasisymmetrical rudder torque. *Forschungszentrum des Deutsch Schiffbaus*, Report No. 152/1984. (in German).

5.131 Shen, Y.T., Jiang, C.W. and Remmers, K.D. A twisted rudder for reduced cavitation. *Journal of Ship Research*, Vol. 41, No. 4, December 1997, pp. 260–272.

5.132 Twisted spade rudders for large fast vessels. *The Naval Architect*, Royal Institution of Naval Architects, London, September, 2004, pp. 49–50.

5.133 Twisted composite rudders ordered. *Warship Technology*, Royal Institution of Naval Architects, London, October 2005, pp. 3–4.

5.134 Jurgens, A.J. Static and dynamic effects of rudder-hull-propeller interaction on fast monohulls. Proceedings of *Eighth International Conference on Fast Sea Transportation, FAST '2005*, St. Petersburg, Russia, June 2005.

5.135 Suhrbier, K.R. An experimental investigation on the propulsive effect of a rudder in the propeller slipstream. *International Shipbuilding Progress*, Vol. 21, No. 234, 1974, pp. 31–39.

5.136 English, J.W. A note on rudder force at zero helm. Proceedings of *20th ITTC*, Vol. 2, 1993, pp. 85–89.

6 Theoretical and numerical methods

6.1 Available methods

Since computational machines first became available to Naval Architects in the late 1950s progressively more complex theoretical methods have been developed to analyse the performance of rudders and control surfaces. The more elementary theoretical approaches such as Glauert's [6.1] lifting-line theory were the first to be used. It comes as a surprise, especially to the modern student brought up in a world of virtual reality and commercial computational fluid dynamics (CFD) packages, how accurate such an approach can be. In essence this is because a large ship rudder is operating at a relatively high Reynolds number. Neglecting viscous effects will only cause small errors in sideforce and steering torque. Likewise, for large ships, rudder drag is typically 2–3% of total resistance so, again, accurate prediction of rudder drag is less important. However, for high-performance systems, as say typified by the appendages on America's cup yachts, where margins of victory can be a few seconds over many hours of racing, such details do become essential and the computational power of CFD becomes essential.

CFD is the use of computational techniques to solve numerically the equations defining fluid flow around, within and between bodies. Principally, the equations solved are numerical approximations to mathematical models describing the physics of fluid flow. There is, therefore, always an inherent level of approximation to reality. It is the level of abstraction of the CFD analysis from physical reality which determines the amount and form of interpretation of analysis required.

Historically, the origins of our ability to mathematically describe the detailed flow around moving objects such as ships came through the work of such luminaries as Newton, Euler, Laplace, Navier and Stokes. For those interested, a number of detailed fluid texts can be recommended [6.2–6.5] that more fully describe the detailed theoretical background to CFD.

A hierarchy of four CFD methods are considered in order of increasing complexity:

- lifting-line methods,
- surface panel or boundary element methods,
- Reynolds-averaged Navier–Stokes (RANS) methods and
- Large Eddy Simulations (LES) and Direct Numerical Simulation (DNS) methods.

However, of the four methods listed above, only the first three are as yet commonly applied to rudder and control surface design. All four methods vary in complexity, and have particular advantages and disadvantages associated with their physical realism and computational cost. A discussion of these four numerical approaches will be given, in the context of finding the best method or combination of methods, for evaluating ship-rudder performance. The most challenging aspect of numerical analysis is to capture the interaction between a propeller and rudder as well as the secondary effects of hull and free surface. As a starting point, the hierarchy of these methods as related to the governing fluid dynamics equations is considered.

6.1.1 The equations

Three major levels of abstraction of the full unsteady Navier–Stokes (N–S) equations used to describe fluid flow around control surfaces are:

(1) Irrotational flow (vorticity $\zeta = 0$): potential theory based on solution of Laplace's equation for potential with the application of appropriate boundary conditions.
(2) Thin boundary-layer theory in which the full Navier-Stokes equations are simplified by systematically following the consequences of the viscous layer thickness being small compared to the streamwise length of the boundary layer.
(3) Direct solutions of the Navier-Stokes equations classified as (i) an assumption that the influence of turbulent behaviour can be captured through Reynolds' averaging; (ii) large-scale eddies are allowed but below a threshold size their effects are modeled at a subgrid scale, LES and (iii) all motions in the fluid are represented, direct numerical simulation (DNS).

(1) and (2) may be combined, using an outer domain irrotational flow past a boundary whose shape is modified to allow for the boundary layer displacement thickness effects, and by calculating the inner domain viscous boundary layer growth and hence displacement thickness in the pressure field created by the irrotational outer flow, Figures 3.8–3.11.

The most complete description of flow is given by the full unsteady Navier-Stokes equations, which are a statement of the conservation of momentum. Associated statements of conservation of mass and conservation of energy complete the description alongside the properties of the fluid (liquid or gas). For the majority of flows of relevance to marine control surface design the flow is treated as incompressible. This gives the statement of incompressible mass conservation as

$$\nabla \cdot q = \frac{\partial u}{\partial x} + \frac{\partial v}{\partial y} + \frac{\partial w}{\partial z} = 0 \tag{6.1}$$

where $q = (u, v, w)^T$ and u, v and w are the corresponding velocity components expressed in a Cartesian (orthonormal) coordinate system $\{x, y, z\}$.

The full Navier-Stokes equations for the conservation of momentum are stated for an incompressible Newtonian fluid, where stress and strain are linearly related by a constant kinematic viscosity v only dependent on the fluid's properties, and in the absence of external forces are given by

$$\frac{Dq}{Dt} = \frac{-1}{\rho} \nabla p + v \nabla^2 q \tag{6.2}$$

Although straightforward to state in compact vector notation for an incompressible fluid, the four primitive variables, pressure p and the three components of velocity u, v, w, are closely coupled and in only a few special cases do closed analytical solutions exist. When expanded into the three components the coupling becomes more obvious:

$$\begin{aligned}
\frac{\partial u}{\partial t} + u\frac{\partial u}{\partial x} + v\frac{\partial u}{\partial y} + w\frac{\partial u}{\partial z} &= \frac{-1}{\rho}\frac{\partial p}{\partial x} + \nu\left(\frac{\partial^2 u}{\partial x^2} + \frac{\partial^2 u}{\partial y^2} + \frac{\partial^2 u}{\partial z^2}\right) \\
\frac{\partial v}{\partial t} + u\frac{\partial v}{\partial x} + v\frac{\partial v}{\partial y} + w\frac{\partial v}{\partial z} &= \frac{-1}{\rho}\frac{\partial p}{\partial y} + \nu\left(\frac{\partial^2 v}{\partial x^2} + \frac{\partial^2 v}{\partial y^2} + \frac{\partial^2 v}{\partial z^2}\right) \\
\frac{\partial w}{\partial t} + u\frac{\partial w}{\partial x} + v\frac{\partial w}{\partial y} + w\frac{\partial w}{\partial z} &= \frac{-1}{\rho}\frac{\partial p}{\partial z} + \nu\left(\frac{\partial^2 w}{\partial x^2} + \frac{\partial^2 w}{\partial y^2} + \frac{\partial^2 w}{\partial z^2}\right)
\end{aligned} \tag{6.3}$$

The above equations can be simplified by considering the curl of the velocity, termed vorticity. The vorticity ζ is a vector quantity and is given by

$$\zeta = \nabla \times q = \begin{Bmatrix} \dfrac{\partial w}{\partial y} - \dfrac{\partial v}{\partial z} \\ \dfrac{\partial u}{\partial z} - \dfrac{\partial w}{\partial x} \\ \dfrac{\partial v}{\partial x} - \dfrac{\partial u}{\partial y} \end{Bmatrix} \tag{6.4}$$

and can be considered as a measure of how much a local fluid element rotates. The vorticity vector has a magnitude twice that of the rate of rotation of the fluid element. A fluid domain in which no element rotates is defined as irrotational and all the right-hand side of (6.4) will be zero. Thus,

$$\begin{aligned}
\frac{\partial w}{\partial y} - \frac{\partial v}{\partial z} &= 0 \\
\frac{\partial u}{\partial z} - \frac{\partial w}{\partial x} &= 0 \\
\frac{\partial v}{\partial x} - \frac{\partial u}{\partial y} &= 0
\end{aligned} \tag{6.5}$$

These equations are satisfied if the velocity field can be specified as the gradient of a scalar potential ϕ:

$$q = \begin{pmatrix} u \\ v \\ w \end{pmatrix} = \nabla\phi = \begin{Bmatrix} \dfrac{\partial \phi}{\partial x} \\ \dfrac{\partial \phi}{\partial y} \\ \dfrac{\partial \phi}{\partial z} \end{Bmatrix} \tag{6.6}$$

Substituting this into (6.1) gives:

$$\nabla \cdot q = \frac{\partial u}{\partial x} + \frac{\partial v}{\partial y} + \frac{\partial w}{\partial z} = \frac{\partial^2 \phi}{\partial x^2} + \frac{\partial^2 \phi}{\partial y^2} + \frac{\partial^2 \phi}{\partial z^2} = 0 \tag{6.7}$$

which is Laplace's equation:

$$\nabla^2 \phi = 0 \tag{6.8}$$

When given appropriate boundary conditions (ϕ or $\partial\phi/\partial n$) over all the surfaces that bound a given fluid flow domain, there is a unique solution to the Laplace equation to define the potential ϕ for the flow. Laplace's equation allows the linear addition of valid solutions to (6.8), e.g., if ϕ_1 and ϕ_2 are both solutions then $\nabla^2(\phi_1+\phi_2) = 0$ is as well. This allows complex flows to be built up by summation of many basic flow elements such as sources, dipoles or vortices.

For the incompressible flow around a rudder outside the thin boundary layer or a region of separated flow, the flow is irrotational and inviscid and Equation (6.8) applies. Within the thin boundary layer Equation (6.3) applies and the zero-slip condition on the rudder surface causes a shear flow and vorticity is convected downstream into the wake. The following two sections, 6.2 and 6.3, describe methods of solving Equations (6.8) and (6.3), respectively. As all the methods are, in effect, a numerical representation of the mathematical equations, it is essential that the inherent limitations associated are well understood. The validation process is described in Section 6.4. Sections 6.5 and 6.6, demonstrate the typical performance of the methods when applied to the free-stream rudder and rudder behind a propeller. The final section considers the challenges associated with computing fully unsteady flow over a rudder in a propeller race.

6.2 Potential flow methods

The methods described are based on Equation (6.8) and the properties of Laplace's equation. A much fuller description of all these methods is given in reference [6.5].

6.2.1 Lifting-line formulation

Glauert [6.1] used a system of bound vorticity to represent the lift (or sideforce) generated by a wing or control surface. Lifting-line theory is a simple method of quickly establishing the performance of a control surface. The three-dimensional (3-D) lifting surface and wake are modeled as a series of horseshoe vortices, known as lifting lines, whose strength is initially unknown.

This method has the advantage that it provides good estimates of spanwise loading, and induced drag, whilst remaining simple to implement and computationally inexpensive. However, it has a number of specific limitations:

- The theory is limited to lifting surfaces of relatively high aspect ratio, and assumes that the wake is aligned in the local flow direction.
- It assumes that the lifting surface is of zero thickness, and therefore neglects any flow effects resulting from section camber and thickness.

- As it is based on potential flow theory, this method neglects viscous flow effects. Hence, it is unable to model directly frictional drag, flow separation and stall.

The following describes how the standard method can be enhanced when considering the typically low aspect ratio of ship rudders [6.6]. A control surface is considered in close proximity to an effectively flat hull so that it can be assumed to have a total span of $2S$. A linear taper ratio TR in root and tip chord is assumed with $c_T/c_R = TR$. The x coordinate is aligned with the hull centreline and y the spanwise ordinate which is replaced by the angle $y = -S\cos\theta$. At any point on the rudder the chord $c = c_R(1 - k\cos\theta)$, where $k = (1 - TR)$. The incidence of the rudder, considered as a lifting line will be

$$\bar{\alpha} = \delta + \beta \pm \lambda \tag{6.9}$$

where δ is the rudder incidence relative to the hull centreline, β the drift angle at the rudder and λ the mean fluid flow inflow angle induced by an upstream propeller.

The local circulation, Γ_L at a point θ on the rudder expressed as a Fourier series is

$$\Gamma_L = 4SV\sum A_n \sin n\theta \tag{6.10}$$

where V is the free-stream velocity. Since, the platform is symmetrical about the midpoint ($\pi/2$), only odd values of n occur in the series. The locally induced local velocity ω_L at a point θ is given by

$$\omega_L = \left(\sum nA_n \sin n\theta\right)V/\sin\theta \tag{6.11}$$

The section experiences a lift force corresponding to 2-D motion at the effective angle of incidence ($\bar{\alpha} - \omega_L/V$) where ω_L/V is the induced downwash angle. The local lift coefficient:

$$C_{LL} = m(\bar{\alpha} - \omega_L/V) \tag{6.12}$$

where m is the 2-D lift curve slope, allowing for thickness and viscosity effects. Hence at any point,

$$\begin{aligned}\Gamma_L &= \frac{L_1}{\rho V} \\ &= \frac{C_{LL}cV}{2} \\ &= \frac{m}{2}cV(\bar{\alpha} - w_L/V)\end{aligned} \tag{6.13}$$

Incorporating the value of induced velocity from Equation (6.11)

$$\Gamma_L = \frac{m}{2}cV\left(\bar{\alpha} - \frac{\left(\sum nA_n \sin n\theta\right)}{\sin\theta}\right) \tag{6.14}$$

Equating (6.10) and (6.14) and setting $\mu = mc_R/8S$ leads to

$$\sum \left[A_n \sin n\theta \left(\mu n + \frac{\sin\theta}{1 - k\cos\theta} \right) \right] = \mu \bar{\alpha} \sin\theta \tag{6.15}$$

The fundamental equation must be satisfied at all points on the rudder to give the local lift coefficient as

$$\begin{aligned} C_{LL} &= L_1 / \tfrac{1}{2} \rho c V^2 \\ &= \frac{8S}{c} \sum A_n \sin n\theta \\ &= \frac{8S}{c_R(1 - k\cos\theta)} \sum A_n \sin n\theta \end{aligned} \tag{6.16}$$

Equation (6.15) must be satisfied at all points along the rudder span. Typically, at least 20 points is found to give satisfactory results. The theoretical total-lift coefficient C_{Lt} is then found as a numerical integration across all the spanwise stations.

Tip trailing vortex corrections: It is assumed [6.7] that the influence of the tip vortex is responsible for the nonlinear component of lift normally exhibited by low aspect ratio lifting surfaces operating in a free stream. The empirical correction equation proposed for the additional downwash due to the tip vortex is

$$\alpha_T(\theta) = 0.645 H_1(\theta) \alpha^2 TR^{1.5} / AR_E \tag{6.17}$$

where $H_1(\theta)$ is the general form of the variation in of downwash across the span at a particular incidence and its distribution is shown in Figure 6.1.

Correction to theoretical lift: The lifting-line analysis when applied to low aspect ratio foils overpredicts lift. Again, as proposed in reference [6.7] and adopted here, the corrected total lift coefficient is modified based on the aspect ratio as follows:

$$C_L = C_{Lt} 1.052 TR^{0.1} \times 0.875 \frac{(1.14AR_E + 2)}{(AR_E + 3.9)} \tag{6.18}$$

6.2.2 Boundary element (surface panel) methods

In the early 1960s, as a consequence of increased computing power, a new numerical approach to the lifting surface problem, known as the panel method or boundary element method began to emerge. This method promised to overcome many of the problems of the early lifting line and the later vortex lattice approach. The technique allowed the treatment of more complex geometries, and actually models the lifting surface itself, allowing the effects of thickness and camber to be calculated. The basic principle of the panel method is based on the linear superposition of source/sinks, vortices and/or doublet elements over the lifting

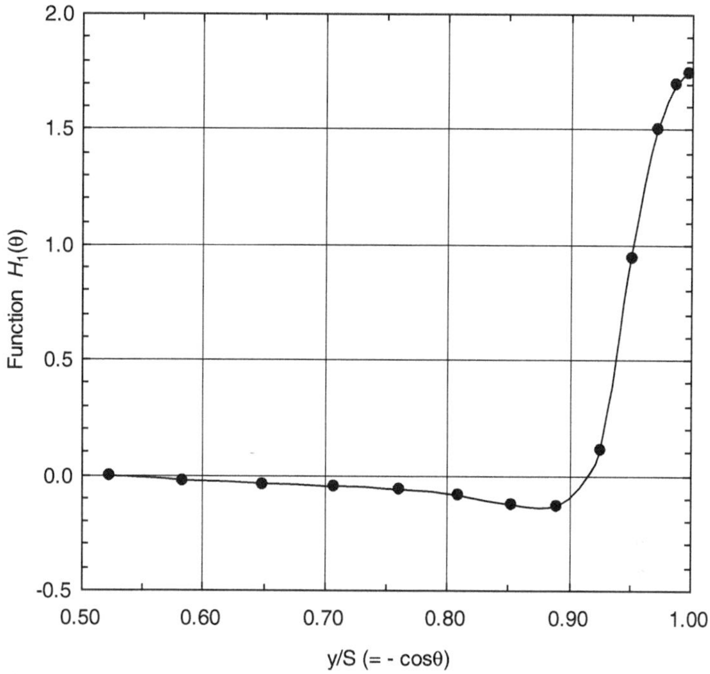

Figure 6.1 General form of function $H_1(\theta)$

surface, such that the boundary conditions are satisfied on the body, across the wake and in the far field.

Extensive research has been carried out in this field. A detailed overview of this method is provided by one of the pioneers of panel methods [6.8].

The advantage of this approach is that it can be used to model actual geometries without requiring further simplification of the geometry. Although panel methods are more complex than lifting-line methods, the computational effort required is still less than that needed for RANS methods. A panel code used to solve a 3-D rudder flow required only 1% of the computational effort needed by a RANS code to solve the same flow problem [6.9]. The advantage of the panel method is that computations are carried out to determine unknowns only on the body, wake, and far field boundary surfaces, and not throughout whole fluid domain.

Panel methods allow considerable freedom in their numerical application so that complex flow features (rotational and viscous) such as wake roll up, separation zones and unsteadiness can be incorporated. However, as a potential flow method they, like the lifting-line method, cannot account directly for frictional drag, separation and stall effects.

Lamb [6.10] showed that a quantity satisfying Laplace's equation (6.8) can be written as an integral over the bounding surface S of a source distribution per unit area σ and a normal dipole distribution per unit area μ distributed over S. If v represents the disturbance velocity field due to the bounding surface (or body) and is defined as the difference between the local velocity at a point and that due to the free-stream velocity, then the disturbance potential can be expressed as a surface integral:

$$\phi = \iint_{S_B} \left[\frac{1}{r}\sigma + \frac{\partial}{\partial n}\left(\frac{1}{r}\right)\mu \right] dS + \iint_{S_W} \frac{\partial}{\partial n}\left(\frac{1}{r}\right)\mu dS \tag{6.19}$$

where S_B is the surface of the body and S_W a trailing wake sheet. In the expression r is the distance from the point for which the potential is being determined to the integration point on the surface, and $\partial/\partial n$ is a partial derivative in the outward normal direction to the local surface. A dipole distribution is used to represent the wake sheet.

The conditions imposed on the disturbance potential are that the [6.8]:

(1) velocity potential satisfies Laplace's equation everywhere outside of the body and wake;
(2) disturbance potential due to the body vanishes at infinity;
(3) normal component of velocity is zero on the body surface;
(4) Kutta–Joukowsky condition of a finite velocity at the body trailing edge is satisfied and
(5) trailing wake sheet is a stream surface with equal pressure either side.

For a steady-state solution the wake dipole strength distribution is uniquely determined by the application of the Kutta condition at the body trailing edge. As conditions (1) and (2) are satisfied as functions of μ and σ, conditions (3) and (4) are used to determine μ and σ on the body. The Kutta condition only applies at the trailing edge and some other relationship has to be used to uniquely determine the distribution of μ and σ over the body. The numerical resolution of this nonuniqueness is referred to as the singularity mix of the lifting-surface method.

The arbitrary selection of the appropriate choice explains the initial diversity of approaches to panel code formulations. The example approach discussed is one of those most commonly adopted for low-order panel codes and was applied by Turnock [6.11] in the surface panel code, Palisupan, used in Sections 6.5 and 6.6.

Lee [6.12] carried out a 2-D investigation into four possible schemes for the solution of Lamb's equation. For lifting surfaces, which have both thin and thick sections (e.g., propeller blades), the perturbation potential method taken from the work by Morino and Kuo [6.13] was found to be the most suitable. The principal advantages of this method are that, because panel potential (scalar) rather than velocity (vector) influence coefficients are calculated, only a third of the memory requirement for the method is needed. Crucially, the perturbation potential influence coefficient is an order less singular.

Morino's numerical procedure is based on representing the body surface by a series of N quadrilateral panels each with an unknown but constant dipole strength per unit area. The vertices of these panels are located on the actual surface of the body. The wake sheet is represented by M panels placed on the stream-surface from the trailing edge of the body surface. Its dipole strength per unit area is related to the difference in dipole potential at the trailing edge. In Morino's work the wake strength μ_W was equated to the difference in potential between the upper and lower surface at the trailing edge: $\mu_W = \phi_u - \phi_l$.

On the body surface the source strength per unit area is prescribed by satisfying the condition for zero normal velocity at the panel centroid:

$$\sigma = \bar{U} \cdot \bar{n} \tag{6.20}$$

where n is the unit normal outward from the panel surface and U the specified inflow velocity at the panel centroid.

The numerical discretisation of Equation (6.19) gives the potential at the centroid of quadrilateral panel i as

$$\begin{aligned} \phi_i = & \frac{1}{2\pi} \sum_{j=1}^{N} ((U_\infty \cdot n_j)S_{ij} - \phi_j D_{ij}) \\ & + \sum_{k=1}^{M} \Delta\phi_k W_{ik} \end{aligned} \tag{6.21}$$

where the following influence coefficients at the centre of panel i are:

S_{ij} is that due to a unit strength source distribution at panel j;
D_{ij} is likewise the dipole influence coefficient due to panel j and
W_{ik} the influence of the unit strength wake strip k extending to infinity.

As there are N independent equations corresponding to the N body surface panel centroids, Equation (6.21) is closed and can be evaluated. Expressed in matrix form it becomes:

$$[D_{ij}]\phi + [W_{ik}]\Delta\phi = [S_{ij}](U_\infty \cdot n) \tag{6.22}$$

For Morino's original trailing edge Kutta condition, which directly relates $\Delta\phi$ to the difference in trailing edge panel potential, the matrix expression (6.22) can then be directly solved to give the vector of dipole potential ϕ. Numerical differentiation of dipole potential along the body surface allows the surface velocity and hence pressures on the surface to be evaluated.

6.2.2.1 *Evaluation of influence coefficients*

At the heart of a lifting surface panel method is the efficient calculation of the potential (or velocity) influence coefficients at a field point due to a particular panel's source or dipole distribution. Newman derived expressions [6.14] for calculating the analytical influence coefficients of a constant or linear strength distribution of sources and normal dipoles over a quadrilateral panel. The approach of Newman was different from that used originally by Hess and Smith [6.15] although the form of the exact source influence coefficient is algebraically similar. As the distance of the field point from the panel centroid increases, approximate expressions can be used for the influence coefficients to reduce computational effort. These ensure that at greater distances from the panel the accuracy of the source and dipole influence coefficient is maintained while at the same time the computational time is reduced. The computational time savings associated with the use of these expressions can be significant as the total number of panels increase, with only a progressively smaller fraction of the panels requiring the use of exact influence coefficients.

6.2.2.2 *Kutta condition*

For a steady-state solution, the dipole strength of the trailing wake sheet has a constant strength in the streamwise direction. This strength is directly related to the circulation around the lifting surface. The original Kutta condition, implemented by Morino, involved setting the trailing wake-sheet dipole strength equal to that of the difference in perturbation potential at the trailing edge. This implies that the pressure difference at the trailing edge would be close to zero. Lee showed that a source term should also be included to ensure that there was zero difference in *total* potential caused by the difference in source strength of the two trailing edge panels. With this additional term, when significant cross-flow occurs at the trailing edge, the upper and lower panels will not necessarily be at the same pressure and a nonphysical trailing edge pressure loading occurs. This was seen by Lee as the need to explicitly equate the upper- and lower-panel pressure using an Newton–Raphson iterative scheme to correct the dipole wake strength based on a factor K multiplying the pressure loading at the trailing edge from the previous iteration. The zeroth order ($k = 0$) approximation for the wake strength is taken to be the original Morino Kutta condition so as ϕ_u and ϕ_l are unknown then the numerical (6.22) is arranged with the unknowns on the left-hand side:

$$[D_{ij} + W_{ik}]\phi = [S_{ij}](U_\infty \cdot n) \tag{6.23}$$

Once the solution vector ϕ is obtained this is used to calculate ΔCp at the trailing edge and a correction to the wake strength is found. This correction vector of known strength is multiplied by the wake strip influence coefficient matrix W_{ik} and applied to the right-hand side of the equation. This modifies Equation (6.23) to

$$\left[D_{ij} + W_{ik}\right]\phi = \left[S_{ij}\right]U_\infty \cdot n_j - \left[W_{ik}\right]\left(\frac{\mathrm{d}\Delta\phi}{\mathrm{d}\Delta p}\Delta p\right)' \tag{6.24}$$

The process is repeated until the pressure loading at the trailing edge has been removed to any significant degree.

6.2.2.3 *Surface pressure*

The numerical solution of Morino's method gives a result vector, which specifies a dipole strength at the centre of each panel. This corresponds to the potential ϕ on the surface of the body. In order to obtain practical engineering information from this surface potential distribution a numerical differentiation is required. The differentiation gives the disturbance velocity tangential to the panel surface. The total velocity at the panel centroid U_t is the vector sum of the tangential disturbance velocity U_d, and the normal component of the relative body surface velocity U_i.

$$U_t = U_d + (U_i - (U_i \cdot n)\,n) \tag{6.25}$$

where n is a unit vector normal to the panel surface.

A second-order finite difference method is used along two parametric directions, s and t as shown in Figure 6.2. The velocities in the t and s directions are then obtained using

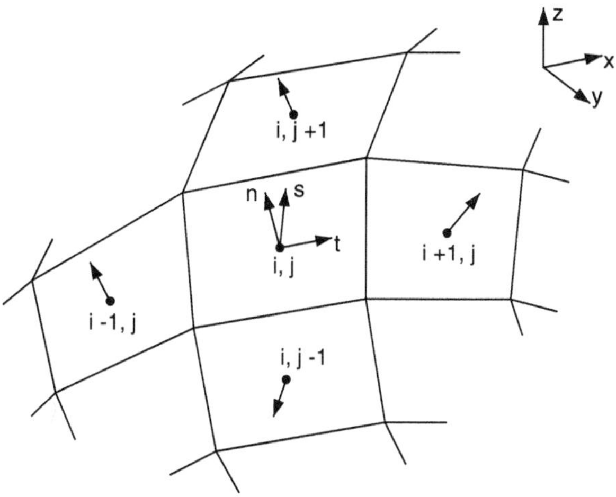

Figure 6.2 Detail of surface panel vectors for evaluation of surface velocity

$$u_t = \frac{\left(t_{i,j} - t_{i-1,j}\right)\left(\phi_{i+1,j} - \phi_{i,j}\right)}{\left(t_{i+1,j} - t_{i-1,j}\right)\left(t_{i+1,j} - t_{i,j}\right)} - \frac{\left(t_{i+1,j} - t_{i,j}\right)\left(\phi_{i,j} - \phi_{i-1,j}\right)}{\left(t_{i+1,j} - t_{i-1,j}\right)\left(t_{i,j} - t_{i-1,j}\right)} \quad (6.26)$$

and

$$v_s = \frac{\left(s_{i,j} - s_{i,j-1}\right)\left(\phi_{i,j+1} - \phi_{i,j}\right)}{\left(s_{i,j+1} - s_{i,j-1}\right)\left(s_{i,j+1} - s_{i,j}\right)} - \frac{\left(s_{i,j+1} - s_{i,j}\right)\left(\phi_{i,j} - \phi_{i,j-1}\right)}{\left(s_{i,j+1} - s_{i,j-1}\right)\left(s_{i,j} - s_{i,j-1}\right)} \quad (6.27)$$

Having determined the surface velocity in the parametric coordinate system a transformation has to be carried out to give the surface velocity components in the overall coordinate system. Unit vector $\boldsymbol{s}$ and $\boldsymbol{t}$ are not necessarily orthogonal and therefore the velocities are first transformed into an orthogonal system with one direction normal to the panel. The u, v and w components can then be found. The combined expression becomes:

$$U_{\mathrm{d}} = \frac{\mathrm{d}\phi/\mathrm{d}t(\boldsymbol{t} - (\boldsymbol{s}\cdot\boldsymbol{t})\boldsymbol{s}) + \mathrm{d}\phi/\mathrm{d}s(\boldsymbol{s} - (\boldsymbol{s}\cdot\boldsymbol{t})\boldsymbol{t})}{||\boldsymbol{s}\ x\ \boldsymbol{t}||^2} \quad (6.28)$$

Knowing the disturbance velocity U_{d} and hence total velocity U_{t} allows the local nondimensional pressure coefficient C_{p} to be found.

$$C_{\mathrm{p}} = 1 - \frac{U_{\mathrm{t}}^2}{U_\infty^2} \quad (6.29)$$

6.2.2.4 *Rudder forces and moments*

The integration of the pressure distribution over the N panels defining the body surface allows the total potential pressure force F on a body to be evaluated as a vector sum, where A_i is the area of the ith panel, n_i the direction of its unit surface normal and N the number of panels on the body.

$$\bar{F}_P = \frac{1}{2}\rho U_\infty^2 \sum_{i=1}^{N} C_{Pi} A_i \hat{n} \tag{6.30}$$

The calculation of the pressure components of the nondimensional body force and moment coefficients requires a further transformation into the correct body coordinate system. For example, for a ship rudder at incidence with the x direction in the free-stream direction and z vertical, the lift is the j component of F. That is,

$$C_L = \frac{F \cdot j}{\frac{1}{2}\rho U_\infty^2 S\bar{c}} \tag{6.31}$$

where S and $\bar{c}$ are the rudder span and mean chord, respectively. The pressure component of drag is correspondingly,

$$C_D = \frac{F \cdot i}{\frac{1}{2}\rho U_\infty^2 S\bar{c}} \tag{6.32}$$

An estimate of the viscous skin friction force acting on a lifting surface can be found by using the panel surface velocity and distance from the leading edge to estimate the skin friction coefficient C_f [6.16]. This gives a viscous force contribution equal to

$$F_{visc} = \frac{1}{2}\rho \sum_{i=1}^{N} C_f A_i (V_T \cdot V_T) v \tag{6.33}$$

where v is a unit vector in the local flow direction. The skin friction coefficient is calculated in terms of local Reynolds number:

$$Re = \frac{Ul}{v} \tag{6.34}$$

where l is the distance to the leading edge. Various expressions may be used for C_f, such as Equations (3.14–3.18) or from Schlicting [6.17]:

$$\begin{array}{ll} Re < 3\times 10^5 & C_f = 0.664\,Re^{-0.5} \\ 3\times 10^5 \le Re < 1\times 10^7 & C_f = 0.074\,Re^{-0.2} - 1050\,Re^{-1} \end{array} \tag{6.35}$$

Combining the viscous and pressure contributions gives the total force F_T acting on the body as

$$\bar{F}_T = \bar{F}_P + \bar{F}_{visc} \tag{6.36}$$

It should be noted that this neglects the change in pressure force due to viscous effects. Similar expressions are derived to give the total moment.

6.2.3 Coupled boundary layer

The surface panel method can be enhanced to include the effect of viscosity through coupling a method for evaluating a solution of the thin boundary layer approximations to the full N–S equations along a series of surface streamlines.

There are two approaches:

(1) The geometry of the body in question is altered by increasing its size in the surface normal direction by an amount equal to the displacement thickness of the local boundary layer, Section 3.2.13 and Figure 3.13.

(2) Rather than the imposed zero normal relative velocity condition on the body surface, a flux of momentum is applied. The magnitude of this flux (transpiration velocity V_t) is proportional to the rate of momentum exchange at the edge of the boundary layer and is given by

$$V_t = \frac{\partial}{\partial s}(V_\delta \delta^*) \tag{6.37}$$

where s is the streamwise direction, V_δ the velocity at the edge of the boundary layer and δ^* the local displacement thickness along a streamline.

The three steps necessary to include the boundary layer growth in the flow solution are:

1. solve the potential flow over the body and obtain the surface pressure distribution;
2. using the pressure distribution, calculate boundary-layer characteristics and
3. modify the surface boundary conditions for the potential flow, and solve for the next iteration.

Considerable effort has been expended in developing accurate methods for solving the thin boundary-layer equations. Particular attention has been given to the prediction of laminar-turbulent transition, attached small separation bubbles and the capture of large zones of separation. In many cases, these methods and their associated mathematical complexity and significant computational effort, are in the process of being superseded by the complete N–S solvers described in Equation (6.3).

Theoretical methods that have the limited objective of predicting overall characteristics of the boundary layer, for example, momentum thickness, displacement thickness and skin friction, rather than details of the actual flow, are more straightforward to apply. For the purpose of improving rudder performance prediction, such methods are all that are needed to modify the potential flow.

The modification is split into a number of components (see Figure 6.3): prediction of the laminar boundary-layer growth from either a sharp edge or stagnation point, estimate of when transition to turbulence occurs, turbulent boundary layer growth and an estimation of whether the flow will separate

$$\frac{1}{2}C_{\mathrm{f}} = \frac{\theta}{V_\delta} \cdot \frac{\mathrm{d}V_\delta}{\mathrm{d}s}(H+2) + \frac{\mathrm{d}\theta}{\mathrm{d}s} \tag{6.38}$$

where the shape factor $H = \delta^*/\theta$, θ is the boundary layer momentum thickness and δ^* the boundary layer displacement thickness.

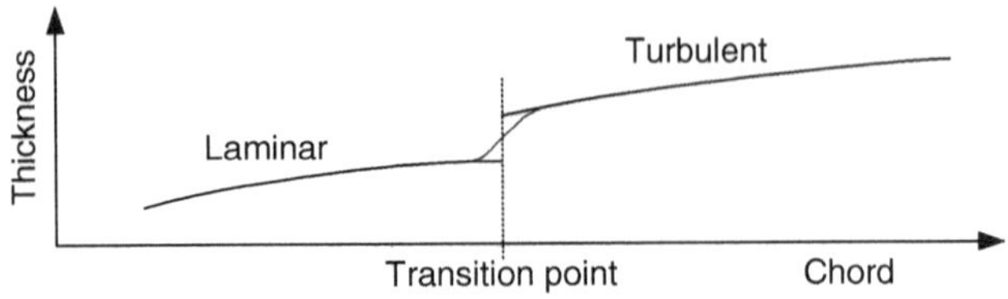

Figure 6.3 Schematic for smoothing the transition point discontinuity

Table 6.1 Predicted flow separation on NACA0020 at 20° angle of attack

H	1.8	1.9	2.0	2.1	2.2	2.3	2.4
% Chord	57.3	64.8	72.1	76.6	79.5	82.5	82.5

6.2.3.1 *Prediction of separation*

Estimates of the likely point of flow separation on the body surface can be made using either the shape factor H or the skin friction coefficient C_f as the criterion for separation. For example, the C_f law given by Ludwieg and Tillmann [6.18];

$$C_f = 0.246 \times 10^{-0.678H} R_\theta^{-0.268} \tag{6.39}$$

which predicts $C_f = 0$ as H tends to infinity. An exact value of H corresponding to separation cannot be specified, but a range between 1.8 and 2.4 has been quoted [6.19]. For a NACA 0020 [6.20] section foil at 20° angle of attack, predictions of the chordwise point of separation, using this range of H values, are given in Table 6.1.

The predicted position varied substantially, between 57% and 83% chord, for the range of H values. Without knowing the precise point of flow separation, the more reliable means is to use the C_f values. In this case, $C_f \to 0$ occurred at 81.7%c, which lies within the quoted range of H.

The starting point for applying such 2-D calculations is the definition of a suitable streamline across the 3-D rudder surface. Methods of varying levels of complexity can be applied to trace the streamline forward to a stagnation point or sharp leading edge, and backward to a point of separation or a trailing edge. Once such a streamline has been found the coordinates of the underlying surface should be used to define a parametric curve with known values of tangential velocity at specified points along its length. The integral boundary layer method can then be applied with an appropriate level of discretisation, which is likely to be different from that used for the surface panels. In some methods, 3-D effects due to streamline curvature normal to the flow direction are used to modify the integral boundary-layer equation. For most rudder-like flows these effects are small except near the tip.

Although small laminar separations can be handled, only the initiation of large zones of separation can be predicted. Such information can provide valuable insight into the limitations of a particular surface panel calculation. If necessary, the edge of the zone of separation can be panelled and the modified potential flow represented.

6.3 Navier–Stokes methods

It is only in the last 10 years that it has become practical to solve the N–S equations, based on Reynolds averaging around 3-D free-stream rudders, with any degree of confidence. Previous work was always limited by the availability of sufficient computational power and memory to define the 3-D computational mesh (or grid) around the rudder and a sufficiently large surrounding domain.

The development of the finite volume method, used by most commercial and research flow solvers, results from the surface integration of the conservative form of the complete N–S equations over a 3-D control volume, as for example, explained by Versteeg and Malalasekera [6.21]. The resulting equations express the exact conservation of the relevant flow properties within the control volume. This relationship between physical conservation and the governing equations forms one of the main attractions of the finite volume method.

6.3.1 RANS equations

Although the complete N–S equations govern both laminar and turbulent flows, they are not suitable for the direct computation of turbulent flows. To do so would require computers estimated to be of the order of at least 10^4 times faster than today's (2006) fastest supercomputer. This requires use of extremely fine grids and over a large number of time steps, in order to capture the turbulent motion at the smallest time and length scales. The statements above assume that accessible computer power has improved by a factor of 10–100 in the 15 years since the values stated by Speziale [6.22].

The method of averaging the behaviour of turbulence was first proposed by Reynolds [6.23]. In this case, the flow variables are resolved into

$$u = U + u' \quad v = V + v' \quad w = W + w' \quad p = P + p' \tag{6.40}$$

where $(u', v', w')^T$ and p are the unsteady time varying flow about a slowly varying mean flow $(U, V, W)^T$ and P. On substitution into the complete N–S equations, and following time averaging, the unsteady RANS equations are obtained. The time-averaged form of the continuity equation is now given by

$$\frac{\partial U}{\partial x} + \frac{\partial V}{\partial y} + \frac{\partial W}{\partial z} = 0 \tag{6.41}$$

The time-averaged N–S momentum equations are given by

$$
\begin{aligned}
&\frac{\partial U}{\partial t}+U\frac{\partial U}{\partial x}+V\frac{\partial U}{\partial y}+W\frac{\partial U}{\partial z}\\
&\quad=\frac{-1}{\rho}\frac{\partial P}{\partial x}+\nu\left(\frac{\partial^2 U}{\partial x^2}+\frac{\partial^2 U}{\partial y^2}+\frac{\partial^2 U}{\partial z^2}\right)-\left(\frac{\partial\overline{u'^2}}{\partial x}+\frac{\partial\overline{u'v'}}{\partial y}+\frac{\partial\overline{u'w'}}{\partial z}\right)\\
&\frac{\partial V}{\partial t}+U\frac{\partial V}{\partial x}+V\frac{\partial V}{\partial y}+W\frac{\partial V}{\partial z}\\
&\quad=\frac{-1}{\rho}\frac{\partial P}{\partial y}+\nu\left(\frac{\partial^2 V}{\partial x^2}+\frac{\partial^2 V}{\partial y^2}+\frac{\partial^2 V}{\partial z^2}\right)-\left(\frac{\partial\overline{u'v'}}{\partial x}+\frac{\partial\overline{v'^2}}{\partial y}+\frac{\partial\overline{v'w'}}{\partial z}\right)\\
&\frac{\partial W}{\partial t}+U\frac{\partial W}{\partial x}+V\frac{\partial W}{\partial y}+W\frac{\partial W}{\partial z}\\
&\quad=\frac{-1}{\rho}\frac{\partial P}{\partial z}+\nu\left(\frac{\partial^2 W}{\partial x^2}+\frac{\partial^2 W}{\partial y^2}+\frac{\partial^2 W}{\partial z^2}\right)-\left(\frac{\partial\overline{u'w'}}{\partial x}+\frac{\partial\overline{v'w'}}{\partial y}+\frac{\partial\overline{w'^2}}{\partial z}\right)
\end{aligned}
\tag{6.42}
$$

where the additional terms (six in total) compared to Equation (6.3) are known as the specific Reynolds stresses and act both in the normal and shear directions.

6.3.2 Turbulence and turbulence modelling

Reynolds averaging of the complete N–S equations gives rise to six additional unknown independent stresses. A process of closure is required to express these stresses in terms of known values. The applied procedure, known as turbulence modelling, provides an engineering approximation of the complex turbulent flow behaviour.

No single turbulence model exists that can be applied universally to any turbulent flow problem. For a turbulence model to be useful in an engineering sense, it must have wide applicability, a known level of accuracy in its application to different flow problems and yet remain easy to implement. A number of turbulence models have been developed over the years, all varying in their complexity and suitability to certain flow situations.

Turbulence models can be roughly divided into four main categories: algebraic (zero-equation), one-equation, two-equation and stress-transport models. A complete discussion of turbulence and turbulence modelling is not appropriate but an excellent description is provided in Wilcox [6.24].

For clarity, a brief description of the theory behind a widely used turbulence model, the two-equation standard $k-\varepsilon$ turbulence model, will be given. This will be used to help explain some of the inherent difficulties when attempting to predict rudder performance either in the free stream or behind a propeller.

6.3.2.1 *Standard k−ε turbulence model*

Its relative simplicity explains both its robustness (a flow solution should be found) and its drawbacks (it is unable to capture key flow behaviour especially near to separation). The $k-\varepsilon$ turbulence model computes the Reynolds stresses based on the Boussinesq [6.25] eddy hypothesis for Newtonian fluids, whereby the Reynolds stresses are related to the mean rate of deformation, turbulent kinetic energy and turbulent viscosity within a fluid, as expressed:

$$\frac{\partial \overline{u'^2}}{\partial x^2} = -v_t\left(2\frac{\partial U}{\partial x}\right) + \frac{2}{3}k \qquad \frac{\partial \overline{v'^2}}{\partial y^2} = -v_t\left(2\frac{\partial V}{\partial y}\right) + \frac{2}{3}k \qquad \frac{\partial \overline{w'^2}}{\partial z^2} = -v_t\left(2\frac{\partial W}{\partial z}\right) + \frac{2}{3}k$$

$$\frac{\partial^2 \overline{u'v'}}{\partial x \partial y} = -v_t\left(\frac{\partial U}{\partial y} + \frac{\partial V}{\partial x}\right) \qquad \frac{\partial^2 \overline{u'w'}}{\partial x \partial z} = -v_t\left(\frac{\partial U}{\partial z} + \frac{\partial W}{\partial x}\right) \qquad \frac{\partial^2 \overline{v'w'}}{\partial y \partial z} = -v_t\left(\frac{\partial W}{\partial y} + \frac{\partial V}{\partial z}\right) \tag{6.43}$$

where υ_t is known as the eddy viscosity and depends on the local turbulence in the flow. This Equation (6.43) is known as the isotropic eddy-viscosity model, as it is assumed that the normal turbulent stresses are the same in all directions, even though this is generally not the case.

The eddy viscosity υ_t is determined from the turbulent kinetic energy,

$$k = \frac{1}{2}\left(\overline{u'^2} + \overline{v'^2} + \overline{w'^2}\right) \tag{6.44}$$

and the rate of turbulent dissipation, ε, as

$$\upsilon_t = c_\mu \frac{k^2}{\varepsilon} \tag{6.45}$$

Physically, turbulent kinetic energy k is the kinetic energy per unit mass of the turbulent fluctuations within a flow. The rate of turbulent dissipation ε is the rate at which turbulent kinetic energy is converted into thermal internal energy within a fluid. Equation (6.45) is derived from dimensional analysis of the velocity and length scales that characterize the turbulent exchange of momentum, where c_μ is an empirical coefficient generally set to 0.09. k and ε are unknowns determined from two transport equations, which require four empirical constants obtained from experiments on a wide range of turbulent flows.

The standard k–ε turbulence model described, has been widely used and shown to perform acceptably well in a variety of applications. In the context of flows around control surfaces, a number of the weaknesses associated with the standard k–ε turbulence model are outlined:

- The turbulent kinetic energy is over predicted in regions of impingement and reattachment, leading to the poor prediction of boundary layer development around leading edges and bluff bodies.
- Flow separation is poorly predicted from surfaces under the action of adverse pressure gradients. The real flow is often found to be much closer to separation or has more separation than the computed flow.
- Flow recovery following the reattachment of a separation zone is often poorly predicted as a result of the implementation of wall functions within the standard k–ε turbulence model.

- Prediction of highly swirling flows is poor, where the turbulent flow field is anisotropic, such as in separated regions of slow flow recirculation and vortex shedding.

It is also unable to capture laminar and transitional flows; this is an important drawback in, for example, performance prediction of rudders or keels on small craft such as yachts which are often operating at a transitional Reynolds number and can have up to 50% laminar flow.

6.3.3 Numerical implementation of RANS techniques

The finite volume method subdivides a specified computational domain into a number of discrete volumes that completely fill the domain. Within each finite volume the numerical implementation attempts to enforce conservation of mass and momentum. Other approaches to solving the RANS equation can be classified as point-solution techniques (earliest applications using finite difference representation of equations) or finite element that represents the variation of flow variables (pressure, velocity) as simple functions across individual elements, which completely fill the whole domain. In practice, the specific classification of the implementation is less important than the overall process. This differs fundamentally from that used in potential flow where equations are solved on the bounding surfaces, to one where equations are now constructed which have to be solved at all locations throughout a specified 3-D fluid domain.

Figure 6.4 is a schematic that illustrates the process of defining the underlying finite volume method. The three dimensional domain requires a prescribed extent with a series of bounding surfaces. The interior of this domain is subdivided into multiple (finite) volumes. Each finite volume (mesh cell) is defined by a number of

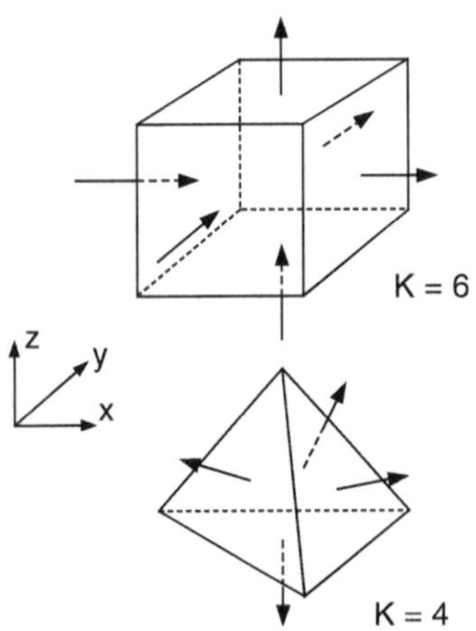

Integral Conservation Equation for a general scalar ϕ (e.g. u, v, or w) with a flux $f(\phi)$ in the direction normal **n** to the local surface S, enclosing a volume Ω.

$$\frac{d}{dt}\bar{\phi}+\frac{1}{\Omega}\oint_S f(\phi)dS = 0$$ or for a polyhedron with k faces is

$$\frac{d}{dt}\bar{\phi}+\frac{1}{\Omega}\sum_k f(\phi_i)A_i = 0.$$

The flux across each face requires use of finite difference approximations.

Figure 6.4 Schematic of finite volume discretisation. This is based on numerical integration of flux of mass and momentum through faces based on finite difference for evaluation of the flux faces consisting of a series of straight edges joining nodes (or vertices). Typically, for an incompressible flow the finite volume is described as either:

- cell centred, with the flow variables associated with the centre of each cell or
- node centred with the variables defined for each node of the domain.

The four scalar unknowns associated with an incompressible flow are the flow velocity and static pressure. In addition, further unknowns are associated with the applied turbulence model. So for a two-equation model six unknowns (or degrees of freedom) are associated with each mesh cell.

The solution of the steady-flow problem requires that the net sum of mass and momentum through each of the cell's faces is zero, i.e., mass and momentum are conserved. This summation requires that the flux of mass and momentum is calculated and this requires evaluation of first and second derivatives of the velocity and the pressure gradient. These quantities are found based on numerical (finite) difference expressions using the velocity and pressure associated with a cell and its surrounding neighbours. The order of accuracy of these flux calculations and the method adopted are an essential part of the overall accuracy of a given finite volume calculation, as will be discussed later and illustrated in Figure 6.5. In essence, the smaller the separation of cell centres (smaller-sized cells) the higher the absolute accuracy of the evaluation of the derivative and hence the fluxes. Hence, the accuracy with which a given flow will be defined is determined by the cell size (density) in a given direction. In a 3-D problem, halving the cell size in all three directions will result in an eightfold increase in the number of cells.

The equations for mass and momentum conservation tightly couple the velocity field and pressure. A number of numerical approaches can be adopted to find a solution that satisfies the local statement of conservation for each cell and the global

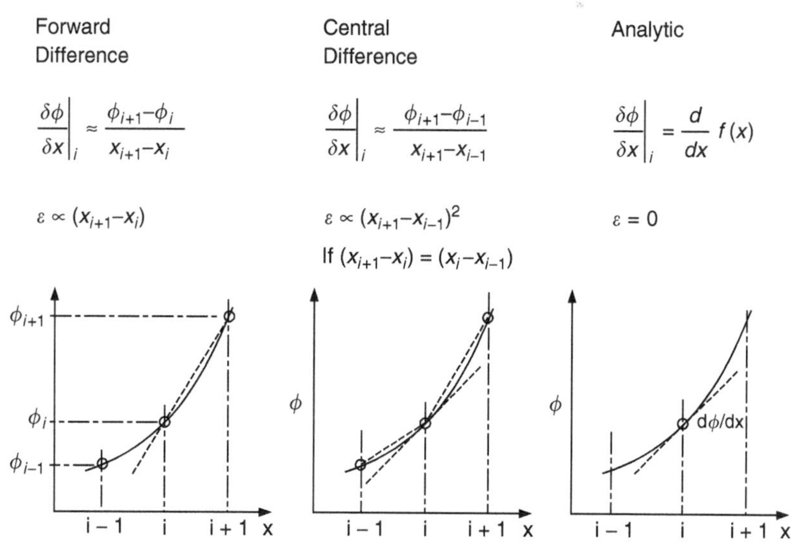

Figure 6.5 Illustration of numerical error for finite difference evaluation of velocity gradient

statement of conservation as expressed by the boundary conditions applied to the surfaces defining the fluid domain. The evaluation of face fluxes couples the solution between adjacent cells. As a result, iterative approaches have to be adopted to find the values of velocity and pressure that satisfy all the defining equations.

6.3.4 Solution process

The process of solving such a problem consists of a number of specific steps, some of which are under the control (and hence expertise) of an individual user and others which are determined by the code developer.

Figure 6.6 illustrates the overall process with the three main stages:

(1) preprocessing (specifying a geometry, computational mesh, initial conditions and domain boundaries);
(2) numerical solution based on some form of iterative procedure to solve the large number of coupled equations associated with each of the individual finite volumes and
(3) post-processing and analysis of the obtained solution.

As a system, the results of whole process can only be used with confidence if an overall validation process is also applied to define the quality and trust that can be associated with the derived results.

6.3.5 Mesh generation

The requirement to subdivide the complete domain volume into contiguous finite volumes has always been one of the critical components required in applying the RANS method. This process of 'grid' or 'mesh' generation is described in most CFD texts.

The mesh-generation process has to balance:

- the need to improve problem resolution through having many finite volumes in regions where flow parameters are changing rapidly;
- the need to define a sufficiently large overall domain around the problem of interest such that the boundary conditions applied will not incorrectly influence the solution; and
- the ability of the individual engineer to actually obtain a solution on a particular computer and within a realistic timescale.

The trade-off between a finite computational resource and solution accuracy has driven many of the misconceptions associated with results obtained from RANS solutions and their associated accuracy.

Development of a 3-D mesh of sufficient quality that a flow solution could actually be obtained could take 6 months only a decade ago. It is in this area that the majority of improvements in solution quality have occurred. Larger computers (faster and more memory for mesh storage) have allowed the user to both generate and manipulate a mesh as well as to solve the problem within a reasonable time-span.

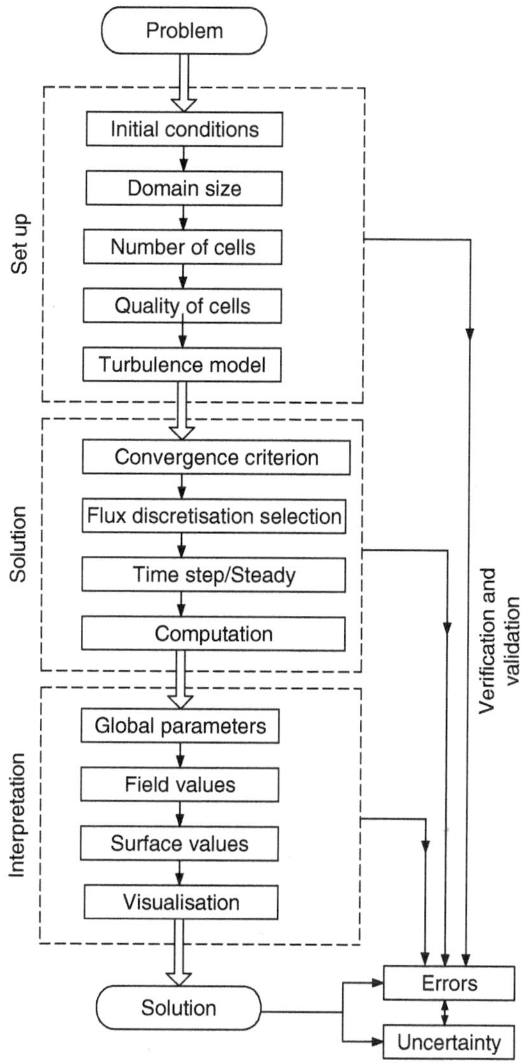

Figure 6.6 Schematic of CFD process

Typically, for the workplace, the ability to start a job one day and obtain a solution the next is deemed a key usability parameter. As an example of large-scale applications, of relevance to this work, the solution of the complete underwater appended hull form of an IACC class yacht with free surface is now routinely solved with of order 120×10^6 million cells. This can be compared, for example, with the early application to the 1987 America's cup campaign when 800 or so surface panels was considered a large computation.

Typically, finite volumes methods use either a six face (hexahedron) or a four face (tetrahedron), although flow solvers that can use arbitrary polyhedra offer greater

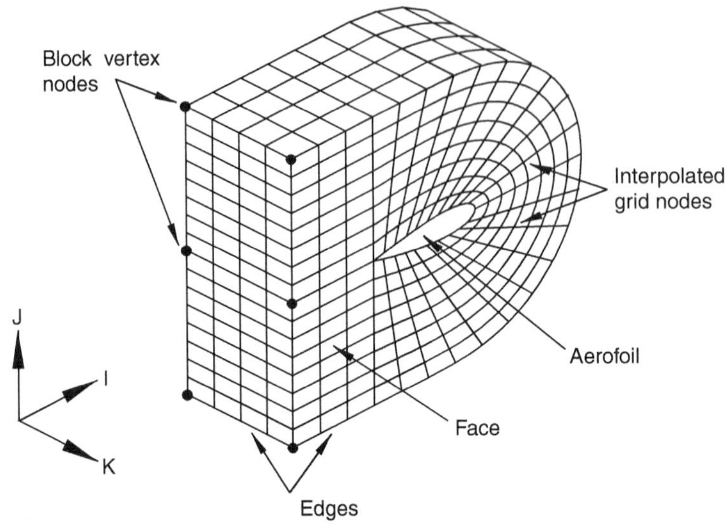

Figure 6.7 Single-block finite volume mesh

flexibility in mesh generation. The correct matching of the finite volume orientation to the flow direction is an important factor in maintaining high-quality flow solutions.

The earliest 3-D RANS flow solvers used a six-faced domain with a hexahedral mesh generated by subdividing each of the three directions shown in Figure 6.7. The neighbouring cells of a given cell were implicitly known from the $\{i, j, k\}$ integer indices locating its components within a 3-D array. How the equations were solved was directly linked to this structured block. Unfortunately, for most 3D problems it is not a straightforward process to map a single six-face domain onto the problem. However, a logical development is to fill the domain of interest with a number of blocks (multiblock) and this gives a much better mapping of the physical space to the computational space. The same flow solver can then work on a block at a time by applying the appropriate boundary conditions between blocks. The drawback is then how to break down the complex domain into a suitable number of blocks. As the number of divisions on a block edge or face propagates into the adjacent block, dense regions of cells can often exist in regions where such density is not required.

An alternative strategy is to chose a basic tetrahedron and fill the complete domain, clustering finite volume cells where required (a challenge in its own right as often the correct location is not known *a priori*). Unfortunately, tetrahedral finite volumes cause significant numerical difficulties in thin boundary layers with their high shear (vorticity). Ideally, the faces of a cell should either be aligned with the local flow direction or be perpendicular to it. This will minimize numerical error when calculating the flux of momentum through a face.

A hybrid approach, which uses prisms in regions of high shear and fills the rest of the space with tetrahedrons, allows for rapid mesh generation and reasonable control of mesh quality. A difficulty then is how to obtain a smooth transition from

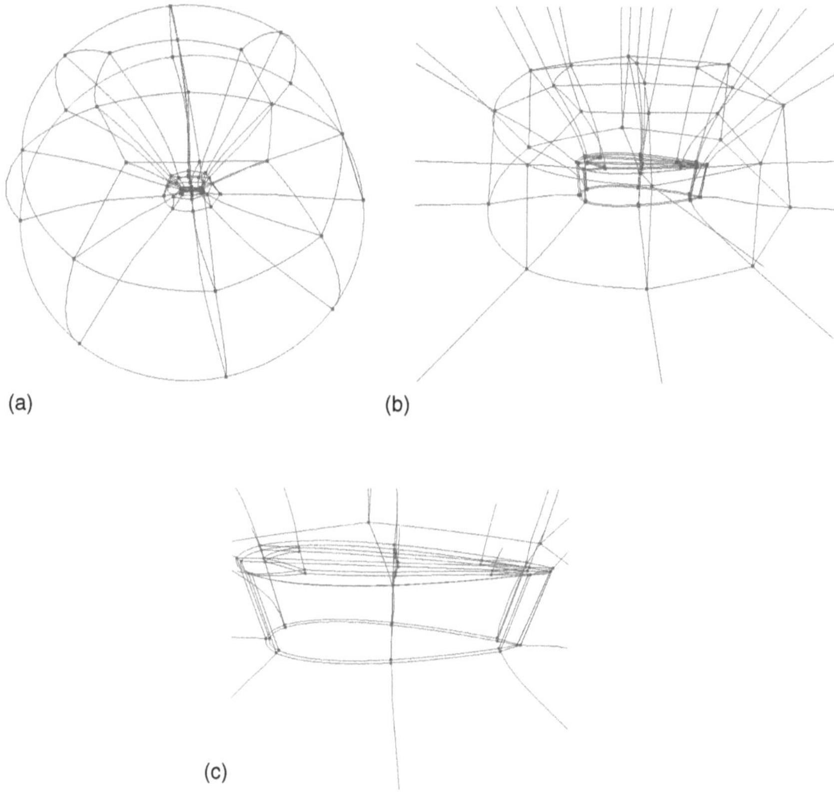

Figure 6.8 Three-layer multiblock mesh topology for control surface CFD analysis, showing block edges. (a) Outer blocks – showing rudder within a hemispherical domain. (b) Middle blocks – showing transition from boundary layer to outer domain. (c) Inner blocks – showing small volume required to capture boundary layer (See Plate 3 of Colour Plate Section)

the last prism onto the first tetrahedral. It is usually a good idea to make sure that this transition occurs well outside the outer limit of the boundary layer.

Figure 6.8 (See Plate 3 of Colour Plate Section) illustrates a typical block mesh strategy that should deliver reliable results. A fine boundary layer is required with the dimensions of the first cell dictated by the turbulence model requirements. A smooth progression in cell size is then applied to the edge of the inner zone. This fine block mesh should extend back from the trailing edge to capture the wake for a number of chord lengths downstream. A relaxed wake solution from a surface panel method can provide a sensible method of *a priori* predicting this location. Likewise, in the vicinity of the rudder tip and downstream a fine mesh is required to capture the correct tip vortex [6.26]. If this is not captured adequately then it is unlikely that the correct sideforce will be predicted. Around all of these regions, which are adapted to the local flow, a further block should be incorporated to provide a reasonably smooth transition to the outer region of blocks within which the large aspect ratio cells and reasonable levels of mesh distortion can be tolerated.

6.3.6 Boundary conditions

In order to obtain correct solutions from the governing equations, it is necessary to define the initial and/or boundary conditions for the dependent variables $(U, V, W, P, k, \varepsilon)^T$ that describe the problem to be solved. The correct selection of these boundary condition is fundamental to obtaining accurate flow solutions.

The use of unrealistic and badly posed boundary conditions can lead to spurious and incorrect flow solutions, or more usually rapid solver divergence. The two most frequently used linear boundary conditions are, the Dirichlet (value specified) and Neumann conditions (gradient of value specified) [6.2]. Typical boundary conditions used for rudder problems are:

6.3.6.1 *Inlet*

The inlet boundary condition is a form of Dirichlet boundary condition. On an inlet boundary, the dependent variables of U, V, W, k, ε are prescribed. The pressure is not set for incompressible flows, as it is extrapolated from downstream. The turbulence quantities k and ε or equivalent for other turbulence models, are often difficult to specify. If the computations are to be compared with experimental data, the inlet turbulence quantities of k and ε should ideally be set according to measured values found from experiment. When these measurements are not available the sensitivity of the flow solution to the selection of k and ε parameters must be carried out.

If measurements of turbulence intensity do exist and the turbulent length scale of the problem is known, then crude estimates of k and ε can be made. The turbulence level or turbulence intensity within a flow is defined by

$$T_i = \frac{\sqrt{\frac{1}{3}\left(\overline{u'^2 + v'^2 + w'^2}\right)}}{U_\infty} \tag{6.46}$$

which reduces to

$$T_i = \frac{\sqrt{\overline{u'^2}}}{U_\infty} \tag{6.47}$$

if all the Reynolds stresses are assumed equal and hence using Equation (6.44),

$$k = \frac{3}{2} T_i^2 U_\infty^2 \tag{6.48}$$

The specification of the rate of dissipation of kinetic energy ε is more difficult. Estimates of ε can be made if measurements of the turbulent length scale l of the problem exist [6.21, 6.24]

$$\varepsilon = c_\mu^{3/4} \frac{k^{3/2}}{l} \tag{6.49}$$

If the flow is known to be free of residual turbulence (which is unusual especially in the case of flows in the marine environment or in wind tunnel experiments), k and ε can be set to zero, the free-stream undisturbed condition. In practice, both k and ε

are set to small values, say 0.0001 to avoid solver convergence problems if k and ε can turn negative. If the upstream inlet boundary is placed far enough upstream, the choice of k and ε is less critical as they tend to dissipate to low values. Such a choice can often come at too high a computational cost.

6.3.6.2 *Wall*

The wall boundary condition requires that the velocity on the wall satisfies the no-slip condition. Also, on the wall k is zero and ε is nonzero. The k and ε turbulence models can only be applied to regions that are fully turbulent and cannot be applied in regions where viscous effects are dominant, such as those found in the laminar sublayer. The wall function approach proposed by Launder and Spalding [6.27] can be used to overcome this problem.

From experimental work, it is known that near-wall flows have a characteristic multilayered structure within the boundary layer as shown in Figures 3.10 and 6.9.

This consists of a laminar sublayer (viscous stress dominated) close to the wall, followed by a buffer layer (viscous and turbulent stress of similar magnitude) and then an outer turbulent core (turbulent stress dominated). Direct methods of resolving the turbulent eddies within this boundary layer require extremely fine grids down to the wall, through the laminar sublayer, which is computationally intensive and very costly. However, most RANS codes can use turbulence models that employ wall functions, based on the universal law of the wall [6.17]. The use of wall functions avoids the need for fine grids in the laminar sublayer, by making use of empirical fits within this region. When dealing with near-wall flows, positions and velocities within the boundary layer are usually considered in nondimensional form, and y^+ is a local Reynolds number, with length scale in the direction perpendicular to the wall y_p as represented by:

$$y^+ = \frac{y_p}{v}\sqrt{\frac{\tau_w}{\rho}} \tag{6.50}$$

and u^+ a velocity based on the wall shear stress,

$$u^+ = \frac{u_p}{\sqrt{\tau_w/\rho}} \tag{6.51}$$

where u_p velocity at distance y_p.

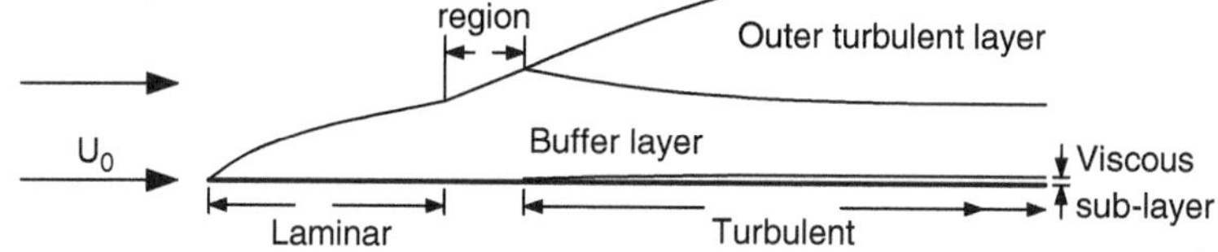

Figure 6.9 Flow regimes for thin laminar-turbulent boundary layer

Specific flow structures within the boundary layer lie within strict bounds of y^+, and these are used in the formulation of wall functions. Positions within the boundary layer, in which $y^+ \leqslant 11.63$, are regarded as laminar in structure and above 11.63 as turbulent. It has been shown that within these two regions, two different functional relationships exist between y^+ and u^+. These are shown as

$$u^+ = y^+ \tag{6.52}$$

$$u^+ = \frac{1}{\kappa} \ln Ey^+ = \frac{1}{\kappa} \ln y^+ + B \tag{6.53}$$

for the laminar linear sublayer and for the turbulent log-law regions, respectively.

The constants in Equation (6.53) are determined from experiment. For hydraulically smooth walls the Von Karman constant, $\kappa = 0.4$, and the log-layer constant, $E = 9.793$. Roughness can be simulated by increasing the value of E. The buffer layer crossover value of 11.63, is found by finding the intersection of the linear laminar sublayer profile (6.52); and the log-law turbulent profile (6.53).

The use of wall function turbulence models places specific requirements on the meshes used in solving turbulent flow problems. When considering turbulent near-wall flows, the most critical mesh parameter is the near-wall grid spacing. It is of paramount importance, that near-wall grid spacing is selected in accordance with the requirements of the wall function. In wall function turbulent calculations, a y^+ of 11.63 usually sets the lower limit for the distance of the first cell, to the wall boundary, with the optimum near-wall position lying somewhere between $y^+ = 30$ and no more than 500. Although the first cell spacing is critical for accurate near-wall flow modelling, enough cells should also be placed through the whole boundary layer to resolve the flow gradients. The use of wall functions, therefore, poses a specific mesh independence problem for near-wall flows.

6.3.6.3 *Mass flow outlet*

The mass flow boundary condition is a form of the Neumann boundary condition. Here the gradients of the dependent field variables $(U, V, W, k, \varepsilon)^T$ normal to the boundary are initially set to zero and, later, $(U, V, W)^T$ is modified to have a constant gradient to maintain global mass continuity. The pressure is extrapolated from upstream. This boundary condition relies on the assumption that the flow is fully developed when it reaches the outlet. Therefore, all outlet boundaries should be placed far enough downstream to ensure that the fluid flow is fully developed, i.e., zero flow variable gradients in the flow direction. Positioning an outlet too close to an area with a flow disturbance may result in solution errors, since the assumed outlet condition of zero-flow gradient will not hold. Bluff body flows have areas of reversed flow and these regions will violate the outlet boundary condition of outward flow if the outlet is placed too close to the body. It is necessary in the case of incompressible flows for global mass continuity to be maintained in order for the pressure correction equation to be well posed. This means that the total flow out of the domain must equal the total flow into the domain at all stages of the solution procedure.

6.3.6.4 *Positioning of boundaries*

It is important that all boundaries are positioned to ensure that they have no demonstrable effect on the flow solution. In any CFD study, where high-accuracy results are required, a sensitivity study should be carried out to demonstrate that the interior flow solution is unaffected by the location of the boundaries. It must also be remembered that it is not good practice to place boundaries at excessive distances from the body. This wastes valuable computational resources, which could be put to better use in resolving areas with high flow gradients.

6.3.7 Obtaining a solution

The actual process of obtaining a flow solution using an appropriate solver technique for the governing equations is a complex process. Typically, for user-friendly codes this complexity is hidden from the user. Even so, obtaining a good quality solution still often remains an iterative process that cannot be reduced to a specified list of tasks. Often, the solver must be run, the results checked and then rerun with slight adjustments to the model, in order to improve the solution. The production of a good simulation usually results from a continual process of trial and error with improvements in productivity arising from developed user expertise.

6.3.7.1 *Differencing schemes*

The accuracy of a flow solution is dependent on the choice of differencing scheme employed in approximating the N–S equations. The type of differencing scheme determines the way in which the partial derivatives in the governing N–S equations are replaced with algebraic difference quotients, Figure 6.4, based on the flow-field variables at the faces of an individual finite volume. A variety of differencing schemes exist, with varying orders of accuracy. Essentially, the greater the number of surrounding control volumes used in formulating the difference quotients, the higher the order of accuracy of the differencing scheme. Higher order schemes can often cause problems near the edge of a fluid domain or in those codes that use overlapping or sliding meshes. Generally, if high gradients are present then domain boundaries should not occur in their vicinity, a task easier stated than achieved and notable for its relevance to coupling say a rotational domain around a marine propeller as an inflow to a downstream rudder.

The choice of differencing scheme used for a particular flow solution depends on the flow type and the degree of accuracy expected from the flow solution. High order accurate differencing schemes have the advantage over the lower order schemes in that they generally need fewer total grid points to obtain comparable overall accuracy, but this is often at the expense of increased computer time. Lower order schemes are often more robust than higher order schemes, and can cause fewer convergence problems, especially when used on grids of poor quality. However, first order schemes suffer from numerical diffusion and good industrial practice [6.28] would indicate that such solutions have little worth other than as a way of providing an initial starting condition for a higher order scheme.

Three common differencing schemes used in finite volume flow codes are: upwind (first order), quadratic upstream interpolation for convective kinetics or QUICK (third order) and hybrid (second order). The upwind differencing scheme forms the difference formula based solely on the value of the upstream control volume face, whereas QUICK uses a three-point upstream-weighted quadratic interpolation. The hybrid interpolation combines an upwind interpolation with a central differencing scheme, which uses one upstream and one downstream face for interpolation.

Similar differencing scheme approaches are applied to transient problems in order to calculate the time-dependent partial derivatives in the N–S equations, in terms of algebraic difference quotients based on the variables at different time levels. Backward and quadratic differencing are two of the most common time marching differencing schemes used by finite volume codes.

6.3.7.2 *Iterative solution process*

The statement of mass and momentum conservation results in an associated variable e.g., $(U, V, W, P)^T$ at the cell centre along with a further variable for the turbulence model. So, for a two-equation single-phase solver this results in $6N$ equations, where N is the total number of cells. Typically, for a higher-order differencing scheme, the fluxes for the faces of a single cell will depend on the state vectors of 1–2 layers of cells, e.g., 26–124 other cells. In practice, these coupled sets of equations can only be solved using either an explicit or more normally an implicit iterative scheme.

This means that an initial solution, normally with all the flow variables set to their free-stream values, is required at the start of the solution process. The numerical equations are then used to produce a more accurate approximation to the numerically correct solution. This is one in which all the variables of each control volume satisfy the governing equations. During this iterative process, the updated solution field variables at the end of an iteration cycle will be different from those at the start of the iteration. Ideally, as the solution proceeds these differences should decrease (convergence).

6.3.7.3 *Convergence stopping criteria*

Most solvers use the residual error from the continuity equation, commonly known as the mass source residual, as the stopping criterion for defining a solution as converged. The mass source residual is the sum of the absolute values (the L_1 norm [6.3]) of the net mass fluxes into or out of every control volume in the flow, and thus has the dimensions of kilogram per second. For the continuity equation to be satisfied this must equal zero. In practice, a small residual tolerance is set by the user, dependent on the accuracy expected from the solution. The specification of this tolerance is problem specific, and can only be determined after running a problem for a number of iterations while studying the convergence history of an important property, like an integrated global parameter such as lift, drag or a field parameter like velocity or pressure at some location within the flow. From this data, the minimum mass source tolerance needed to obtain a converged solution, in terms of the flow property of interest, can be found.

6.3.7.4 *Iterations on inner equations*

A number of solvers use a pressure coupling process that has an inner iterative loop to balance the momentum equations and mass conservation equations on each cell directly. In some cases, increasing this usually fixed number of inner loops can often help the convergence of difficult problems. Poor convergence of the inner pressure correction iteration, in particular, can lead to loss of mass conservation and solver divergence. Therefore, increasing the number of iterations on pressure can sometimes help. It must be noted that increasing the number of inner iterations can greatly increase the solution time, so the smallest number of inner iterations should always be sought.

6.3.7.5 *Solution troubleshooting*

Obtaining a valid converged solution is often a continual process of trial and error. Even though much of the solver detail is hidden from the user, many parameters still exist for the user to define, all of which can influence the solution process.

If the problem has been specified correctly, is physically reasonable and has a well defined stable mathematical solution, yet still fails to converge satisfactorily, then numerical difficulties may be suspected. Many of these problems can be rectified by the selection of more conservative solution parameters. Some likely causes and possible solutions to poor solver convergence are given:

- *Incorrect initial and/or boundary conditions.* A common cause of failure to converge is the prescription of incorrect initial and/or boundary conditions. If the errors in the boundary conditions are not easily identifiable, it is useful to visualise the data as this often clarifies the source of error in the initial model. If the initial conditions are suspected it can often prove advantageous to run a lower order accurate solution and restart the more complex problem using the field values generated from the solution.
- *Poor mesh quality.* The definition of a poor mesh is one whose control volumes differ in shape greatly from a cuboid. Nonorthogonal grids can cause convergence problems, due to inaccurate calculation of fluxes through the control volume faces. Smoothing the grid and making it more orthogonal by altering the block shape and/or structure, can often help overcome this. Nonorthogonal grids are often unavoidable, especially around complex geometries. Rapid changes in cell size/aspect ratio can also cause solver divergence and must generally be avoided, especially in areas where rapid changes in flow variables are expected.
- *False diffusion.* False diffusion, or numerical diffusion, results from the truncation errors associated with discretisation. Depending on the differencing scheme used, its effect is most prominent in flows where the flow is not aligned with the mesh. It results in the distributions of the transported properties becoming smeared. Most prominent false diffusion occurs when upwind schemes are used. The degree of false diffusion also increases with Reynolds number. False diffusion can be minimized by using higher-order differencing schemes, such as hybrid or QUICK. A comprehensive study into the effect of false diffusion on flow solutions, for various differencing schemes, is discussed by Versteeg and Malalasekera [6.21].

- *Turbulence model problems.* The use of a turbulence model can bring its own convergence problems. Two specific problems are: slow or nonconvergence, due to incorrect first cell size on wall boundaries or due to divergence for negative values of k and ε from the turbulence transport equations. For example, ensuring for law-of-the-wall turbulence models that the first cell size satisfies at least the minimum criteria of $y^+ > 11.63$ usually resolves this problem. Negative calculated values of k and ε often arise when higher order upwind and QUICK differencing schemes are applied to the turbulence equations, or when the shear layer is not adequately resolved by the grid. It is often recommended, even when using high order differencing schemes on the velocity terms, that lower order schemes such as upwind or hybrid should be used for the k and ε terms, thus ensuring positive results. Although the upwind and hybrid schemes are of lower order, the overall accuracy of the flow solution is usually unaffected, since k and ε are dominated by production and dissipation.

6.3.8 DNS and LES

DNS and LES computations of marine flows are as yet uncommon. DNS involves the direct solution of the unsteady N–S equations, and are thought to be capable of resolving even the smallest eddies and time scales of turbulence within a flow. Although the DNS method does not require any additional closure equations (as in the case of the RANS method), very fine grids and extremely small time steps need to be used, in order to obtain accurate solutions. This method is currently confined to simple flow problems at relatively low Reynolds numbers. DNS computations of the fully turbulent high Reynolds number flows associated with ship flows await major advances in computational hardware. Although DNS solvers are limited to solving low Reynolds flows, they are seen as playing a role in further RANS code turbulence model development.

Like DNS codes, the use of LES solvers is still mainly as a research tool. LES [6.24] is a method that can be used to predict accurately the large scale turbulent structures within a flow, requiring a subgrid scale model to represent the smaller scale eddies. Although only the large scale eddies are resolved individually, this still requires the use of extremely fine grids, making solutions expensive and demanding on present computer resources. For example, this method has been successfully utilized in solving numerous high Reynolds number problems, like the turbulent flow over a NACA0012 aerofoil, carried out by Creismeas [6.29].

6.4 Interpretation of numerical analysis

The described hierarchy of control surface analysis tools provides a choice as to the most appropriate tool for a specific problem. This choice requires knowledge as to the likely quality (how close they are to reality?) of the results and hence the trust (how much uncertainty is associated with the value obtained?) that can be placed in them.

The key question is how such data can be interpreted for inclusion within the rudder design process. Interpretation is taken to be the process whereby the results of a CFD analysis are modified to take account of all the underlying assumptions, implicit and explicit, in order that they can be applied to the real full-scale rudder design.

The easy availability of results from complex computational analysis often fosters the belief that when it comes to data more detail implies more accuracy. Hence, greater reliance is placed in the results than is actually justified. An oft assumed, and usually not stated, belief that small changes in input lead only to small changes in output [6.30] cannot be guaranteed for the complex, highly nonlinear nature of viscous flow. Typical examples of relevance to rudder design that violate this assumption include: laminar-turbulent transition, flow separation, cavitation and breaking waves.

6.4.1 Validation of CFD calculations

In the same way in which there is always error (or more correctly a degree of uncertainty) in the acquisition of experimental data, numerical modelling gives rise to uncertainty in the answer obtained. The process of validation can be seen as an attempt to eliminate or at least quantify these uncertainties.

The process of validation can be seen as a series of stages:

(1) *Verification of the code implementation against the underlying mathematical formulation.* This is to ensure the code is free of error due to mistakes in expressing the mathematics in the particular computer language used. Ideally, the comparison should be made against an analytic solution although often the comparison can only be made with other numerical codes.

(2) *Investigation of the independence of the solution from numerical parameters.* The most common form of dependence is on the density of the grid of points at which the governing equations are solved. For iterative techniques, which use a convergence criterion, the dependence of solution on this value and any other has also to be investigated.

(3) *Direct comparison of numerical and corresponding experimental data.* Computational fluid dynamic codes are an approximation to the actual physics of the flow so there will always be differences between the experimental and numerical results. Experimental data should also have a specified accuracy. This should then allow the difference between experimental result and numerical prediction to be quantified. In many codes some degree of empiricism is used to adjust the numerical model to fit specific experimental data. The extent to which such an empirically adjusted model can be said to be valid for cases run at different conditions requires careful consideration.

As yet there is no standard method for evaluating numerical uncertainty, accepted by the CFD community, hence it is a subject of much debate, together with the exact definitions of verification and validation. However, the simple definition of verification and validation given by Blottner [6.31], provides a broad definition which encompasses many of the accepted interpretations:

> ***Verification is solving the equations right, and validation is solving the right equations.***

Stern *et al.* [6.32] discuss many of the verification and validation issues surrounding CFD. A valuable guide to the validation process for both surface panel and finite volume CFD can be found in reference [6.28].

6.4.2 Error and uncertainty estimation

Identification of the error and uncertainty resulting from the use of the underlying finite difference approximations of the governing RANS equations is an essential component of a numerical analysis. A spatial discretisation or grid independence study involves obtaining solutions on successively refined grids. Depending on the number of meshes computed, estimates of discretisation error and order of accuracy can be made. A number of systematic approaches to the quantification of uncertainty and order of accuracy have been proposed. Stern *et al.* [6.32] advocate error estimates based on a form of Richardson extrapolation [6.33] using three model grids systematically refined in the three coordinate directions. Eça and Hoekstra [6.34] argue that this method can be unreliable.

Ideally, asymptotic convergence of an important integrated global parameter such as lift, drag or field parameter like velocity or pressure at some important location in the flow, should occur as the computational mesh is refined. In practice, asymptotic convergence can be hard to achieve due to numerical problems associated with the solver used. The process of finding a grid independent solution can be a complex one, especially when 3-D grids are considered, as the grid properties in each dimension are often interrelated with regard to the flow-field variables.

Researchers who follow the approach of Stern *et al.* [6.32] to discretisation error estimation, successively refine their grids in all the dimensions by the same refinement factor. This approach can result in meshes that are over-refined in some areas and under-refined in others. For instance, Date [6.35] showed that highly refined grids can produce solutions which violate the asymptotic convergence criteria. It should be noted that the degree of grid independence for a particular CFD model should be related to the degree of accuracy needed in the final solution.

Time-accurate flow computations should include a temporal time step independence study, whereby the temporal time accuracy of the flow solution is assessed. In the same way as the grid independence study, the time step size used should be varied and its effect on the solution investigated.

Mesh extrapolation: A convergence study can be used to extrapolate a mesh-independent solution. At least three meshes are necessary to estimate the convergence behaviour. These should be refined in all coordinate directions, i, j, k, each time using the same refinement factor

$$r_{\mathrm{G}}^{i,j,k} = \frac{\Delta X_{\mathrm{A}}}{\Delta X_{\mathrm{B}}} \tag{6.54}$$

ΔX_{A} and ΔX_{B} are the grid line distances for the base and refined mesh. Mesh convergence is estimated using the ratio of changes in the solution:

$$R_{\mathrm{G}} = \frac{\varepsilon_{21}}{\varepsilon_{32}} = \frac{S_2 - S_1}{S_3 - S_2} \tag{6.55}$$

$S_{1,2,3}$ describes the solutions (a scalar of a chosen property) on the fine (1), medium (2) and coarse (3) grid. The following cases occur:

- Converging condition: $0 < R_{\mathrm{G}} < 1$
- Oscillatory condition: $R_{\mathrm{G}} < 0$
- Diverging condition: $R_{\mathrm{G}} > 1$

According to the Richardson extrapolation [6.3, 6.33] the first order error on the fine grid is

$$\delta_{\mathrm{RE,G1}} = \frac{\varepsilon_{21}}{r_{G}^{P_{\mathrm{G}}} - 1} \tag{6.56}$$

and gives the difference between the computation at high resolution and the grid independent solution, neglecting all higher-order terms and r_{G} is the grid refinement factor. The estimated order of accuracy is defined as

$$P_{\mathrm{G}} = \frac{\ln(\varepsilon_{31}/\varepsilon_{21})}{\ln r_{\mathrm{G}}} \tag{6.57}$$

A correction factor

$$C_{\mathrm{G}} = \frac{r_{\mathrm{G}}^{P_{\mathrm{G}}} - 1}{r_{\mathrm{G}}^{P\mathrm{th}} - 1} \tag{6.58}$$

is introduced, using Pth for the theoretical order of accuracy of the applied method. For $C_{\mathrm{G}} \approx 1$ the obtained solutions are within asymptotic range. In this case sign and magnitude of the error are given by

$$\delta_{\mathrm{G1}} = C_{\mathrm{G}} \delta_{\mathrm{RE,G1}} \tag{6.59}$$

The uncertainty can be estimated by

$$U_{\mathrm{G1}} = |(1 - C_{\mathrm{G}} \delta_{\mathrm{RE,G1}})| \tag{6.60}$$

For $C_G > 1$ Equation (6.60) under-predicts the error, $C_G < 1$ indicates an over-prediction. If C_G is far from one, a lack of confidence in Equation (6.58) and (6.59) is justified. The uncertainty should then be calculated from

$$U_{G1} = |C_G \delta_{RE,g1}| + |(1 - C_G)\delta_{RE,g1}| \qquad (6.61)$$

In practice, this procedure can only predict a grid-independent solution in case of asymptotic grid convergence. For oscillatory behaviour, more results are necessary to resolve the grid-dependent oscillations and prove possible oscillatory convergence. Typically, this requires large computational effort and even then it is not clear if the oscillatory behaviour can be extrapolated reliably to a grid independent solution [6.34].

Parametric mesh independence studies: An alternative to the above procedure is that of mesh-independence studies, focusing on the set of independent mesh parameters. This approach is more flexible, especially when it comes to complex grid configurations. With respect to unstructured hybrid grids it is more practical to apply this approach.

The variation of mesh parameters is carried out to find a parameter independent solution. It is useful to work on the parameters of decreasing importance, e.g., resolve in surface normal direction to capture the boundary layer first. In contrast to a single grid-refinement factor, no grid extrapolation is carried out. The procedure is rather regarded as a way of efficient grid optimisation. By parametric studies the mesh can be adjusted and it should be possible to generate and optimize a mesh, avoiding cells in areas where further refinement is unnecessary.

6.5 Free-stream rudders

Many applications, from underwater vehicles with control surfaces mounted upstream of the propulsor to fully appended yachts, require a knowledge of the free-stream manoeuvring forces and control surface loadings. The following sections illustrate how well each of the previously described methods capture the free stream performance of control surfaces.

6.5.1 Lifting line

The low aspect ratio corrected lifting line is applied to the free-stream experimental tests given in Chapter 5. The method is applied to a semi-balanced skeg rudder with a geometry as shown in Figure 6.10. In addition to the tip vortex correction discussed in 6.2.1, a further line vortex is required to represent the mid vortex at the gap between all-movable and skeg rudder. A schematic of the downwash influence is given in Figure 6.11.

As an example of the performance using the semi-empirical corrections for the strength of the imposed line vortices, Figure 6.12 shows the net spanwise loading distribution C_N for a range of skeg angles and rudder angles between 0° and 30°. A close comparison is obtained with the experimentally measured pressure integration.

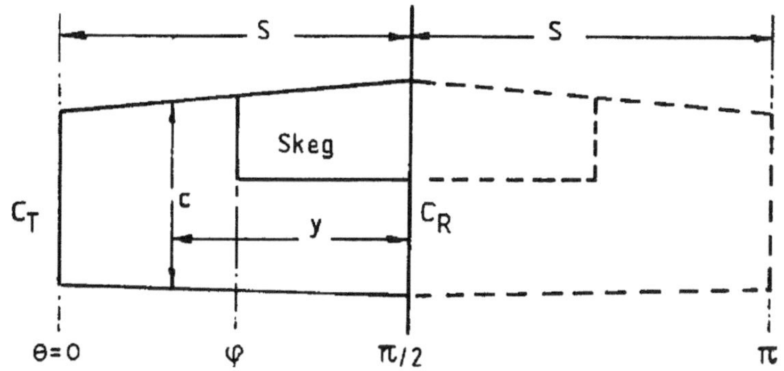

Figure 6.10 Geometry used in lifting-line analysis

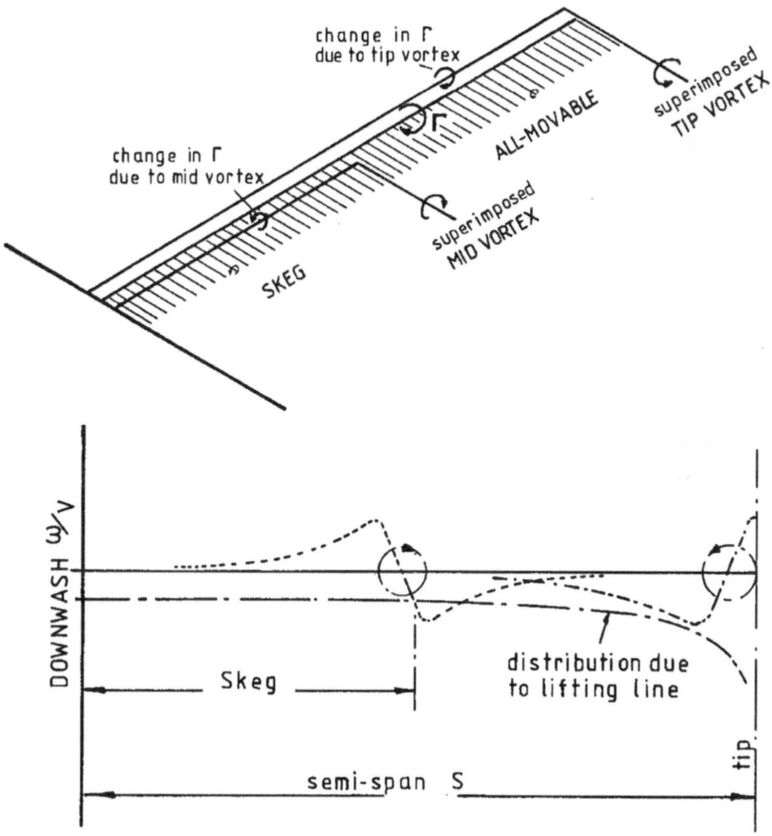

Figure 6.11 Vortex model and assumed downwash distribution

The lower load over the skeg, where it acts as a flapped foil, can be seen as can the strong tip vortex present at high rudder angles.

Figure 6.13 gives the total rudder forces estimated by the lifting line code compared to the wind tunnel measurements. Excellent comparison is obtained. It should be

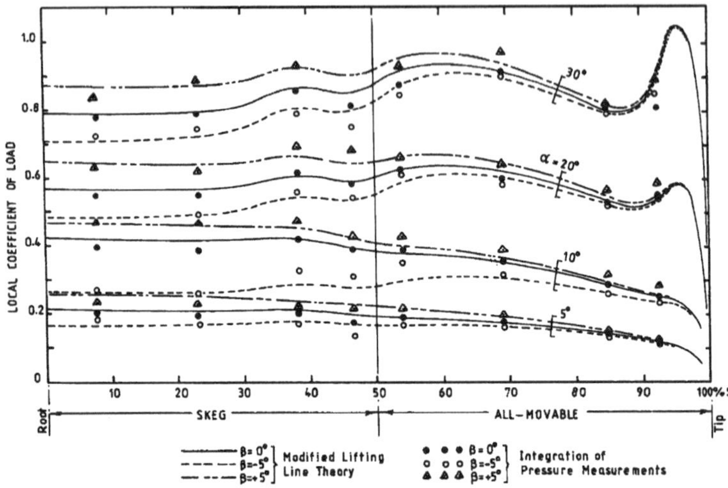

Figure 6.12 Comparison of theoretical and experimental spanwise loadings, semi-balanced Skeg Rudder No.1, Figure 5.39

noted that although *CPs* is calculated directly, *CPc* is obtained using an empirical estimate at each spanwise station for the location *CPc*.

The drop in sideforce as the skeg gap behaviour becomes important beyond 10° is captured through the mid-vortex model.

Such a calculation requires negligible computational resources and, as a result, is ideal for direct inclusion within a ship simulator. However, the excellence of comparison is restricted to the range of experimental geometries tested and used to develop the semi-empirical corrections.

6.5.2 Surface panel

The prediction of free-stream performance of all-movable rudders is a straightforward task for a surface panel code. This will be illustrated using the surface panel code, Palisupan, previously described. Similar results can be expected with commercial surface panel codes. Rudder No. 2, Figure 5.55, will be taken as the benchmark case. Figure 6.14(a) illustrates the process of meshing the simple ruled geometry. A series of spanwise sections are defined and a bidirectional cubic spline interpolation is used to generate the quadrilateral panels. It should be apparent that in areas of high curvature, e.g., near the leading edge, more and thus smaller panels will be required to minimize the error in capturing the actual geometry. For geometries of $t/c < 20\%$, although the correct application of the theory requires a completely closed body, if the square tip is left open it only has a small effect on the developed forces.

For a control surface adjacent to a hull, or as in the case of the wind tunnel tests the working section floor, an image system is required to capture the effective aspect ratio effect (Section 5.5).

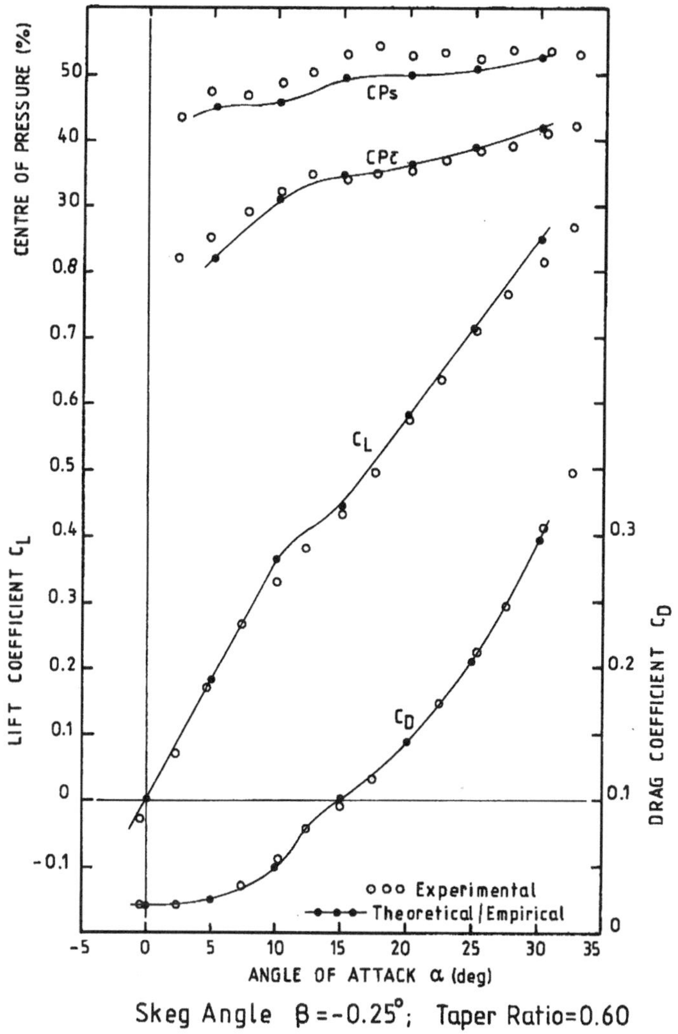

Figure 6.13 Comparison of theoretical and experimental results, semi-balanced skeg Rudder No.1, Figures 5.39 and 5.41(d), rudder plus skeg

Although the zero pressure loading condition on the rudder wake implies that it should follow the local flow if the bisector of the trailing edge is used to prescribe the wake direction (line 1, Figure 6.15) only small changes are found in the developed forces compared to the more realistic shape (line 2, Figure 6.15), whereas, line 3 causes a considerable error. The shape of line 2 was generated using a simple blending function between the foil T.E. bisector and the free-stream direction imposed over a distance of 1.5*c*.

The length of the meshed wake needs to extend a sufficient distance such that adding further panels only makes small changes to the developed forces. In many codes a semi-infinite wake model removes the need to do this. However, when the

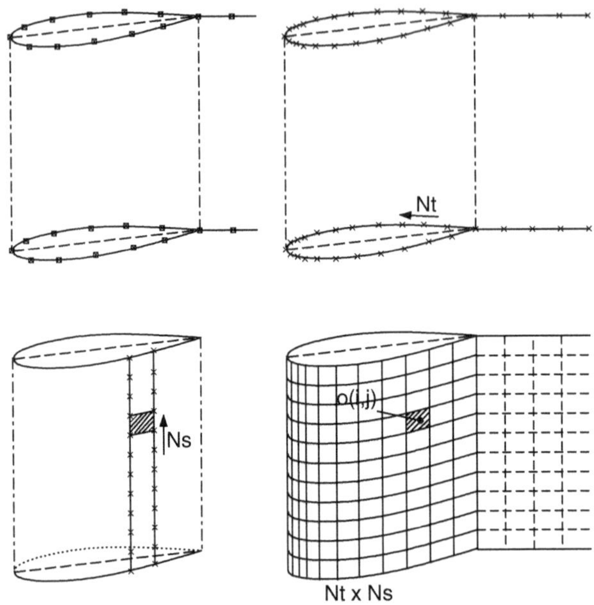

Figure 6.14 (a) Development of surface panel mesh using bicubic spline surface

rudder is in the presence either of other control surfaces or other bodies the panelled wake must extend a reasonable distance beyond their downstream limit in order to satisfy the implicit conditions imposed on the semi-infinite wake. Close proximity of a series of trailing wake panels with another lifting surface can often cause numerical difficulties and requires considerable attention to the size and location of panels on both the wake and control surface.

In some codes, the wake is allowed to 'relax' and follow the actual stream surface near to the trailing edge. In this case, the wake tip starts to roll-up as occurs in the real flow. Unfortunately, as in the far distance all of the vorticity in the wake concentrates into a line vortex, the roll-up will become unstable for a finite number of wake panels. Most panel codes impose modifications that attempt to limit these effects.

The order of accuracy of a particular code will be controlled by that used to describe: the geometry, flat panels or curved surfaces; the surface variation in potential, constant, linear or a higher order polynomial variation and finally the method used to find the surface potential. In code development there is a trade-off [6.36] between complexity of coding for higher order applications, the time required to calculate the influence coefficients and the number of panels used. In general, low order panel methods are typically more robust.

The numerical discretisation used in the surface panel methodology results in a scheme where the results are dependent on:

- the size of the panels in a given direction compared to the surface curvature;
- the aspect ratio of the panels;
- the relative size and proximity of panels; and
- how much the panels are skewed.

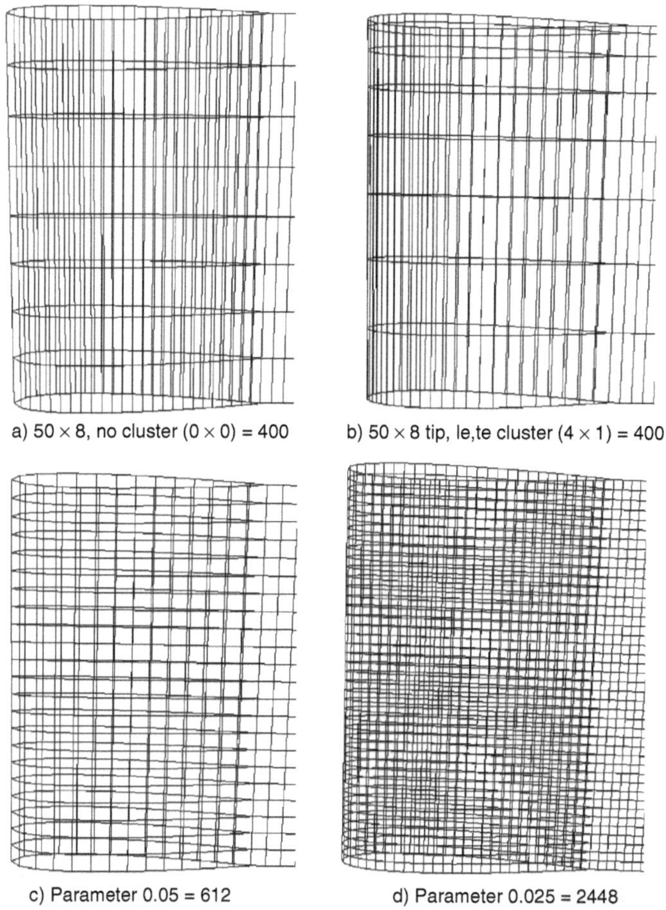

Figure 6.14 (b) Four examples of various panel mesh densities and distributions used for sensitivity study (Table 6.2)

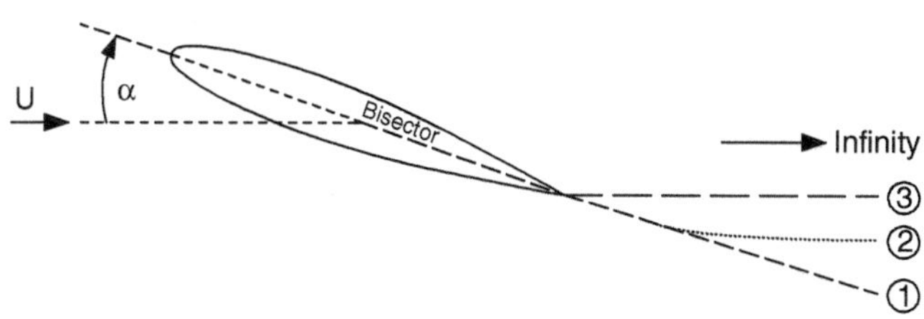

Figure 6.15 Possible directions for wake sheet from trailing edge

A systematic study should be carried out to investigate these dependencies. The following guidelines can be used as a starting point in mesh generation.

- Typically, high curvature is confined to the chordwise direction so a greater density of panels per unit length should be used but limiting the increase so that aspect ratio of the spanwise to chordwise dimension is no more than 3.
- The chordwise size of the trailing edge panels should be similar to those at the start of the wake.
- The distance between two panel centroids should not be closer than 10% of the largest diagonal of the larger panel. If this occurs the overall panel density will need to be increased. This problem becomes more important as the foil t/c reduces and also for thin (cusped) trailing edges.
- For standard foil sections, progressively closer agreement in surface pressure is found as the number of chordwise panels going all the way round the foil rises from 25 to 50. Beyond this, the gains in accuracy probably do not outweigh the extra computational time, especially for high aspect ratio foils.

Figures 6.16–6.18 illustrate the performance of Palisupan for a comparison of surface pressure and sectional lift at +9.6° and for C_L, C_D and CPc, CPs across a range of rudder incidence.

It is worth noting that the tip vortex behaviour is not captured as this is a rotational flow feature not represented directly in potential flow. This is seen in the local lift rise towards the tip and in the corresponding change in surface pressure at the outer span. Such strong tip vortices are a feature of higher thickness/chord ratio. For high aspect ratio rudders used on yachts such behaviour would not be as important.

The overall forces and moments are well predicted at low incidence with progressive increase in error beyond 15°. This corresponds to the gradual initiation of stall across more of the rudder span (not captured in surface panel codes). This is reflected in the movement aft of CPc and towards the tip for CPs.

Figure 6.14b and Table 6.2 show the results of a mesh sensitivity study carried out for Rudder No. 2 at +9.6°. The performance of the panel code is investigated with respect to the evaluation of the nondimensional lift and drag found from the numerical integration of the normal pressure (Equation (6.31)) and tangential shear (Equation (6.33)). Two types of mesh generation strategies were applied. The first used two parameters to specify the number of panels in the chordwise (Nt) and spanwise (Ns) directions, respectively. The second used a single mesh size parameter (m) that ensures a uniform panel size with an aspect ratio of one across the whole rudder surface.

In each case the mesh refinement factor (Equation (6.54)) is 1.5. In addition to the uniform mesh refinement, two examples of the influence of varying the relative size of panels are given with clustering just at the leading edge (4×0) and clustering at the leading edge and rudder tip (4×1). It can be seen that, for progressive refinement (increase in panel numbers), the values are converging for all lift and drag components. The rate of convergence is slower for the drag values with the viscous estimate and pressure integration responding both to the better representation of the foil shape and the surface area. However, as these changes are in opposition,

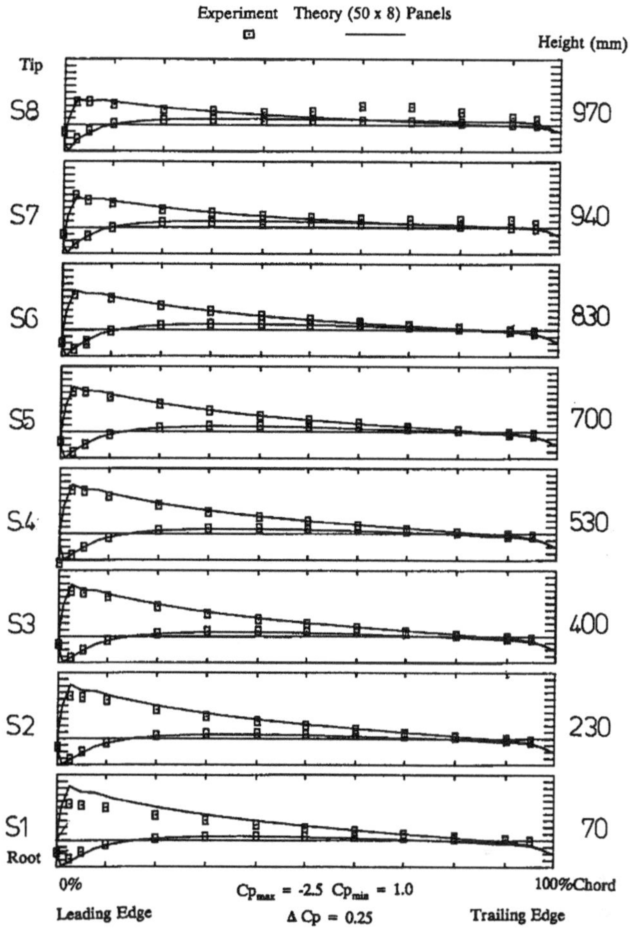

Figure 6.16 Comparison of numerical and experimental chordwise pressure distributions at eight spanwise positions; Rudder No. 2, $\alpha = +9.6°$

the actual coefficient of variance (COV) for all values given in Table 6.2 is 3% whereas for the two components it is 14%.

The differences between the two-mesh strategies and with the clustering of panels at the leading edge and tip indicate the importance of selecting correct mesh strategies to minimise computational effort. The strategy that selects less panels in the spanwise direction and clusters panels at the leading edge requires (74 × 12, 4 × 0) only 888 panels compared to the constant aspect ratio choice with a comparable pressure drag, C_{Dp} of 2,944. This first strategy corresponds to a reduction in computational effort by a factor of 11.

Finally, Figure 6.19 (See Plate 4 of Colour Plate Section) illustrates the surface pressure and viscous stress distribution for coupled boundary layer and surface panel calculations.

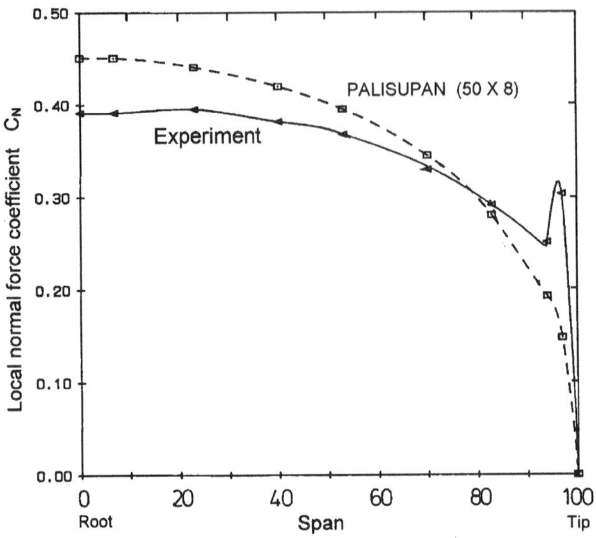

Figure 6.17 Comparison of numerical and experimental spanwise pressure distribution, Rudder No. 2, $\alpha = +9.6°$

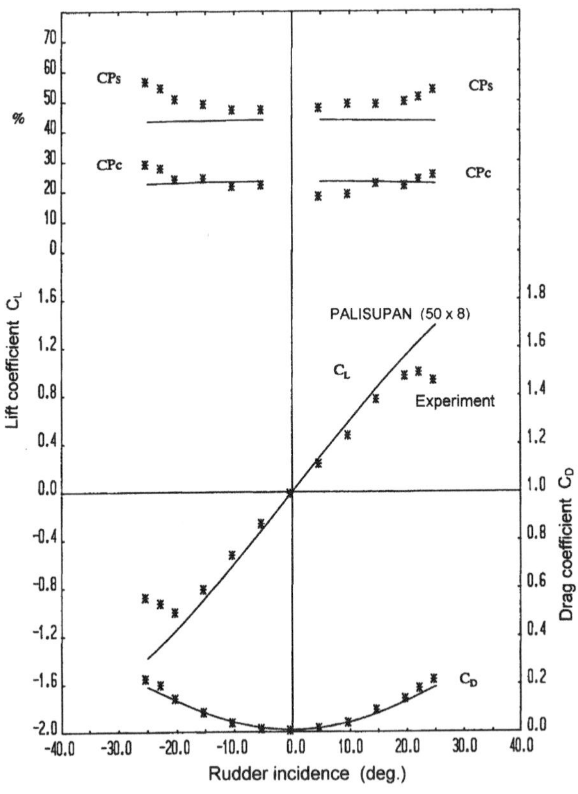

Figure 6.18 Comparison of lifting surface and experimental lift, drag and centre of pressure characteristics; Rudder No. 2 in free stream

Table 6.2 Sensitivity of lift and drag prediction to panel number and distribution at +9.6°

Panel										
Nt	*Ns*	*Dist.*	C_{Dp}	C_{Dv}	C_D	C_{Lp}	C_{Lv}	C_L	C_L/C_D	$dC_L/d\alpha$
50	8	0×0	0.034802	0.020781	0.05558	0.540831	−0.002261	0.53857	9.689473	0.056101
74	12	0×0	0.032956	0.022794	0.05575	0.533541	−0.002531	0.53101	9.524843	0.055314
112	18	0×0	0.032079	0.025357	0.05744	0.528571	−0.002836	0.52574	9.153406	0.054764
74	12	4×0	0.032765	0.022664	0.05543	0.533267	−0.002295	0.53097	9.579318	0.05531
74	12	4×1	0.034359	0.022314	0.05667	0.543203	−0.002268	0.54094	9.544845	0.056347
Parameter	*N*									
0.125	220		0.046595	0.018237	0.06483	0.534351	−0.001936	0.53242	8.212225	0.05546
0.075	490		0.042291	0.01986	0.06215	0.537606	−0.002098	0.53551	8.616241	0.055782
0.05	1,100		0.039277	0.021867	0.06114	0.527007	−0.002345	0.52466	8.58076	0.054652
0.0333	2,336		0.036553	0.024096	0.06065	0.522856	−0.002633	0.52022	8.577602	0.05419
0.0222	2,944		0.032973	0.026608	0.05958	0.521856	−0.002911	0.51895	8.709908	0.054057
		Mean μ	0.036465	0.0224578	0.058923	0.5323089	−0.0024114	0.529898	9.018862	0.055198
		Standard deviation σ	0.005231	0.0033277	0.001995	0.0069839	0.0003948	0.007346	0.190519	0.000765
		COV σ/μ %	14.35	14.82	3.39	1.31	16.36	1.39	2.11	1.39

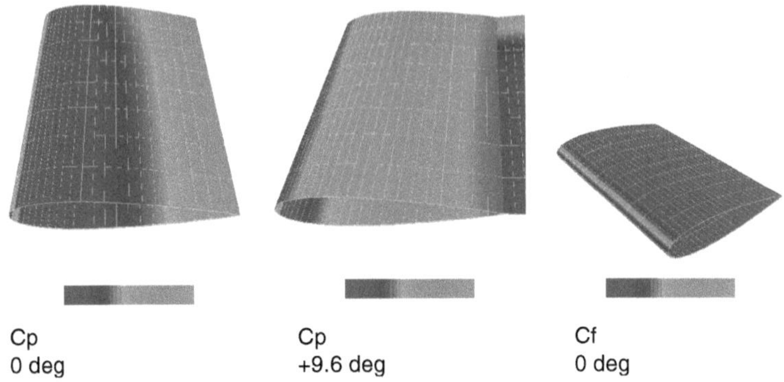

Figure 6.19 Visualisation of surface pressure and skin friction for coupled boundary layer calculation on Rudder No. 2 (See Plate 4 of Colour Plate Section)

6.5.3 Reynolds averaged Navier–Stokes

As an example of the use of a RANS code for free-stream rudder performance prediction a validation study [6.37] is presented. A commercial flow solver was used to investigate the behaviour of the all-movable Rudder No. 2, Figure 5.55. This work compared various mesh strategies and their influence on solution accuracy. Hybrid meshes were generated. In 2-D the outer mesh consists of either triangles or unstructured quadrilaterals. The inner boundary mesh uses only quadrilateral cells to resolve the boundary layer flow. The 3-D meshes are generated by extruding the 2-D base grids into the third dimension, defining the 3-D node distribution by meshing the volume edges.

The coordinate system (x, y, z) is defined on the trailing edge with the x-axis pointing downstream in the direction of the chord, Figure 6.20. For the tunnel cases the rudder is turned within the solution domain around the pivot at $0.3c$ from the leading edge. The origin is defined on the point of the trailing edge at $\alpha = 0°$, the x-axis pointing downstream along the tunnel symmetry line.

In order to investigate the blockage of the wind tunnel, the rudder is modelled within the tunnel as well as in the free stream. The meshes used are shown in Figure 6.21 for the tunnel and free-stream boundaries.

For the 3-D case it is necessary to refine the grid around the tip with respect to the tip vortex and next to the root, which is mounted as in the original wind tunnel tests as a no-slip wall. This requires expansion ratios going out from the tip in high and low z-direction, and from the root of the rudder (tunnel floor). The mesh ratios used are given Table 6.3.

In 3-D the tip face of the rudder also has to be meshed. It is subdivided into two faces, serving as base faces for the inner blocks. Here the advantage of unstructured meshing can be exploited as shown in Figure 6.22. Along the larger front face of the tip a boundary layer mesh is created, adjacent to the outer boundary layer mesh. This provides a smooth transition from the outer mesh, across the outer

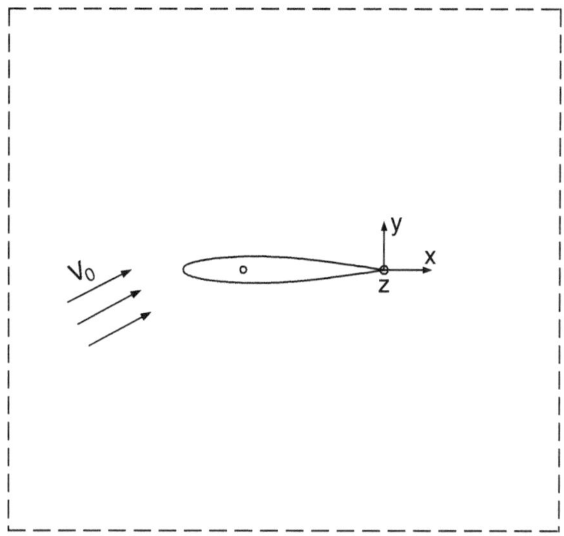

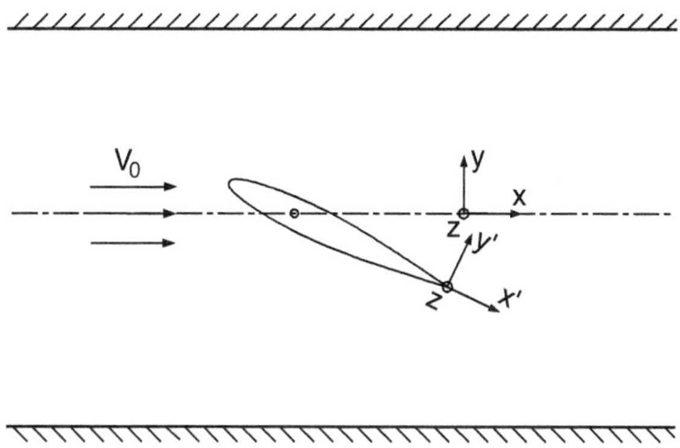

Figure 6.20 Coordinate system for RANS calculation of Rudder No. 2: (a) free stream, (b) wind tunnel

boundary layer mesh, across the inner boundary layer mesh, to the interior area. The inside is filled up by unstructured quadrilaterals. The aft triangle is defined as an extra block and meshed separately. Here cell aspect ratios of the adjacent outer mesh are relatively small so that a smooth transition, without an inner boundary layer mesh, can be realised. Within structured C-grids especially, this aft part of the tip face causes problems of high cell skewness.

Boundary Conditions: In the free stream case any disturbing influences of the boundaries are to be minimised. Following the recommendations of reference [6.38], the boundaries are located ten chord-lengths horizontally and six chord-lengths

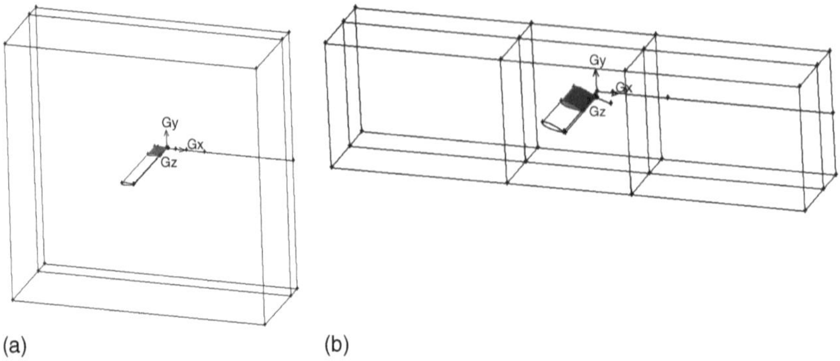

Figure 6.21 Block topology applied to rudder: (a) free stream, (b) wind tunnel

Table 6.3 Three-dimensional grids, NACA 0020, Re $= 8.0 \times 10^5$

No. of nodes	*Free stream*	*Wind tunnel*
Base mesh	20,420	18,970
Main tip mesh	1,830	1,785
Aft tip mesh	186	193
z-distribution ($z_{floor}...z_{tip}$)	80	80
z-distribution ($z_{tip}...z_{roof}$)	60	50
$\sum$ Total	2,979,760	2,565,000

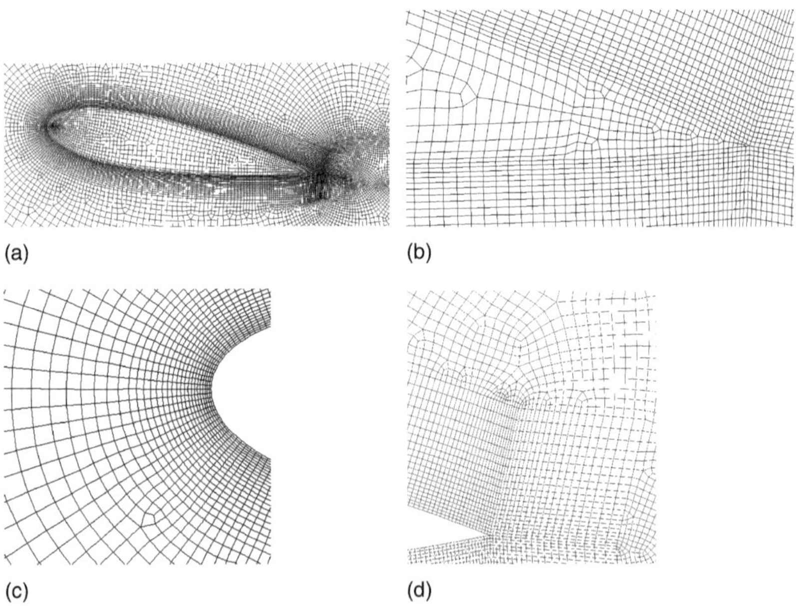

Figure 6.22 Details of mesh at tip ($z = 1.0$): (a) whole rudder (tip-plane z = 1.0) tip showing inner structured mesh, (b) trailing edge, (c) leading edge, (d) influence on block boundaries near trailing edge

vertically away from the rudder. A velocity inlet is applied, defining the angle of incidence by the velocity components of the flow. Downstream a pressure outlet is defined. The side faces are set to a velocity inlet or pressure outlet, assuming the same definitions of flow variables as for the upstream inlet and the downstream outlet, respectively. For the zero angle of incidence both side faces are defined as pressure outlets. The ground face ($z = 0$), the rudder is attached to, is set to a no-slip wall. Its viscous boundary layer is modelled by wall functions, using a nondimensional wall distance $y^+ \sim 40$. The top of the domain is modelled as a pressure outlet. The rudder surfaces are no-slip walls.

For the wind tunnel case the solution domain extends 10c up- and downstream, going out from the trailing edge. The upstream length of the domain will also affect the thickness of the boundary layers on the tunnel walls. Its floor $z = 0$ and side-walls are modelled as no-slip walls with $y^+ \sim 40$. The tunnel roof is defined as a slip wall; this saves resources as the viscous boundary layer on the wall is neglected and does not need resolution. Upstream a velocity inlet is defined, and downstream a pressure outlet. The angle of incidence is realised by turning the rudder, hence the direction of the inlet velocity is constant for all tunnel cases. Figure 6.23 shows a plot of the 3-D wind tunnel mesh.

Convergence: For the 3-D investigation the automatic convergence criterion is defined by the mass residual, reducing by four orders of magnitude. It can be noted that this criterion causes a numerical overhead as forces converge much earlier. Figure 6.24 shows the convergence histories for the free stream case at $\alpha = 10°$, using the $k-\varepsilon$ RNG turbulence model.

3-D Mesh refinement study: The mesh generation scheme, used to extrude the 2-D base mesh, allows an independent refinement study of node distribution in the z-direction. Parameter convergence is estimated on the base of the integral rudder forces as well as on the spanwise distribution of the normal force coefficient.

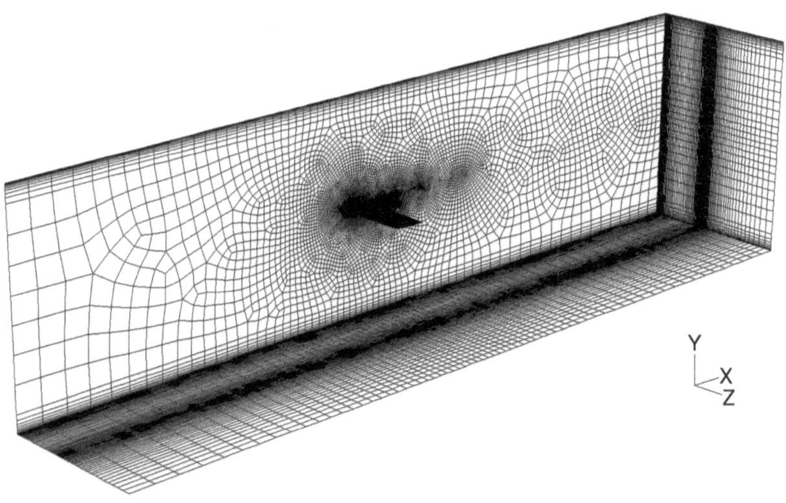

Figure 6.23 Mesh on bounding surface for rudder in wind tunnel

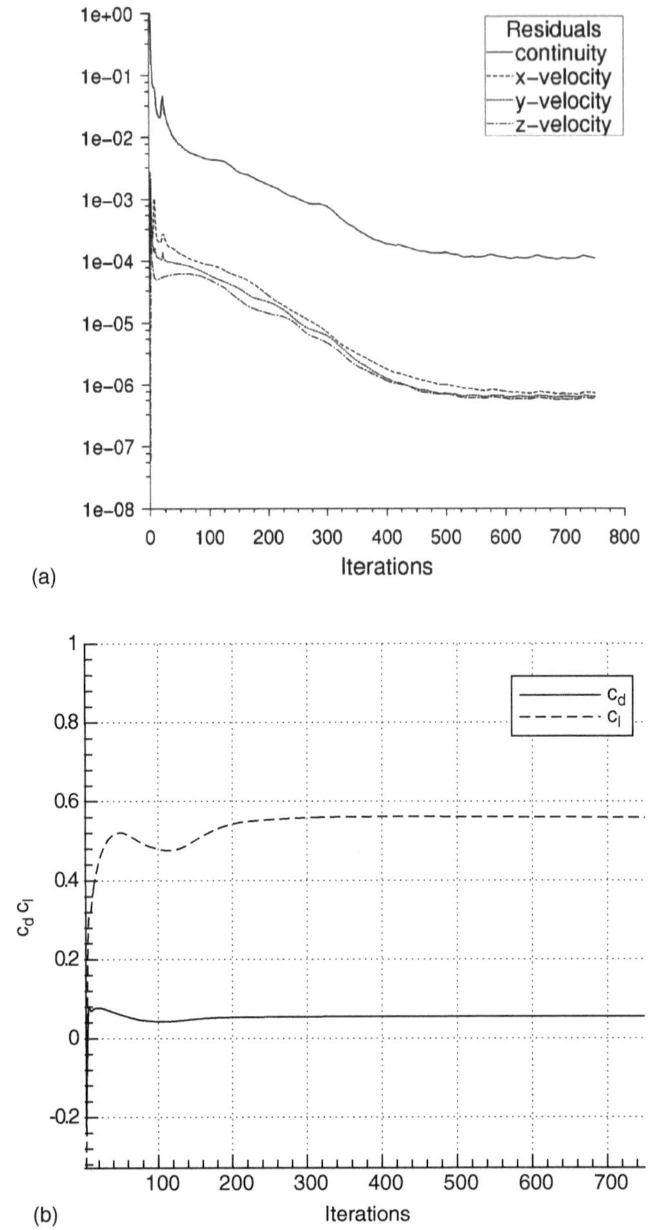

Figure 6.24 Typical convergence history for RANS calculation of rudder flow; NACA0012, $Re = 8 \times 10^5$, $\alpha = 10°$, k–ε RNG. (a) Scaled residuals, (b) forces

The tip flow is expected to be sensitive to the spanwise node distribution. This investigation is based on three free-stream grids, successively increasing the number of nodes in the z-direction by $\sqrt{2}$, while the first cell size for the nondimensional wall distance y^+ is kept constant at $z = 0$, floor and $z = 1m$ tip. The base grids are identical to previously refined 2-D free-stream cases.

The z-refinement study is carried out using the $k-\varepsilon$ RNG turbulence model at $\alpha = 10°$. In this case the prediction of lift and drag appear independent of the variation in spanwise node distribution (Table 6.4).

Also, the spanwise plot of the normal force coefficient, Figure 6.25, shows little difference. While the first cell sizes are kept constant, only the expansion ratios in z-direction profit from the increasing number of nodes. These ratios are sufficiently small, as the results are almost constant. Except for the nondimensional wall distance y^+, the first cell size is also kept constant for the cell geometry around the tip. The constant cell size provides a homogeneous, almost cubic cell shape around the tip.

With respect to computations at higher angles of incidence, the medium grid is selected for the following calculations.

Table 6.4 Mesh parameters and results of z-refinement study, NACA 0020, Re = 8.0×10^5, α = 10°

	Coarse	*Medium*	*Fine*	*Exp.*
$n_{\text{nodes,z}}$	99	140	198	–
$n_{\text{nodes,tot}}$	2,106,252	2,979,760	4,241,520	
C_d	0.0545	0.0540	0.0542	0.0440
C_l	0.5280	0.5304	0.5288	0.5107
$\Delta C_d(\%)$	+23.8	+22.8	+23.2	–
$\Delta C_l(\%)$	+3.4	+3.9	+3.6	–

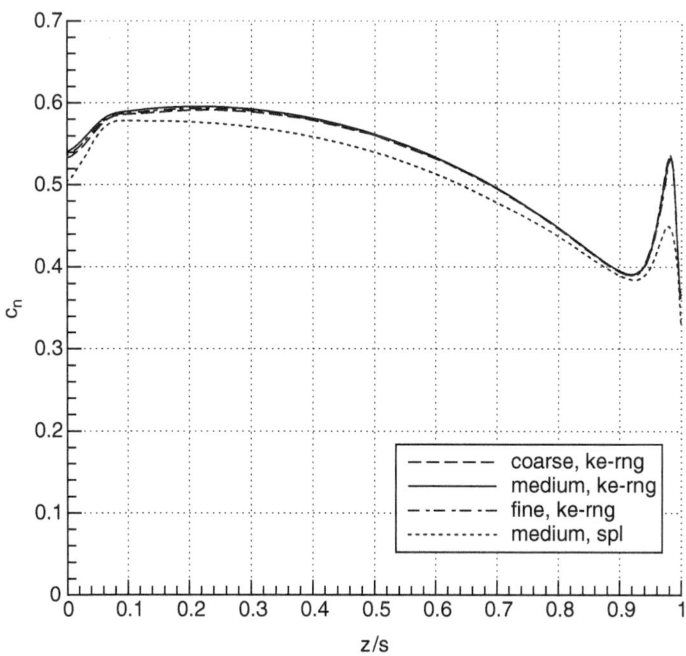

Figure 6.25 Spanwise normal force distribution for different mesh resolutions, NACA0020, $Re = 8 \times 10^5$, α = 10°, free stream

Integral forces: Calculations are carried out for free-stream and tunnel boundary conditions. Both cases are investigated using the $k-\varepsilon$ RNG and the Spalart–Allmaras (S–A) turbulence models [6.24]. Results are plotted in Figure 6.26.

The lift prediction qualitatively captures the blocking effect of the modelled wind tunnel, hence the lift increases due to an increasing effective Reynolds number. This behaviour can be observed for both turbulence models. The $k-\varepsilon$ RNG computations predict a higher lift than the corresponding S–A case. It is noticed that the latter turbulence model tends to predict a decrease in lift earlier. Over the considered range of incidence the tunnel case, which geometrically comes closest to the real experimental case, over-predicts the lift. Therefore, the closer agreement between the free stream and the experimental curves is explained by the superposition of two counter effects: The lower free-stream velocity decreases the lift relatively, which is compensated by the tendency of over-predicting the lift. This over-estimation is observed for both turbulence models and was also noticed during 2-D investigations of the NACA0012 [6.35,6.37].

Figure 6.26 shows that the tunnel drag is predicted higher than found in the free stream. In 3-D the increase in frictional and viscous pressure drag due to blockage is expected to be partly compensated by a reduced induced drag. Compared to the free-stream case the closer tunnel roof restricts the tip flow, and with it the drag component induced by the tip vortex. For low angles of incidence the S–A model calculates the higher drag, whilst at $\alpha = 10°$ both turbulence models calculate almost the same values and for $\alpha = 20°$ S–A predicts a higher drag. At $\alpha = 20°$ the results are dominated by the different predictions of the viscous pressure drag as S–A can cope with detached flow much better than $k-\varepsilon$. The relative over-prediction

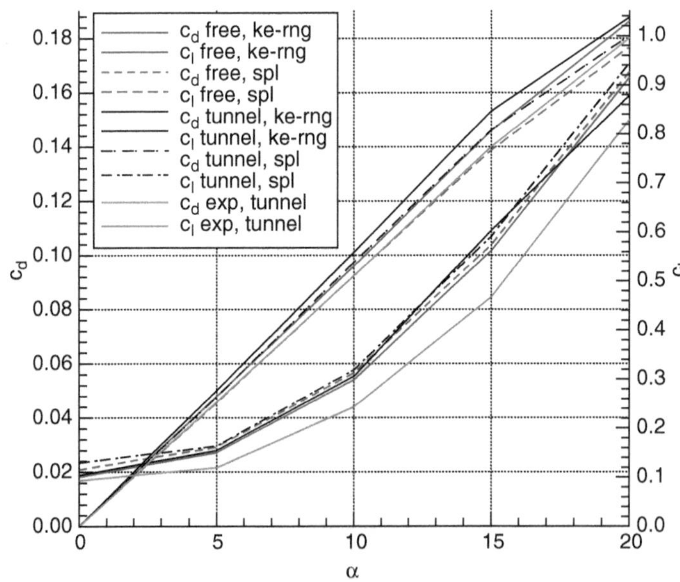

Figure 6.26 Influence of boundary conditions and various turbulence models on rudder force prediction using a commercial RANS code

of drag is higher and less sensitive to changes in boundaries and turbulence modelling than is the case for lift. It must be kept in mind that the pressure drag is caused by small differences fore and aft in the pressure-field.

The spanwise distribution of the normal force coefficient C_n is shown in Figure 6.27. At the root C_n is reduced as the rudder acts within the boundary layer of the tunnel floor, whilst near to the tip it is affected by the tip vortex.

Figures 6.25 and 6.27 demonstrate how the C_n prediction depends on turbulence modelling. Generally, less lift is predicted by S–A, but particularly the tip peak only reaches half the $k-\varepsilon$ RNG height.

The influence of boundary modelling on the tip peak, Figure 6.27, shows the expected behaviour. While the normal force distribution over most of the span is higher for the tunnel case, its tip peak is predicted lower than in free stream. This shows that moving the tunnel roof closer to the tip slightly decreases the development of the tip vortex.

Confirming the results of the previous sections, plots of the pressure fields and tip vortices help giving a more detailed view of the solutions. The vortex plots, Figure 6.28 (See Plate 5 of Colour Plate Section), display a selected surface of constant axial vorticity,

$$\zeta = \frac{\partial v}{\partial z} - \frac{\partial w}{\partial y} = -500 \tag{6.62}$$

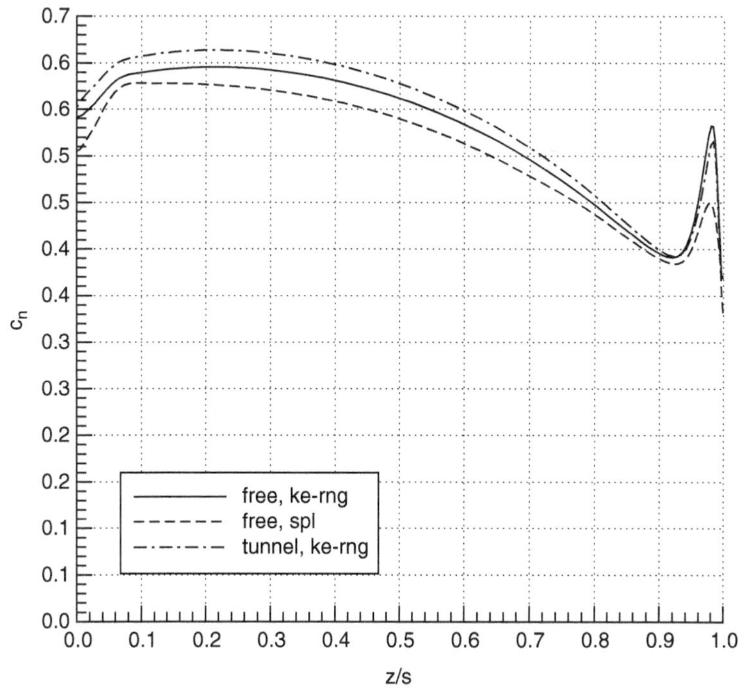

Figure 6.27 Spanwise normal force distribution for free stream and wind tunnel; NACA0020, $Re = 8 \times 10^5$, $\alpha = 10°$

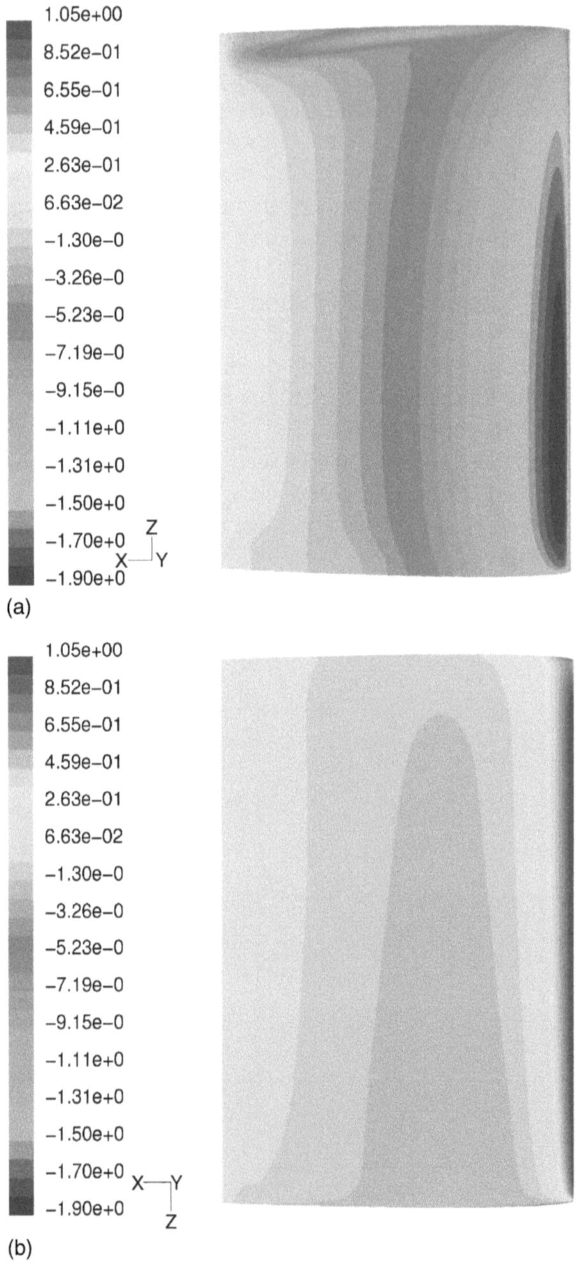

Figure 6.28 Pressure distribution on rudder surface; NACA0020, $Re = 8 \times 10^5$, $\alpha = 10°$, k–ε RNG. (a) Pressure side, (b) suction side (See Plate 5 of Colour Plate Section)

as well as the geometrical rudder surfaces. The plots are coloured according to the distribution of the pressure coefficient C_p.

Tip vortex development is shown in Figure 6.29 (See Plate 6 of Colour Plate Section). At $\alpha = 0°$ the vortices separate from the trailing edge and on either side of

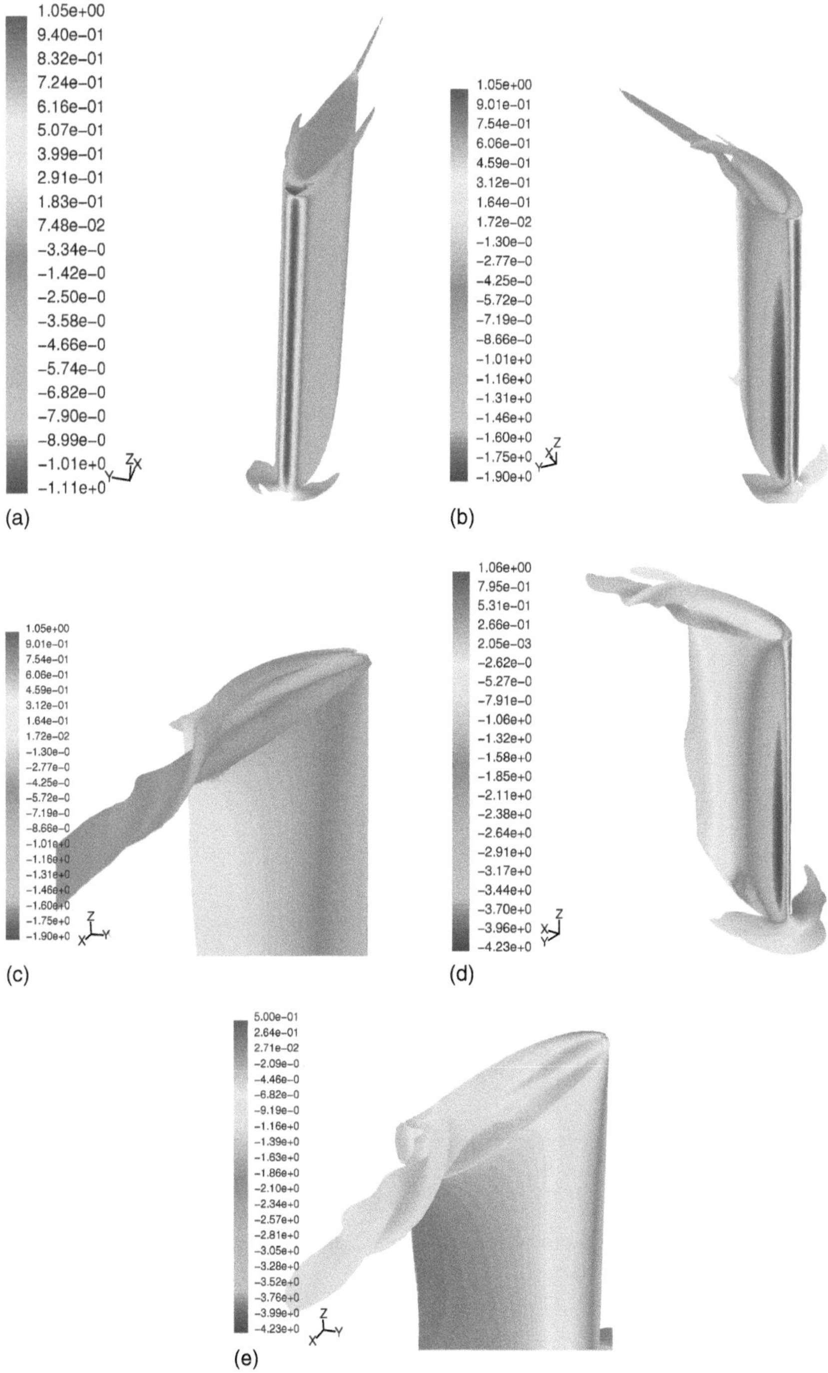

Figure 6.29 Tip vortex development captured using a constant vorticity surface (500), NACA0020, $Re = 8 \times 10^5$, $\alpha = 10°$, $k-\varepsilon$, RNG. (a) $\alpha = 0°$, (b) front view, $\alpha = 10°$, (c) rear view, $\alpha = 10°$, (d) front view, $\alpha = 20°$, (e) rear view, $\alpha = 20°$ (See Plate 6 of Colour Plate Section)

the tip. They are simply caused by the displacement effect of the rudder. The plot for $\alpha = 10°$ shows the reunion of the suction and pressure side vortices to one tip vortex behind the rudder. At $\alpha = 20°$ the vortices grow large due to the higher circulation and are also predicted to coalesce earlier. More detailed discussion of the influence of mesh on vortex capture and its important influence on rudder force prediction can be found in Pashias [6.26].

Figure 6.30 shows at selected spanwise positions the chordwise distribution of the pressure coefficient C_p: the maximum lift position, the local minimum before the tip, and the tip maximum itself; Additionally, the 2-D distribution for the NACA0020 section is provided. Comparing the pressure distributions on the rudder surfaces, Figure 6.30, and the development of the tip vortex, it becomes clear how the pressure field on the suction side is influenced by the tip vortex, converging in the negative z-direction. It finally leaves the rudder, coalescing with the vortex of the pressure side, which is of very limited influence on the tip peak in the C_n distribution. This can also be observed in the chordwise pressure plot for the maximum C_n.

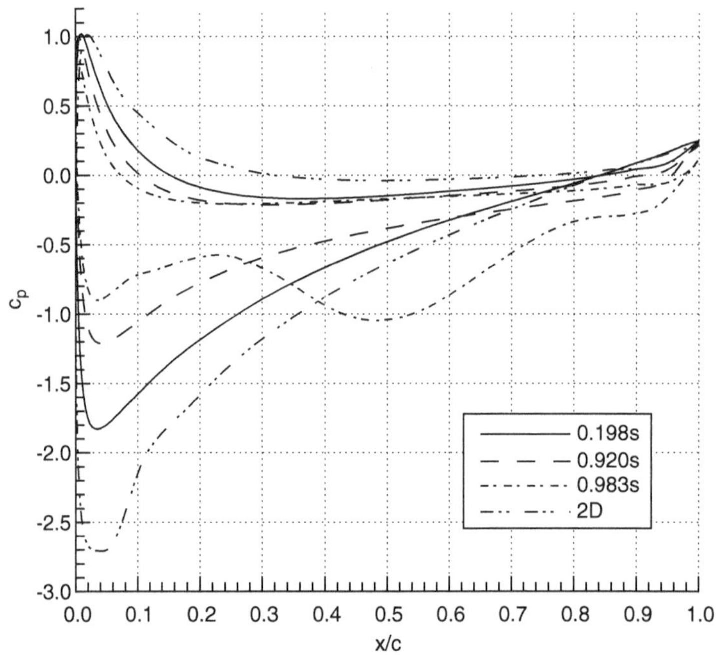

Figure 6.30 Chordwise pressure distributions at selected positions of span; NACA0020, $Re = 8 \times 10^5$, $\alpha = 10°$, free stream

6.6 Rudder–propeller interaction

The core problem associated with investigating the influence of an upstream propeller on the behaviour of a control surface is the induction of significant swirl velocity and axial acceleration. Prediction of the magnitude of the swirl and local velocity is the key to the successful analysis. The induced flow across the rudder is unsteady but dominated by the circumferential mean flow [6.39,6.40].

6.6.1 Lifting line/BEM

The circumferential mean flow influence can be captured by modifying the inflow to the lifting-line method described in Section 6.5.1 [6.41]. In this the work the propeller race flow is estimated using blade element-momentum theory. As described in Chapter 3 and demonstrated in Chapter 5, the rudder both blocks and diverts the flow through the propeller. This effect is included through use of theoretical estimates based on surface panel calculations.

The blade element-momentum output determines the axial and rotational inflow factors a, a' which are then multiplied by the Goldstein K factor [6.42] to give the mean-induced velocity and angle at each radius. The flow in the propeller race accelerates as it contracts. A simple correction, using a Gutsche [6.43]-type approach, based on the rudder distance from the propeller (X/D), gives a correction factor, using Equation (3.37):

$$K_{\mathrm{R}} = 1.0 + 1/(1 + 0.15/(X/D)) \tag{6.63}$$

and the induced axial V_{A} and tangential velocity V_{T} components become

$$\begin{aligned} V_{\mathrm{A}} &= V(1 + K_{\mathrm{R}}a) \\ V_{\mathrm{T}} &= \mathrm{K}_{\mathrm{R}}a'\Omega r \end{aligned} \tag{6.64}$$

The slipstream contraction is estimated by applying continuity between the propeller and rudder and using, as a first approximation, the axial velocity changes induced by the Gutsche correction

$$D_{X/D}/D = [(1 + a)/(1 + K_{\mathrm{R}}a)]^{1/2} \tag{6.65}$$

The upstream blockage effect of rudder on the propeller is included through a reduction to the applied advance ratio seen by the propeller J.

The resultant modification to the rudder lifting line is expressed as a change to the onset flow and an additional upwash/downwash dependent on the direction of rotation of the propeller and whether the particular rudder spanwise location is above or below the propeller axis.

$$\begin{aligned} \frac{V_{\mathrm{R}}}{V} &= \sqrt{\frac{\pi^2 x^2 \left(K_R d\right)^2}{J^2} + (1 + K_{\mathrm{R}}a)^2} \\ \lambda &= \tan^{-1}\frac{\pi x}{J}\cdot\frac{K_{\mathrm{R}}d}{(1 + K_{\mathrm{R}}a)} \end{aligned} \tag{6.66}$$

Figure 6.31 shows the performance of the lifting-line approach for predicting the spanwise load distribution for all-movable Rudder No. 2. Good agreement can be seen and this is also reflected in the predictions of rudder forces given in Figure 6.32.

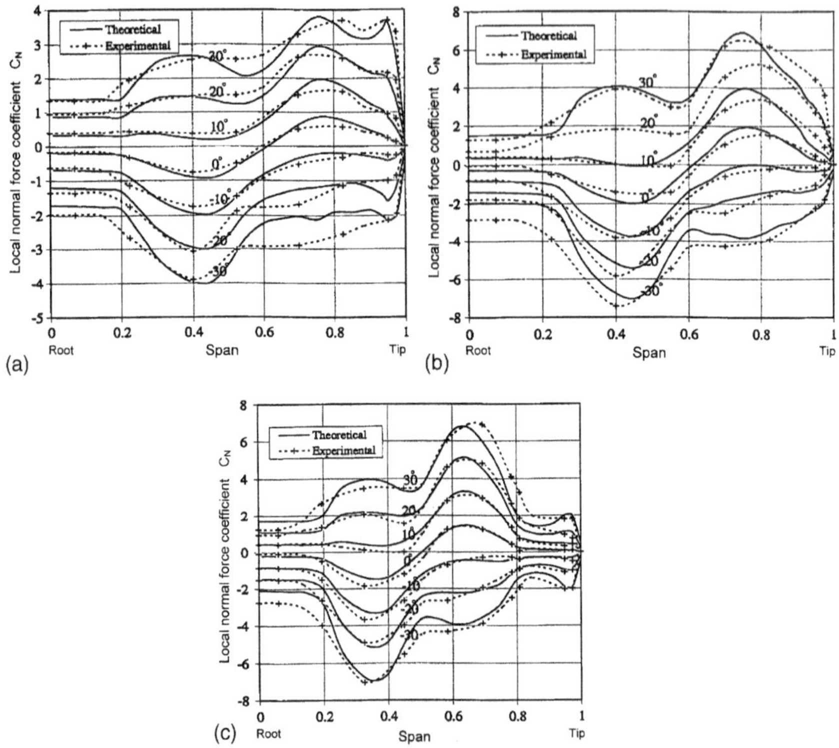

Figure 6.31 (a) Spanwise load distributions; comparison of lifting-line theoretical predictions and experimental results; Rudder No. 2, $X/D = 0.39$, $J = 0.51$; (b) Spanwise load distributions; comparison of lifting-line theoretical predictions and experimental results; Rudder No. 2, $X/D = 0.39$, $J = 0.35$; (c) Spanwise load distributions; comparison of lifting-line theoretical predictions and experimental results; Rudder No. 3, $X/D = 0.39$, $J = 0.35$

Although a number of empirical adjustments are required to provide accurate predictions these are based on physical effects.

6.6.2 Surface panel

It is possible to incorporate the effect of the propeller within a panel or lifting surface method [6.44–6.50] and recently a team at INSEAN [6.51] have all used a variety of boundary element methods to model ship rudders operating under both free stream and propeller flow conditions. The methods vary in their detail and complexity which they use to represent the structure of the propeller race. The ongoing work at INSEAN includes a complete, unsteady calculation of the wake shape evolution even in the presence of the rudder. The challenge in such approaches is in how to deal with the difficulty of controlling the numerical instabilities associated with wake roll-up while

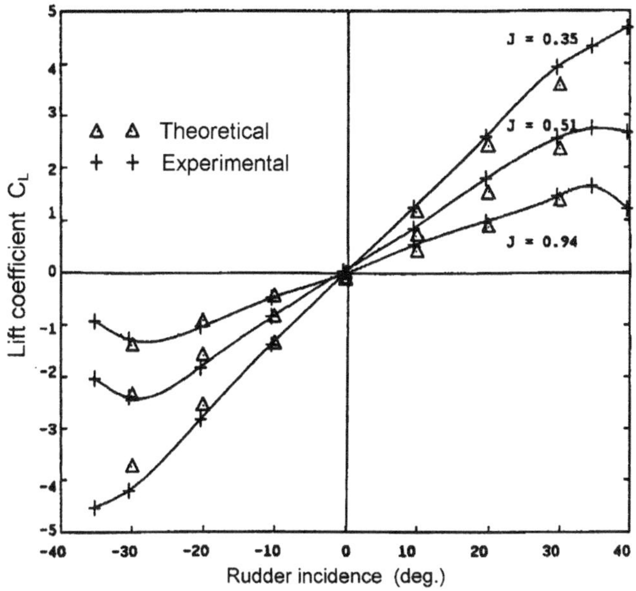

Figure 6.32 Comparison of lifting-line theoretical total lift predictions and experimental results; Rudder No. 2, $X/D = 0.39$

at the same time allowing the vortex filaments to stretch correctly as they arrive at the rudder leading edge and sweep along either side of the rudder.

The simpler approach is to start by considering just the circumferential mean influence of the propeller race. The problem can then be separated into two separate calculations, for example, reference [6.40] uses this approach for wing/propeller interaction. The interaction effects are then captured through a modification of the respective inflow conditions.

Turnock [6.47,6.51,6.52] developed a surface panel method based on the perturbation potential formulation with the aim of capturing the interaction between the rudder and propeller. This program, Palisupan, captures multiple body interaction by splitting the flow solution into multiple domains within which a number of bodies are placed. The interaction between each domain is captured through the flowfield modifications induced by the bodies within one domain on all others.

Using this interaction velocity field (IVF) method the bodies are not all modelled in one numerical pass thus creating a further iteration loop around the solution method described earlier. Take the example of rudder–propeller interaction, first the propeller flow model is solved to get a velocity influence upon the rudder. The rudder flow is then solved with the modified inflow velocity field to get a subsequent field on the propeller. This process is repeated with the starting point on the propeller being the velocity influence solved in the previous numerical pass. The procedure repeats until the difference in the results of body forces have iterated down to a minimum required value. Figure 6.33 is a flow chart of the overall velocity interaction process.

The interaction velocity field approach allows the available number of panels for a given body to be maximised to the computer memory available and hence allow higher quality grids to be created on the individual bodies. The approach is useful when two bodies are in close proximity as large dipole influence coefficients D_{ij} between panels on different bodies are replaced by a modification to the boundary condition on the right-hand side for an M domain problem. For example, for panels on body 1, U_i in Equation (6.25) becomes:

$$U_i = U_\infty + \sum_{l=2,M} U_{IVF-l} \tag{6.67}$$

and U_{IVF-1} is the net interaction (disturbance) velocity induced at the panel centre due to all the panels in body l.

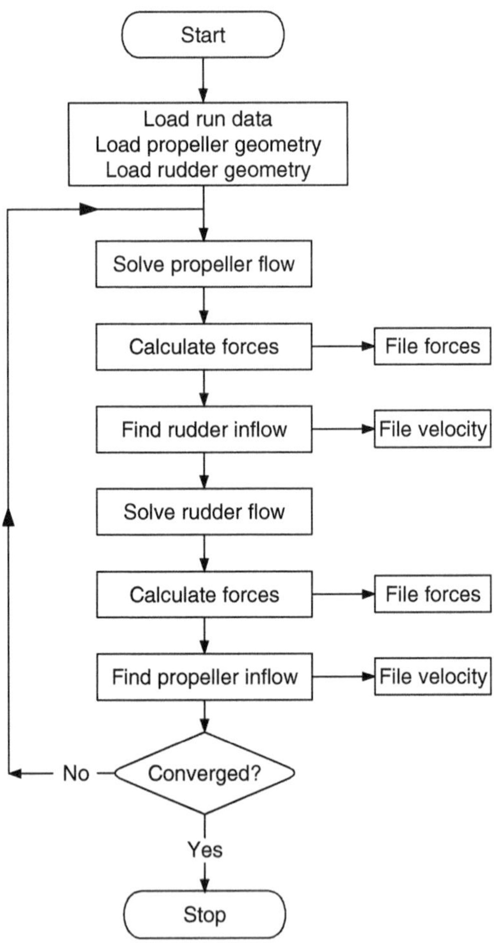

Figure 6.33 Flow chart of interaction velocity field algorithm

As an example of this approach the IVF method is applied to the modified Wageningen four bladed propeller and Rudder No. 2. A frozen propeller wake was used to determine the typical circumferential mean velocity field seen by the rudder, Figures 6.34 and 6.35, for the propeller advance ratio J of 0.35. The effect of the discontinuity of velocity field due to the propeller wake can be clearly seen. At these locations the large changes were filtered out for the calculation of circumferential mean with a velocity field value required every 3°.

The similar mean upstream effect of the rudder on the propeller is given in Figure 6.36 for variable X/D.

Figure 6.37 shows the variation of axial and tangential components of velocity field U_i, at the rudder stock compared to the values found using blade element-momentum theory. The two different curves show the influence of using a frozen propeller wake with a constant (fixed) radial value of far field wake pitch and with a radial variation (variable) pitch based on individual propeller blade section loading.

The number of complete iteration cycles for convergence is shown in Figure 6.38 where the difference in rudder lift is 0.1% after four cycles.

The performance of the method across a range of rudder incidence between $-30.4°$ and 29.4° is shown in Figure 6.39 against experimental measurements for the variation of spanwise loading C_n at a propeller advance ratio J of 0.51. Excellent agreement is observed for rudder incidence between $\pm 10°$. Above this value the influence of the rudder on the direction of propeller race needs to be included in the interaction process.

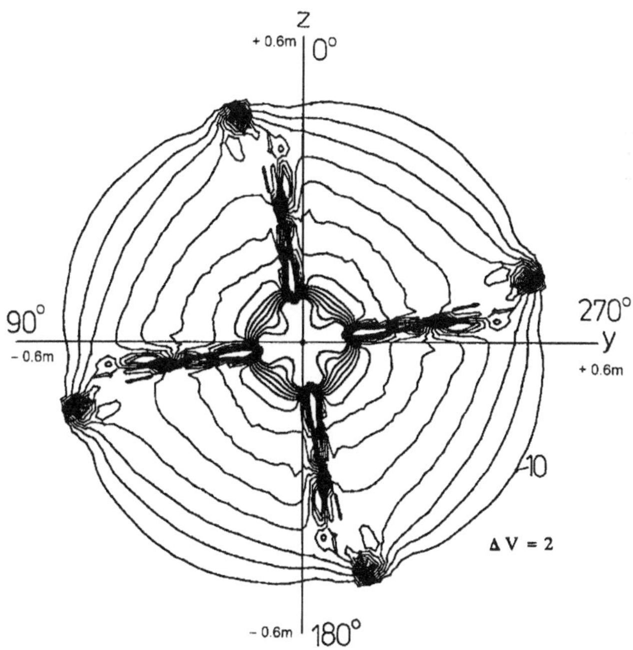

Figure 6.34 Contour plot of total velocity for vertical plane downstream of propeller; $X/D = 0.39$, $J = 0.35$

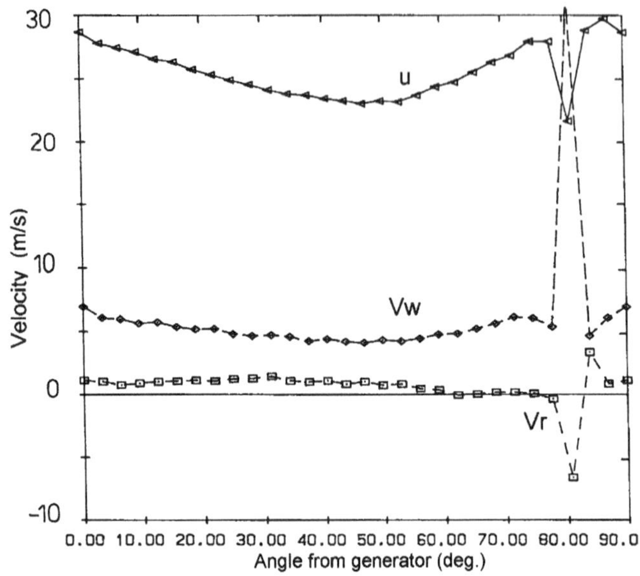

Figure 6.35 Circumferential variation of axial, radial and swirl velocity; $X/D = 0.39$, $r/D = 0.35$, $J = 0.35$

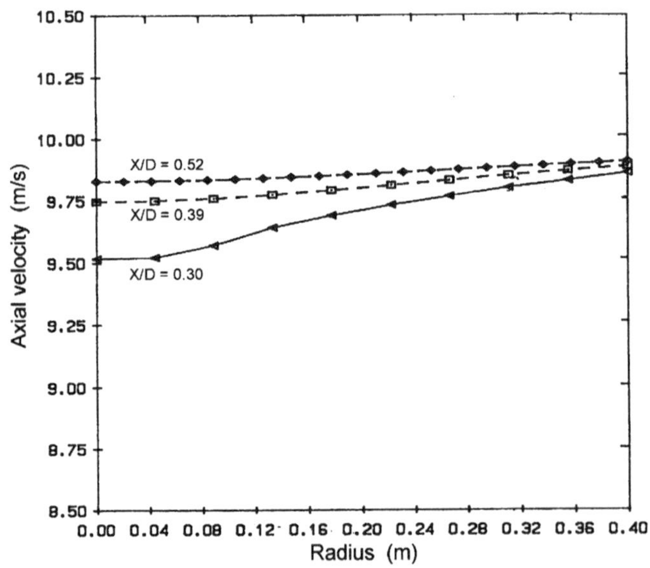

Figure 6.36 Effect of longitudinal separation on circumferentially averaged velocity upstream of Rudder No. 2

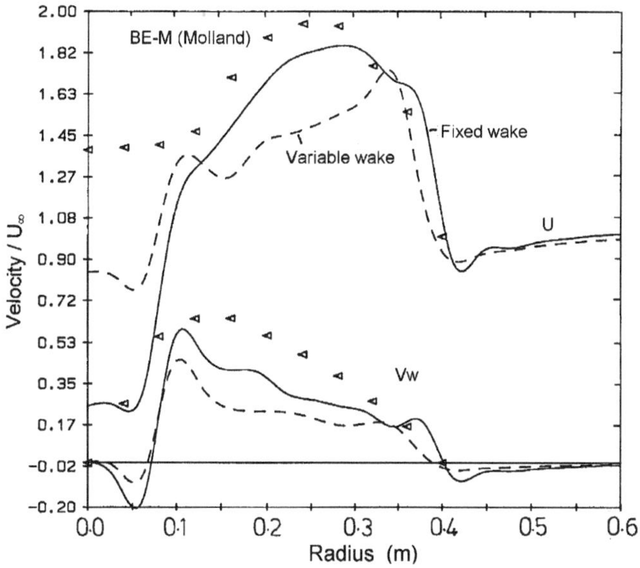

Figure 6.37 Radial variation of axial and swirl velocity; $X/D = 0.39$, $J = 0.35$

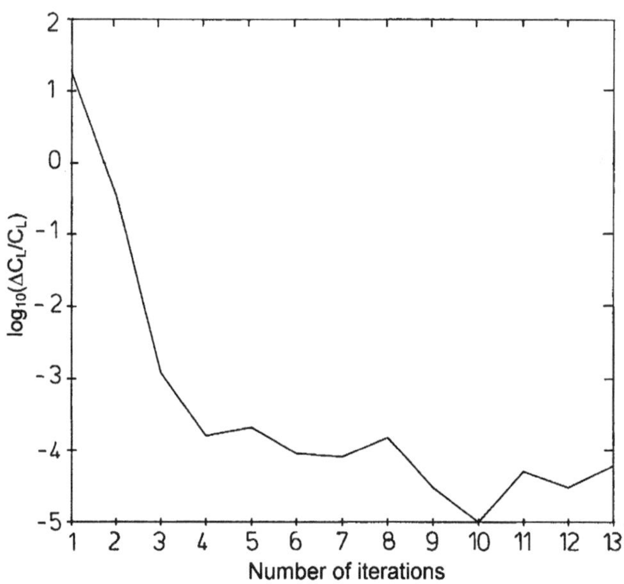

Figure 6.38 Convergence of rudder lift with number of IVF iteration cycles

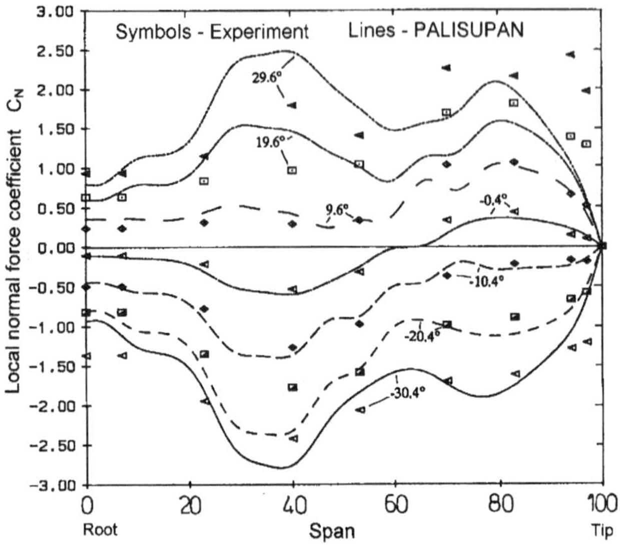

Figure 6.39 Spanwise load distributions; comparison of lifting-surface predictions and experimental results; Rudder No. 2, $J = 0.51$

The corresponding surface pressure values are given in Figure 6.40 for a rudder incidence of $-0.4°$ and at three advance ratios. The overall shape and values agree well except in the region where the propeller hub flow impact on the rudder (S4, S5). This hub flow is likely to be a combination of separated flow from the hub and the hub vortex system that will be poorly captured by the frozen propeller wake.

Finally, Figure 6.41 presents the force characteristics across the angle range and for the three advance ratios. Overall, good agreement is found across the range of incidence and different propeller thrust loadings. The good agreement arises from the effective cancellation of the overprediction of freestream performance due to a surface panel code with the missing effects of the unsteady spatial variations in velocity field excluded from the circumferential mean. The combination of the good overall force prediction with the variability in C_n distribution at high rudder incidence is another indication that the propeller race has an integrating effect. The important measure to achieve good comparison is the capturing of the correct amount of energy associated with the axial and tangential flow fields.

The drag prediction is reasonable at angles below 10° but underpredicts by 70% at higher incidence. This is not surprising given the approximation of the frictional estimate.

6.6.3 RANS

At the current time, it is not possible routinely to use RANS methods to predict the full unsteady interaction of the hull, propeller and rudder, to the level required for

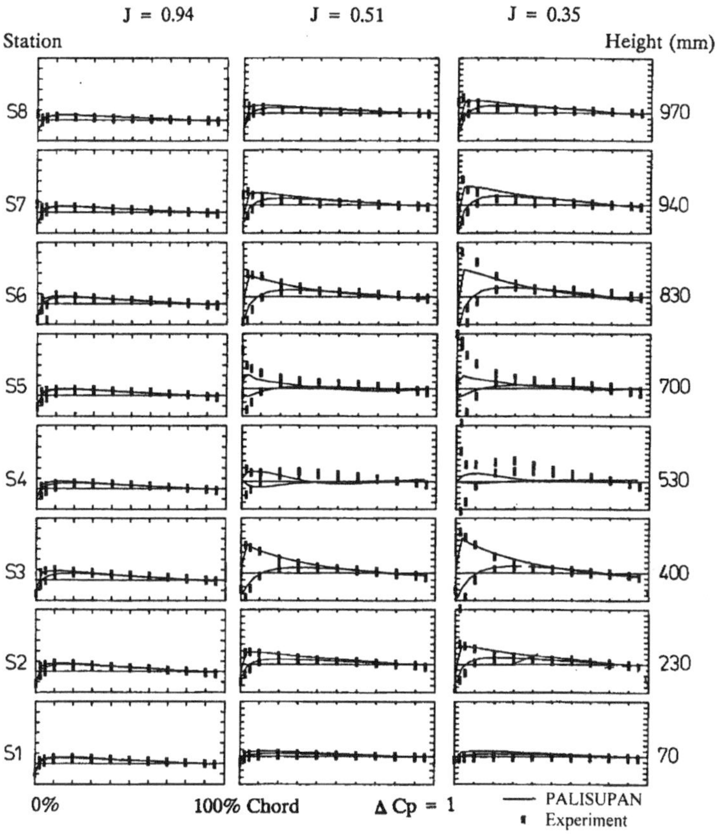

Figure 6.40 Chordwise pressure distributions; comparison of lifting-surface predictions and experimental results; Rudder No. 2, $J = 0.94$, 0.51 and 0.35, $\alpha = 0°$

rudder designers [6.54–6.56]. Propeller effects can be incorporated within the RANS method using one of two approaches. The simplest way is to model the propeller as an actuator disk. This method involves applying body forces, i.e., the source terms in the momentum equations, to the cells located within the propeller disk, such that the flow is accelerated in the same way as a propeller with an infinite number of blades, with the required thrust and torque. This actuator disk approach was proposed by Schetz and Favin [6.57]. However, this method only accounts for the axial and tangential forces, and neglects any radial force components, which would be present in the real flow. For simplicity, the effect of the propeller is usually represented as circumferentially averaged body forces, input into the steady RANS momentum equations, hence neglecting any unsteady effects. It is perfectly feasible for unsteady body forces to be included in unsteady RANS momentum equations. However, due to the high computing overheads associated with time-accurate simulations, these computations are uncommon. Examples of investigations using the circumferentially averaged body forces, and time varying body force approaches are found in references [6.9,6.54,6.55,6.58,6.59]. Various degrees of success have

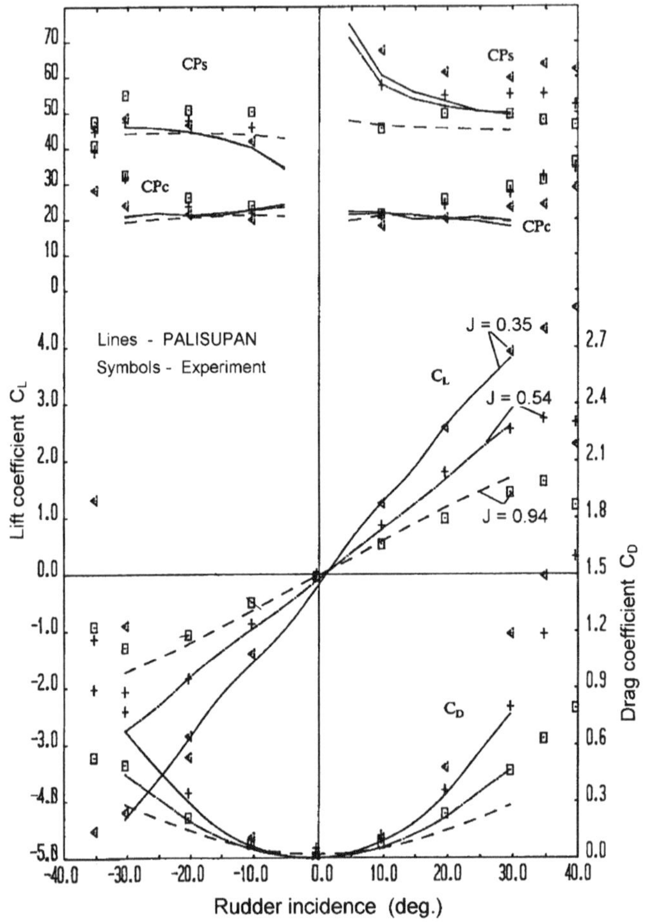

Figure 6.41 Comparison of lifting-surface and experimental force characteristics; Rudder No. 2, $J = 0.94, 0.51, 0.35$

been obtained using this body force method, with qualitative results comparing more favourably than the quantitative results.

The second, and more complex, way of incorporating propellers within the RANS model, is to compute the actual unsteady flow over the real rotating propeller geometry. This method is complicated, requiring the generation of complex nonmatching grids around the hull, propeller and rudder geometries, with fixed and rotating frames of reference. In addition, this approach needs to take into account the different time scales in the flow, as the propeller flow requires a much smaller time step than the ship flow. This approach requires extremely large computing resources. Limited examples of such calculations are available [6.60]. Improvements in the mesh generation capabilities and more suitable turbulence models are required before the full potential of this method can be exploited. A good overview of capabilities in application of CFD to hull–propeller–rudder systems is given in references [6.61,6.62].

Notwithstanding the lack of progress in this area, it can be expected that progressively more success will be achieved as computing power reduces in cost and increases in availability. Future advances will be reported at SIMMAN workshop to be held in Copenhagen in October 2007 where international research groups will use CFD to predict dynamic ship manoeuvres with rudder, propeller and free surface.

6.7 Unsteady behaviour

A ship rudder can be located in the race of a propeller, which in turn is located within the wake of a hull; all of these are subject to a greater or lesser extent by the presence of the free surface and the motion of the ship. Such unsteady flows as seen by the rudder are still a particularly challenging area for the application of time accurate RANS equations. One of the difficulties is whether the assumption that the time period of unsteady turbulent fluctuations is sufficiently distinct from that of the variations due to the propeller race.

The highly turbulent, periodic and interactive wake produced by a propeller, gives rise to rudder performance characteristics which differ significantly from those experienced in a free stream. Date [6.35], imposed periodic flow conditions, representative of the flow produced in the wake of a propeller, to both 2-D NACA0020 and high lift sections. It was expected that a better understanding of the performance of rudders operating in propeller wakes could be achieved. The response prediction of 2-D rudder sections subjected to periodic flow conditions can be regarded as a necessary first step towards understanding the requirements for full periodic 3-D rudder computations.

6.7.1 Transient flows and time stepping

For transient problems, it is often difficult to determine a suitable value of the time step necessary to obtain a convergent solution, as it is not possible to obtain an analytic stability criterion. The choice of time step depends on the time scales of the important flow features that need to be resolved. Using too large a time step can often result in resolution of nonphysical flow behaviour. Although from a numerical stability point of view the time step used by implicit solvers does not have to satisfy the Courant–Friedrichs–Lewy (CFL) condition (Equation (6.68)), it is often advisable for relatively small time steps close to the CFL limit to be used initially,

$$C = c\frac{\Delta t}{\Delta x} < 1 \tag{6.68}$$

where C is the Courant Number, c is the speed of propagation of some important flow feature, Δt is the time step and Δx is the grid spacing in the direction of propagation.

Using this (CFL) criterion, an estimate of the time step needed in transient problems can be made. Typically, a time step of the order of magnitude of the residence time of a fluid particle passing through a control volume is used. The residence time is the time it would take a fluid particle to move through a cell from one face to the opposite face. For example, if a fluid particle moves in the x-direction with a velocity u, the residence time and hence the time step necessary to capture this movement would be given by

$$\Delta t = \frac{\Delta x}{u} \tag{6.69}$$

This calculation is carried out on a control volume, which is known to be located at a point with the greatest flow instability, such as the vortex street in bluff body flows. However, care must be taken to ensure that the time step used is not so small that the Reynolds-averaging assumptions are violated.

For steady-state problems with convergence problems, these can be run as an unsteady problem with an initial time step set close to the Courant limit and slowly increased by means of adaptive time stepping, whereby the time step is increased by a fixed factor at each new time step. This approach can also be applied to transient flows when seeking an optimum time step.

6.7.2 Rudder performance in periodic flows

Considering a propeller and rudder in isolation, i.e., neglecting the wake field produced behind a ship hull, the rudder is subjected to a periodically varying flow field. The helical and nonaxisymmetric nature of this flow field means that the velocities within the propeller wake vary as a function of space and time.

A simple way of visualising what is happening to a rudder located in a propeller race is to consider the 2-D flow, i.e., ignoring rotational or cross flow effects, at a spanwise section. The section effectively experiences periodic variations in both flow incidence and speed as the propeller rotates. Downstream of a propeller, the variation in axial V_a and tangential V_t velocities with propeller-blade angular position is similar to that in Figure 6.35. As can be seen, an almost sinusoidal variation in the axial and radial velocity components occurs, as the propeller rotates. Figure 6.42 emphasises how large the changes in local flow incidence angles produced in the propeller race can be.

Now consider the same spanwise section, operating within a 2-D propeller wake, experiencing a time varying effective flow incidence of α_E, as shown in Figure 6.43. The resolved axial $A(t)$ and normal $N(t)$ forces are related to the lift and drag forces by

$$\begin{aligned} N(t) &= L(t)\cos(\alpha + \alpha_E(t)) + d(t)\sin(\alpha + \alpha_E(t)) \\ A(t) &= -L(t)\sin(\alpha + \alpha_E(t)) + d(t)\cos(\alpha + \alpha_E(t)) \end{aligned} \tag{6.70}$$

If $\alpha + \alpha_E = 0$, the lift and drag forces act in the normal and axial directions, respectively. However, if the lift/drag ratio becomes large enough, a net reduction in drag can be achieved. It is also theoretically possible for a net propulsive thrust to be

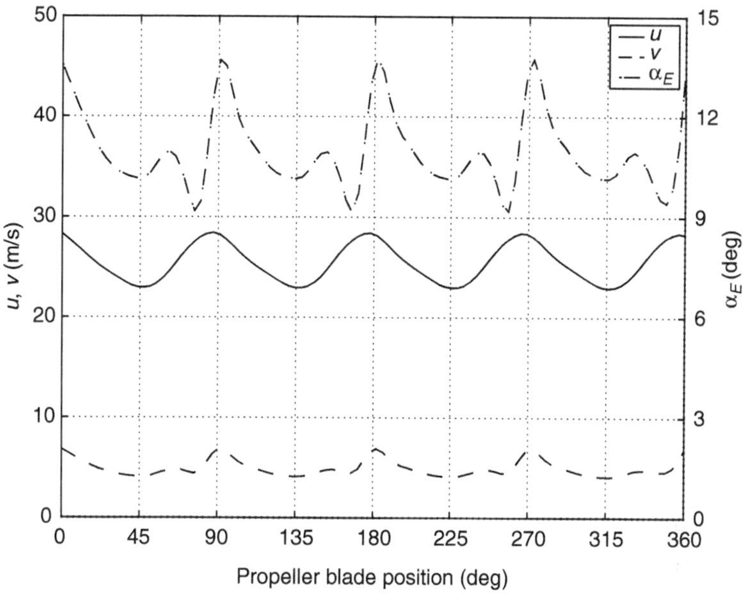

Figure 6.42 Variation of *u*, *v* and α in the wake of a four-bladed propeller; $X/D = 0.39$, $2r/D = 0.70$, $J = 0.35$

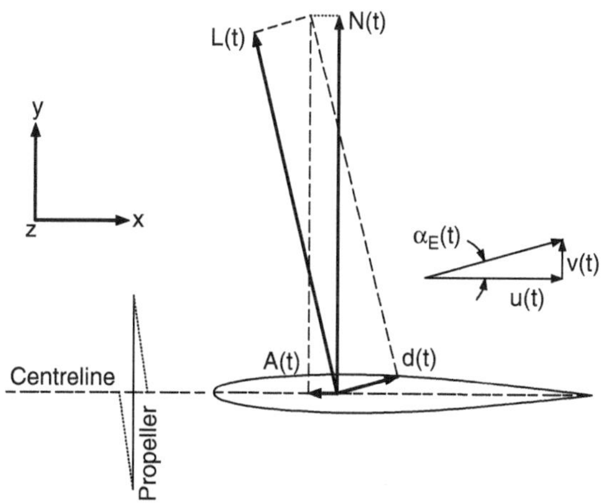

Figure 6.43 Resolution of lift and drag forces in the normal and axial directions

produced, in the same way as a sail produces the forward thrust on a yacht when sailing upwind, as shown in Figures 6.43 and 3.33(b) and discussed in Section 3.5.1. For the flow over the whole rudder, the asymmetry of the propeller wake, above and below the propeller shaft line, would result in the cancellation of the normal forces. The axial force components above and below the rudder shaft line are additive. So, over one complete revolution of the propeller, it is possible for a net thrust force to be produced. This thrust force combined with the stator effect of the rudder, can result in improvements in propulsive efficiency. This advantage may be significant enough that a ship may not require any increase in propulsive power [6.63], over what would be required if the rudder were not located in the propeller race.

The problem is, however, not quite that simple since the magnitude of lift and drag forces acting on the rudder are periodic, and highly dependent on the circular passage frequency of the propeller. For a propeller with B number of blades and rotating at n revolutions per second, the circular passage frequency ω of the propeller is given by

$$\omega = 2\pi nB \tag{6.71}$$

For periodic flows, it is more common to characterise the fluid dynamic phenomena with respect to the nondimensional or reduced frequency parameter k given by

$$k = \frac{\omega c}{2U} \tag{6.72}$$

6.7.3 Periodic aerofoil performance prediction

The response of a rudder to the periodic flow conditions produced by a propeller is governed by unsteady aerofoil and boundary layer theory. The main motivation for research in this field is focused on how the undesirable effects of vibration, buffeting, gust response, dynamic stall and flutter caused by unsteady flows can be reduced. However, some attention has also been focused on ways of optimising the beneficial effects of flow unsteadiness, such as improvements in propulsive efficiency and stall delay. Assessment of these effects, requires the prediction of both the magnitude and time lag of the unsteady fluid dynamic loads [6.64].

Paterson and Stern [6.65] overview the historic approach to solving the unsteady foil problem based either on fundamental analysis of unsteady boundary layers or unsteady lifting flows. When boundary layer effects have been the primary driver, viscous unsteady boundary layer methods have been implemented; when lift performance has been the driver, unsteady inviscid flow methods have been applied.

The pioneers of early theoretical unsteady aerofoil performance prediction were Theodorsen [6.66], Von Karman [6.67] and Sears [6.68]. Using linear potential theory, these researchers investigated various periodic incompressible flows, over flat plate aerofoils of infinite span. By considering only small disturbances to the steady flow, they found that it was possible to linearise the flow about a uniform parallel mean flow, and thus uncouple the time-dependent component of the flow completely from the steady-state flow characteristics.

Unsteady classical aerofoil theory is best explained by considering the change in developed circulation around an aerofoil, as it undergoes unsteady motion. For every change in circulation about the aerofoil, resulting from the unsteady behaviour of the flow or aerofoil, vorticity of opposing sign must be shed into the wake and carried away by the mean flow. The vortices shed into the wake represent the time history of the unsteady flow about the aerofoil, and induce a velocity field which is proportional to the vortex circulation, and inversely proportional to the distance away from the vortex centre. In essence, the wake of the aerofoil acts as a memory of the previous aerofoil unsteady flow, and the total velocity field therefore depends on the entire history of the airfoil motion.

Sears' [6.68] approach involved the derivation of a lift (Sears) function for a rigid aerofoil subjected to periodic transverse gusts. Similar functions have also been derived by other researchers such as Horlock [6.69], for fluctuations parallel and at an angle to the mean flow direction.

Linear potential theory has its limitations, specifically small-amplitude and low-frequency oscillations, due to the use of the steady Kutta condition. Experimental work by Poling and Telionis [6.70] has shown that the steady Kutta condition, on which classical linear potential theories are based, is never satisfied for reduced frequencies above $k = 2$. Linear potential theory also neglects any viscous effects, such as the interaction of the outer unsteady flow with the mean flow boundary layer development. These nonlinear effects, found in real unsteady aerofoil flows, cause a departure from the predicted performance. Nonlinear methods have been developed which account for the unsteady interaction and distortion of the travelling gust. However, like classical linear theory, nonlinear potential theory has reduced validity for unsteady flows of high reduced frequency.

There is scope for the application of unsteady RANS flow methods to overcome these limitations. Using a time-accurate RANS approach, implementing the Baldwin–Lomax turbulence model, Paterson and Stern [6.65] investigated the response of a propeller-blade geometry operating under unsteady flow conditions. The aims of their research were; the validation of time-accurate solutions obtained using the RANS approach and investigation of the response of turbulent propeller blade boundary layers and wakes to external-flow travelling waves. Validation was carried out using the results obtained from the Massachusetts Institute of Technology (MIT) flapping foil experiments [6.71,6.72] as a benchmark. They showed that high-frequency lifting surface flows display a complex response, which is significantly different from classical and fundamental boundary layer and potential theories. Viscous-inviscid interaction was also found to be an important mechanism of lifting surface response. Despite implementing the Baldwin–Lomax turbulence model in a quasi-steady manner, the unsteady velocity profiles over the propeller blade were found to show a close correlation with the experimental data. Quantitative details of the flow, such as lift and drag, were found to be highly dependent upon frequency, geometry and waveform. Agreement between the CFD and the experimental unsteady pressure response was found to be poor. The CFD indicated upstream and downstream travelling pressure waves over the foil and in the wake. Detailed analysis showed that distortion of the external flow wave, particularly on the suction side of the propeller blade, was significant and may have been partially responsible for the complex wake structure.

6.7.4 Investigation of propeller race on rudder section performance

A study of small and large-amplitude gusts related to those created by a propeller race were applied to a NACA0020 section with an incidence angle of 0° [6.35]. The typical reduced frequency of the wake produced by a four-bladed propeller of a VLCC tanker is approximately $k = 3$, based on a propeller r.p.m. of 80, rudder chord of $c = 5$ m and mean propeller race speed of $u = 25$ m/s. The results presented are subject to the limitations of a quasi-steady application of the standard k–ε turbulence model.

The NACA0020 sections was modelled at a $Re = 4.19 \times 10^7$. The magnitude of the u velocity component was held fixed and the v velocity component was sinusoidally varied:

$$v = v_0 \cos\left(\omega\left(t - \frac{x_V}{U_\infty}\right)\right) \tag{6.73}$$

where v_0 is the amplitude of the transverse gust moving past the aerofoil, and x_V the vertex node location on the inlet boundaries. Small and large-amplitude gusts of 0.2 and 2.5 m/s were investigated for four different reduced frequencies of $k = 0.5$, 1.0, 2.0 and 5.0. Time steps of $\Delta t = T/50$ s were used for all the NACA0020 computations. QUICK differencing was used for the spatial terms and hybrid for the turbulence quantities. Quadratic second-order time differencing was used in conjunction with fixed time stepping for all transient calculations. Pressure correction was carried out using the SIMPLE algorithm. The mass source residual stopping convergence criterion, was set at 1.0×10^{-4} kg/s in all computations. The upper and downstream boundaries were located 2 and 6 chord lengths respectively, away from sections. This was done in order to reduce dissipation of the imposed gust in upstream areas of insufficient grid resolution. The periodic gust boundary conditions were implemented by means of a user FORTRAN subroutine.

Figures 6.44 and 6.45 show the response histories for the NACA0020 section subjected to small-amplitude transverse gusts of 0.2 m/s, at k of 0.5 and 5.0, respectively. As can be seen, both frequency response histories show a first-harmonic response, consistent with the imposed first-harmonic inflow boundary conditions. It is evident that lower frequencies produce larger C_l amplitudes; with a 10-fold increase in frequency, resulting in a 8% reduction in C_l amplitude. The C_D response histories show a 100% reduction in amplitude with a 10-fold increase in frequency. It is interesting to note that the drag response for a k of 0.5 oscillates about a mean drag equal to the free stream (zeroth harmonic) while for $k = 5$ the mean drag is 4% higher.

Figure 6.46 compares the C_l response amplitude for small-amplitude transverse gusts of $v_0 = 0.2$ m/s against the linear theory of Sears [6.68], over a range of reduced frequencies. A general trend of reducing C_l amplitude with increasing frequency can be seen. For low frequencies of around $k = 0.5$, the C_l response amplitude shows reasonable agreement with the classical linear theory. This confirms the correct implementation of the periodic boundary conditions. However, at higher frequencies, departure from classical linear theory is seen to occur.

Figure 6.47 shows the pressure distributions for gusts of 0.2 m/s at the maximum and minimum transient response values of $C_l = 0$ and 0.056, at $k = 0.5$. For comparison, the corresponding free-stream pressure distributions are shown. Both of the pressure distributions for the maximum and minimum response values are nearly identical to their free-stream counterparts. Although not shown, the pressure distributions at the minimum and maximum C_l values for $k = 5.0$ were identical to the free-stream pressure distribution at $C_l = 0$. This behaviour can be understood since at low frequency the section responds as a slowly varying incidence whereas

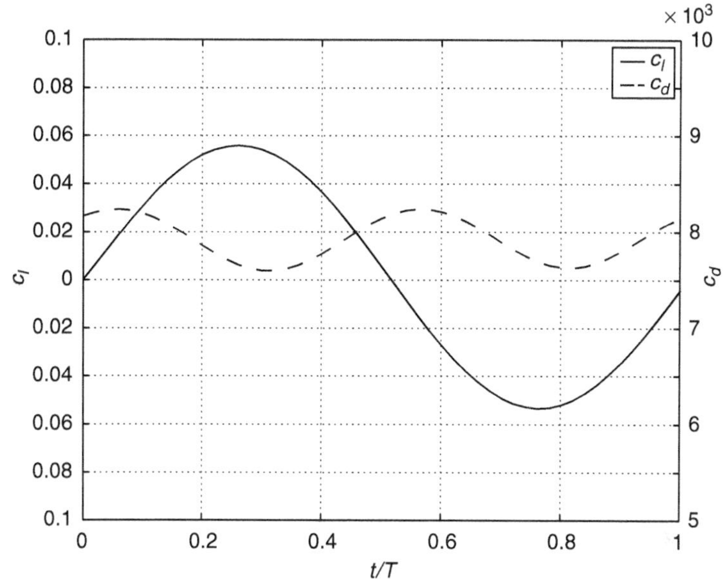

Figure 6.44 Response of C_L and C_D for NACA0020, u = 10 m/s, v_0 = 0.2 m/s, k = 0.5

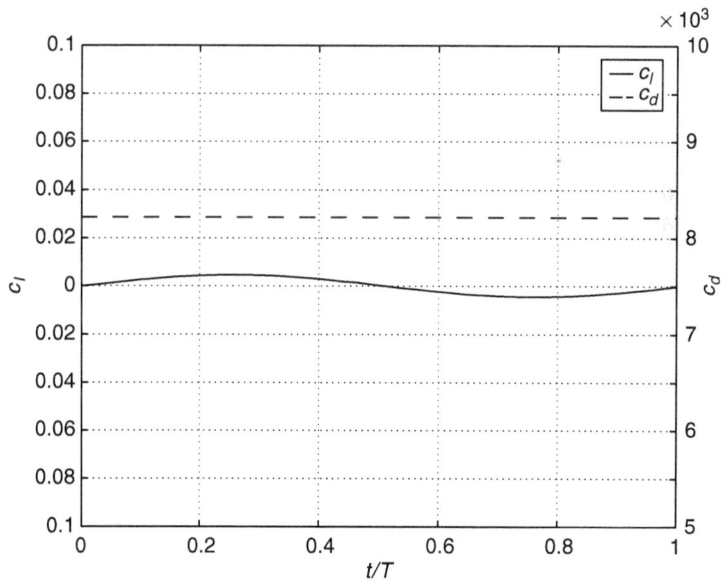

Figure 6.45 Response of C_L and C_D for NACA0020, u = 10 m/s, v_0 = 0.2 m/s, k = 5.0

at high frequency the section response diminishes because the section is experiencing a flow disturbance for a very small period of time. However, between these two limiting frequencies complex flow interactions occur between the travelling gust and the flow about the section.

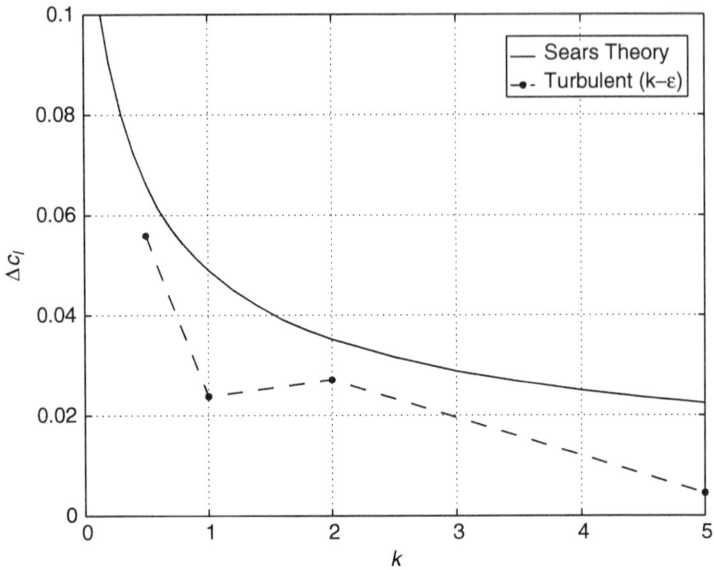

Figure 6.46 Amplitude of response of C_L for NACA0020, $u = 10$ m/s, $v_0 = 0.2$ m/s

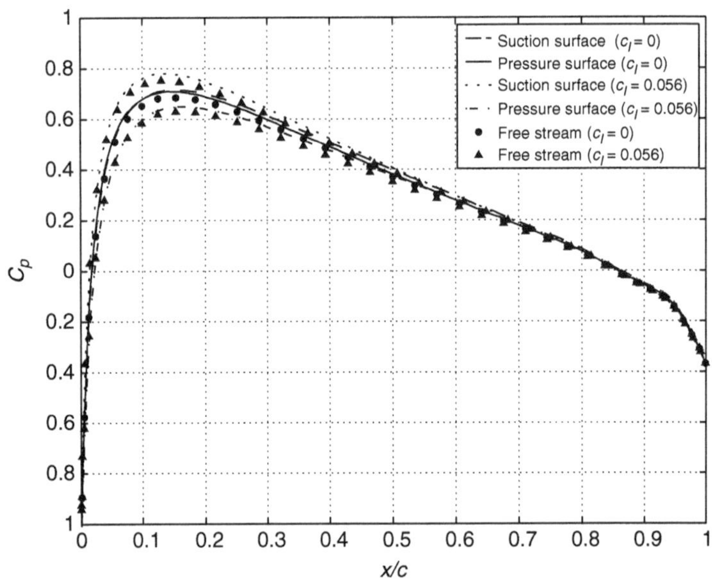

Figure 6.47 Chordwise pressure distributions for periodic inflow; NACA0020, $u = 10$ m/s, $v_0 = 0.2$ m/s, $k = 0.5$

Figures 6.48 and 6.49 show the response histories for the NACA0020 section subjected to large-amplitude transverse gusts of 2.5 m/s, at $k = 0.5$ and 5, respectively. A similar form of behaviour is seen but with an increased amplitude of response. It is interesting to note that the response C_D for $k = 0.5$ shows that an appreciable thrust force is being generated by the rudder.

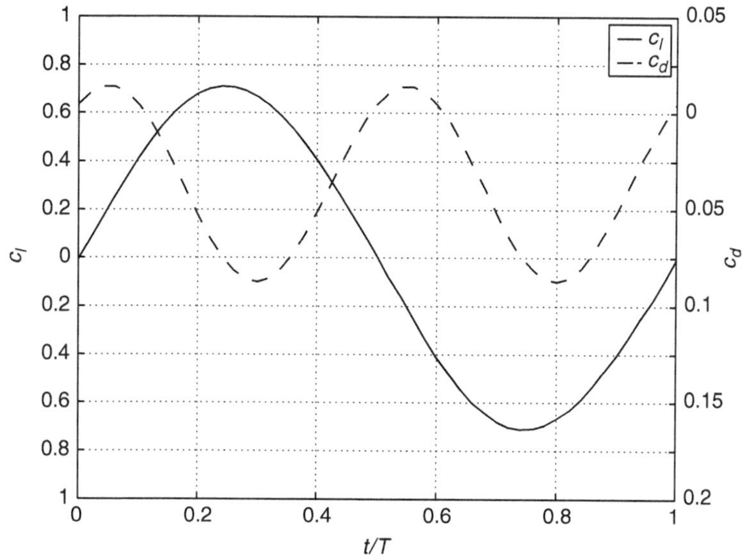

Figure 6.48 Response of C_L and C_D for NACA0020, $u = 10\,\text{m/s}$, $v_0 = 2.5\,\text{m/s}$, $k = 0.5$

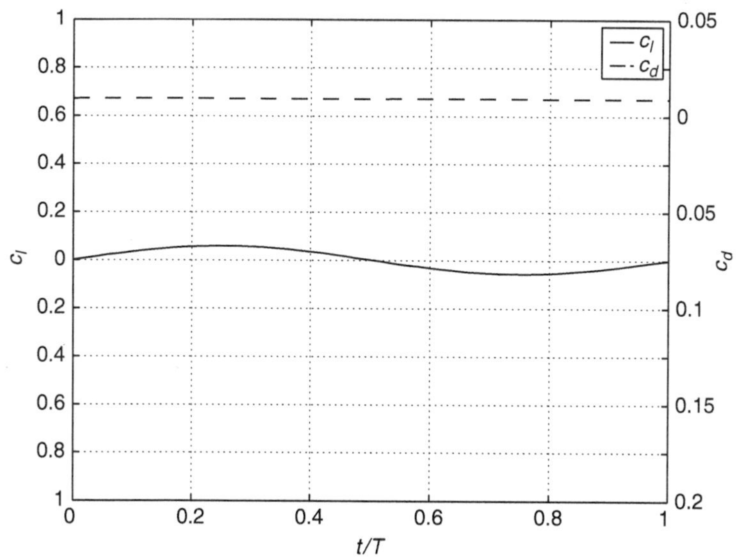

Figure 6.49 Response of C_L and C_D for NACA0020, $u = 10\,\text{m/s}$, $v_0 = 2.5\,\text{m/s}$, $k = 5.0$

Figure 6.50 compares the C_l response amplitude for the NACA0020 section subjected to large-amplitude transverse gusts of 2.5 m/s. Like the small-amplitude case, a general trend of reducing C_l amplitude with increasing frequency is evident.

Figure 6.51 shows the corresponding pressure distributions for large-amplitude transverse gusts of 2.5 m/s at the maximum and minimum transient response values of $C_l = 0$ and $C_l = 0.71$, at $k = 0.5$. The pressure distributions for both the minimum and maximum C_l response values show a different behaviour than that found in the low-amplitude investigation. Looking first at the suction surface for the $C_l = 0$ it is apparent that the high-amplitude periodic inflow conditions result in a reduction in the suction pressure from the leading edge back as far as the $x/c = 0.4$ location. Aft of this the suction surface pressure is coincident with the free-stream

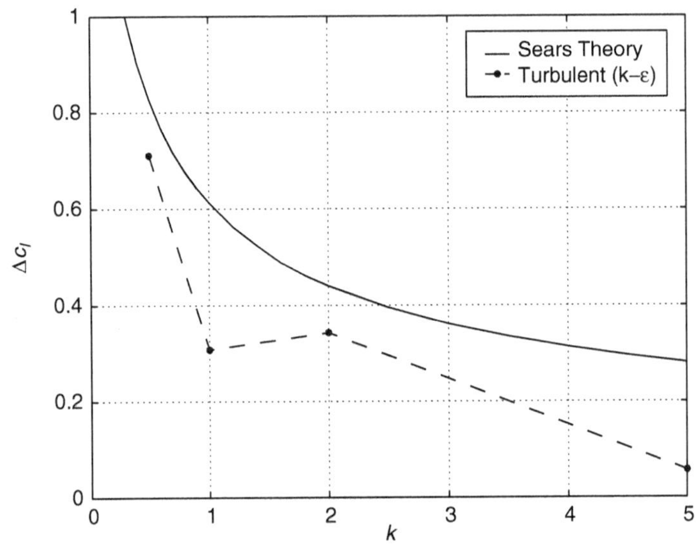

Figure 6.50 Amplitude of response of C_L for NACA0020, $u = 10$ m/s, $v_0 = 2.5$ m/s

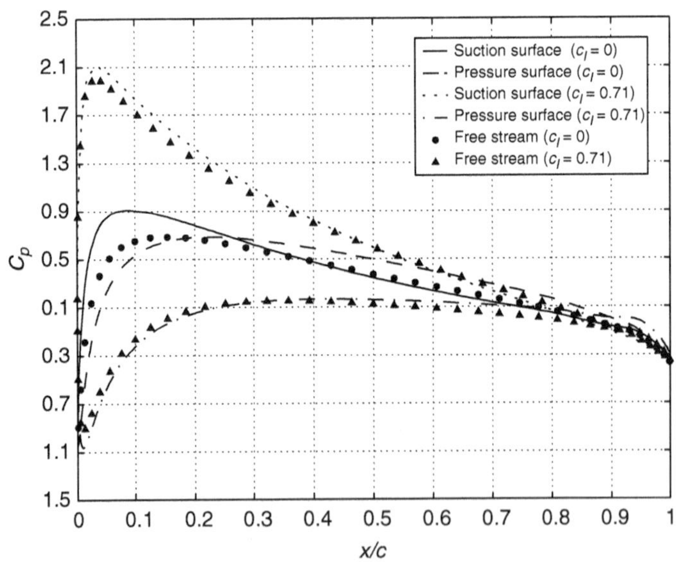

Figure 6.51 Chordwise pressure distributions for periodic inflow; NACA0020, $u = 10$ m/s, $v_0 = 2.5$ m/s, $k = 0.5$

data. The pressure surface, on the other hand, undergoes an increase in the leading edge pressure back as far as the $x/c = 0.2$ location, and a reduction in pressure aft of this location. This pressure distribution explains the origin of the mean thrust force identified previously. The pressure plots for the $C_l = 0.071$ case show a similar response to the $C_l = 0$ case, although less marked. The high frequency, $k = 5$ response shows little deviation from the steady surface pressure distribution.

6.8 Future developments

It is likely that as computing costs continue to reduce in real terms then progressively more of the rudder design process will be carried out using computational fluid dynamics. Over the timescale of thirty plus years during which the authors have been involved in marine rudder analysis, computational methods have moved from developing empirically corrected lifting-line techniques, through surface panel techniques to the present date when there are groups applying LES techniques, to others carrying out unsteady manoeuvres of ships with a representative propeller and rudder all solved using RANS equations. A similar rate of development can still be expected but it is likely that, for the foreseeable future, all of the techniques described in this chapter will still find a place in the hierarchy of design analysis tools for the Naval Architect.

A convincing solution for an unsteady RANS calculation has still not been achieved for an all-movable rudder operating in the race of a propeller. The inherent difficulties required to capture and preserve the strong tip vortex, let alone the complexities of the hub vortex system when they impact the rudder, still present too much of a challenge for the available turbulence models and mesh generation strategies. It is to be expected, however, that progress in this area will be made over the next decade. A major requirement is the ability to simulate cavitation processes as a complete multiphase calculation or as a micro-physical model embedded within a RANS framework.

These applications will require major computational resources and will not be suitable for routine use in concept design. It is in this area that the surface panel techniques for prediction of detailed loadings (Section 7.2) and blade element/ lifting line for complete ship-manoeuvring simulations will find their use.

For many tasks the practitioner will be faced with a choice as to what tool to use. This choice will have to be made based on the blend of an individual's experience of a particular tool, alongside the project timescales and costs involved.

References

6.1 Glauert, H. *The Elements of Aerofoil and Airscrew Theory*, 2nd edition. Cambridge University Press, Cambridge, 1983.

6.2 Anderson, J.D. *Computational Fluid Dynamics: The Basics with Applications*. McGraw-Hill International Editions, New York, 1995.

6.3 Ferziger, J.H. and Peric, M. *Computational Methods for Fluid Dynamics.* Springer-Verlag, 2004.
6.4 Cebeci, T., Shao, J.P., Kafyeke, F. and Laurendeau, E. *Computational Fluid Dynamics for Engineers: From Panel to Navier–Stokes Methods with Computer Programs.* Springer, Berlin 2005.
6.5 Katz, J. and Plotkin, A. *Low-Speed Aerodynamics: From Wing Theory to Panel Methods*, McGraw-Hill, New York, 1991.
6.6 Molland, A.F. *The free-stream characteristics of ship skeg-rudders.* PhD Thesis, University of Southampton, 1981.
6.7 Molland, A.F. A method for determining the free-stream characteristics of ship skeg-rudders. *International Shipbuilding Progress*, Vol. 32, No. 370, 1985, pp. 138–150.
6.8 Hess, J.L. Panel methods in computational fluid dynamics. *Annual Review of Fluid Mechanics*, Vol. 22, 255–274. 1990.
6.9 Turnock, S.R. and Wright, A.M. Directly coupled fluid structural model of a ship rudder behind a propeller. *Marine Structures*, Vol. 13, 2000, pp. 53–72.
6.10 Lamb, H. *Hydrodynamics*, 6th edition. Cambridge University Press, Cambridge, 1932.
6.11 Turnock, S.R. Technical manual and user guide for the surface panel code Palisupan. *University of Southampton, Ship Science Report No. 100*, 1997.
6.12 Lee, J.-T. *A potential based method for the analysis of marine propellers in steady flow.* PhD Thesis, M.I.T., 1987.
6.13 Morino, L. and Kuo, C.-C. Subsonic potential aerodynamics for complex configurations: a general theory. *AIAA Journal*, Vol. 12, No. 2, 1974.
6.14 Newman, J.N. Distribution of sources and normal dipoles over a quadrilateral. *Journal of Engineering Mathematics*, Vol. 20, 1986, pp. 113–126.
6.15 Hess, J.L. and Smith, A.M.O. Calculation of non-lifting potential flow around arbitrary three-dimensional bodies. *Journal of Ship Research*, Vol. 8, No. 2, 1964.
6.16 Maitre, T.A. and Rowe, A.R. Modelling of flow around a propeller using a potential based method. *Journal of Ship Research*, Vol. 35, No. 2, 1991, 114–126.
6.17 Schlichting, H. *Boundary Layer Theory.* McGraw-Hill, New York, 1968.
6.18 Ludwieg H and Tillmann W. Investigations of the wall shearing stress in turbulent boundary layers. *NACA Report TM 1285*, 1949.
6.19 Curle N. *The Laminar Boundary Layer Equations.* Oxford University Press, Oxford, 1962.
6.20 Cebeci T and Bradshaw, P. *Momentum Transfer in Boundary Layers.* Mcgraw-Hill, New York, 1977.
6.21 Versteeg, H.K. and Malalasekera, W. *An Introduction to Computational Fluid Dynamics.* Longman Scientific and Technical, Harlow, 1995.
6.22 Speziale, C.G. Analytical methods for the development of Reynolds-stress closures in turbulence. *Annual Review of Fluid Mechanics*, Vol. 23, 1991, pp. 107–157.
6.23 Reynolds, O. On the dynamical theory of incompressible viscous fluids and the determination of the criterion. *Philosophical Transactions of the Royal Society*, Series A, Vol. 186, 1895, pp. 123–164.

6.24 Wilcox, D.C. *Turbulence Modelling for CFD.* D.C.W. Industries, La Canada, CA, 1998.
6.25 Boussinesq, J. Essai Sur La Theorie Des Eaux Courantes. *Mem. Present'es Acad. Sci.*, Vol. 25, 1877.
6.26 Pashias, C. *Propeller tip vortex simulation using adaptive grid refinement based on flow feature identification.* PhD Thesis, University of Southampton, 2005.
6.27 Launder, B.E. and Spalding, D.B. The numerical computation of turbulent flows, *Computer Methods in Applied Mechanics and Engineering*, Vol. 3, 1974, pp. 269–289.
6.28 WS Atkins Ltd. Best practice guidelines for marine applications of computational fluid dynamics, https://pronet.wsatkins.co.uk/marnet/publications/bpg.pdf, accessed Nov. 2006.
6.29 Creismeas, P. Application de la LES Implicite a Letude d'un Ecoulement Autour D'un Profil NACA0012, 7th *Journees De L'Hydrodynamique*, Marseille, France, 8–10 March, 1999.
6.30 Gleick, J. *Chaos, the Amazing Science of the Unpredictable.* William Heinemann, London, 1988.
6.31 Blottner, F.G. Accurate Navier–Stokes results for the hypersonic flow over a spherical nosetip. *Journal of Spacecraft and Rockets*, Vol. 27, No. 2, 1990, pp. 113–122.
6.32 Stern, F., Coleman, H.W., Wilson, R.V. and Paterson, E.G., Verification and validation of CFD simulations. Proceedings of the *3rd ASME/JSME Joint Fluids Engineering Conference.* San Francisco, California, July 18–23, 1999.
6.33 Richardson, L.F. The deferred approach to the limit. *Transactions of the Royal Society, London Series A*, Vol. 226, 1927, pp. 229–361.
6.34 Eça, L. and Hoekstra, M. On the numerical verification of ship stern flow calculations. *MARNET-CFD, 1st Annual Worshop*, Barcelona, 18–19 November, 1999.
6.35 Date, J.C. *Performance prediction of high lift rudders operating under steady and periodic flow conditions.* PhD Thesis, University of Southampton, 2002.
6.36 Margesson, R.J., Kjelgaard, S.O., Sellers, W.L., Morris, C.E.K., Walkley, K.B. and Shields, E.W. Subsonic panel methods – a comparison of several production codes. Proceedings of *AIAA 23rd Aerospace Science Meeting*, Reno, Jan 1985.
6.37 Stück, A., Turnock, S.R. and Bressloff, N.W. An evaluation of the RANS method for the prediction of steady rudder ship rudder performance compared to wind tunnel measurements. *University of Southampton, Ship Science Report No. 130*, 2004, pp. 1–64.
6.38 El Moctar, O. A. M. Berechung von ruderkraften. *Institut für Schiffbau der Universitat Hamburg*, Bericht 582, 1997.
6.39 Tsakonas, S., Jacobs, W.R. and Ali, M.R. Application of the unsteady lifting-lifting-surface theory to the study of propeller–rudder interaction. *Journal of Ship Research*, Vol. 14, No. 3, 1970, pp. 181–194.
6.40 Cho, J. and Williams, M. Propeller-wing interaction using a frequency domain panel method. *Journal of Aircraft*, Vol. 27, No. 3, 1990.

6.41 Molland, A.F. and Turnock, S.R. A compact computational method for predicting forces on a rudder in a slipstream. *Transactions of The Royal Institution of Naval Architects*, Vol. 138, 1996, pp. 227–244.

6.42 Eckhard, M.K. and Morgan, W.B. A propeller design method. *Transactions of Society of Naval Architects and Marine Engineers*, Vol. 63, 1955.

6.43 Gutsche, F. Die induction der axialen strahlzusatzgeschwindigheit in der umgebung der schraubenebene. *Schiffstechnik*, Heft 12/13, 1955.

6.44 Li, D.-Q. Non-linear method for the propeller-rudder interaction with the slipstream deformation taken into account. *Computer Methods in Applied Mechanics and Engineering*, Vol. 130, No. 1–2, 1996, pp. 115–132.

6.45 Guo, C. and S. Huang Study of the rudder with additional thrust fins used by non-linear vortex lattice method. *Journal of Huazhong University of Science and Technology*, Vol. 34, No. 6, 2006, pp. 87–89.

6.46 Willis, C.J., Crapper, G.D. and Millward, A. A numerical study of the hydrodynamic forces developed by a marine rudder. *Journal of Ship Research*, Vol. 38, No. 3, 1994, pp. 182–192.

6.47 Turnock, S.R., Molland, A.F. and Wellicome, J.F. Interaction velocity field method for predicting ship rudder–propeller interaction. Proceedings of *SNAME Propeller/Shafting Symposium*, Paper 18, pp. 1–14, Sept., Virginia Beach USA, 1994.

6.48 Tamashima, M., Matsui, S., Yang, J., Mori, K. and Yamazaki, R. The method for predicting the performance of propeller–rudder system with rudder angle and its application to rudder design. *Transactions of the West-Japan Society of Naval Architects*, Vol. 85, 1993.

6.49 Söding, H. Limits of potential theory in rudder flow predictions. *Ship Technology Research*, Vol. 45, No. 3, 1998, pp. 141–155.

6.50 Laurens, J.-M., Grosjean, F. Numerical simulation of the propeller–rudder interaction. *Ship Technology Research*, Vol. 49, No. 1, 2002, pp. 3–12.

6.51 Felli, M., Greco, L., Colombo, C., Salvatore, F., Di Felice, F. and Soave, M. Experimental and theoretical investigations of propeller–rudder interaction phenomena. Proceedings of *26th Symposium on Naval Hydrodynamics*, Rome, Sept. 2006.

6.52 Turnock, S.R. *Prediction of ship rudder–propeller interaction using parallel computations and wind tunnel measurements.* PhD Thesis, University of Southampton, 1993.

6.53 Turnock, S.R. A transputer based parallel algorithm for surface panel analysis. *Ship Technology Research Schiffstechnik*, Vol. 41, No. 2, 1994, pp. 93–104.

6.54 Laurens, J.-M. Unsteady hydrodynamic behaviour of a rudder operating in the propeller slipstream. *Ship Technology Research*, Vol. 50, No. 3, 2003, pp. 141–148.

6.55 Simonsen, C.D. and Stern, F. RANS manoeuvring simulation of ESSO OSAKA with rudder and a body-force propeller. *Journal of Ship Research*, Vol. 49, No. 2, 2005 pp. 98–121.

6.56 Abdel-Maksoud, M. and Karsten, R. Unsteady numerical investigation of the turbulent flow around the container ship model (KCS) with and without propeller. Proceedings of *Gothenburg 2000, A Workshop on Numerical Ship Hydrodynamics*, Chalmers University of Technology, 2000.

6.57 Schetz, J.A. and Favin, S. Numerical solution for the near wake of a body with propeller. *Journal of Hydronautics*, Vol. 11, No. 10, 1977, pp. 136–141.
6.58 Stern, F., Kim, H.T., Zhang, D.H., Toda, Y., Kerwin, J. and Jessup, S. Computation of viscous flow around propeller–body configurations: series 60 $C_B = 0.6$ ship model. *Journal of Ship Research*, Vol. 38, No. 2, 1994, pp. 137–157.
6.59 Tzabiras, G.D. A numerical study of additive bulb effects on the resistance and self-propulsion characteristics of a full ship form. *Ship Technology Research*, Vol. 44, 1997, pp. 98–108.
6.60 McDonald, H. and Whitfield, D.L. Self-propelled manoeuvring underwater vehicles. *21st Symposium on Naval Hydrodynamics*, Trondheim, Norway, June 24–28, pp. 478–489, 1996.
6.61 Hino, T. (ed.) Proceedings of *CFD Workshop*, Tokyo, 2005.
6.62 Report of Resistance Committee. Proceedings of *24th ITTC*, Vol. I, pp 17–72, 2005.
6.63 Lewis, E.V. *Principles of Naval Architecture*, SNAME, New York, 1988.
6.64 McCroskey, W.J. Unsteady airfoils. *Annual Review Fluid Mechanics*, Vol. 14, 1982, pp. 285–311.
6.65 Paterson, E.G. and Stern, F. Computation of unsteady viscous marine-propulsor blade flows. Part 1: Validation and analysis. *Journal of Fluids Engineering*, Vol. 119, 1997, pp. 145–154.
6.66 Theodorsen, T. General theory of aerodynamic instability and the mechanism of flutter. *NACA Report 496*, 1935.
6.67 Von Karman, T. and Sears, W.R. Airfoil theory for non-uniform motion. *Journal of the Aeronautical Sciences*, Vol. 5, No. 10, 1938, pp. 379–390.
6.68 Sears, W.R. Some aspects of non-stationary airfoil theory and its practical application. *Journal of the Aeronautical Sciences*, Vol. 8, No. 3, 1941, pp. 104–108.
6.69 Horlock, J.H. Fluctuating lift forces on aerofoil moving through transverse and chord-wise gusts. *Journal of Basic Engineering*, December, 1997, pp. 494–500.
6.70 Poling, D.R. and Telionis, D.P. The response of airfoils to periodic disturbances – the unsteady Kutta condition. *AIAA Journal*, Vol. 24, No. 2, 1986, pp. 193–199.
6.71 Rice, J.Q. *Investigation of a two-dimensional hydrofoil in steady and unsteady flows*. S.M. Thesis, Department of Ocean Engineering, M.I.T., 1991.
6.72 Delpero, P.M. *Investigation of flows around a two-dimensional hydrofoil subject to a high reduced frequency gust loading*. S.M. Thesis, Department of Ocean Engineering, MIT, 1992.

PART THREE

Design Strategy and Methodology

7 Detailed rudder design

7.1 Background and philosophy of design approach

The forces developed by a rudder or control surface depend fundamentally on its area, profile shape and aspect ratio (ratio of span to chord), section shape, the square of the inflow velocity, the density and viscosity of water and the rudder angle of attack or incidence α. These parameters are described in Section 3.4.2 and, for a precise estimate of the rudder forces all of the parameters have to be taken into account.

Classical approaches to rudder design have tended to concentrate mainly on the derivation of rudder torque and a suitable stock diameter. Rudder force and its centre of pressure were derived using empirical equations that were based on relatively limited model- and full-scale data. Derived equations are of the form:

$$\text{Force } N = kf_1(\alpha)AV^2$$

$$CPc = f_2(\alpha)AV^2$$

$$\text{Torque } Q_R = N \times \text{lever of stock position to } CPc$$

In early formulae, $f_1(\alpha)$ is a function of the rudder angle only and does not take account of aspect ratio, although later formulae would allow for this in the constant k. Many alternatives for $f_1(\alpha)$ are listed by Kinoshita [7.1], and a popular choice was $\sin\alpha$. Another assumption was to take $CPc = 0.375\bar{c}$ at 35° rudder angle.

Jöessel [7.2] carried out tests on plate rudders in 1873 in the Loire river and derived empirical formulae for rudder torque and centre of pressure as follows:

$$\text{Torque } Q_R = 405.8AV^2 \sin\alpha\bar{c}$$

$$CPc = (0.195 + 0.305 \sin\alpha)\bar{c}$$

and

$$\text{Force } N = 405.8AV^2\left[\frac{\sin\alpha}{0.195 + 0.305\sin\alpha}\right]$$

These formulae were used extensively for many years, with different values of the coefficient developed for various ship and rudder types. Mention should also be made of the extensive early experimental work carried out on rudders by Denny [7.3], Baker and Bottomley [7.4–7.9], Abell [7.10] and Gawn [7.11] and their proposals for empirical equations for force and centre of pressure. Many of the results of Baker, Bottomley, Abell and Denny are summarised and discussed by van Lammeren *et al.* [7.12], together with the results of several rudder tests carried out in the Wagengingen test tank. Equations due to Baker, Bottomley and Gawn include:

For twin screw-twin rudders, Gawn [7.11]:

$$\text{Force } N = 21.1\ AV^2\alpha$$

For single screw-single rudder, Baker and Bottomley [7.4]:

$$\text{Force } N = 18.0\ AV^2\alpha$$

Attwood and Pengelly [7.13] give a good description of the application of such empirical formulae for forces and centre of pressure. For nonrectangular rudders, such as Figure 7.1(a), the forces on each rectangle can be calculated as though they acted independently and the forces then combined to give the total force. For the centre of pressure of nonrectangular rudders such as Figure 7.1(b), the recommended method is to divide the rudder into a number of horizontal strips, each considered as a rectangle with its own *CPc*. Using quadrature across the span, the moments of area of each strip are summed to give the total moment; this is divided by the total area to give the centre of pressure. Applying the total force from a formula (such as above), multiplied by the appropriate lever from the centre of pressure to the position of the stock, provides the total torque. Attwood and Pengelly emphasise the need to include frictional losses in the rudder bearings and to check the astern torque.

A practical approach using such data and methods is described by Lamb and Cook [7.14].

A thorough review of the hydrodynamic aspects of appendage design is presented by Mandel [7.15] including, in particular, the important aspects concerned with the design of rudders.

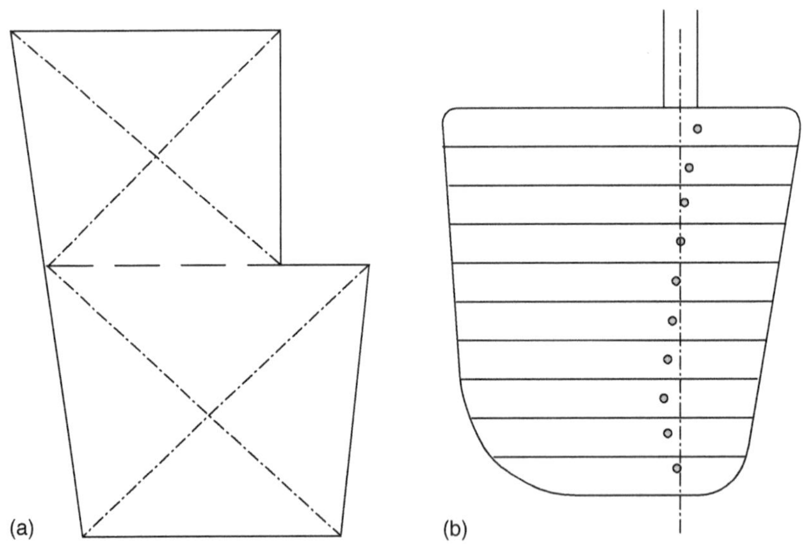

Figure 7.1 Nonrectangular rudders [7.13]

Jaeger [7.16] presents methods and calculations for rudder torque prediction that include the use of wind tunnel data and the incorporation of the influence of aspect ratio. Estimates of propeller-induced velocities are included based on propeller apparent slip values.

Gover and Olsen [7.17] present a method for predicting the torque of semi-balanced centreline rudders on multiple screw ships, that is with the rudder not in a propeller slipstream. The rudder is broken down into two parts A_1 and A_2, Figure 7.2, each with *CPc* distance x_1 and x_2 from the leading edge. Curves are provided for normal force coefficient C_N and centre of pressure *CPc* based on the aspect ratio of each portion. The method shows reasonable agreement with the experimental results of Hagen [7.18].

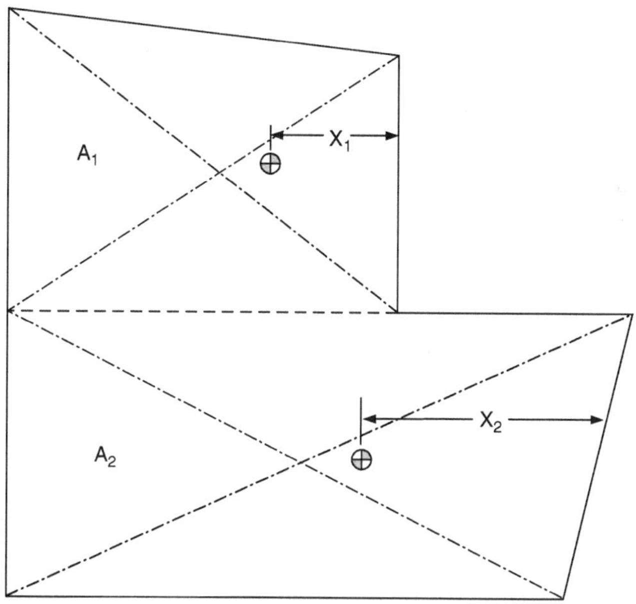

Figure 7.2 Semi-balanced centreline rudder [7.17]

A significant change in the level of understanding of the behaviour of control surfaces, with relatively low aspect ratio suitable for marine applications, occurred with the publication of the extensive free-stream tests of Whicker and Fehlner [7.19]. An early publication using the results of Whicker and Fehlner, and a more rigorous approach to rudder forces, torques and moments, was that of Taplin [7.20]. He includes example calculations to illustrate the methodology. Other publications of note through that period include those of Romahn and Thieme [7.21], Thieme [7.22] and Okada [7.23].

Harrington [7.24] presents an extensive review of rudder torque prediction. He makes use of the Whicker and Fehlner data, also comparing the results with the Jöessel method. He includes estimates of the frictional losses in the rudder bearings and compares his results with full-scale measurements.

In the field of yachts and small powercraft, relevant publications on rudder design include those of Millward [7.25] and Molland [7.26].

The manual of Brix [7.27] presents a wide range of detailed information concerning the design of rudders of many types. It also includes a thorough review of other manoeuvring devices.

Son *et al.* [7.28] and Kresic [7.29] present methods for estimating the torque of semi-balanced skeg, or horn, rudders. Son *et al.* use the modified lifting line analysis of Molland [7.30] for the prediction of forces and *CPc* and carried out a regression analysis of the results. Satisfactory agreement with full-scale data was obtained. Kresic developed a program for detailed estimates of torque for skeg rudders. He compared his results with model data, including the free-stream work of Goodrich and Molland [7.31].

It can be seen that the publication of the free-stream characteristics of various rudder series, including all-movable, flapped and semi-balanced skeg rudders, has enabled a more rigorous analysis and physically correct approach to be used for the design of marine rudders and control surfaces. Consequently, the most common rudder design and performance prediction method currently employed entails the use of free-stream or open-water characteristics for a particular rudder or control surface. The free-stream characteristics of the rudder represent its performance in the absence of the hull, propeller, or appendages. In order to take account of the presence of the hull and/or propeller, corrections are applied to the rudder aspect ratio, the inflow velocity to the rudder and to the rudder angle of attack to yield their effective values. These effective values are then used to enter the appropriate free-stream characteristic curves, and hence to compute the rudder forces and moments. The rudder in the free-stream condition and the modifying effects on the free stream of the hull and propeller are, therefore, treated as individual components of the complete system. This approach is discussed further in Section 7.2.

A number of mathematical models of rudder–propeller interaction using the individual component approach have been developed over the years, generally using actuator disc theory (Section 3.6), to model the rudder axial inflow velocity. This approach, together with a relatively large number of empirical modifications, can achieve reasonable predictions in simulations. The method of using free-stream characteristics with correction factors is, however, deficient in that it does not correctly account for the actual physical interaction between the various components including, for example, the asymmetric performance of a rudder downstream of a propeller, the spill over effects when the propeller slipstream is not completely covering the rudder span, or the significant increase in stall angle when the rudder is downstream of a propeller. Consequently, test data have been derived in various investigations for the rudder–propeller combination working as a unit. In this case, the rudder plus propeller is modelled as a combination in isolation, taking account of the governing parameters described in Section 3.5.2. The influences of an upstream hull and drift angle β are then applied in the form of velocity and flow straightening inputs to the basic isolated model of the rudder–propeller combination. The feasibility of this approach has been demonstrated through experimental work, Molland and Turnock [7.32], which has indicated, for example, that a systematic change in drift angle applied to the rudder–propeller combination leads to an effective shift in the sideforce characteristics of the combination by an angular offset, Figure 5.93, Section 5.4.2.7. Thus if these stages in the procedure are modelled in the manner described, with sufficient detail and adequate accuracy, then a versatile and more physically correct model of rudder action in the presence of a propeller can be established. Example applications using this approach are included in Chapter 11. A design methodology using the rudder–propeller interaction data of Molland and Turnock is presented by Smithwick [7.33] and Molland *et al.* [7.34].

7.2 Rudder design process

In the process of designing a rudder it is necessary to identify the performance and design requirements, choose and apply a rudder with appropriate geometric parameters and estimate its performance characteristics for the given flow conditions. The overall rudder design process may then be summarised as follows:

Input rudder parameters

(i) Number of rudders
(ii) Rudder type
(iii) Area
(iv) Aspect ratio
(v) Profile shape: taper ratio and sweep
(vi) Chordwise section shape and thickness
(vii) Position of stock, balance
(viii) Rudder location relative to hull
(ix) Rudder location relative to propeller

Input flow conditions

(i) Effective inflow velocity
(ii) Effective rudder incidence

Output data

(i) C_L over range of incidence
(ii) C_D over range of incidence
(iii) C_{Lmax}
(iv) α_{stall}
(v) Centre of pressure
(vi) Pressure (load) distribution

Outcomes

The output data are used to derive rudder torque and bending moments to size the rudder stock diameter, size the steering gear, estimate rudder scantlings from the load distributions and provide lift and drag data for coursekeeping and manoeuvring simulations. An outline of the overall rudder design flow path is shown in Figure 7.3.

The following section discusses the topics within the design process:

7.2.1 Rudder parameters

Number. The number of rudders will depend on the ship type and service, or yacht or boat size and purpose. In motor-propelled vessels, the number of rudders

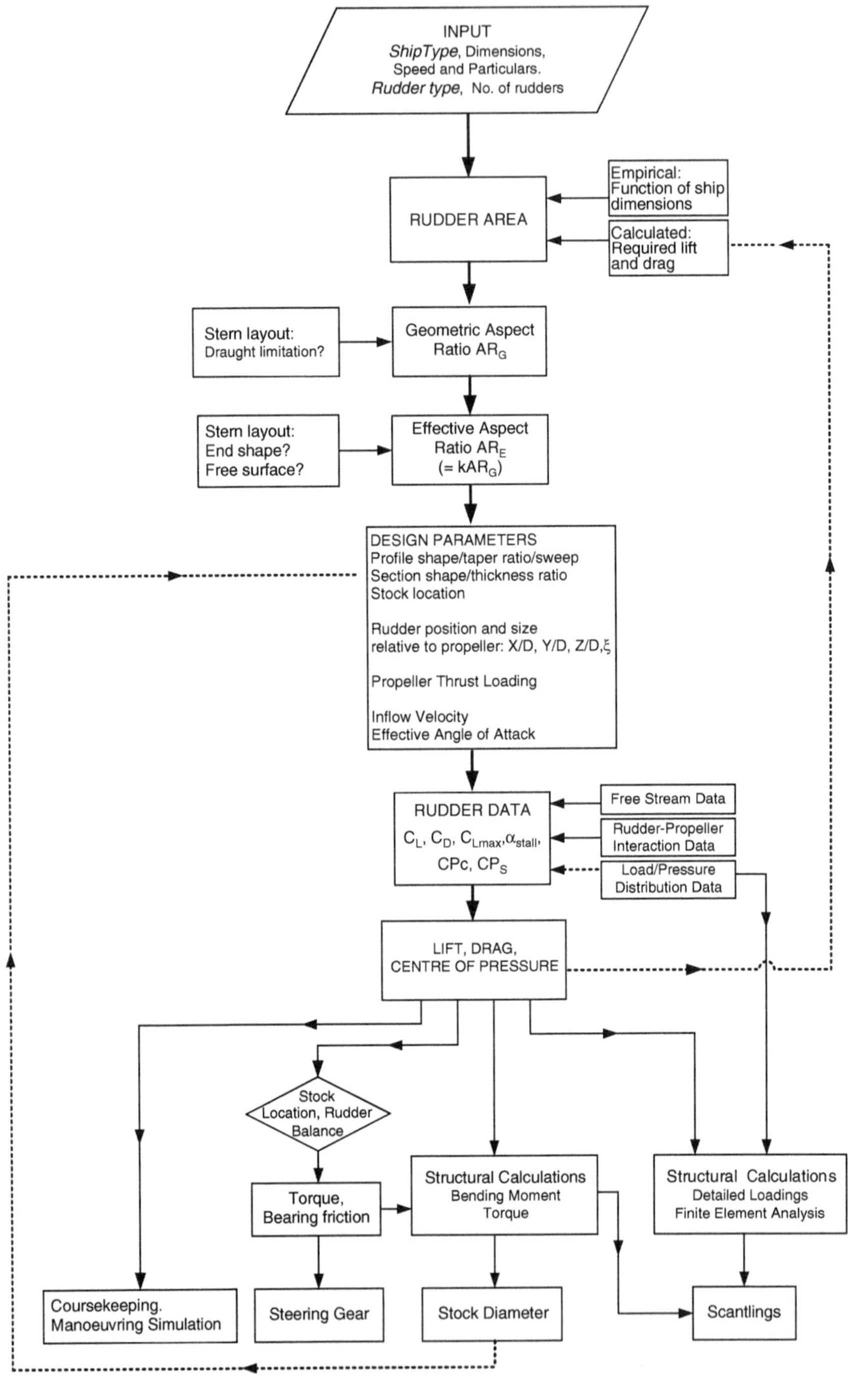

Figure 7.3 Rudder design flow path

will generally follow the number of propellers. In sailing yachts, the number will depend on the required total rudder area and performance requirements.

Type. Typical rudder types are shown in Figure 2.2, Section 2.2. The rudder type chosen will often be related to the ship type and stern arrangement. There are, however, circumstances where alternatives may be available, such as the choice between an all movable spade-type rudder, a full-skeg rudder, or a semi-balanced skeg rudder. Typical reasons for choosing the alternatives include hydrodynamic performance, structural design, layout and maintenance.

Area. Rudder area would ideally be estimated using a coursekeeping and manoeuvring simulation that would indicate the size of the rudder necessary to provide a certain level of steering performance. In practice, this is generally not possible at the preliminary design stage when the stern arrangement, propeller and rudder layouts are being decided. An alternative, and often used procedure, is to estimate the area, generally based on a proportion of the immersed lateral area, from the area used for similar ships with satisfactory steering properties (see Section 5.6). This has been found to be a satisfactory procedure for existing ship types and aft end layouts. Care must however be exercised if radical changes to the aft end layout are applied, when more fundamental investigations may be necessary including model tests and simulations.

Aspect ratio. Aspect ratio may be deemed the most important parameter as far as hydrodynamic performance is concerned, with increase in aspect ratio leading to an increase in overall hydrodynamic efficiency of the control surface. In merchant ships, aspect ratio tends to evolve as a result of the rudder–propeller layout. For example, if there are any draught limitations, then maintaining a required rudder area will lead to an increase in rudder chord length and decrease in aspect ratio. Such low aspect ratios can be seen on shallow draught inland waterway vessels. The shape of the hull above the rudder is important in that it affects the aspect ratio used for the performance predictions. It should be noted that an increase in aspect ratio for a spade rudder can lead to conflicting outcomes since it will lead to an increase in rudder root bending moment, increase in root thickness for structural reasons and a consequent decrease in hydrodynamic performance. Such a scenario in the case of a sailing yacht is addressed in example application 13 in Chapter 11.

Profile shape. Profile shape tends not to have a significant influence on hydrodynamic performance (see Section 5.3.2). Small amounts of taper and sweep tend to be the norm. Further adjustments to shape may occur to suit particular stern arrangements.

Section shape. The choice of chordwise section shape will follow design requirements for hydrodynamic performance. Standard aerofoil type sections are used in most cases, but specialised sections may be employed where increased lift curve slope, delayed stall, low drag or the avoidance of cavitation is sought (see Sections 5.3.2 and 5.8.4). Numerical methods can be usefully employed in the design of section shapes (Chapter 6). Section thickness generally results from structural requirements.

Balance. Balance can be fundamental to the rudder design since it influences tiller forces and steering gear size. However, as the centre of action of the forces (centre of pressure) moves aft with increase in incidence, it is generally not possible to fully balance a rudder or control surface over a range of incidence. The location of the stock will depend on whether the centre of pressure should always be aft of the stock, which would lead to a trailing rudder in the event of a tiller or a steering gear malfunction, but with relatively high torques at large incidence, or a compromise where some negative torque is accepted at small angles in order to lower peak torques at large angles. In this case, the rudder will flop over and increase in angle of attack in the event of a tiller or steering gear malfunction. In order to limit an excessive size of steering gear, this tends to be the practice for large merchant ships. In the case where astern operation is important, such as for some ferries and warships, a compromise stock position may have to be adopted to achieve peak ahead and astern torques at broadly the same level. Example applications in Chapter 11 illustrate the effects of change in the position of the stock on torque.

Rudder-hull. The rudder location relative to the hull can be important as it may influence the end effect between the rudder and the hull, the effective aspect ratio and, consequently, hydrodynamic performance (see Section 5.5).

Rudder-propeller. The rudder location relative to the propeller influences the performance of the rudder depending on the relative longitudinal locations, the amount of asymmetry in the propeller race and the proportion of the rudder within the propeller race, as described in Sections 5.4 and 5.9. The whole propeller diameter will, where possible, be within the rudder span to utilise fully the accelerated flow from the propeller. In practice, this may not always be achievable, such as the case of some small twin-screw ships, Figure 4.8. A discussion of the influences of the relative positions of the rudder and propeller on performance is included as example application 9 in Chapter 11.

7.2.2 Flow conditions

Velocity. In the case of a sailing craft, or a twin-screw motor ship with a rudder not in way of the propellers, the effective inflow velocity will be estimated by taking into account the slowing down effect of the hull (Section 3.7). In the case where the rudder is operating downstream of a propeller, this will amount to estimating the slowing down effect of the hull on the propeller, together with the accelerating effect of the propeller on the flow into the rudder, Sections 3.5 and 3.6. Where performance data are available for the rudder operating downstream of a propeller, such as the experimental data presented in Section 5.4, the data are entered at the appropriate propeller thrust loading, K_T/J^2. Alternatively, the propeller induced velocity, Sections 3.5, 3.6, is applied directly to the free-stream rudder data, Section 5.3.

Incidence. The effective inflow incidence on a control surface is likely to be different from the set incidence. For a sailing craft, the rudder can often be operating in the downwash of the keel. For a ship or a boat just entering a turn, the hull develops a drift or leeway angle which decreases the effective rudder angle, whilst the hull and the propeller have flow straightening effects, which increase the effective angle (see Section 5.4.2.7). These factors will be taken into account in manoeuvring simulations and may be considered in the preliminary rudder design process.

7.2.3 Output data

For given effective inflow velocity and incidence, performance data will be obtained and applied for the control surface type and size under consideration. The sources of an extensive range of such performance data are described in Chapters 5 and 6.

7.2.4 Outcomes

The output data can be used in a systematic way to estimate the forces acting on the rudder, to estimate the diameter of the rudder stock and to size the steering gear. The data will broadly be applied in the following manner:

(A) *Forces, torques, moments*: Levers of centres of centre of pressure, Figure 5.1:

$$\bar{x} = \left[\frac{CPc}{100} \times \bar{c}\right] - x_1$$

$$= \left[\frac{CPc}{100} - \frac{x_1}{100}\right] \times \bar{c} \qquad (7.1)$$

$$\bar{y} = \frac{CPs}{100} \times S \qquad (7.2)$$

Force data:

$$C_N = C_L \cos\alpha + C_D \sin\alpha \qquad (7.3)$$

where α is the effective rudder incidence

Normal force $$N = C_N \times 0.5\,\rho A V^2 \qquad (7.4)$$

Rudder torque $$Q_R = N \times \bar{x} \qquad (7.5)$$

Typical curves of CPc, C_N and Q_R are shown in Figure 7.4. In the example shown, the rudder has some balance and the stock axis has been chosen whereby there is some negative torque at low angles of attack, leading to a lower maximum (positive or negative) torque.

Resultant force coefficient: $$C_R = \sqrt{C_L^2 + C_D^2} \qquad (7.6)$$

and resultant force $$R = C_R \times 0.5\,\rho A V^2 \qquad (7.7)$$

Root bending moment (spade rudder case)

$$M = R \times \bar{y} \qquad (7.8)$$

Equivalent bending moment

$$BM_E = \frac{M}{2} + 0.5\sqrt{M^2 + Q_R^2} \qquad (7.9)$$

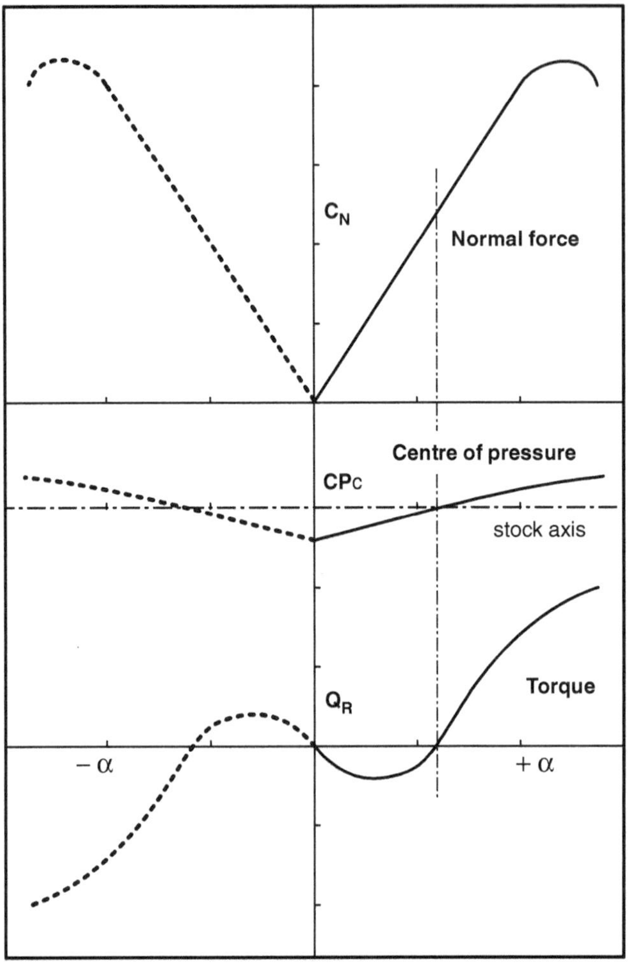

Figure 7.4 Typical curves of rudder normal force C_N, centre of pressure *CPc* and torque Q_R

Equivalent torque

$$Q_{RE} = M + \sqrt{M^2 + Q_R^2} \tag{7.10}$$

Diameter of rudder stock

$$D = \sqrt[3]{(BM_E \times 32)/\pi \times \sigma} \qquad \text{or} \qquad = \sqrt[3]{(Q_{RE} \times 16)/\pi \times \sigma} \tag{7.11}$$

where σ is the allowable stress in the stock material.

For an equivalent tubular rudder stock, the outside and inside diameters d_1 and d_2 have to satisfy the equation:

$$D = \sqrt[3]{(d_1^4 - d_2^4)/d_1} \tag{7.12}$$

Use of these formulae is illustrated in the example applications in Chapter 11.

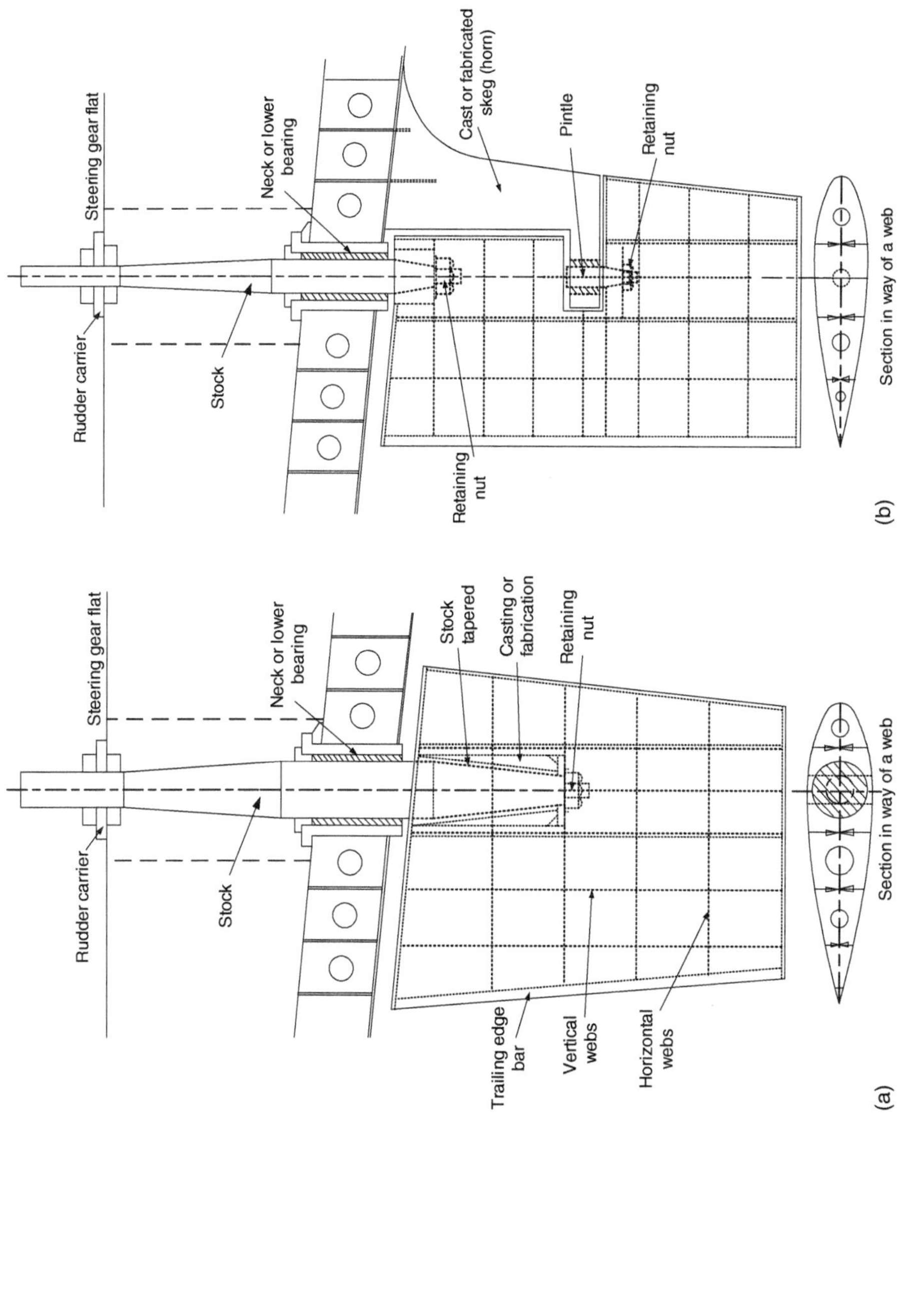

Figure 7.5 (a) Structural layout: spade rudder; (b) Structural layout: skeg rudder

(B) *Load distributions and scantlings*: Examples of outline structural layouts for a spade rudder and a semi-balanced skeg rudder are shown in Figure 7.5.

The scantlings for the rudder webs and plating will normally be obtained or checked using the rules of classification societies and standards such as references [7.35–7.40]. Direct calculations may also be carried out. In this case, the load distributions can be established from available experimental pressure distributions such as those described in Section 5.4.2.3, or one of the numerical methods described in Chapter 6. These load distributions may then be linked to a finite element analysis, FEA, which will determine the structural response of the rudder to the prescribed load distribution. This will allow appropriate structural scantlings for the rudder to be determined. Derived rudder scantlings, together with the derived stock diameter, will then normally be compared with the requirements of the various classification societies and standards. The application of an FEA package to a prescribed distribution of rudder loading is illustrated in example application 8 in Chapter 11.

(C) *Steering gear*: The rudder characteristics determine the hydrodynamic torque, Q_H, which is effectively Q_R in Equation (7.5) and Figure 7.4. The frictional torque Q_F due to friction in the rudder bearings also has to be overcome. The total torque Q_T to be provided by the steering gear is

$$Q_T = Q_F \pm Q_H \tag{7.13}$$

The sign in equation (7.13) depends on whether the rudder angle is being displaced or restored. The effect of friction and whether the rudder angle is being displaced or restored is shown schematically in Figure 7.6. This shows the basic hydrodynamic torque Q_H, say Q_R from Figure 7.4, together with Q_F to give the total torque Q_T. As the rudder is displaced, with increasing angle, the effect of the frictional

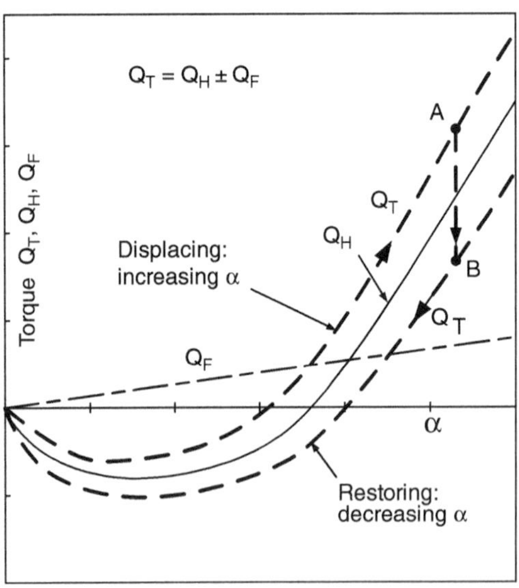

Figure 7.6 Effect of friction in bearings on rudder torque

torque Q_F is to decrease the initial negative torque and increase the positive torque at larger angles. When the rudder action is reversed, say at point A, the torque drops to point B on the restoring curve. When the rudder is restoring, with decreasing angle (rudder effectively driving the steering gear), the effect of Q_F is to decrease Q_T at positive torque and increase Q_T at negative torque. Note that, in reality, there will be a complete reversal of sign for the restoring Q_T at point B, which for schematic purposes, is not shown in Figure 7.6.

The frictional torque Q_F is derived from the reactions in the rudder stock bearings due to the total rudder normal force N. The frictional torque at each bearing is then the resultant force multiplied by the coefficient of friction multiplied by the bearing turning radius. For the case of the spade rudder, Figure 7.7:

$$Q_F = \mu_B R_1 N \left[\frac{y_1 + y_2}{y_3} \right] + \mu_B R_2 N \left[\frac{y_1 + y_2 + y_3}{y_3} \right] \qquad (7.14)$$

where μ_B is the coefficient of friction of the bearing material and R_1 and R_2 are the radii of the bearings. Typical values of μ_B for sleeve bearings are 0.1 for metals and 0.2 for synthetic materials, Taplin [7.20] and Harrington [7.24].

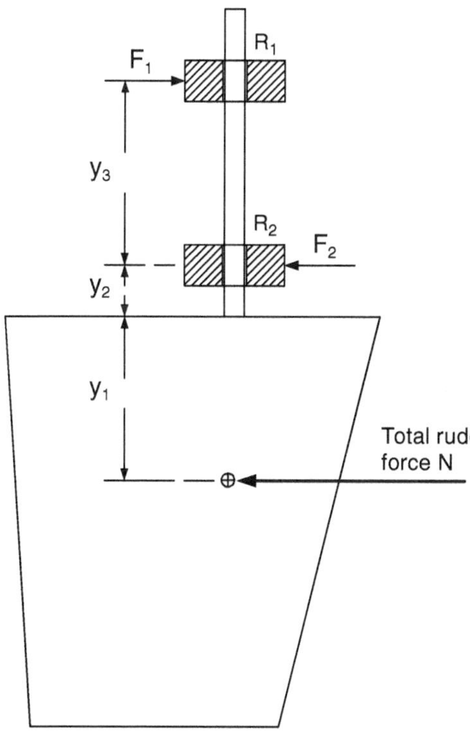

Figure 7.7 Resolution of forces at bearings

Harrington [7.24] uses a number of worked examples to illustrate the derived hydrodynamic and frictional torques for spade and semi-balanced skeg rudders. Taplin [7.20] includes calculations for frictional torque. The frictional torque Q_F is assumed linear with angle, Figure 7.6, and at 30–35° may be about 5–10% of the total torque Q_T. At 10–20°, when hydrodynamic torque Q_H can be very low or tending to zero (but the normal force is still present, Figure 7.4), Q_F will represent a much higher proportion of the total torque. The effect of friction in the rudder bearings is included in example application 2 in Chapter 11.

(D) *Rudder rate*: The steering gear torque may be further influenced by the rate of change of rudder angle. The time to change heading and abilities such as zig–zag overshoot characteristics are affected by rudder deflection rate, although the rate has no influence on the diameter of the steady turn. Minimum rudder rates of 2⅓°/s are required by regulatory bodies and classification societies. That is typically putting the rudder over from 35° one side to 35° the other side in 30 s, or 35° one side to 30° the other side in 28 s. Such rudder rates are generally acceptable for most ship types. Ideally, a fast rudder rate is called for initially, with rudder angle rising to just below stall. As drift angle develops, the rudder angle would be increased such that the effective angle remains just below stall. As pointed out by Mandel [7.15], this ideal goal is generally not possible.

Mandel [7.15] investigated the influence of changes in rudder rate, based on the time to change the heading by 30°, resulting in a trend shown schematically in Figure 7.8. It was found that increases in rudder rate provided greater improvements in shorter ships than larger ones. It was also deduced that, based on the results of the investigation, a good rudder rate to select for most ships would be a rate that corresponds to about 15° rudder deflection per ¼ ship length of travel, and indicates this would amount to a rudder rate of about 5°/s for a 122 m, 20 knot ship. This is in the area where the curve in Figure 7.8 flattens and further increases in rate are not very effective. Mandel concluded that most ships will have a rate somewhat lower than 5°/s. The effect of rudder rate was examined in some detail by Eda and

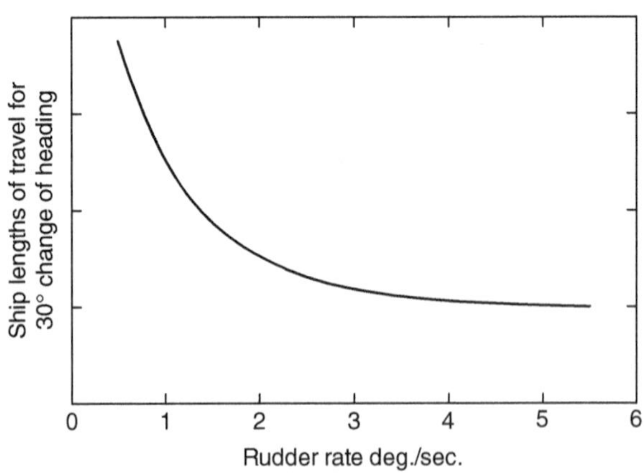

Figure 7.8 Influence of rudder rate

Crane [7.41] and Eda [7.42, 7.43]. Their results indicate that beyond about 3°/s, further improvements become very small. It should be finally noted that increased rudder rate will increase the torque to a certain extent and increase the load on the steering gear.

(E) *Full-scale tests*: In making predictions for full-scale rudder performance, note will be made of the relevant Reynold's number and whether any correction for scale effects should be included. Effects of Reynold's number are discussed in Section 5.3.2.

Many difficulties can arise when attempting to measure full-scale rudder forces and torques. Ideally, forces and torque would be measured using suitably sited strain gauges on the rudder and rudder stock, and such an approach was used in the trials reported by Becker and Brock [7.44]. Very few results of such detailed measurements have been published elsewhere. More normally, torque has to be measured indirectly using the pressures in the hydraulic rams of the steering gear. Further corrections then have to be made to allow for mechanical friction in the steering system and in the rudder stock bearings. The difficulties arising from this approach are discussed by Hagen [7.18], Taplin [7.45] and Harrington [7.24]. Son *et al.* [7.28] compare, for a large tanker, the sea-trial torque from the steering gear with a regression model of rudder torque.

In making a comparison of full-scale measurements with those predicted by model, care has to be taken with the measurement or prediction of the actual full-scale rudder angle of attack and inflow speed, allowing for wake, propeller slipstream, hull drift angle and change in rudder angle of attack as the ship turns. With careful consideration of all the corrections, reasonable correlation between model and full-scale can be achieved. The extended discussion to Harrington's paper covers most aspects of the problems associated with full-scale predictions and tests.

A further indirect approach to checking overall rudder forces is to compare the full-scale manoeuvring trial results, such as in reference [7.46], with those from a mathematical manoeuvring simulation that adequately models the rudder, such as in reference [7.47]. It has to be noted that, as this approach models the hull, propeller and rudder, interaction effects need to be modelled correctly, otherwise acceptable overall manoeuvring predictions may be obtained for the wrong reasons.

Full-scale tests on roll stabilisers can be more productive. The ship can be force rolled (in calm water) using the fins and the resulting roll angle compared with the design waveslope capacity of the fins (see Section 9.1.2).

7.3 Applications of numerical methods

Chapter 6 describes in some detail the application of numerical methods suitable for the design of control surfaces. Techniques reviewed include potential flow and Navier–Stokes methods, together with unsteady behaviour. Particular applications include the design of section shape, for example to delay stall or cavitation, together with the ability to determine surface pressures and load distributions for a wide range of shapes in the freestream and behind a propeller.

7.4 Guidelines for design

When considering the overall design process and in deciding the most appropriate rudder or control surface for a particular task, three complementary areas of knowledge can be used:

(1) Empirical knowledge derived from in-service experience but, more usually these days, from the results of model-scale experimentation of varying levels of complexity and expense. These are discussed in Chapter 5.
(2) Theoretical investigation, using dimensional analysis, that allows the categorisation of the appropriate flow regime and then adoption of a suitable mathematical approximation, as discussed in Chapter 6.
(3) The use of numerical methods to solve the many fluid dynamic equations required to discretise a complete domain, as discussed in Chapter 6.

The most effective design strategy will be one that permits an appropriate synthesis of the three areas in a suitable blend.

It is difficult to develop a universal guideline. As in the majority of designs, the quality and sophistication of the final product will depend on the resources made available, both in terms of expenditure associated with model testing (computational or experimental) and, most importantly, the amount of time available for work on the project.

References

7.1 Kinoshita, M. and Okada, S. On the twisting moment acting upon a ship's rudder stock. *International Shipbuilding Progress*, Vol. 2, No. 9, 1955, pp. 230–240.
7.2 Jöessel, Rapport sur des experiences relatives aux gouvernails. *Memorial du Genie Maritime*, Rapport 9, 1873.
7.3 Denny, M.E. The design of balanced rudders of the spade type. *Transactions of the Royal Institution of Naval Architects*, Vol. 63, 1921, pp. 117–130.
7.4 Baker, G.S. and Bottomley, G.H. Manoeuvring of ships. Part I – Unbalanced rudders of single screw ships. *Transactions of the Institute of Engineers and Shipbuilders in Scotland*, Vol. 65, 1921/22, pp. 522–583.
7.5 Bottomley, G.H. Manoeuvring of ships. Part II – Unbalanced rudders of twin screw ships. *Transactions of the Institute of Engineers and Shipbuilders in Scotland*, Vol. 67, 1923/24, pp. 509–559.
7.6 Bottomley, G.H. Manoeuvring of ships. Part III – Unbalanced rudders behind single screw ships, effect of varying fullness of form. *Transactions of the Institute of Engineers and Shipbuilders in Scotland*, Vol. 70, 1926/27, pp. 463–500.
7.7 Bottomley, G.H. Manoeuvring of ships. Part IV – Unbalanced rudders behind twin screw ships, effect of varying fullness of form. *Transactions of the Institute of Engineers and Shipbuilders in Scotland*, Vol. 74, 1930/31, pp. 94–123.
7.8 Bottomley, G.H. Manoeuvring of ships – Semi-balanced rudders of twin screw ships. *Transactions of the North East Coast Institution of Engineers and Shipbuilders*, Vol. 48, 1931/32, pp. 97–114.

7.9 Bottomley, G.H. Manoeuvring of single screw ships – The effect of rudder proportions on manoeuvring and propulsive efficiency. *Institution of Civil Engineers*, Selected Engineering Papers, No. 175, 1935, pp. 1–18.

7.10 Abell, T.B. Some model experiments on rudders placed behind a plane deadwood. *Transactions of the Royal Institution of Naval Architects*, Vol. 78, 1936, pp. 135–144.

7.11 Gawn, R.W.L. Steering experiments, Part I. *Transactions of the Royal Institution of Naval Architects*, Vol. 85, 1943, pp. 35–73.

7.12 Van Lammeren, W.P.A., Troost, L. and Koning, J.G. *Resistance, Propulsion and Steering of Ships*. The Technical Publishing Co. H. Stam-Haarlem, Holland, 1948.

7.13 Attwood, E.L. and Pengelly, H.S., revised by A.J. Sims. *Theoretical Naval Architecture*. Longmans, Green and Co. London, 1953.

7.14 Lamb, B.J. and Cook, S.B. A practical approach to rudder design. *Shipbuilding and Shipping Record*. September 1961, pp. 334–336.

7.15 Mandel, P. Some hydrodynamic aspects of appendage design. *Transactions of the Society of Naval Architects and Marine Engineers*, Vol. 61, 1953, pp. 464–515.

7.16 Jaeger, H.E. Approximate calculation of rudder torque and rudder pressures. *International Shipbuilding Progress*, Vol. 2, No. 10, 1955, pp. 243–257.

7.17 Gover, S.C. and Olsen, C.R. A method for predicting the torque of semi-balanced centreline rudders on multiple-screw ships. *DTMB Rep. 915*, 1954.

7.18 Hagen, G.R. A contribution to the hydrodynamic design of rudders. Ministry of Defence, *Third Ship Control Systems Symposium*, Bath, September 1972.

7.19 Whicker, L.F. and Fehlner, L.F. Free stream characteristics of a family of low aspect ratio control surfaces for application to ship design. *DTMB Report 933*, December 1958.

7.20 Taplin, A. Notes on rudder design practice. *DTMB Report 1461*, October 1960.

7.21 Romahn, K. and Thieme, H. On the selection of balance area for rudders working in the slipstream (in German). *Schiffstechnic*, No. 21, 1957.

7.22 Thieme, H. Design of ship rudders. *Jarbuch der Schiffbautechnischen Gesellschaft*, Vol. 56, 1962 or DTMB Translation No. 321, 1965.

7.23 Okada, S. On the performance of rudders and their designs. *The Society of Naval Architects of Japan*, 60th Anniversary Series, Vol. 11, 1966, pp. 61–137.

7.24 Harrington, R.L. Rudder torque prediction. *Transactions of the Society of Naval Architects and Marine Engineers*, Vol. 89, 1981, pp. 23–90.

7.25 Millward, A. The design of spade rudders for yachts. *Southampton University Yacht Research Report*, SUYR No. 28, 1969.

7.26 Molland, A.F. Rudder design data for small craft. University of Southampton, *Ship Science Report* No. 1/78, 1978.

7.27 Brix, J. (ed.) *Manoeuvring Technical Manual*, Seehafen Verlag, Hamburg 1993.

7.28 Son, D.I., Ahn, J.H. and Rhee, K.P. An empirical formula for steering gear torque of tankers with a horn rudder. Proceedings of *the 8th International Symposium on Practical Design of Ships and Mobile Units, PRADS'2001*, Shanghai, 2001.

7.29 Kresic, M. Estimating hydrodynamic force and torque acting on a horn-type rudder. *Marine Technology*, Vol. 39, No. 2, April 2002, pp. 118–136.

7.30 Molland, A.F. A method for determining the free-stream characteristics of ship skeg-rudders. *International Shipbuilding Progress*, Vol. 32, No. 370, June 1985 pp. 138–150.

7.31 Goodrich, G.J. and Molland, A.F. Wind tunnel investigation of semi-balanced ship skeg-rudders. *Transactions of the Royal Institution of Naval Architects*, Vol. 121, 1979, pp. 285–307.

732 Molland, A.F. and Turnock, S.R. Flow straightening effects on a ship rudder due to upstream propeller and hull. *International Shipbuilding Progress*, 49, No. 3, 2002, pp. 195–214.

7.33 Smithwick, J.E.T. *Enhanced design performance prediction method for rudders operating downstream of a propeller*. Ph.D. Thesis, University of Southampton, UK, 2000.

7.34 Molland, A.F., Turnock, S.R. and Smithwick, J.E.T. Design studies of the manoeuvring performance of rudder–propeller systems. Proceedings of *the 7th International Symposium on Practical Design of Ships and Mobile Units, PRADS '98*, The Hague, The Netherlands, September 1998, pp. 807–816.

7.35 Lloyds Register of Shipping. *Rules and Regulations for the Classification of Ships*, Ship Control Systems, Part 3, Chapter 13, July 2005.

7.36 DNV *Rules for Classification of Ships*, Part 3, Chapter 3, Hull Equipment and Safety, Det Norske Veritas, Oslo, January 2006.

7.37 American Bureau of Shipping. *Guide for building and classing Motor Pleasure Yachts*. 1990.

7.38 American Bureau of Shipping. *Guide for building and classing Offshore Racing Yachts*. 1994.

7.39 Germanischer Lloyd. *Rules for Classification and Construction, 2006*.

7.40 ISO *Draft International Standard: Hull Construction – Scantlings – Rudders*. ISO/DIS 12215–12218, 2005.

7.41 Eda, H. and Crane, C.L. Steering characteristics of ships in calm water and in waves. *Transactions of the Society of Naval Architects and Marine Engineers*, Vol. 73, 1965, pp. 135–177.

7.42 Eda, H. Notes on ship controllability. *Society of Naval Architects and Marine Engineers Bulletin* No. 1-41, April 1983.

7.43 Eda, H. Shiphandling simulation study during preliminary ship design. Proceedings of *Fifth CAORF Symposium*, Kings Point, New York, May 1983.

7.44 Becker, L.A. and Brock, J.S. The experimental determination of rudder forces during trials of USS Norfolk. *Transactions of the Society of Naval Architects and Marine Engineers*, Vol. 66, 1958, pp. 310–344.

7.45 Taplin, A. Sea trials for measuring rudder torque and force. Proceedings of *Fourth Ship Control Systems Symposium*. Royal Netherlands Naval College, The Hague. Vol. 5. 1975, pp. 99–115.

7.46 Clarke, D., Patterson, D.R. and Woooderson, R.K. Manoeuvring trials with the 193,000 Tonne deadweight tanker 'Esso Bernicia'. *Transactions of the Royal Institution of Naval Architects*, Vol. 114, 1973, pp. 89–109.

7.47 Molland, A.F., Turnock, S.R. and Wilson, P.A. Performance of an enhanced rudder force prediction model in a ship manoeuvring simulator. Proceedings of *International Conference on Marine Simulation and Ship Manoeuvrability, MARSIM '96*, Copenhagen, Denmark, 1996, pp. 425–434.

8 Manoeuvring

8.1 Rudder forces

The basic requirements of a rudder and the forces acting during a manoeuvre are described in Chapter 4 and Figure 4.2. The manoeuvring performance of a ship is controlled by the performance of its rudder and it is therefore necessary to be able to estimate rudder forces at any stage in the manoeuvre. At a point in the manoeuvre, for a given rudder design and arrangement, it is necessary to estimate the effective rudder incidence and the effective rudder velocity. Knowing the incidence and velocity, the rudder performance data such as that reviewed in Chapter 5, can be used to estimate the forces produced by the rudder.

In a manoeuvre, the total sideforce will be made up of:

(i) the contribution from the rudder,
(ii) the sideforce due to the propeller in oblique flow and
(iii) the sideforce developed on the hull due to the rudder–propeller combination.

The sideforce due to the propeller when the ship is in a turn can be significant and will depend on drift angle and speed. Guidance on the likely levels of propeller sideforce may be derived from research on the performance of propellers on inclined shafts, such as that of Gutsche [8.1], Hadler [8.2] and Peck and Moore [8.3].

From the aspect of propulsion when in a manoeuvre, as well as the basic thrust and torque characteristics of the propeller, account has to be taken of:

(i) Changes in rudder drag or thrust ΔK_R due to the influence of the propeller and,
(ii) Changes in propeller thrust ΔK_T and torque due to the presence of the rudder.

An overall algorithm showing the derivation of the total manoeuvring sideforce and the net propulsive force is shown in Figure 8.1.

Estimates of the effects of ΔK_R and ΔK_T and the net propulsive force are discussed further in Chapter 10.

8.2 Hull upstream

The hull upstream of a rudder, or rudder–propeller combination, can have a significant influence on the rudder forces and the production of total sideforce.

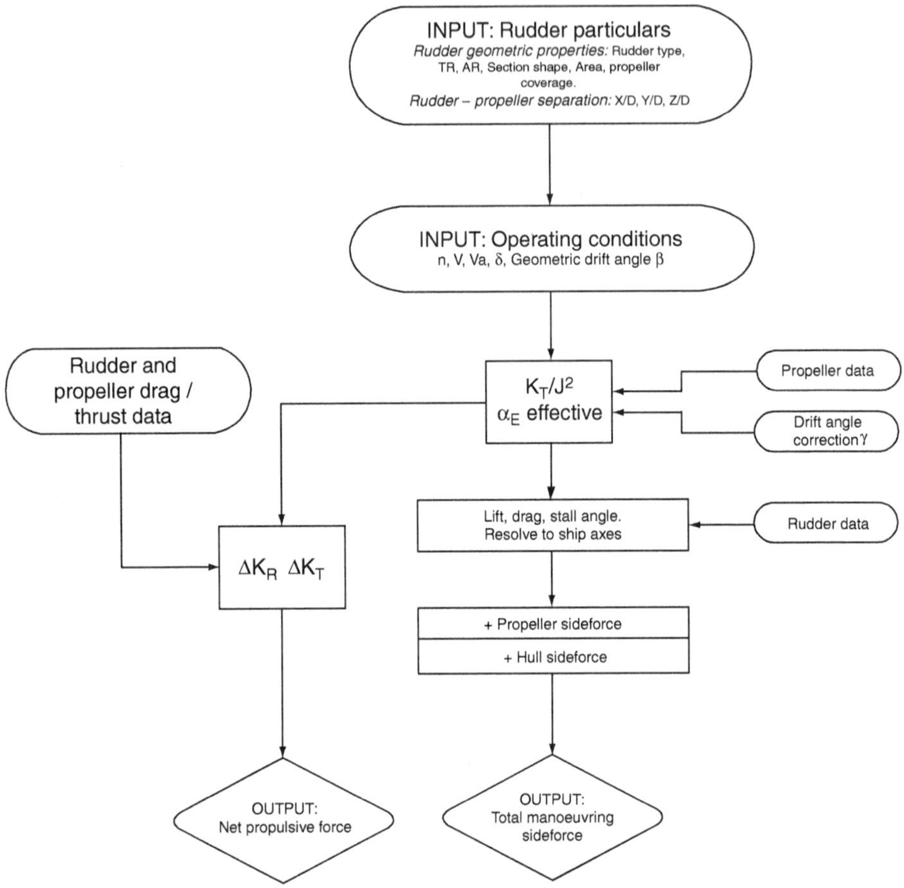

Figure 8.1 Development of total manoeuvring sideforce

(i) The hull slows down the flow speed into the rudder, or the rudder–propeller combination. The average wake speed in the vicinity of the propeller can be estimated using a suitable wake fraction, as described in Section 3.7.
(ii) The hull has a flow straightening effect when the ship is on a turn, as described in Section 5.4.2.7.
(iii) The hull contributes to the production of sideforce, due to pressure changes on the adjacent hull induced by the rudder, Figure 8.2.

The total sideforce (ship axis) may be written as

$$F_T = F_Y(1 + a_H) \tag{8.1}$$

where F_T = total sideforce and F_Y = rudder sideforce.

$(1 + a_H)$ is the contribution due to the hull, which will depend on the distance between the rudder and hull and the overall hull–propeller–rudder arrangement. Typical values of $(1 + a_H)$ are from 1.10 to 1.30. a_H may be expressed as a function of block coefficient as suggested by Hirano [8.4], indicating values of $a_H = 0.15$ at $C_B = 0.55$ up to $a_H = 0.30$ at $C_B = 0.80$, and to reflect some dependence on speed.

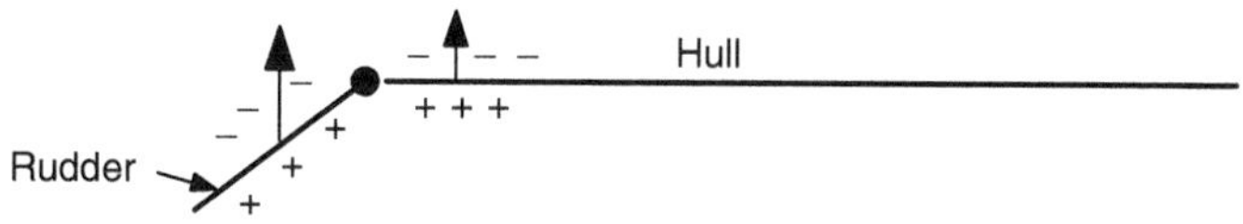

Figure 8.2 Pressures around rudder and hull

Results of tests by Gong et al. [8.5] for a tanker with $C_B = 0.73$ and large $B/T = 3.77$ indicate values of a_H ranging from 0.12 at low J(0.30) up to about 0.20 at high J(0.90).

8.3 Influence of drift angle

When rudder angle is applied, the ship develops a drift angle β, Chapter 5, Figure 5.37, which leads to a cross flow at the stern and a geometric drift angle at the rudder β_R, which is larger than β. The net effect is a decrease in the effective rudder incidence, although the flow-straightening effects of the propeller and hull, described in Section 5.4.2.7 and Figure 5.91, lessen this decrease. The final effective rudder incidence will therefore result from the effects of drift angle and flow straightening. Typical values for flow straightening are contained in Section 5.4.2.7.

An approximate value for β for single-screw merchant ships is proposed in reference [8.6], based on the results of Shiba [8.7] and Mandel [8.8], as

$$\beta = 22.5\ L/R\ \text{(deg.)} \tag{8.2}$$

where R is the steady turning radius.

Actual β will depend on the ship form and aft end hull–propeller–rudder arrangement and Mandel suggests β values between 18 L/R and 22.5 L/R + 1.4.

For a rudder located $L/2$ aft of G, β_R at the rudder, Figure 5.37, is related to the ship drift angle β at G as

$$\tan \beta_R = \tan \beta + L/2R\cos \beta \tag{8.3}$$

As drift angle is developed on a turn, the effective incidence on the rudder decreases and the rudder helm will be increased to compensate. It is apparent that large rudder angles may need to be applied, say up to 60°, to reach the full effectiveness (e.g. stall) of the rudder. For example, for a ship on a steady turn with a diameter of 5 ship lengths, or $L/R = 0.4$, then from Equation (8.2), approximate $\beta = 9°$ and from Equation (8.3) approximate $\beta_R = 19.8°$. From Figure 5.96, a typical flow-straightening angle is about 10° and a rudder with a set helm of 50° would therefore see an effective incidence of about 40°.

8.4 Low and zero speed and four quadrants

8.4.1 Methodology

Although most requirements for ship-manoeuvring capabilities are defined at service speed, it is crucial that a ship manoeuvres well at low speed and has a known performance in all four quadrants of operation, Figure 5.85, i.e.

(i) Ship ahead, propeller ahead
(ii) Ship ahead, propeller astern
(iii) Ship astern, propeller astern
(iv) Ship astern, propeller ahead.

The experimental database, described in Sections 5.4.2.5 and 5.4.2.6, include tests at zero, low speed and in all four quadrants.

The rudder characteristics in the four quadrants are discussed in Section 5.4.2.6 and a methodology for accessing the data proposed. It was shown that a useful presentation of the data in the four quadrants case is

$$C_L^{**} = \frac{L}{\frac{1}{2}\rho A[V^2 + K_T n^2 D^2]} \qquad C_D^{**} = \frac{d}{\frac{1}{2}\rho A[V^2 + K_T n^2 D^2]}$$

and are presented in terms of the propeller advance angle $\psi = \tan^{-1} J/0.7\pi$ across the four quadrants for different values of rudder incidence δ, Figure 5.89.

A flow chart showing the principal features of the approach, using curve fits and look-up tables for the database is given in Figure 8.3. In this example, for the first quadrant data, the standard and low speed presentations are used, namely:

$$C_L = \frac{L}{\frac{1}{2}\rho A V^2} \qquad C_D = \frac{d}{\frac{1}{2}\rho A V^2}$$

$$C_L^* = \frac{L}{\frac{1}{2}\rho A K_T n^2 D^2} \qquad C_D^* = \frac{d}{\frac{1}{2}\rho A K_T n^2 D^2}$$

Low and zero speed data are reviewed in Section 5.4.2.5, and presented in terms of C_L^* and C_D^*. It is noted that, working downstream of a propeller, stall angle is delayed and significant sideforce can be generated by a rudder in the static $J = 0$ condition, an attribute utilised in the low speed handling of ships. In Example application 16 of Chapter 11, it is seen that, at very low speed, the sideforce produced by the rudder on a tug is almost half as much as the thrust produced by the propeller. For low speed work, it is common practice to produce polar or vector diagrams of the net manoeuvring force for various rudder angles, as shown in Figure 8.4. Such a diagram can be applied to manoeuvring simulations and can also be used to compare alternative rudder types, configurations and manoeuvring devices. In Figure 8.4, the propeller thrust has been reduced by the effect of thrust deduction, discussed in Chapter 10.

Slow-speed manoeuvring is discussed in some detail by Brix [8.9]. Various simulation models of low speed and backing manoeuvres have been proposed, including those of Oltman and Sharma [8.10], Abkowitz [8.11] and Shouji *et al.* [8.12]. Kang *et al.* [8.13] describe a methodology for predicting the manoeuvring of full-form ships with low speed. A mathematical model is developed and a regression analysis carried out on the results of several model tests to provide a database for manoeuvring simulations. Semi-balanced skeg, Schilling and flap rudders were investigated. It was concluded that special rudders could significantly improve the performance of ships that had poor manoeuvrability.

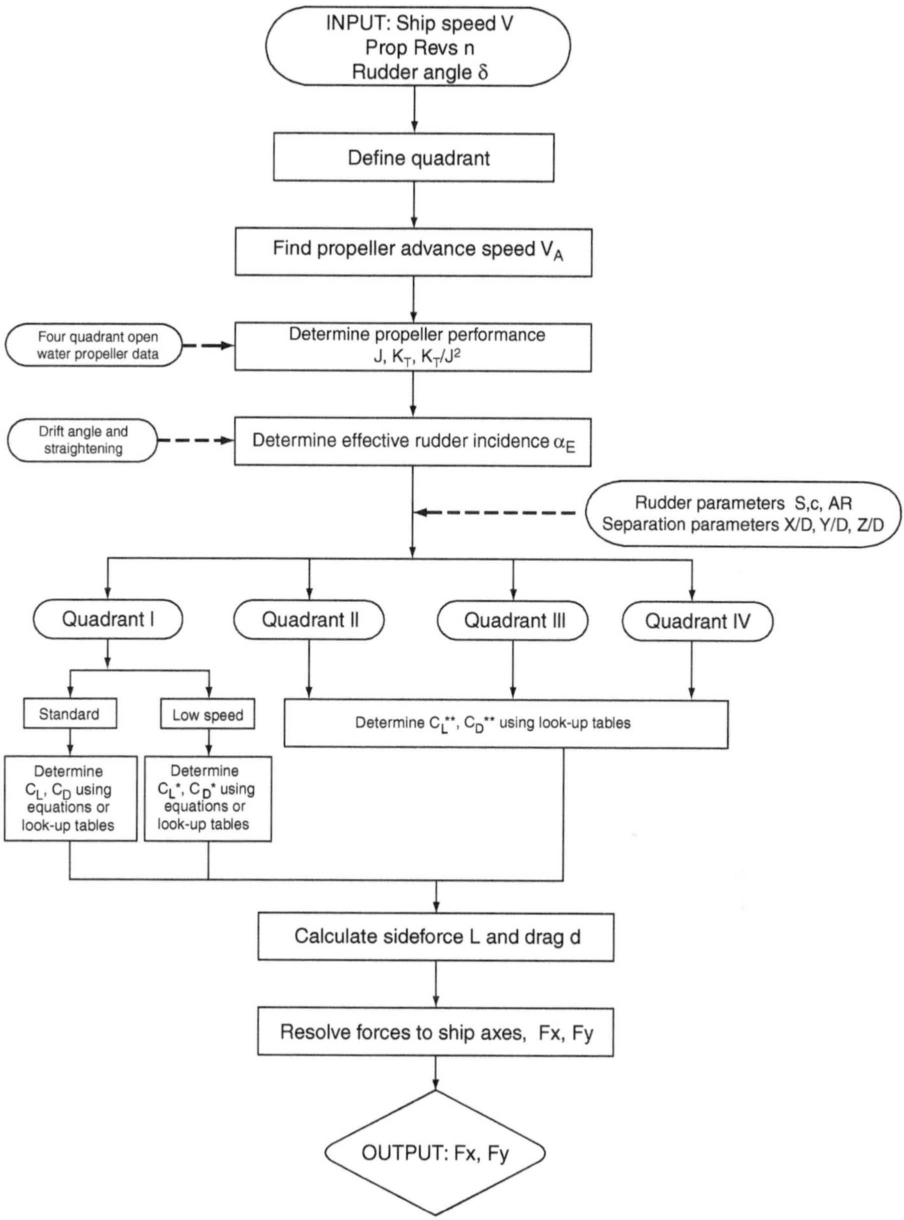

Figure 8.3 Flow chart for four-quadrant rudder force prediction algorithm

8.4.2 High lift rudders and control surfaces

A number of high lift rudders and control surfaces have been proposed and employed, primarily to enhance low-speed manoeuvring. The main objective of such devices is to extend the rudder angle before stall and increase the maximum lift achievable.

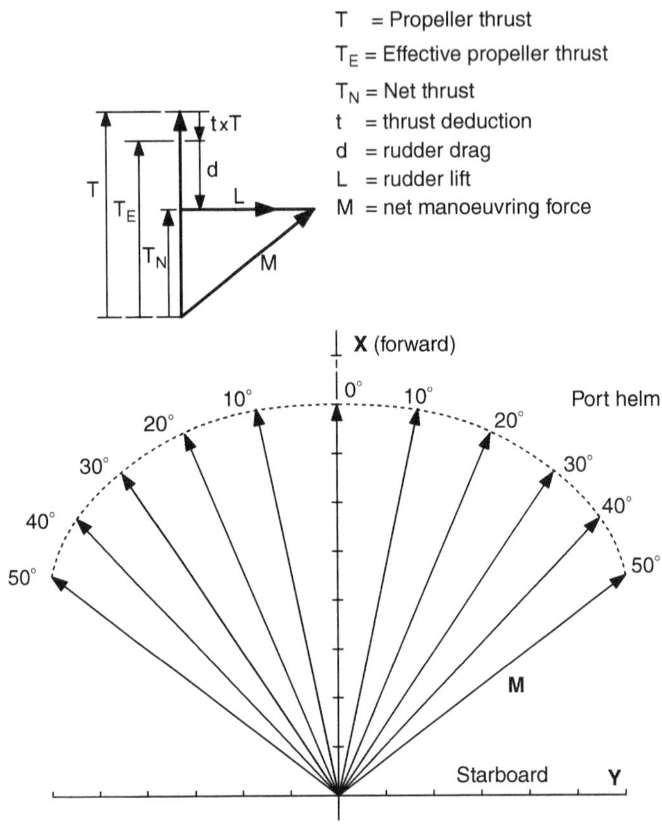

Figure 8.4 Polar plot of net manoeuvring force

They often result from the modification and/or addition to the ordinary aerofoil type section. A brief review is made of some of the various devices that have been proposed for high lift purposes, as shown in Figure 8.5.

(a) *Flapped aerofoil*, Figure 8.5(a). The flapped rudder is discussed and performance data reviewed in Sections 5.3.2 and 5.3.3. The concept is used to increase the lift curve slope, delay stall and increase C_{Lmax}. The increased lift curve slope, giving a faster response for a given helm, can be utilised for coursekeeping and for other control surfaces requiring a fast response such as the fin stabiliser. It is the concept used in the Becker high-lift rudder [8.9].

(b) *Schilling rudder*, Figure 8.5(b). Has a special section designed to delay stall and increase C_{Lmax}. It is described in Section 5.3.2 and by Bingham and Mackey [8.14].

(c) *Wedge at tail*, Figure 8.5(c). Designed to increase lift curve slope and C_{Lmax}. Can be used in situations requiring a fast response, such as the fin roll stabiliser. The concept is discussed and data reviewed in Sections 5.3.2 and 5.3.3.

(d) *Gurney flap*, Figure 8.5(d). The Gurney flap amounts to a small flat plate attached at the trailing edge at right angles to the chord. Designed originally to be fitted to one side of an asymmetrical section, it could equally be applied to a symmetrical movable control surface. It induces an effective camber

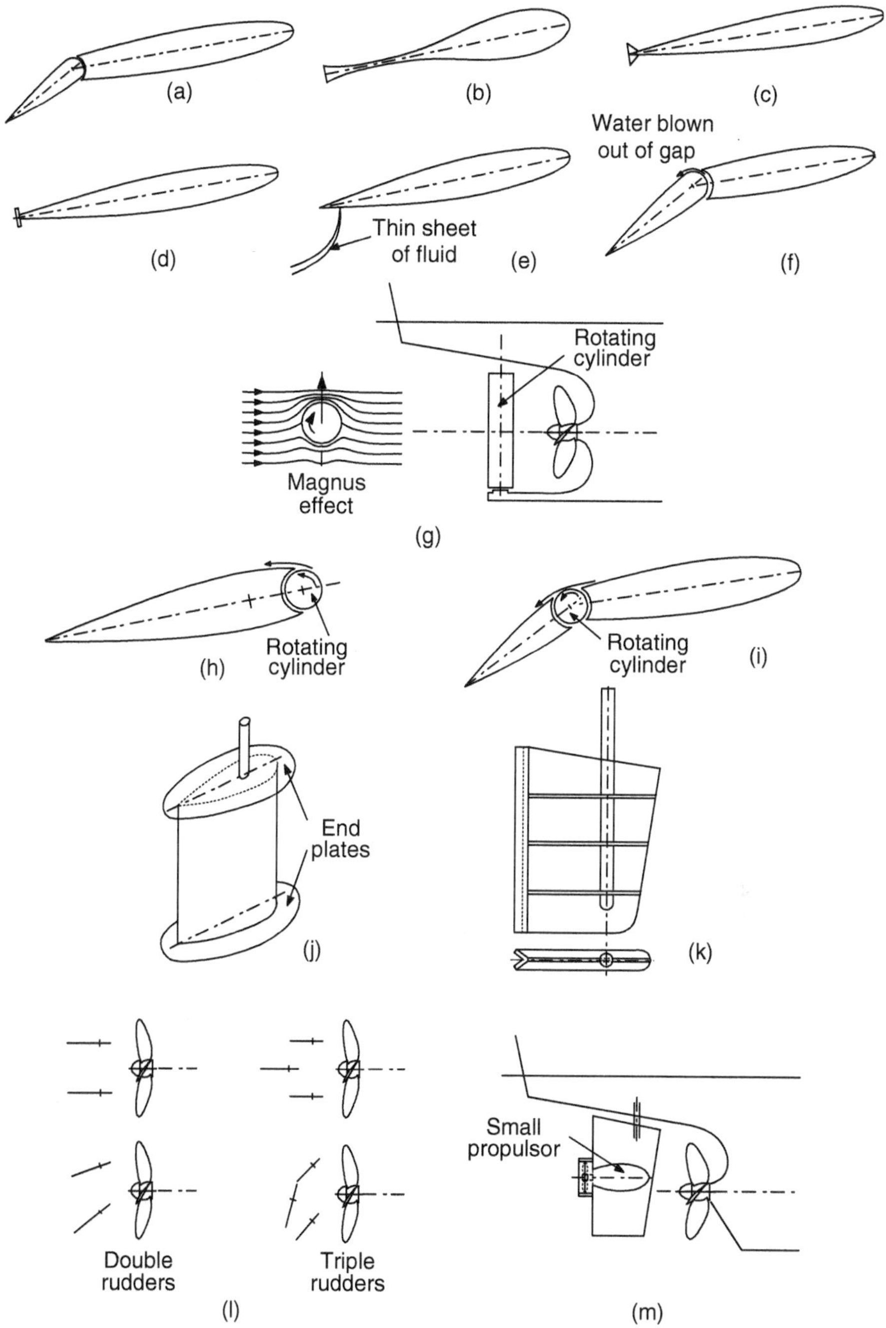

Figure 8.5 Various high-lift rudders

disproportionate to its size, increasing lift for a given incidence. Whilst increasing lift, a penalty is an increase in zero-lift drag and a reduced *L/D* at low to moderate values of lift. An investigation into this concept is included in the work of Date [8.15].

(e) *Jet flap*, Figure 8.5(e). The jet flap has origins in the aircraft industry. A thin sheet of fluid is discharged from the trailing edge (either side) at an angle to the chord of the foil. The emerging jet has the effect of increasing the circulation around the foil and foil lift. English *et al.* [8.16] describe the operation of the jet flap and report on the results of tests carried out in a water tunnel. With the jet flap working, they found an improvement in lift, a rearward movement in *CPc* and a reduction in drag, in spite of the increased lift. The advantage of this type of flap is that it does not have the mechanical complexities of a conventional hinged flap. Whilst meant for zero and low speed manoeuvring, it has attractions also for coursekeeping, where the rudder would be fixed amidships and the jet flap operated for coursekeeping.

(f) *Blown flap gap*, Figure 8.5(f). When a water jet is blown tangential to the suction surface of a flapped rudder, extra lift force is induced by delaying stall and increasing circulation, especially at large angles of attack. A rudder employing this concept was designed and tested by Choi *et al.* [8.17]. Conventional rudder lift slope was improved with an ordinary flap by 35–64%, depending on flap angle/rudder angle ratio. With the addition of jet injection, there were further large increases in lift. The model test results were used in a manoeuvring simulation and it was concluded that the use of a blown flap rudder is an effective way of improving the ship's tuning ability.

(g) *Rotating cylinder in isolation*, Figure 8.5(g). The rotating cylinder in isolation produces lift due to the Magnus effect. Research has been carried out on the application of such a concept as a ship rudder/low speed manoeuvring device [8.18]. The concept, using the cylinder alone, does not seem to have had many practical applications.

(h) *Rotating cylinder in association with rudder*, Figure 8.5(h). A thorough review of the design of control surfaces with rotating cylinders is carried out by Cordier [8.19]. With a rotating cylinder at the leading edge of the rudder, the cylinder imparts energy into the boundary layer. The boundary layer can be controlled and the flow on the back low-pressure side maintained up to very large rudder angles. For example, with this rudder type, angles up to 80° have been achieved without stalling. From the tests reported by Brix [8.9], it was found that with optimum rotational speed/forward speed, increases in rudder lift of up to 100% could be achieved at large angles. Work on this concept has been carried out at NPL [8.18] who also consider the use of a rotating cylinder at the leading edge of a flap. McGeough and Millwood [8.20] carried out water tunnel tests on a rotating cylinder rudder (cylinder at fore end of rudder). With the cylinder rotating, the stall angle was delayed from about 20° (without rotor) to about 50°, with an increase in C_L from 0.65 to 1.46. The use of a blown trailing edge cylinder is mentioned by English [8.16].

(i) *Rotating cylinder in association with flap*, Figure 8.5(i). The rotating cylinder is located at the leading edge of the flap. The concept provides flow control over the flap and was considered by Steele and Harding [8.18].

(j) *End plates*, Figure 8.5(j). End plates have been used over many years to increase the effective aspect ratio of a control surface and to enhance its lift performance. In straight-line flow, there will be an increase in rudder drag due to the frictional drag on the end plates. In oblique flow, which is far more likely in practical situations, there can be significant increases in drag due to separated flow across the plates and shed vortices from the edges of the plates. For this reason, the use of end plates tends to be limited to rudders used mainly in low-speed manoeuvring situations. At low speeds, with the influences of a propeller slipstream, the use of end plates on a high lift section, such as the Schilling section, can lead to very high incidence and lift values before stall [8.14].

(k) *Robust simple rudder*, Figure 8.5(k). It may be important to design a rudder where robustness and reliability are the key design features, such as for the rudder on a vessel working mainly in harbours and coastal waters on manoeuvring and towing duties. As drag is generally not a problem at these low speeds, a flat-plate rudder can be used and the design can concentrate on the method of construction and strength, rather than hydrodynamics. The hydrodynamic properties for flat plates are given in Section 5.3.2. Plate rudders, such as that shown in Figure 8.5(k), have been employed for such vessels and construction is typically from 14 mm to 20 mm steel plate, with an oversize stock diameter. Horizontal stiffeners may be used as necessary and a wedge tail (from flanged plate) may be used at the aft end to improve the strength and increase the stall angle. Recommended scantlings for plate rudders are provided by the classification societies, such as references [7.35–7.39].

(l) *Use of double/triple rudders*, Figure 8.5(l). Twin rudders have been used to vector the slipstream from the propeller in various directions. Such a concept is used with a twin Schilling rudder installation [8.14,8.21] with a controller that allows differential operation of the two rudders. Guarino [8.22] reports on the use of three rudders in a differential manner (Figure 8.5(l)), whereby large steering forces are generated.

(m) *Active rudder*, Figure 8.5(m). This concept consists of a submerged electric motor contained in a streamlined casing, set in a normal rudder with a ducted propeller at the aft end [8.9]. Large rudder angles may be employed. The unit also offers some auxiliary propulsion. There tends to be little detrimental effect on the overall cruising efficiency [8.23].

8.5 Shallow water/bank effects

The effectiveness of the rudder helm is influenced by the presence of the upstream propeller and the wake of the hull. The proximity of the seabed and/or banks will in turn alter these effects. Studies into these effects include those by Kijima *et al.* [8.24, 8.25] who use a parametric approach based on extensive model tests and theoretical analysis. Kobayashi [8.26] describes a method for evaluating the manoeuvring efficiency in deep and shallow water. The MMG manoeuvring model [8.27], was adapted for simulations in shallow water and was found to show satisfactory agreement with free-running and captive model tests that were carried out in shallow water. Turnock and Molland [8.28] investigated shallow water effects using a surface panel code. The methodology is described in Chapter 6.

A thorough review of ship performance in confined waters is presented by Brix [8.9] and Dand [8.29] reports on an extensive series of experiments involving the behaviour of steered ship models in shallow water close to a bank.

References

8.1 Gutsche, F. The study of ships' propellers in oblique flow. *Schiffbauforschung*, 3, ¾, 1964, 97–122. Translation from German, DRIC Translation No.4306.

8.2 Hadler, J.B. The prediction of power performance of planing craft. *Transactions of The Society of Naval Architects and Marine Engineers*, Vol. 74, 1966.

8.3 Peck, J.G. and Moore, D.H. Inclined-shaft propeller performance characteristics. Paper G, *Society of Naval Architects and Marine Engineers, Spring Meeting*, April 1973.

8.4 Hirano, M.A. Practical calculation method of ship manoeuvring motion at initial design stage. *The Society of Naval Architects of Japan*, Naval Architecture and Ocean Engineering, Vol. 19, 1981, 68–80.

8.5 Gong, I-Y., Kang, G-G., Kim, Y-G., You, Y-B., Huh, D. and Kim, Y-H. The influence of rudder area on the manoeuvrability of a ship with large beam-to-draught ratio. Proceedings of *The Sixth International Symposium on Practical Design of Ships and Mobile Units*, PRADS '95, *Society of Naval Architects of Korea*, Seoul, Korea, 1995.

8.6 Lewis, E.V. (ed.) *Principles of Naval Architecture*. The Society of Naval Architects and Marine Engineers, New York, 1989.

8.7 Shiba, H. Model experiments about the manoeuvrability and turning of Ships. *First Symposium on Ship Manoeuvrability*, DTMB Report 1461, October 1960.

8.8 Mandel, P. Some hydrodynamic aspects of appendage design. *Transactions of The Society of Naval Architects and Marine Engineers*, Vol. 61, 1953, pp. 464–515.

8.9 Brix, J. (ed.) *Manoeuvring Technical Manual*, Seehafen Verlag, Hamburg, 1993.

8.10 Oltman, P. and Sharma, S.D. Simulation of combined engine and rudder manoeuvres using an improved model of hull–propeller interactions. *15th ONR Symposium on Naval Hydrodynamics*, Hamburg, 1984.

8.11 Abkowitz, M.A. A manoeuvring simulation model for large angles of attack and backing propellers. Proceedings of *19th ITTC*, Vol. 2, Madrid, 1990.

8.12 Shouji, K., Ishiguru, T. and Mizoguchi, S. Hydrodynamic forces by propeller and rudder interaction at low speed. Proceedings of *MARSIM and ICSM*, Society of Naval Architects of Japan, 1990.

8.13 Kang, G-G., Gong, I-Y., Kim, S-Y., Yum, D-J., Kim, H-C., Ha, M-K., Sen, B-R. and Ahn, S-P. The manoeuvrability of full form ships with low speed. Proceedings of *International Symposium on Forces Acting on a Manoeuvring Vessel*, MAN'98, Val de Reuil, France, 1998.

8.14 Bingham, V.P. and Mackey, T.P. High performance rudders with particular reference to the Schilling rudder. *Marine Technology*, Vol. 24, No. 4, October 1987, pp. 312–320.

8.15 Date, J.C. *Performance prediction of high lift rudders operating under steady and periodic flow conditions.* Ph.D. Thesis, University of Southampton, UK, 2001.
8.16 English, J.W., Rowe, S. and Bain, D. Some manoeuvring devices for use at zero and low ship speed. *Transactions of NECIES*, Vol. 88, 1972, pp. 31–50.
8.17 Choi, B., Park, H., Kim, H. and Lee, S.H. An experimental evaluation on the performance of high lifting rudder under Coanda effect. *9th Symposium on Practical Design of Ships and Other Floating Structures*, PRADS'2004, Lubeck-Travemunde, Germany, 2004, pp. 329–336.
8.18 Steele, B. and Harding, M. The application of rotating cylinders to ship manoeuvring. *National Physical Laboratory*, Ship Report 148, 1970.
8.19 Cordier, S. Design of control surfaces with rotating cylinders. Proceedings of Conference, *Manoeuvring and Control of Marine Craft*, MCMC '92, Computational Mechanics Publications, 1992, pp. 503–518.
8.20 McGeough, F.G. and Millward, A. The effect of cavitation on the rotating cylinder rudder. *International Shipbuilding Progress*, Vol. 28, No. 317, 1981, pp. 2–9.
8.21 The Schilling rudder – 10 years on. *The Naval Architect*, The Royal Institution of Naval Architects, London, April 1985, p. E185.
8.22 Guarino, S.J. The slot augmented, flap effect rudder. *8th International Tug Convention.* Organised by Thomas Reed Publications Ltd., Singapore, 1984.
8.23 High performance rudders for improved shiphandling. *The Naval Architect*, The Royal Institution of Naval Architects, London, March 1979, pp. 53–54.
8.24 Kijima, K., Katsuno, T., Nakiri, Y. and Furakawa, Y. On the manoeuvring performance of a ship with the parameter of loading condition. *Journal of the Society of Naval Architects of Japan.* Vol. 168, 1990, pp. 141–148.
8.25 Kijima, K., Tanaka, S., Matsunga, M. and Hori, T. Manoeuvring characteristics of a ship in deep and shallow waters as a function of loading condition. Proceedings of Conference, *Manoeuvring and control of marine craft*, MCMC'92, Computational Mechanics, 1992, pp. 73–86.
8.26 Kobayashi, E. The development of practical simulation system to evaluate ship manoeuvrability in shallow water. Proceedings of *The Sixth International Symposium on Practical Design of Ships and Mobile Units*, PRADS '95, Society of Naval Architects of Korea, Seoul, Korea, 1995.
8.27 Kose, K. On a new mathematical model of manoeuvring motions of a ship and its applications. *International Shipbuilding Progress*, Vol. 29, No. 336, 1982, pp. 205–220.
8.28 Turnock, S.R. and Molland, A.F. The effects of shallow water and channel walls on the low-speed manoeuvring performance of a mariner hull with rudder. *International Symposium and Workshop on Forces Acting on a Manoeuvring Vessel*, MAN '98, Val de Reuil, France, 1998.
8.29 Dand, I.W. On ship–bank interaction. *Transactions of the Royal Institution of Naval Architects*, Vol. 124, 1982, pp. 25–40.

9 Other control surfaces

9.1 Fin stabilisers

9.1.1 Applications

Fin stabilisers are used on several ship types to provide roll reduction. These include passenger and car ferries, larger passenger ships, warships and some cargo ships. The fins are fitted at about the turn of bilge and are generally retractable, although they may be nonretractable on warships and some smaller installations. Whilst a stabiliser fin may be seen as behaving like a horizontal rudder, as Allan [9.1] points out, the stabiliser fin requires attention to detailed design because it is usually restricted in area and location, and also has to operate out to out in about 1–2 s compared with a rudder in 25–30 s. At the same time, most of the free-stream data for all-movable, flapped, and other foils described in Section 5.3 are directly applicable to stabiliser fins as well as to rudders. There are a number of references that provide a background to the use, design and operation of fin stabilisers, including Allan [9.1], Conolly [9.2], Gunsteren [9.3], Lloyd [9.4,9.5,9.9], Cox and Lloyd [9.6], Fairlie-Clarke [9.7] and Dallinga [9.8].

9.1.2 Design procedure

The basic consideration is to provide the necessary fin force and couple to oppose the rolling moments applied by the waves to the ship. The stabilising moment should be equal to the wave heeling moment and opposite in phase. A parameter suitable for use at the design stage for checking fin area and specifying stabiliser power is the *waveslope capacity* [9.7].

If a ship is heeled to an angle ϕ, the restoring moment is $\Delta gGZ = \Delta gGM \cdot \phi$, where Δ is the ship displacement and GM the metacentric height, Figure 9.1.

The stabilising moment of the fins is $2L_F R$, Figure 9.2.

Thus, if ϕ is seen as the wave slope, Figure 9.3, and $\Delta gGM \cdot \phi$ the couple applied by the wave, then the stabilising moment has to match this couple and

$$2L_F R = \Delta gGM\phi$$

and

$$\phi = 2L_F R \times 57.3/\Delta gGM \qquad (9.1)$$

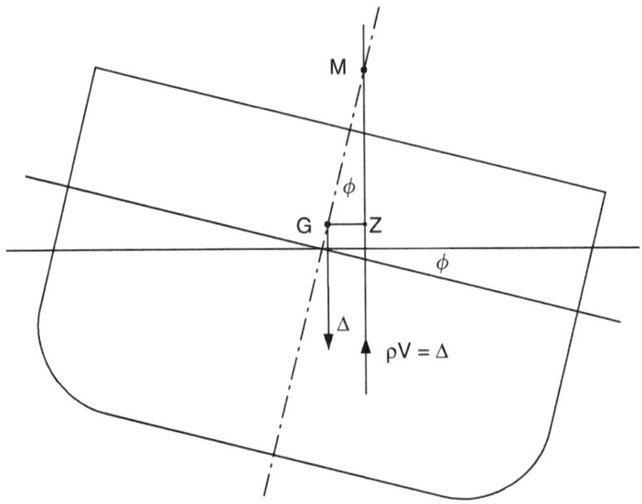

Figure 9.1 Equivalent wave couple

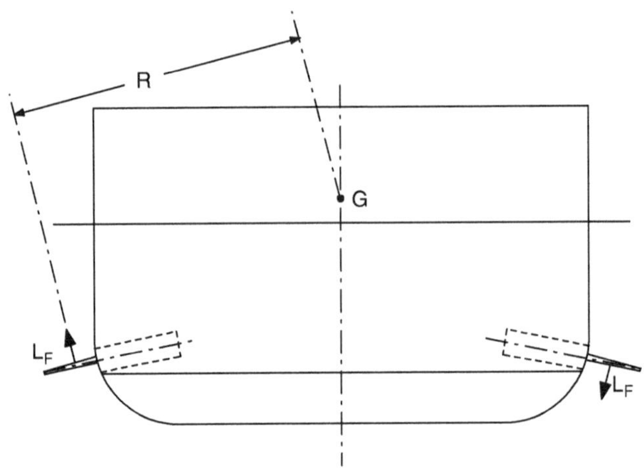

Figure 9.2 Stabilising moment

where ϕ, expressed in degrees, is commonly referred to as the *waveslope capacity* of the stabiliser. This criterion represents the approximate maximum sea state in which the stabiliser can operate effectively. It can also be considered as the angle of steady heel that the fins can create with the ship moving ahead in calm water at a given speed.

The required waveslope capacity will typically lie between 3° and 5°, depending on size and type of ship and its roll period [9.7]. Gunsteren [9.3] suggests 3°–4° for large ships and 4°–5° for small ships. Approximate waveslopes, for the Pierson–Moskowitz wave spectrum, for given ship size, are given in Table 9.1 which was assembled from data in Reference [9.7]. Fairlie-Clarke points out that other spectra may have higher average waveslopes and that the ship will not respond in full to waveslopes such

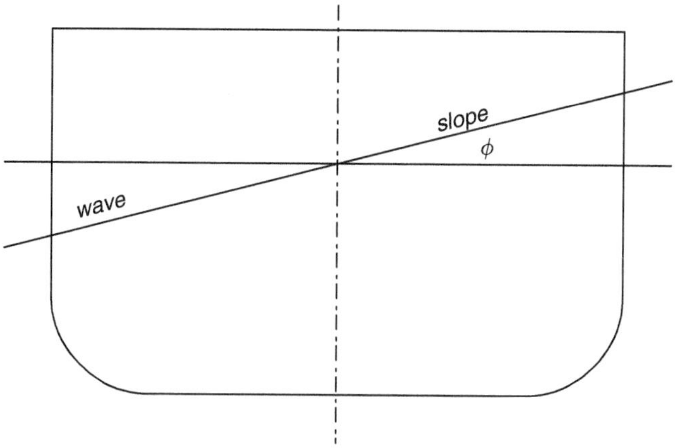

Figure 9.3 Wave slope

Table 9.1 Approximate average RMS wave slopes (degrees)

Ship breadth (m)	*Sea state*				
	3	*4*	*5*	*6*	*7*
10	2.0	3.5	4.8	5.6	6.5
20	0.9	2.2	3.5	4.4	5.2
30	0.5	1.5	2.6	3.5	4.5

as those in Table 9.1 due to dynamic effects and entrained water which will reduce the static heeling force by about 20–25% for a ship with bilge keels.

Equation (9.1) illustrates the importance of *GM* in the stabiliser fin design process. For a given design waveslope, a larger *GM* will require a larger restoring moment and fin force. This will generally mean larger fins and fin power. Thus from the point of view of stabiliser fin design, a small *GM* is desirable, subject to the requirements of overall ship stability and safety.

9.1.3 Design data and process

Given the dimensions and displacement of the ship, its *GM* and the position of the stabilisers then, for an assumed wave slope ϕ, the required force per fin F_L can be found using Equation (9.1). The geometry and working incidence of the fin will be designed to deliver this lift. Free-stream lifting surface data suitable for fin design are described in Section 5.3. A worked example using the stabiliser design process is given in Chapter 11, example application 17.

9.1.3.1 *Location*

Stabiliser fins should preferably be located near amidships to avoid breaking the surface when pitching, although within 20% L of amidships is also likely to be

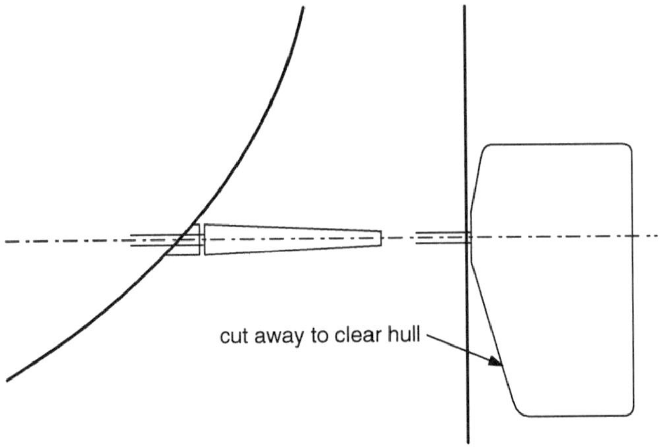

Figure 9.4 Root gap effects with horizontal fin axis

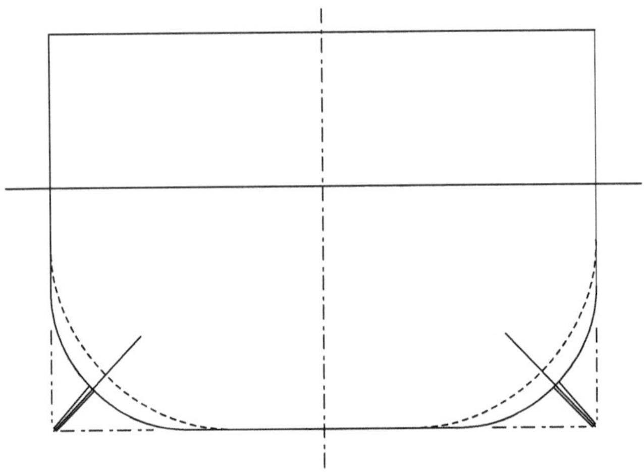

Figure 9.5 Possible stabiliser size limits (nonretractable)

acceptable. The fin should preferably be at right angles to the local hull to avoid shaping the root of the stabiliser fin to clear the hull when at incidence, Figure 9.4, leading to significant root gap effect. For reasons of cost, maintenance and convenience a number of small ships have nonretractable fins. Warships tend to have nonretractable fins as they have greater immunity to damage from shock and explosion [9.9]. For safety, docking and port operation nonretractable fins need to be confined to a rectangle formed by the baseline and ship breadth, Figure 9.5. This will restrict the span of the fin and, for a given required area, lead to a smaller aspect ratio. Lloyd [9.9] points out that with limited span, there is little advantage in increasing the chord to decease the effective aspect ratio to less than about 1.0. With such limited spans, the effective aspect ratio is typically from about 1.0 to 1.3. Further, lift can only be achieved by fitting more pairs of fins. Retractable fins do not suffer this restriction and larger aspect ratios with a higher hydrodynamic efficiency can be used. Retractable fins may be mounted higher and be nearly horizontal, typically at about 15° to the horizontal, without much loss in lever, Figure 9.2. In this case, hull shape and root gap losses can also be minimised.

The required level of restoring moment, or limits on lift due to aspect ratio, may be such that more than one pair of fins is necessary. It should be noted that interference effects occur between fins in line, one behind another. The downwash and upwash from the upstream fin influences the effective incidence of the downstream fin and the lift forces produced. This aspect is addressed further by Allan [9.1] and Lloyd [9.5,9.9].

It has been pointed out by Lloyd [9.4] and Dallinga [9.8] that when measured experimental data have been less than predicted, this may be due to possible effects of the free surface, cavitation, body–hull interference, boundary layer and root gap.

9.1.3.2 *Design speed*

Design speed will normally be the service speed. There may, however, be cases where the design speed is taken as the maximum speed achievable, or some lower cruise speed at which the ship may spend much of its time.

9.1.3.3 *Influence of the ship boundary layer*

The presence of the boundary layer leads to a reduction in the effective fin area. This reduction will clearly be larger with smaller aspect ratio fins.

An estimate of the boundary layer thickness at the position of the stabilisers may be made using Equation (3.17), for the thickness of a turbulent boundary layer:

$$\delta = x \times 0.370\ Re_x^{-1/5}$$

where x is the distance from the fore end and Re_x is the Reynolds number based on x, and using Equation (3.20) for the displacement thickness:

$$\delta^* = x \times 0.0463\ Re_x^{-1/5}$$

A velocity correction can be made across the boundary layer thickness based on a power law turbulent velocity distribution, Figure 3.11. The stabiliser lift L_F (or drag) may then be calculated over the boundary layer region as

$$\begin{aligned} L_F &= C_L \tfrac{1}{2}\rho c \int u^2 \mathrm{d}\delta \\ &= C_L \tfrac{1}{2}\rho c \int [V(y/\delta)^n]^2 \mathrm{d}\delta \qquad (9.2) \end{aligned}$$

An alternative approximate approach is to use the displacement thickness to correct the span, area and aspect ratio and to apply these to the database and coefficients.

9.1.3.4 *Influence of hull and waves on flow speed*

The water speed in the outer flow (outside boundary layer) changes as it flows around the hull with speeds up to 5–10% higher than ship speed around the bilge region where the stabilisers are located. Such speeds have been assessed for calm-water conditions, but the changes in flow speed and direction will be considerably more complicated when the ship is rolling.

In a similar manner, the sub-surface orbital motion of the waves producing the roll will lead to changes in the flow speed and direction. Again, these are complicated and difficult to quantify.

9.1.3.5 *Influence of adjacent hull shape and root gap*

This can be assessed in a similar manner to the derivation of effective aspect ratio for rudders, Section 5.5. Equation (5.25) is an approximate equation for the aspect ratio factor k, covering a range of effective hull shapes, Figure 5.104, and is defined as

$$k = 2 - 0.016\,\alpha$$

Hence at say 15° incidence, $k = 1.76$ and from Figure 5.100 for say $AR = 4.0$, the lift curve slope is 94% of the reflection plane lift curve slope, leading to an approximate loss in lift of 6%. At 25°, the loss in lift is about 10%.

From Figure 5.101, with say a gap/chord ratio of 0.01 (a gap of 10 mm for a chord of 1.0 m), $k = 1.90$ and from Figure 5.100, the loss in lift is only about 2.5%.

9.1.4 Section design

9.1.4.1 *Section shape*

Sections used for fin stabilisers tend to be all-movable, or high lift such as a flapped foil or a foil with a wedge at the trailing edge, Figure 9.6. The all-movable section

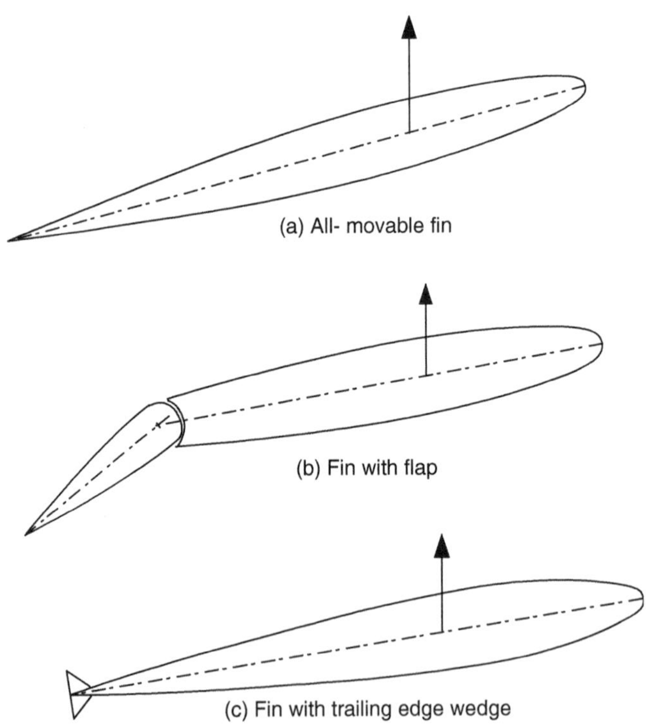

Figure 9.6 Fin stabiliser section types

is normally used for nonretractable fins. For retractable or folding fins the high-lift devices tend to be employed in that they provide a higher lift curve slope, hence faster response to change in incidence, a delay in stall and larger stall angle and a larger maximum lift coefficient. The differences between the performance of the flapped foil and foil with trailing edge wedge tend to be marginal. The flapped foil tends to be a little more efficient in terms of lift/drag ratio, whilst the foil with trailing edge wedge has no moving parts to manufacture and maintain.

9.1.4.2 *Lift and drag*

Lift curve slope, stall angle and maximum lift for a given fin aspect ratio and incidence can be derived from the database described in Section 5.3. Data are available for the various section types in Figures 5.9–5.19 and Tables 5.2–5.5. Corrections to aspect ratio can be carried out using Equation (5.17).

Drag on the fins may also be derived from the data in Figures 5.9–5.19 and Tables 5.2–5.5. The drag of the fins is important in that it can lead to a loss in ship speed.

Minimising the drag of nonretractable fins is important in that they produce a parasitic drag even in calm water. Typically, all-movable fins with NACA00 type sections will be employed and the thickness/chord ratio will be minimised within the limits of stock diameter and structural integrity.

As mentioned earlier, retractable fins tend to be of the high-lift type including flapped foils and sections with a wedge at the trailing edge. These both tend to have higher drag at a given incidence than the all-movable equivalent (see Table 5.5). When not in use, drag is not a problem as the fins are retracted or folded away. The actual drag on the fins can be quantified in the normal way, using drag coefficients from the database. The influence of this drag on ship speed is difficult to quantify as the stabilisers reduce roll and hence the ship resistance due to roll. Thus, the drag due to the stabilisers is offset to a certain extent by the reduction in ship resistance due to roll. These aspects of speed loss due to stabilisers are discussed in more depth by Allan [9.1] and Dallinga [9.8].

9.1.4.3 *Centre of pressure, torque and stock diameter*

CPc and *CPs* for the relevant section and aspect ratio and incidence can be obtained from the data in Figures 5.9–5.19 and Tables 5.2–5.5. In order to minimise torque, but preclude negative torques, the stock will normally be located a little forward of the estimated forwardmost position of *CPc*, Figure 5.15. Knowing the lift, drag and centre of pressure, the torque, bending moment and stock diameter can be calculated using Equations (7.1)–(7.11).

9.1.5 Cavitation

A cavitation check can be carried out using the principles applied to rudders (Chapter 5, Section 5.8). Allan [9.1] carried out cavitation tunnel tests on stabiliser sections and the results and limiting data are, as expected, similar to the rudder data.

The cavitation check entails the calculation of the appropriate cavitation number and the use of cavitation inception curves for a given lift coefficient, Figures 5.139 and 5.141. Cavitation number σ is defined as

$$\sigma = \frac{(P_{AT} + \rho g h - P_V)}{0.5\rho V^2} \tag{9.3}$$

It should be noted that the depth of immersion h is likely to be less than that for rudders, particularly when taking into account the reduction in $\rho g h$ due to roll.

From the results of the example application 17 (Chapter 11), for the roll stabiliser fin ($\sigma = 2.68$, $C_L = 0.922$) it is seen that cavitation is unlikely to be present at ship speeds less than about 20 knots. Overall, the maximum design and operational lift may be limited by cavitation in larger faster vessels, whilst in slower speed craft it will be limited by stall. Section shapes that delay the onset of cavitation (Chapter 5, Section 5.8), may also be considered.

9.1.6 Operation

9.1.6.1 *Control*

Maximum fin angle will normally be limited by the occurrence of stall and/or cavitation. A maximum lift may be specified based on stall, cavitation, torque, or materials and structure. The fin controller will be programmed in such a way that this maximum is not exceeded. The controller, which may for example include feedback from a force measurement on the stock, will limit fin angle as necessary to avoid stall, cavitation, or some prescribed maximum lift. Some fundamentals of stabiliser control are discussed in [9.1,9.2] and more recently by Perez [9.10].

9.1.6.2 *Sway and yaw effects*

It can be seen from Figure 9.2 that the stabiliser force L_F will also have vertical and horizontal components. Whilst the vertical components effectively cancel, there is a net horizontal sway force that will induce a yaw motion. This is to port in Figure 9.2 but will oscillate from port to starboard as the fin incidence is reversed. This yawing effect is generally not a problem on most ships, but will depend on the location of the fins relative to the ship LCG, a large lever increasing the yaw moment, and the directional stability of the ship, Section 4.1.2.

9.1.6.3 *Dynamic effects*

The static approach to the design of fin stabilisers described neglects the dynamic behaviour of the ship and stabilisers including, for example, the roll-induced angle of attack on the fins and the influence of nonstationary motion on the lift characteristics. For a more detailed estimate of overall performance, a more rigorous and complete approach can be used which will include the application of the equation of motion for roll, with the stabiliser fins increasing the motion damping term. Such an approach is described in [9.2]–[9.4].

9.1.7 Roll stabilisation with rudders

The heel angle produced by the rudder was described in Section 4.1.4. This effect can be used to produce a stabilising moment in roll and the rudder used as a roll stabiliser. A number of investigations have been carried out to determine the efficacy of such an approach, such as those described by Cowley and Lambert [9.11], Carley [9.12], Cowley and Lambert [9.13] and Lloyd [9.14]. The use of a lateral force estimator (LFE), using the lateral acceleration, to control the rudder as a stabiliser is described by Tang and Wilson [9.15]. A thorough review of the use of the rudder as a roll stabiliser together with suitable controllers is given by Perez [9.10]. Estimates of the likely level of heeling/righting moment due to a rudder are included in example application 18 in Chapter 11.

Restricted rudder angles generally have to be applied to avoid the effects of coupling roll, sway and yaw. It is suggested by Lloyd [9.5] that using the rudder may be acceptable for lower-speed merchant ships but may be unsuitable for high-speed ships, as in low frequency following and quartering seas the rudder might be expected to amplify the roll motion. The use of the rudder as a roll stabiliser seems to have received only limited practical application.

9.1.8 Roll stabilisation at rest

An interesting development is stabilisation by fins whilst the vessel is at rest [9.16]. Available fin incidence is increased to 40° from the more normal maximum of 20°–25°, with fin action responding to roll sensors/controllers. Such systems are finding applications in smaller vessels such as large motor yachts, small ferries and offshore vessels.

9.2 Hydroplanes

9.2.1 Applications

Hydroplanes are horizontal control surfaces used to control the vertical motions of submarines and other underwater vehicles, Figure 2.5. The shape, area and location of these control surfaces will depend on the size, speed and operational requirements of the vehicle. Owing to possible compression of the hull with increase in immersion and a reduction in buoyancy, a submarine is unstable for motions in the vertical (heave) direction and it is difficult to obtain precise equilibrium between weight and buoyancy, particularly when fuel or stores are being consumed during operation. Thus the control surfaces must be able to exert an upward or downward force on the submarine and to provide a pitching moment. Two pairs of control surfaces are therefore required, being normally situated near the fore and aft ends to provide the largest moments. Such an arrangement allows the control of heave and pitch independently. Underwater bodies may be short/fat or long/thin, Figure 9.7.

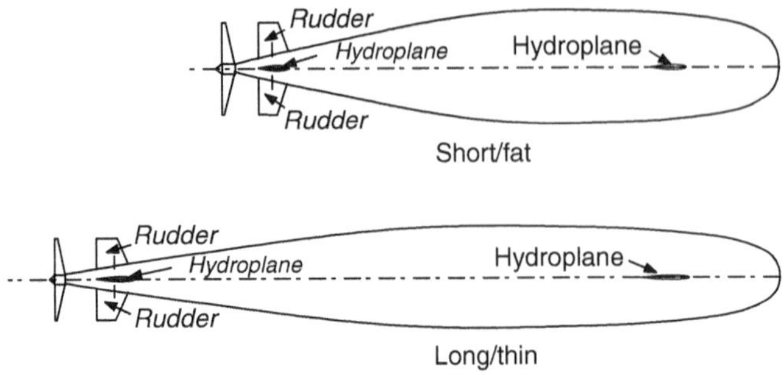

Figure 9.7 Submarines with different length/diameter ratio

They will display different hydrodynamic characteristics, the short/fat body having greater directional instability but good manoeuvrability. The shorter body limits the lever arms of the control surfaces and larger control surfaces may be required.

A useful background to the stability and control of submarines and other underwater vehicles is given by Rawson and Tupper [9.17] and Burcher and Rydill [9.18].

9.2.2 Design procedure and data

For a submarine, the centre of buoyancy B lies above the centre of gravity G. When trimmed by an angle θ there is a restoring moment $\Delta BG\theta$. This moment contributes to trim or pitch corrections at lower speeds, but at speeds higher than above about 10 knots [9.17], the control surfaces dominate since their moments are a function of V^2. In the case of the submarine, the normal design approach is to use simulations of typical manoeuvres, adjusting the size of the control surfaces to provide suitable forces and moments. The simulations will often be supported by model tests.

Hydroplanes will typically be all-movable or flapped with the fixed forward part of about 20–30% of the chord providing structural support. In the case of the submarine, their dimensions will be such that they do not exceed the maximum diameter of the submarine. Having determined the required forces on the control surface, and assumptions made for local flow velocity, actual dimensions, section shape and operational angles can be determined from data for all-movable and flapped control surfaces, such as those reviewed in Chapter 5. The vertical force produced by the hydroplanes can be compared with the displacement of the submarine, indicating the ability to cope with any differences between vessel weight and buoyancy.

It should be noted that tip vortices shed from the forward hydroplanes can interfere with the after hydroplanes, leading to a change in effective incidence on the after hydroplanes. This is similar to the interference between roll stabilisers fins (Section 9.1.3).

9.2.3 Operation

Cavitation is unlikely to be a problem with control surfaces on submarines or underwater vehicles, particularly when deeply immersed. Incidence is therefore likely to be limited by the stall angle. The motions and control of submarines and other underwater vehicles can be complex. An insight into these requirements is provided by Burcher and Rydill [9.18].

9.3 Pitch damping fins

9.3.1 Applications

The energy in pitch is much greater than the energy in roll and the required forces and moments to stabilise pitch are much larger than those required for roll. Hence pitch damping or stabilisation is generally not a practical proposition for large ships. For example, Conolly and Goodrich [9.19] carried out full-scale sea trials on fixed antipitching fins fitted to a coastal minesweeper. They concluded that it was doubtful whether fixed fins would ever provide sufficient attenuation of motion to justify their installation in any ship. Successful stabilisation in pitch tends to be limited to smaller faster semi-displacement vessels, and some success with ride control has been achieved using variable incidence lifting foils situated near the fore end, variable incidence stern flaps (trim tabs), adjustable interceptors and combinations of these various devices.

9.3.2 Design procedure and data

Pitch-damping foils tend to take the form of a flapped foil, Figure 9.8, where the forward part can act as a fixed part and provide a suitable connection for one or two vertical supporting struts. The design procedure is similar to that for the roll stabiliser fin and, again, use can be made of the extensive database of performance characteristics for all-movable and flapped control surfaces such as those reviewed in Chapter 5.

An approximate approach is to relate the required stabilising moment to the longitudinal hydrostatic restoring moment. The hydrostatic restoring moment for a trim angle θ is $\Delta g GM_L \theta = \Delta g GM_L \times \text{trim}/L$.

[In a similar manner, the moment to change trim 1 cm is $\Delta g GM_L/100L$].

The stabilising moment of the fins is $L_{FP} \times R_P$, Figure 9.9, where $L_{FP} = C_L \times \frac{1}{2}\rho AV^2$.

If θ is considered to be the pitch angle and $\Delta g GM_L \theta$ the couple applied by the pitch motion, then the stabilising moment must match this couple and

$$L_{FP} \times R_P = \Delta g GM_L \theta \tag{9.4}$$

θ can be considered as the angle of steady pitch that the foil(s) can create with the ship moving ahead in calm water. This approach tends to suggest the need for a very large control surface area.

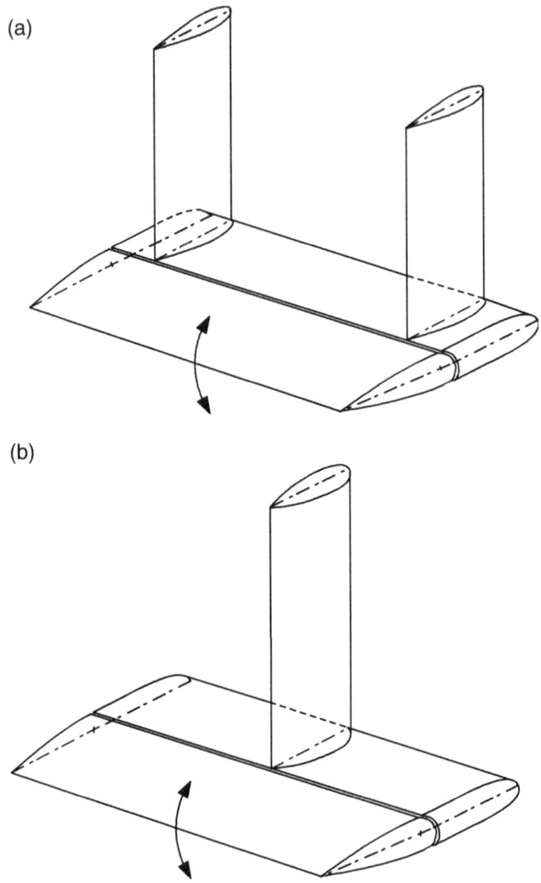

Figure 9.8 Pitch-damping foils

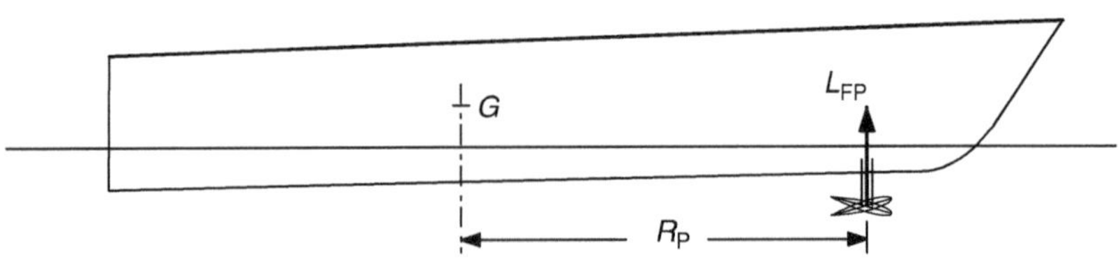

Figure 9.9 Pitch-stabilising moment

An alternative approach is to carry out a ship motion simulation using say strip theory [9.9], and to determine a level of damping that will have a useful effect in decreasing the pitch motion and vertical accelerations. The pitch-damping foils then need to be sized whereby the forces and moments provide an adequate level of damping.

Haywood and Benton [9.20] describe the use of lifting foils and their application also to ride control of a high-speed trimaran. The lifting foils, situated at amidships, lift approximately one third of the mass of the vessel. They are also capable of changing angle of attack due to vessel motion and form part of the ride control system. The transom also has an active transom mounted tab, linked to the ride control system. Full-scale trials demonstrated a high level of damping and good seakeeping performance. Davis *et al.* [9.21] describe the theoretical prediction of motions and full-scale measurements on a high-speed catamaran ferry fitted with T-foils and stern flaps. Various investigations into ride control are reported in references [9.22]–[9.26].

9.3.3 Operation

Like the roll stabiliser fin, the incidence of the pitch-damping foil may be limited by stall or cavitation and the fin controller will be programmed in such a way that this maximum is not exceeded. Both these characteristics may be examined in a manner similar to that for a roll stabiliser fin, Section 9.1.

References

9.1 Allan, J.F. The stabilisation of ships by active fins. *Transactions of the Royal Institution of Naval Architects*, Vol. 87, 1945, pp. 123–159.

9.2 Conolly, J.E. Rolling and its stabilisation by active fins. *Transactions of the Royal Institution of Naval Architects*, Vol. 111, 1969, pp. 21–48.

9.3 Gunsteren, F.F. Van. Analysis of roll stabiliser performance. *International Shipbuilding Progress*, Vol. 21, No. 237, 1974, pp. 125–146.

9.4 Lloyd, A.R.J.M. Roll stabiliser fins: A design procedure. *Transactions of the Royal Institution of Naval Architects*, Vol. 117, 1975, pp. 233–254.

9.5 Lloyd, A.R.J.M. Recent developments in roll stabilisation. *The Naval Architect*, published by RINA, London, March 1977, pp. 44–45.

9.6 Cox, G.G. and Lloyd, A.R.J.M. Hydrodynamic design basis for navy ship roll motion stabilisation. *Transaction of the Society of Naval Architects and Marine Engineers*, Vol. 85, 1977, pp. 51–93.

9.7 Fairlie-Clarke, A.C. Fin stabilisation of ships. *The Naval Architect*, Published by RINA, London, January 1980, pp. 10–11.

9.8 Dallinga, R.P. Hydromechanic aspects of the design of fin stabilisers. *Transactions of the Royal Institution of Naval Architects*, Vol. 136, 1994, pp. 189–200.

9.9 Lloyd, A.R.J.M. *Seakeeping: Ship Behaviour in Rough Weather*. Published by the Author, Gosport, UK, 1998.

9.10 Perez, T. *Ship Motion Control. Course Keeping and Roll Stabilisation using Rudder and Fins*. Springer, 2005.

9.11 Cowley, W.E. and Lambert, T.H. The use of the rudder as a roll stabiliser. *Third Ship Control Systems Symposium*, Bath, UK, 1972.

9.12 Carley, J.B. Feasibility study of steering and stabilising by rudder. *Fourth Ship Control Systems Symposium*, The Hague, 1975.

9.13 Cowley, W.E and Lambert, T.H. Sea trials on a roll stabiliser using the ship's rudder. *Fourth Ship Control Systems Symposium*, The Hague, 1975.

9.14 Lloyd, A.R.J.M. Roll stabilisation by rudder. *Fourth Ship Control Systems Symposium*, The Hague, 1975.

9.15 Tang, A and Wilson, P.A. LFE Stabilisation using the rudder. Proceedings of Conference, *Manoeuvring and Control of Marine Craft*, MCMC '92, Computational Mechanics Publications, 1992, 475–492.

9.16 Stabilisation at rest: a new solution. *The Naval Architect*, Royal Institution of Naval Architects, London, September 2006, p. 96.

9.17 Rawson, K.J. and Tupper, E.C. *Basic Ship Theory*, 5th ed. Butterworth Heinemann, Oxford, 2001.

9.18 Burcher, R. and Rydill, L. *Concepts in Submarine Design*. Cambridge Ocean Technology Series 2, Cambridge University Press, Cambridge, 1998.

9.19 Conolly, J.E. and Goodrich, G.J. Sea trials of anti-pitching fins. *Transactions of the Royal Institution of Naval Architects*, Vol. 112, 1970, pp. 87–100.

9.20 Haywood, A.J. and Benton, H.S. The integration of lifting foils into ride control systems for fast ferries. Proceedings of *Eighth International Conference on Fast Sea Transportation*, FAST '2005, St. Petersburg, Russia, June 2005.

9.21 Davis, M.R., Watson, N.L. and Holloway, D.S. Wave response of an 86 m high speed catamaran with active T-Foils and stern tabs. *Transactions of the Royal Institution of Naval Architects*, Vol. 145, 2003, pp. 87–106.

9.22 Davis, M and Holloway, D. Effect of sea, ride controls, hull form and spacing on motion and seasickness incidence for high speed catamarans. Proceedings of *Seventh International Conference on Fast Sea Transportation*, FAST '2003, Ischia, Italy, October, 2003.

9.23 Xi, H. and Sun, J. Effects of actuator dynamics on stabilisation of high-speed planing vessels with controllable transom flaps. Proceedings of *Eighth International Conference on Fast Sea Transportation*, FAST '2005, St. Petersburg, Russia, June 2005.

9.24 Katayama, T., Suzuki, K. and Ikeda, Y. A new ship motion control system for high speed craft. Proceedings of *Seventh International Conference on Fast Sea Transportation*, FAST '2003, Ischia, Italy, October, 2003.

9.25 Folsø, R., Nielsen, U.D. and Torti, F. Ride control systems – Reduced motions on the cost of increased sectional forces. Proceedings of *Seventh International Conference on Fast Sea Transportation*, FAST '2003, Ischia, Italy, October, 2003.

9.26 Doctors, L.J. Theoretical study of the tradeoff between stabiliser drag and hull motion. *Transactions of the Royal Institution of Naval Architecture*, Vol. 146, 2004, pp. 289–298.

10 Propulsion

10.1 Propeller–rudder interaction

With a rudder operating downstream of a propeller, the rudder blocks the flow upstream in way of the propeller, whilst the propeller accelerates and rotates the flow onto the rudder. Rotational energy in the slipstream is recovered and the overall propulsive characteristics may be improved. Interaction effects depend primarily on the size of the separation between the propeller and the rudder and the level of propeller thrust loading, K_T/J^2. The basic changes in propulsive effects due to propeller–rudder interaction are described in Section 5.9, and the development of the net propulsive force is shown in Figures 8.1 and 8.4.

10.2 Propeller effects

The net propeller thrust is based on the open-water thrust together with any corrections arising from interaction with the hull and rudder. The open-water thrust

$$T = K_T \rho n^2 D^4$$

When situated behind the hull, the propeller accelerates the flow ahead of itself, thereby (a) increasing the rate of shear in the boundary layer and hence increasing the frictional resistance of the hull and, (b) reducing pressure over the rear of the hull and hence increasing the pressure resistance. Hence the action of the propeller is to alter the resistance of the hull (usually to increase it) by an amount that is approximately proportional to thrust. This means that the thrust must exceed the naked resistance of the hull. This may be seen as an augment to the hull resistance or, as is more common, as a reduction in thrust, defined by a thrust deduction fraction as

$$t = (T - R)/T = 1 - R/T$$

or

$$T = R/(1 - t) \tag{10.1}$$

where R is the hull resistance.

The thrust deduction fraction t is obtained from a model self-propulsion experiment [10.1], and will take account of the hull–propeller–rudder layout.

In course-keeping or manoeuvring algorithms, when matching the propeller thrust to the resistance R of the ship, it is usual to write the effective propeller thrust T_E as

$$T_E = T(1 - t) = (1 - t)K_T \rho n^2 D^4 \quad (10.2)$$

The conclusions of Kracht [10.2] and discussion to [10.3] are that the overall effects of the rudder are to increase the effective wake fraction and decrease the thrust deduction fraction, leading to an increase in the hull efficiency η_H, where $\eta_H = (1 - t)/(1 - w_T)$, and a decrease in the delivered power.

The actual effects of the rudder on the physical performance of the propeller are to change the effective propeller inflow velocity and to change the propeller thrust and torque coefficients by ΔK_T and ΔK_Q. Typical changes in K_T and K_Q are shown in Figures 5.144–5.148. It is seen that significant changes in ΔK_T and ΔK_Q can occur with change in thrust loading (J) and rudder–propeller separation (X/D, Y/D, Z/D).

10.3 Rudder effects

The net effect of the propeller on the rudder is to increase the flow velocity (Section 3.6) and impart an effective angle of attack on the rudder (rudder set at zero incidence), Section 3.5. With an increase in propeller thrust loading K_T/J^2 (decrease in J), the level of energy recovery increases. This is seen as a decrease in rudder drag. With a high thrust loading (low J) and relatively large separation between rudder and propeller, then thrust (or negative drag) may be produced by the rudder, Figure 5.143. The force vectors in this situation are shown in Figure 3.33(b). Although the rudder is often considered only as a component of resistance, it is apparent that it can contribute to a drag reduction.

10.4 Overall effects

It is clear that the siting of the rudder relative to the propeller has an impact on the performance of both the rudder and propeller and should, if possible, be considered at an early design stage. This is discussed in Example application 5 in Chapter 11. Further possible overall improvements can be made in propulsive efficiency by using a twisted rudder, Section 5.8 and Example 6 in Chapter 11, and/or a faired hub behind the propeller which continues into the rudder.

References

10.1 Carlton, J.S. *Marine Propellers and Propulsion.* Butterworth Heinemann, Oxford 1994.

10.2 Kracht, A.M. Rudder in the slipstream of a propeller (in German). *VWS Report No. 1178/90*, 1990.

10.3 Molland, A.F., Turnock, S.R. and Smithwick, J.E.T. Design and performance characteristics of semi-balanced ship skeg rudders. *Transactions of the Royal Institution of Naval Architects*, Vol. 142, 2000, pp. 230–247.

PART FOUR

Design Applications

11 Applications

11.1 Background

A number of worked examples are presented to illustrate typical applications of the rudder data and some of the methodologies used for design and investigation purposes. Although grouped broadly into large ship and small craft categories, the methodologies in most of the examples are applicable to either category.

11.2 Large ships

11.2.1 Example application 1. Spade rudder: Free stream (e.g., twin screw, single centreline rudder)

Consider a spade rudder with the following particulars:

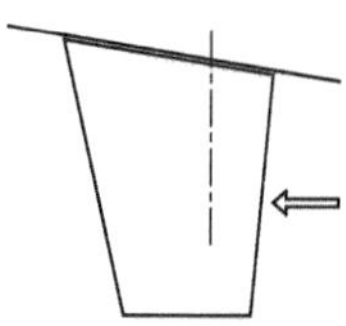

Span: 1.74 m
Chord: 1.16 m
AR_G: 1.5
TR: 0.6
Sweep: zero
Effective inflow speed: 10 m/s ahead, 4 m/s astern

Assume the rudder to be working adjacent to a relatively flat hull, and the effective aspect ratio to be $2 \times AR_G = 3.0$

Determine the forces and torque over a range of incidence for ahead and astern operation.

Assume the initial position of the stock to be 19% aft of the leading edge (LE). Investigate the effect on torque of moving the stock to 22% aft of LE.

Use the Whicker and Fehlner data, Table 5.2 and Equations (5.6)–(5.10) for the control surface with appropriate particulars.

The data are applied to Equations (7.1)–(7.5) and example calculations are shown in Table 11.1. The results are plotted in Figure 11.1.

From the data in Table 5.2 it is noted that the rudder stalls at about 23° when going ahead and about 16° when going astern.

Table 11.1 Examples of calculations for two rudder angles: stock position 19% aft LE

	α (degree)	C_L	C_D	CPc	C_N	N (kN)	$\bar{x}$ (m)	Torque (kNm)
Ahead	10	0.55	0.049	0.20	0.546	56.45	0.010	0.56
Ahead	20	1.11	0.165	0.21	1.109	113.98	0.022	2.49
Astern	10	0.45	0.077	0.12	0.457	7.56	0.80	6.05
Astern	15	0.59	0.152	0.15	0.609	10.08	0.77	7.76

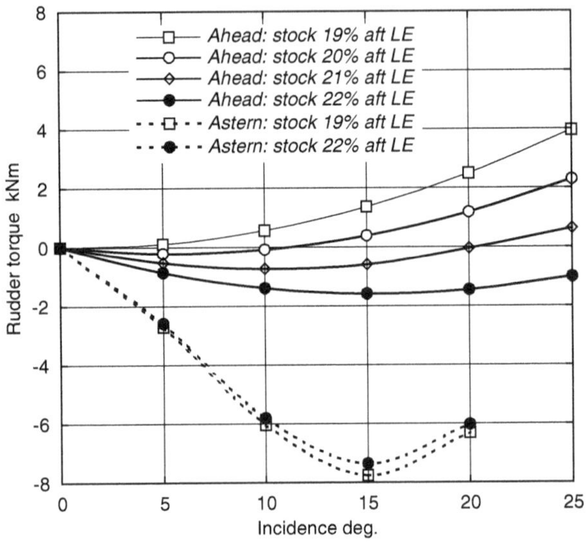

Figure 11.1 Rudder torque: Example application 1

From the results, it is seen that the ahead torque is always positive when the stock is 19% aft of LE and always negative when 22% aft. When the stock is 20% and 21% aft of LE the torque is initially negative, followed by positive torque.

For astern operation, at 40% of ahead speed, the astern torque rapidly exceeds the maximum ahead torque, suggesting that the astern speed or rudder angle would need to be limited.

11.2.2 Example application 2. Container ship: Spade rudder

Consider an outline rudder design for a single-screw container ship having the following particulars:

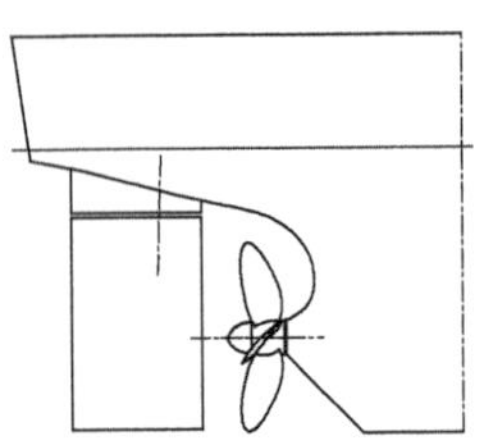

LBP: 140 m
B: 22 m
T: 9.3 m
Service speed: 20 knots = 10.29 m/s
Wake fraction at service speed: 0.255
Propeller diameter: 6.3 m
Propeller revolutions at service speed: 130 rpm

Rudder design data: The rudder is to be a spade type of rectangular profile, with area = 2% of lateral area ($L \times T$) (see Section 5.6) and $AR_G = 1.5$ to suit stern layout. Assume in the first instance no loss in effective aspect ratio due to the shaped hull above rudder and effective aspect ratio $AR = 3.0$. Take the top (root) of the rudder to be 250 mm below the bottom of the lower bearing.

Use the Molland and Turnock data, Figure 5.58 and/or Tables A1.4–A1.6 (Appendix 1) for Rudder No. 2 (rectangular with $AR_G = 1.5$).

Summary of calculations: Area = 2% of immersed lateral area = $0.02 \times (140 \times 9.3) = 26.04\,m^2$. With $AR_G = 1.5$ this leads to a span = 6.25 m and chord = 4.17, say 4.2 m, and final area $6.25 \times 4.2 = 26.25\,m^2$

$$\text{Re} = V\bar{c}/\nu = (10.29 \times 4.17/1.19) \times 10^6 = 3.6 \times 10^7$$

In the service speed condition,

$$J = Va/nD = 20(1 - 0.255) \times 0.5144 \times 60/130 \times 6.3 = 0.561$$

From propeller chart for assumed ship propeller at $J = 0.561$, $K_T = 0.200$ and $K_T/J^2 = 0.635$.

Using Figure 5.58 or Tables A1.4–A1.6, interpolate between $K_T/J^2 = 0.88$ and $K_T/J^2 = 0.05$ for Rudder No. 2.

The data are applied to Equations (7.1)–(7.11) and example calculations for three rudder angles are shown in Tables 11.2(a) and (b). The calculations are repeated for stock positions of 15, 20, 25 and 30% aft of LE. The results are plotted in Figures 11.2 and 11.3. It is seen that the stock position may be adjusted until the negative and positive torques are about the same, hence minimising the steering gear torque. This does however lead to negative torques at low angles of attack, which means that the rudder would not trail (at zero incidence) in the case of a tiller or steering-gear malfunction. From the data graphs and tables, it is seen that at this propeller thrust loading the rudder stalls at about 35°–40° at positive incidence and 30°–35° at negative incidence.

Table 11.2(a) Examples of calculations for torque for three rudder angles: stock position 25% aft LE

α (degree)	C_L	C_D	CPc	C_N	N (MN)	$\bar{x}$ (m)	Torque (MNm)
9.2	0.808	0.111	0.137	0.816	0.640	−0.471	−0.302
19.2	1.719	0.351	0.196	1.739	1.365	−0.225	−0.307
35.0	2.888	1.208	0.335	3.059	2.400	0.354	0.850

Table 11.2(b) Examples of calculations for root bending moment for three rudder angles

α (degree)	C_L	C_D	CPs	C_R	R (MN)	$\bar{y}$ (m)	BM (MNm)
9.2	0.808	0.111	0.617	0.816	0.640	3.85	2.46
19.2	1.719	0.351	0.569	1.755	1.377	3.56	4.90
35.0	2.888	1.208	0.547	3.131	2.457	3.42	8.40

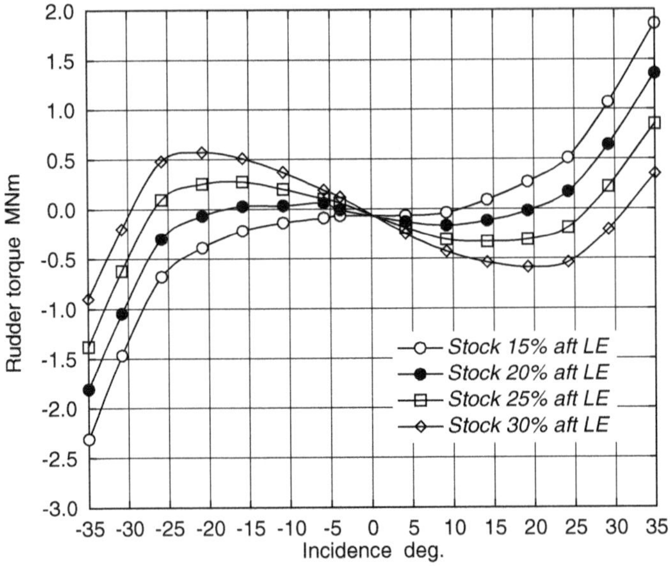

Figure 11.2 Rudder torque: Example application 2

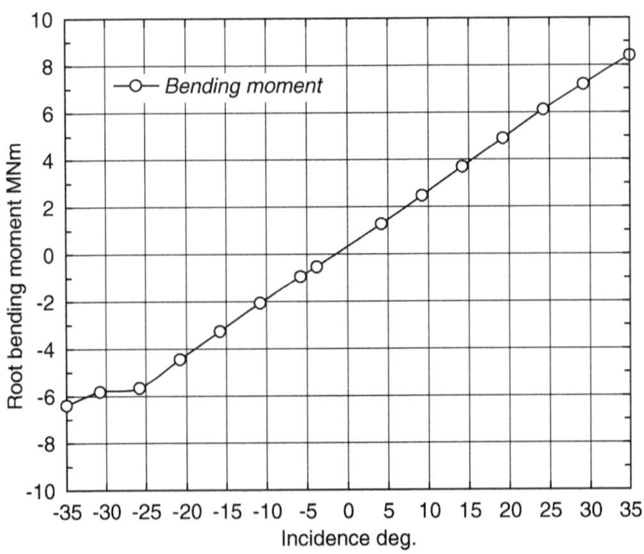

Figure 11.3 Rudder root bending movement: Example application 2

Stock diameter: Assume 30° rudder angle as the maximum *effective* incidence (due to drift angle the set rudder angle will be higher than this)

From Tables and graphs at 30°:

Normal force $N = 2.06$ MN (interpolating from Table 11.2(a))

Torque $Q_R = 0.50$ MNm (balancing positive and negative torques for say stock 25% aft LE)

Bending moment at rudder root $M = 7.40$ MNm
Lever to root = 3.47 m (interpolating from Table 11.2(b))
Lever to lower bearing = 3.47 + 0.25 = 3.72 m
Bending moment at bearing = 7.40 × 3.72/3.47 = 7.93 MNm
Using Equation (7.9), Equivalent bending moment

$$BM_E = 7.93/2 + 0.5\,(7.93^2 + 0.55^2)^{0.5} = 7.94\,\text{MNm}$$

Using Equation (7.11), Stock diameter

$$D = ((7.94 \times 1{,}000 \times 32)/(\pi \times 118 \times 1{,}000^2))^{1/3} = 882\,\text{mm}$$

(where the assumed maximum stress = 118 N/mm^2, being approximately half of the yield stress for steel).

It can be noted that in the case of the spade rudder the bending moment dominates the equivalent bending moment. Reduction in torque will decrease the size of the steering gear but will have little effect on the stock diameter.

If an aspect ratio factor k of say 1.65 is applied to allow for the shape of the hull above the rudder, see Figure 5.104, then $AR = 1.5 \times 1.65 = 2.5$. Using Equation 5.7a or Figure 5.100, this leads to a 10% decrease in the lift curve slope. Assuming this decrease is reflected in the forces, then BM_E is reduced to 7.15 MNm. This in turn reduces the required stock diameter to 852 mm, a reduction in diameter of about 3.5%.

Stock diameter using Classification Society rules: Using the rules of Lloyd's Register or Det Norske Veritas (DNV) [7.35,7.36],

Rudder force $P_L = 1.77$ MN, torque $M_T = 0.74$ MNm and bending moment $M_B = 6.60$ MNm.

Using a permissible equivalent stress

$$\sigma_e = (\sigma_b{}^2 + 3\tau^2)^{0.5} < 118\,\text{N/mm}^2$$

where σ_b = bending stress = $(10.2 \times M_B/\delta_S^3) \times 10^3$ and τ = torsional shear stress = $(5.1 \times M_T/\delta_S^3) \times 10^3$

Final stock diameter to meet these stress requirements $\delta_S = 831$ mm.

This is about 2.5% less than the diameter derived from the above detailed calculations, but typically within the error bands of the assumptions made for the forces and levers.

Friction in rudder bearings: The friction in the rudder bearings and the frictional torque QF may be estimated using Equation (7.14) and Figure 7.7.

Taking the case of 19.2° rudder angle:
Normal force $N = 1.365$ MN (from Table 11.2(a))
Using Equation (7.14),

$$Q_F = \mu_B \times R_1 \times N[(y_1 + y_2)/y_3] + \mu_B \times R_2 \times N[y_1 + y_2 + y_3)/y_3]$$

Assume the coefficient of friction of the bearing material $\mu_B = 0.10$ (see Section 7.2).
In Figure 7.7,

$y_1 = 0.569 \times 6.25 = 3.56$ m (from Table 11.2(b))
$y_2 = 0.250$ (as before)
$y_3 = 3.0$ m (assumed)

Radius R_2 at lower bearing = 0.831/2 = 0.416 m and assume radius R_1 at tiller also = 0.831/2 = 0.416 m.

Then

$$\begin{aligned} Q_F &= 0.1 \times 0.416 \times 1.365[(3.56 + 0.25)/3.0] \\ &\quad + 0.1 \times 0.416 \times 1.365[(3.56 + 0.25 + 3.0)/3.0)] \\ &= 0.201 \text{ MNm} \end{aligned}$$

This can be compared with the hydrodynamic torque at 19.2° of 0.307 MNm.

11.2.3 Example application 3. Container ship: Skeg rudder

Same requirements as example 2 but using a semi-balanced skeg rudder

Rudder span = 6.25 m, mean chord = 4.2 m, area = 26.25 m^2

$K_T/J^2 = 0.635$

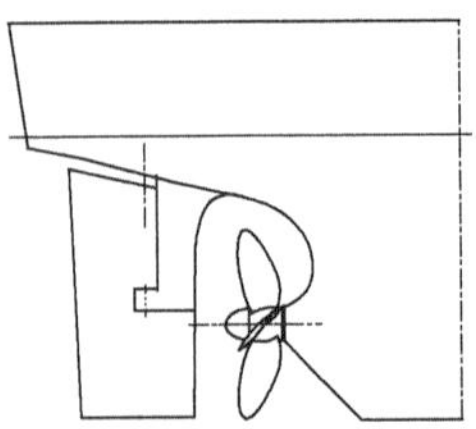

Use the Molland and Turnock data, Figures 5.67 and 5.68 or Tables A1.7–A1.12 for the Skeg rudder (No. 0) ($AR_G = 1.5$, taper ratio = 0.8 and stock = 35.5% aft of LE of mean chord). Interpolate between $K_T/J^2 = 0.88$ and $K_T/J^2 = 0.05$ for Rudder No. 0

The data are applied to Equations (7.1)–(7.5) and example calculations are shown in Table 11.3. The results are plotted in Figure 11.4.

From the results it is seen that torque at the maximum angle of ±35° is less than the all-movable spade rudder (Example application 2) namely 0.706 and 0.785 MNm compared with a mean of 0.860 MNm for the spade rudder. It should however be noted that the torque is produced by the movable area of the skeg rudder which is 79% of the total area. The total sideforce produced by the the skeg rudder (rudder plus skeg) is also less than that produced by the all-movable rudder. A more realistic comparison using the total resultant force (R) for each case is shown in Figure 11.5. The sideforce produced by the movable rudder alone is about 90% of the total rudder plus skeg value at positive rudder angles and about 70% at negative angles. The total rudder plus skeg resultant force is about 85–90% of the spade rudder values at higher angles. This in broad agreement with the conclusions drawn in Section 5.3.5 of Chapter 5 and illustrates the superior performance characteristics of the spade rudder compared with the skeg rudder.

Table 11.3 Examples of calculations for torque for three rudder angles: skeg rudder with stock position 35.5% aft LE of mean chord

α (degree)	C_L	C_D	CPc	C_N	N $\bar{x}$ (MN)	$\bar{x}$ (m)	Torque (MNm)
9.8	0.764	0.064	0.299	0.764	0.604	−0.235	−0.142
19.8	1.368	0.290	0.368	1.385	1.094	0.055	0.060
34.8	1.891	0.833	0.460	2.026	1.601	0.441	0.706

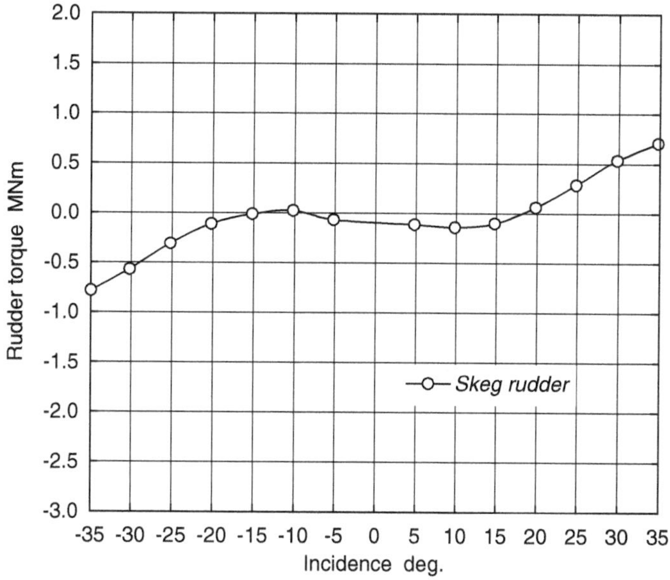

Figure 11.4 Skeg rudder torque: Example application 3

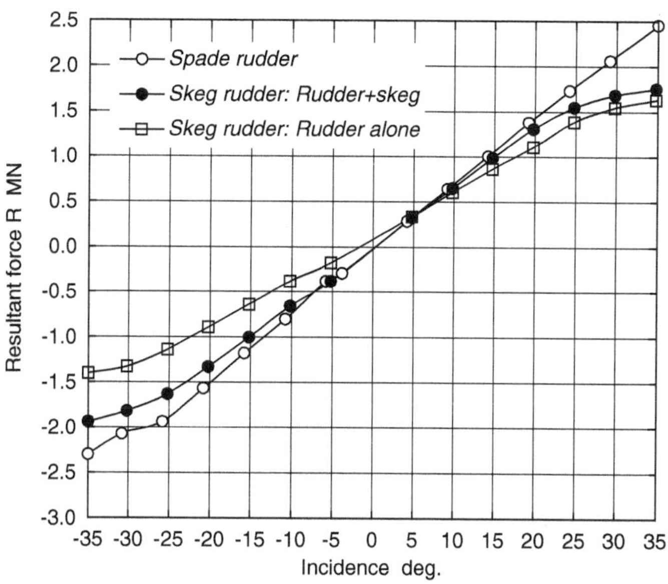

Figure 11.5 Comparison between all-movable spade rudder and skeg rudder: Example application 3

11.2.4 Example application 4. Container ship: spade rudder

Same requirements as example 2, but using free-stream data and estimate of propeller induced velocity.

In example 2, $AR_E = 3$, $X/D = 0.39$ and $K_T/J^2 = 0.635$
For comparative purposes, use 10°.
From Table 11.2(a), interpolating for 10°, $C_L = 0.881$
For $AR_E = 3$, free-stream data using Equation (5.6) or Figure 5.13,

$$dC_L/d\alpha = 0.0548, \text{ and at } 10°, \; C_L = 0.548$$

from Equation (3.37),

$$K_R = f(X/D) = 0.861$$

and from Equation (3.38),

$$V_R/V_0 = f(K_T/J^2) = 1.532$$

at 10°,

$$C_L \text{ corrected for propeller induced velocity} = 0.548 \times (1.532)^2 = 1.286$$

i.e., 46% higher than $C_L = 0.881$ derived above using the rudder–propeller interaction data.

As mentioned in Section 3.6, significant attenuation of Equation (3.38) is required to bring the theoretical estimate of velocity into line with experimental data. In the present example, a reduction factor of 0.83 would need to be applied.
i.e.,

$$V_R/V_0 = f(K_T/J^2) = 0.83 \times 1.532 = 1.272$$

and

$$C_L \text{ corrected for propeller induced velocity} = 0.548 \times (1.272)^2 = 0.887$$

It should also be noted that the *CPc* using the free-stream data approach (Table 5.2) is about 0.25*c* at stall, compared with about 0.335*c* (Table 11.2(a)) for the rudder downstream of a propeller. This must be taken into account when estimating torques.

A calibration of Equation (3.38) can be carried out using the rudder–propeller interaction data, such as that in Figure 5.58, or Tables A1.4–A1.6 for Rudder No. 2

$$X/D = 0.39, \quad \text{and} \quad K_R = 0.861 \text{ as before}$$

Using Tables A1.4–A1.6, and interpolating for 10°, the results may be tabulated as in Table 11.4.

In Table 11.4, **0.559 has been reduced to 0.504 to represent the free-stream value, with $RF = 1.0$

Table 11.4 Calculated reduction factors

rpm	800	1,460	2,100
K_T/J^2	0.05	0.88	2.30
C_L @ 10°	0.559	0.922	1.326
Equation (3.38) V_R/V_0	1.053	1.689	2.393
Calculated C_L	0.504**	1.437	2.886
Reduction factor (RF)	1.0	0.80	0.67

i.e., $0.504 = 0.559 \times 1/1.053^2$ and $1.437 = 0.504 \times 1.689^2$
A suitable fit to the above data is

$$RF = \left[1 - 0.215\left(K_T/J^2\right)^{1/2}\right] \tag{11.1}$$

This relationship is approximate, but suitable for preliminary design and simulation purposes. The experimental data would suggest that this relationship should be suitable for a wide range of rudder angles.

11.2.5 Example application 5. High-lift flapped rudder/all-movable rudder comparison

A comparison is made between an all-movable rudder and two flapped rudders, one with a 20% chord flap and one with a 30% chord flap. The data in Kerwin *et al.* [5.6], reviewed in Section 5.3.3, are used for the study.

$$AR_E = 2.8,\ TR = 0.6,\ Re = 1.2 \times 10^6$$

C_L, C_D and L/D were derived over a range of angles for each case and the results are shown in Figure 11.6.
The flapped rudders had stalled by about 20°, so only data up to 15° are included.

In comparing the flapped rudders with the all-movable rudder in Figure 11.6, it is interesting to note the significant increase in lift curve slope with the 20% flap,

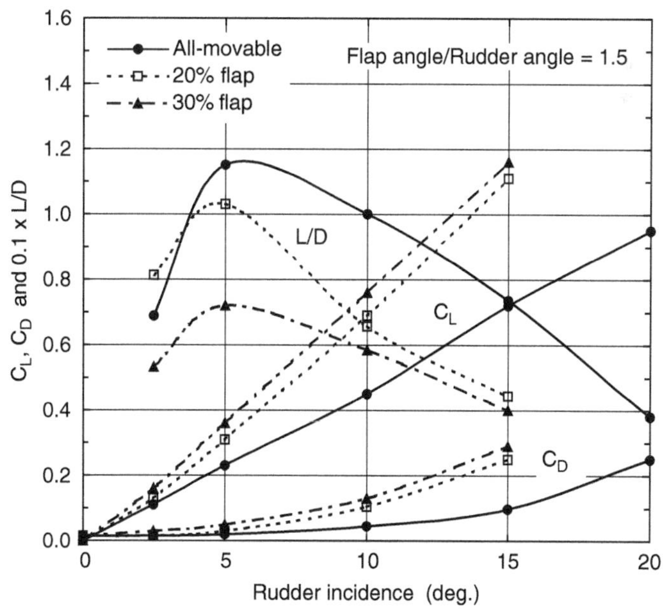

Figure 11.6 Comparison between all-movable and flapped rudders

but with a decrease in L/D ratio. With the 30% flap, the further increase in lift curve slope is relatively small and there is a significant fall in L/D.

As far as rudder efficiency is concerned, measured by L/D, the all-movable rudder is superior over the whole range of angles. There is, however, a significant increase in lift curve slope with the flapped rudder or foil, which may be utilised where space may be limited, such as for a stabiliser fin, or for a rudder required to produce a good manoeuvring performance. In both of these cases, drag is not likely to be a problem. It is apparent that further increases in flap ratio may not significantly improve these attributes.

11.2.6 Example application 6. Twisted rudder design

The reasons for adopting a twisted rudder have been outlined in Sections 5.8.3 and 5.8.4. The concept basically amounts to adapting the local rudder incidence across the span to the effective inflow angle resulting from the rotation of the propeller slipstream. The purpose of applying a suitable twist, for a set rudder angle of zero, is to have zero effective incidence and load across the span. This leads to a decrease in induced drag (at zero rudder angle), see Section 3.4.1, and an improvement in overall propulsive efficiency. An effective incidence of zero across the span (at zero rudder angle) also decreases the vulnerability of the rudder to cavitation at zero and low rudder angles.

The design process basically requires an estimate of the likely effective inflow angle across the span due to the propeller. This will depend on the position of the rudder relative to the propeller (X/D, Y/D and Z/D) and the thrust loading of the propeller (K_T/J^2) for the chosen design condition.

Estimates of the required distribution of inflow angle, hence twist, may be made experimentally or theoretically. An experimental approach is described in Section 5.8, which uses pressure measurements, the local C_N and the inferred required local incidence, as shown in Figures 5.125 and 5.126. Theoretical estimates of twist may be made using lifting-line analysis, panel methods, or Reynolds-Averaged Navier–Stokes (RANS) solvers. The lifting-line analysis, described in Chapter 6 and Appendix 2, can produce useful guidance on the distribution of twist. The results for C_N for zero incidence when compared with experiment in reference [11.1] demonstrate the capability to predict the C_N distribution and from it the required local incidence and twist.

Reference [11.2] reports on the application of a panel method, described in Chapter 6, to investigate required levels of rudder twist for a particular application. The results are summarised to demonstrate the approach. The condition examined was one of the cases used in the wind tunnel rudder–propeller interaction tests reported in Section 5.4. The rudder is based on Rudder No. 2 as given in Table 5.10 and the propeller on a modified Wageningen B4.40 as described in Table 5.12. A propeller thrust loading $J = 0.51$ ($K_T/J^2 = 0.88$) was used in the investigation.

Figure 11.7 shows the inflow velocities to the rudder and Figure 11.8 the adopted initial twist, which was derived from the axial and circumferential components of velocity. A very similar level and distribution of twist is obtained using the lifting line code described in Appendix 2. Figure 11.9 shows a comparison of C_N between

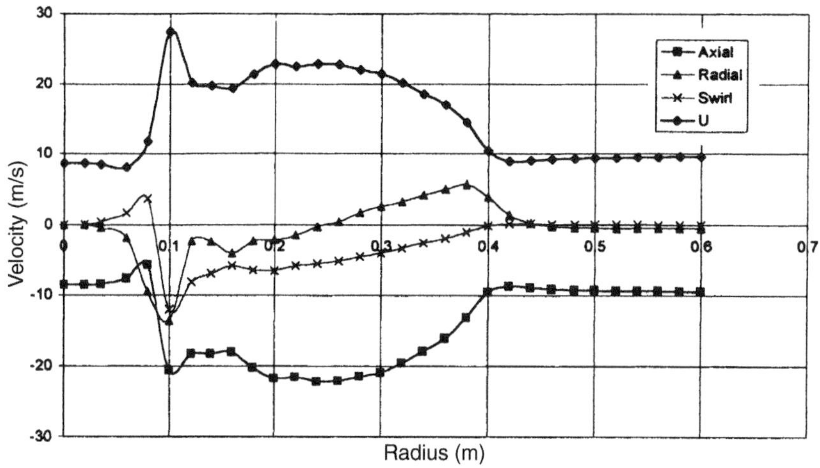

Figure 11.7 Inflow velocities

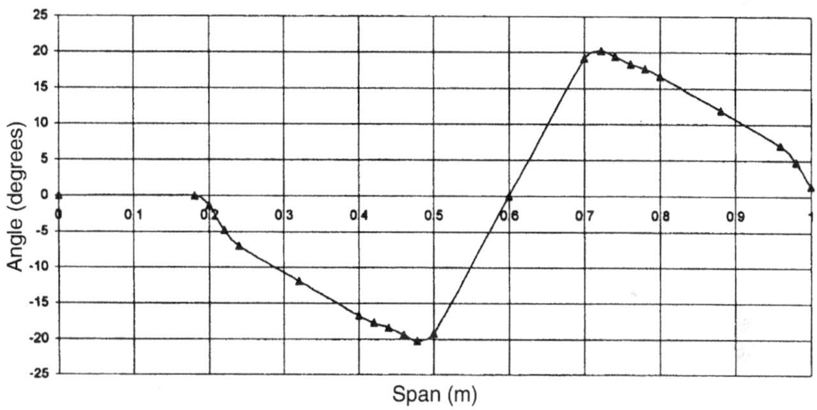

Figure 11.8 Adopted initial twist distribution

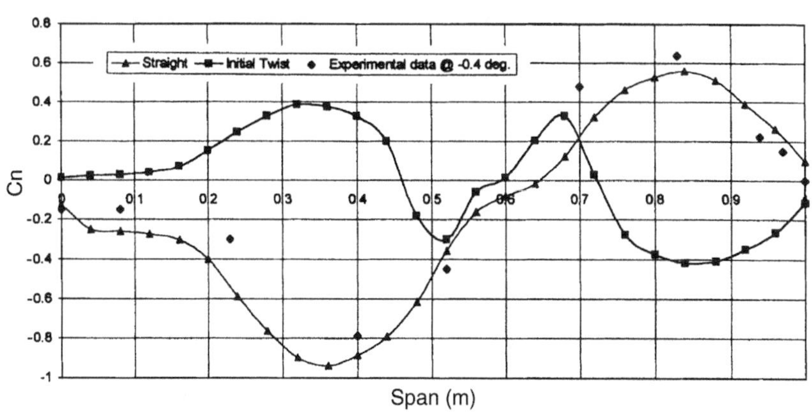

Figure 11.9 Comparison for C_N for straight and twisted cases

the initial assumption for twist and the original straight (or zero twist) case. It is seen that the assumed twist overcompensates for the propeller-induced velocity and angle. This is probably due to the use of what is effectively the 2-D inflow velocity and angle, which will in reality be less due to the downwash on the actual 3-D rudder, Section 3.4.2 and Figure 3.28. It is also noted that the twist distribution derived experimentally in Figure 5.126 (which is for the same rudder–propeller and propeller load case) is much lower at up to ±9°. A systematic variation in twist was then investigated, as shown in Figure 11.10, and the results are shown in Figure 11.11. It was found that a hybrid between the 70% twist and 80% twist was the most suitable. The peaks around the boss are due primarily to the radial velocity components in this region. The use of twist will not eliminate this component of lift, although optimisation of local twist may reduce the effect.

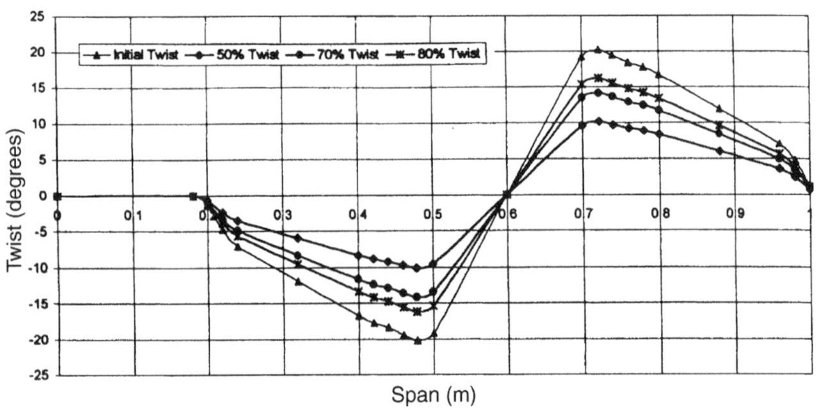

Figure 11.10 Twisted distributions

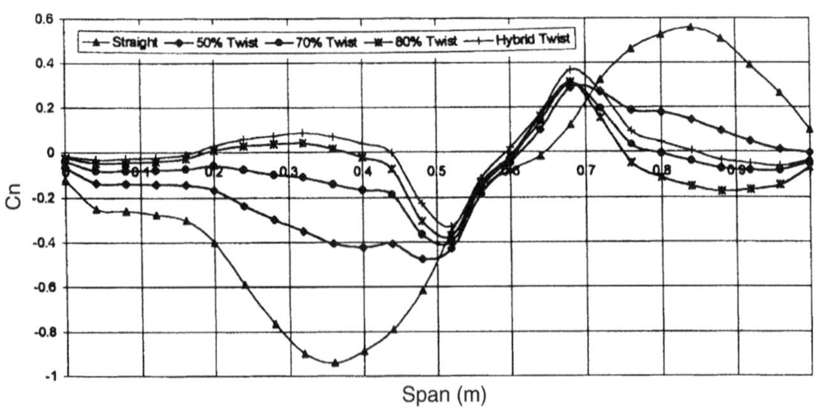

Figure 11.11 C_N distribution for different levels of twist

The use of cambered sections was also investigated, where the sections are cambered such that the ideal angle of attack, with zero lift, is attained without twist. Different levels and distributions of camber were investigated but, in the main, the results were not as successful as when twist was applied. It was, however, concluded that this approach deserved further investigation.

Comparisons of C_N between the original straight, the twisted and cambered sections are shown in Figure 11.12. The twisted rudder yields a more elliptical loading over the span and has nearly symmetrical loading for positive and negative rudder angles. The cambered form was less successful. Peak (negative) pressures are given in Table 11.5 where it is seen that the twisted rudder shows a significant decrease in minimum pressure, with the potential for delaying the onset of cavitation. The cambered form did not perform well due to pressure peaks near the leading edge.

The investigation demonstrates the ability of the panel method approach to estimate suitable levels of twist. The approach also allows further studies in optimising the twist distribution and the performance of the rudder in off-design situations. For practical construction considerations, it may be necessary to adopt a constant mean value of twist for the upper and lower halves of the rudder. Alternatively, if such a rudder is built using composite materials, then a continuous change in twist can be easily incorporated.

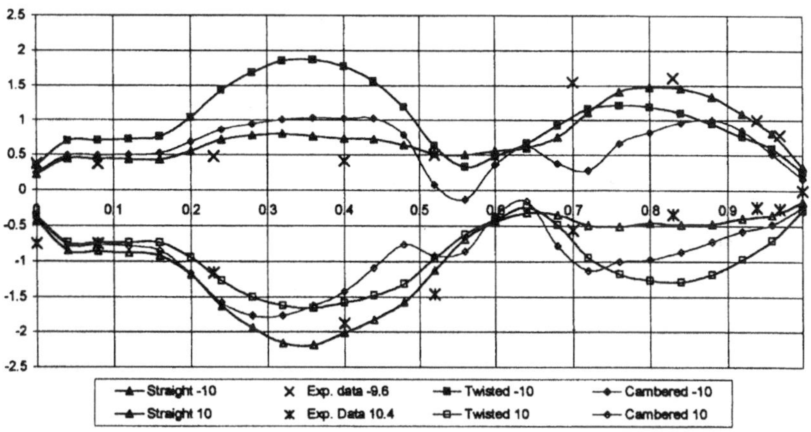

Figure 11.12 C_N distribution for ±10° for straight, twisted and cambered cases

Table 11.5 Minimum values of pressure coefficient C_P

	Incidence		
Rudder type	*−10°*	*0°*	*10°*
Straight	−14.78	−9.59	−17.89
Twisted	−11.02	−6.63	−11.51
Cambered	−27.75	−20.81	−32.13

11.2.7 Example application 7. Cavitation check: Spade rudder on a single-screw container ship

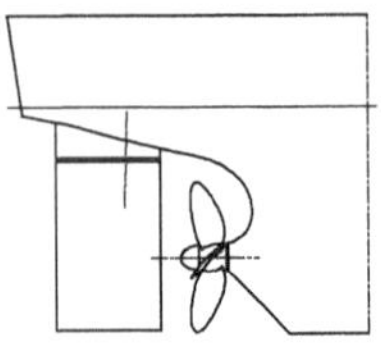

Container ship, 7,000 TEU, draught = 14 m, service speed = 25 knots.

Propeller diameter = 9.5 m

Propeller revolutions at service speed = 110 rpm

Assume geometric aspect ratio $AR_G = 1.5$ and, at small rudder angles, take the effective aspect ratio $AR_E = 2 \times AR_G = 2 \times 1.5 = 3.0$.

Block coefficient $C_B = 0.60$ and from Equation (3.46) wake fraction $w_T = 0.248$.

At the service speed, $J = V_A/nD = 25(1 - 0.248) \times 0.5144 \times 60/110 \times 9.5 = 0.555$.

From propeller chart for assumed ship propeller at $J = 0.555$, $K_T = 0.220$ and $K_T/J^2 = 0.714$.

Wake speed $V_A = 25(1 - 0.248) \times 0.5144 = 9.67\,\text{m/s}$

The propeller-induced axial speed at the rudder V_R can be obtained using Equations (3.37) and (3.38).

Assume $X/D = 0.38$, hence from Equation (3.37), $K_R = 0.858$

And from Equation (3.38), where $V_0 = V_A = 9.67\,\text{m/s}$, $V_R = 1.582 \times V_0 = 15.30\,\text{m/s}$.

Take the depth of immersion under consideration to be $h = 7.0\,\text{m}$, being about 40% below the top of the rudder and in way of the main propeller-induced velocity:

$$\begin{aligned}\text{Cavitation number } \sigma &= (P_{AT} + \rho g h - P_V)/\tfrac{1}{2}\rho V^2 \\ &= 101 \times 10^3 + 1{,}025 \times 9.81 \times 7.0 - 3 \times 10^3/\tfrac{1}{2} \times 1{,}025 \times 15.30^2 \\ &= 1.40\end{aligned}$$

Using the cavitation envelopes for a NACA 0020 section in Figure 5.139, C_L limits to avoid cavitation are approximately $C_L = \pm 0.4$. With an aspect ratio of 3.0 and a lift curve slope $\mathrm{d}C_L/\mathrm{d}\alpha = 0.052$ from Figure 5.13, then limiting rudder angles to avoid cavitation are about ±7.7°. Based on the discussion in Section 5.8 and Chapter 6, it is not expected that propeller-induced angles and helm changes for course-keeping will exceed this limit for any extensive period of time.

If the calculations are repeated taking a depth of immersion $h = 5\,\text{m}$, say in a ballast condition, then $\sigma = 1.23$, limiting $C_L = \pm 0.34$ and limits on rudder angle are about ±6.5°. This just about within the region where propeller-induced angles and helm changes for coursekeeping may come into play.

If the speed were increased to 28 knots, and assuming the same J and K_T/J^2, then $V_R = 17.14\,\text{m/s}$, $\sigma = 1.11$ (for the14 m full draught condition), limiting $C_L = \pm 0.23$ and limiting rudder angle is about ±4.5°.

These calculations provide a broad insight into the likelihood of cavitation on the rudder. They show, for this case, a vulnerability to cavitation, particularly with less immersion at a lower ballast draught or if higher design service speeds were to be considered.

For a rudder working downstream of a heavily loaded propeller on a high-speed ship, consideration may also be given to using a section with a wider cavitation inception envelope, Figure 5.141, or employing a twisted rudder, Sections 5.8.4 and 11.2.6. In the case of the heavily loaded propeller, the influence of the shed propeller vortices will also require consideration.

11.2.8 Example application 8. Rudder structural analysis

An outline is given of the structural analysis of a rudder, including derivation of scantlings using classification society rules, loads and stresses by finite element analysis (FEA) and a description of a coupled fluid-structure model.

The rudder for Example application 2 is used for the study and has the following particulars:

Spade rudder with span 6.25 m, chord 4.20 m and with a NACA 0020 section constant across the span, Figure 11.13.

Classification society rules: In Example application 2 the rudder torque was estimated and a suitable stock diameter derived. This was compared with the 831 mm derived from the rules of Lloyd's Register and DNV. The rules were also used to

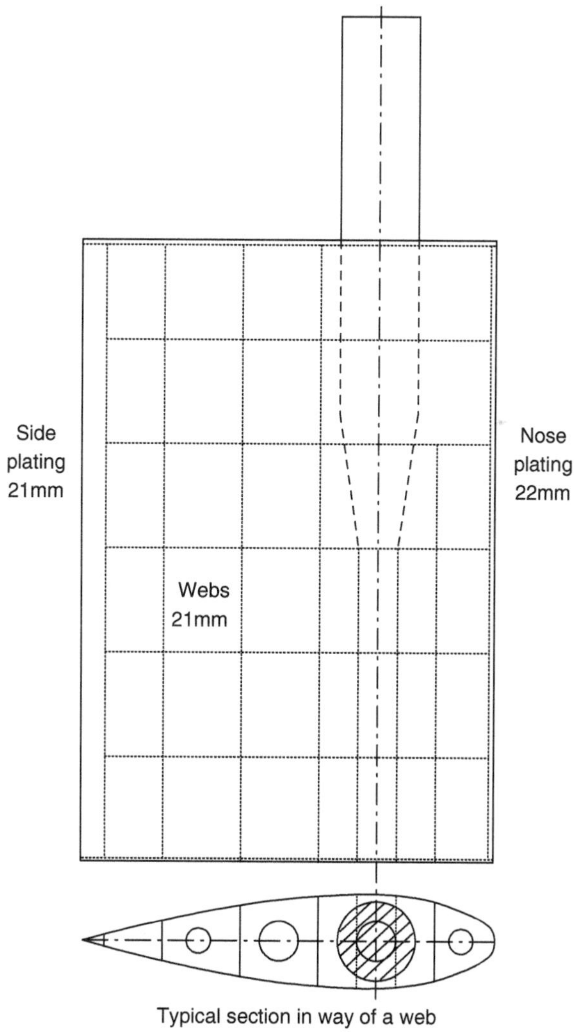

Figure 11.13 Rudder structural layout

obtain preliminary scantlings for the side plating and webs. These are derived from formulae and the results are shown in Figure 11.13.

Finite element analysis: In order to illustrate the methodology, a simplified model of the rudder was used with two horizontal webs at 33% and 67% of the span from the root. The side plating and webs were all assumed to have a thickness of 20 mm.

A realistic model of the load distribution over the rudder was established using experimental pressure data for a rudder angle of 30° and drift angle −7.5°. The method is described in Section 5.4.2.3 and data from the experimental database is used, as described in Appendix 1. These data are similar in format to the 20° example pressure data given in Table A1.14 and Figure 5.72. As noted in Chapters 5 and 6, the pressure distribution over the rudder may also be estimated using theoretical methods such as potential, Euler, or RANS flow solvers.

ANSYS version 5.4 was used for the FEA. All structural members were modelled with four-node elastic shell elements which contain both bending and membrane capabilities. The FEA model of the structure is similar to that shown in Figure 11.14.

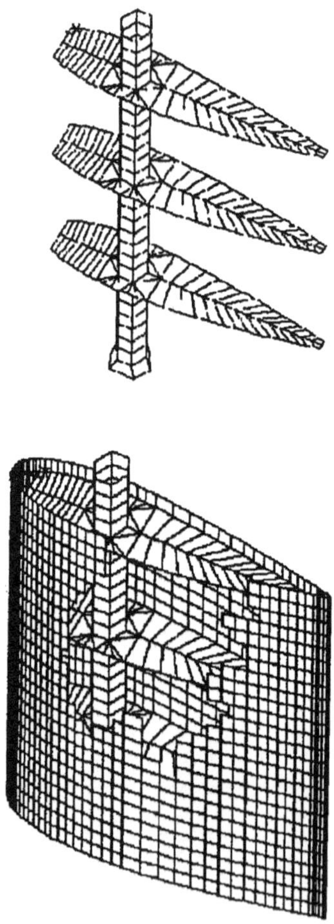

Figure 11.14 Mesh for ANSYS model

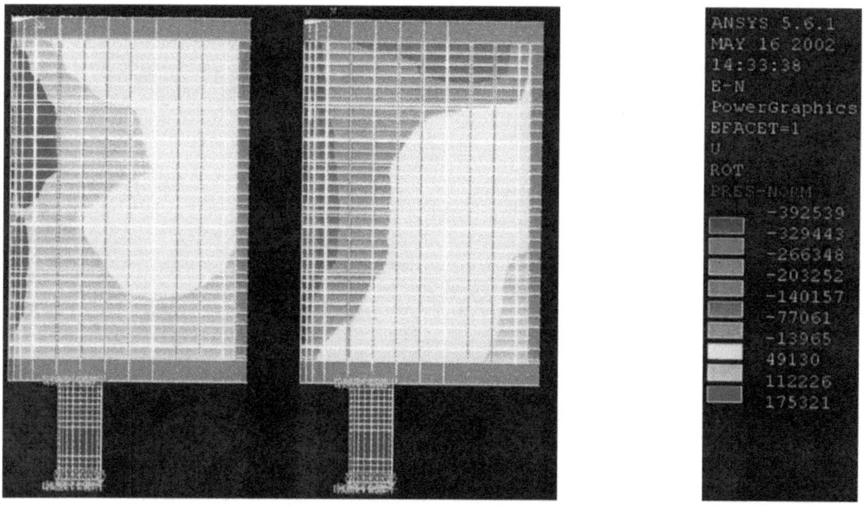

Figure 11.15 Total pressure distribution (See Plate 7 of Colour Plate Section)

Larger stresses were expected at the webs and mid-way between the webs, and the elements in these areas were made smaller to account for this. It is assumed that the stock acts as a cantilever and is fixed where it meets the lowest bearing. The loads transmitted through the webs can be modelled as point loads. The stock within the rudder was modelled with finer elements around the intersections with the webs since higher stress gradients were expected in these areas.

The pressure coefficients were converted to forces and distributed over the rudder. The hydrostatic (ρgh) pressures were then superimposed. The pressure (force) distribution taking into account hydrostatic and hydrodynamic effects is shown in Figure 11.15 (See Plate 7 of Colour Plate Section). Back suction pressure is on the left and face pressure (compression) is on the right. Solution of the model yielded deflections and stresses which are shown in Figures 11.16 (See Plate 8 of Colour Plate Section), 11.17 (See Plate 9 of Colour Plate Section) and 11.18 (See Plate 10 of Colour Plate Section). The largest deflections are at the tip of the rudder, Figure 11.18. Local deflections can be seen as local minima between the webs. The deflection at the trailing edge is greater than that at the leading edge since its distance from the support (stock) is greater. The stresses in the stock in the *Z* direction are shown in Figure 11.19 (See Plate 11 of Colour Plate Section) and, as expected, the largest stress occurs near the supported end of the stock.

The stress at any point within the structure must not exceed a given limiting value (e.g., the yield stress of the material). The total magnitude of stress can be calculated from $\sigma_T = (\sigma_x^2 + \sigma_y^2 + \sigma_z^2)^{1/2}$ and compared locally with the design limiting stress.

The *Z* component of stress has high values in the skin at the location of the internal webs and near the root. In this example, the total stress in the skin exceeds the yield stress of mild steel where it interacts with the webs.

It is clear that the FEA has identified areas of high stress concentration. This would indicate the number of webs to be added, together with appropriate increases in the scantlings.

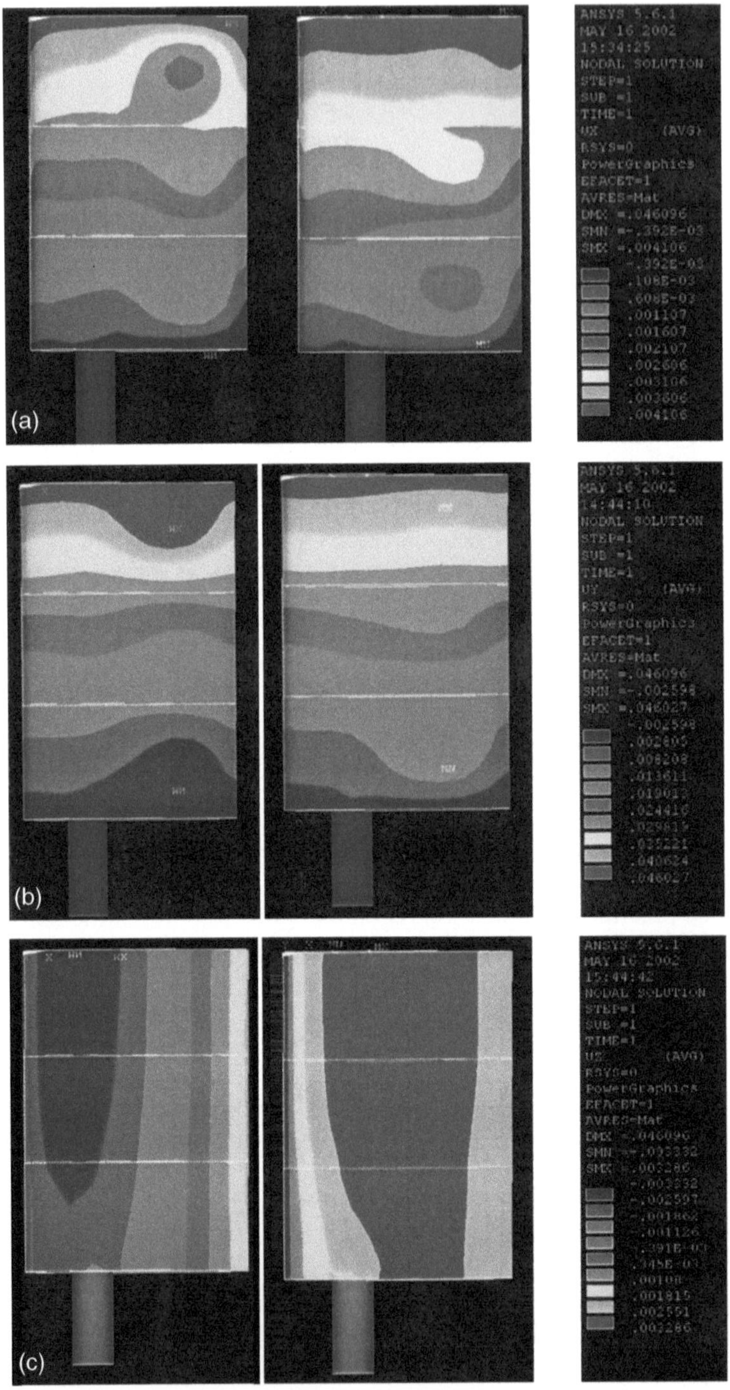

Figure 11.16 (a) Deflection in *x* direction; (b) Deflection in *y* direction; (c) Deflection in *z* direction (See Plate 8 of Colour Plate Section for Part(a))

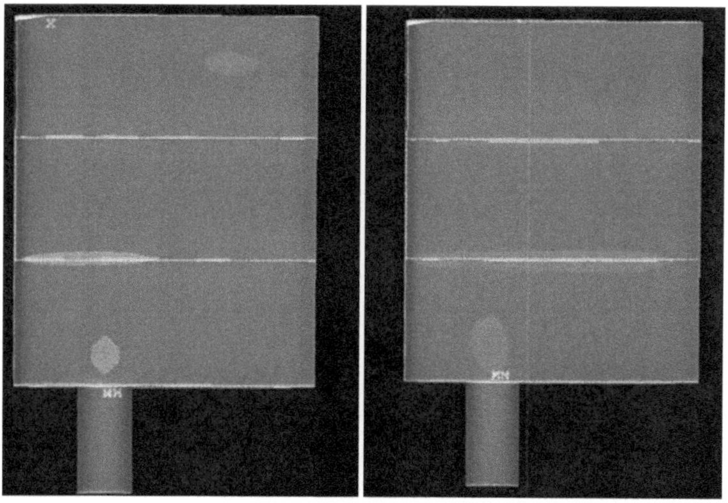

Figure 11.17 *Z*-component stresses (See Plate 9 of Colour Plate Section)

Coupled fluid-structure model: The major shortcoming of the experimental loading-finite element approach described in the previous section is that, with deflections, the hydrodynamic properties of the rudder change. As the experimental data are for a rigid body, it is not possible to model these effects.

An alternative to the experimental data is to derive the pressure distribution using one of the theoretical methods described in Chapter 6. A property of the theoretical methods is that the body shape can be changed and the influence on the hydrodynamic forces monitored. This property can be exploited in the analysis of structures and a coupled fluid-structural model developed, Figure 11.20. In the broader context this model resides within the domain of hydroelasticity. In principle, the pressure loading resulting from the fluid analysis is passed to the structural code, and the deflected shape solution from the structural code passed to the fluid code. The process continues until an acceptable convergence is achieved. A significant problem can lie in the differences required in the discrete meshes of the fluid and structural codes. An appropriate approach is to use a virtual surface, Figure 11.20, which allows a variety of shapes to be modelled and can accommodate large disturbances in the structural and fluid models. An example using such an approach is described by Turnock and Wright [11.3]. An important merit of the coupled model is its flexibility and, in this application, its ability to respond to changes in rudder pressure loading due to changes in upstream propeller loading.

11.2.9 Example application 9. Rudder and propeller layout – design implications

The geometric parameters *X/D*, *Y/D* and *Z/D* that affect rudder–propeller interaction are shown in Figures 4.8–4.10 and are summarised in Figure 11.21. Changes in these

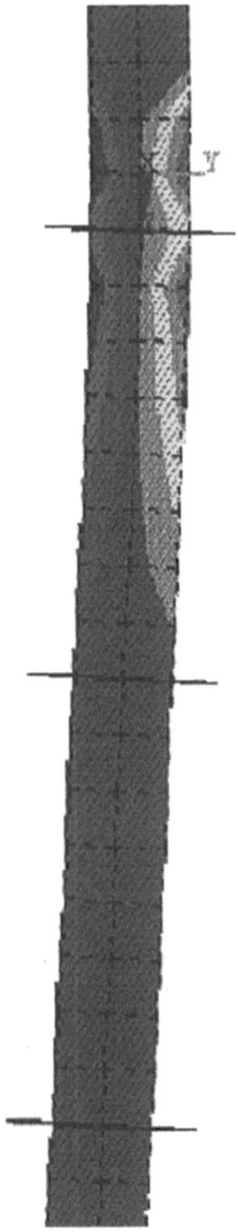

Figure 11.18 Bending stresses (See Plate 10 of Colour Plate Section)

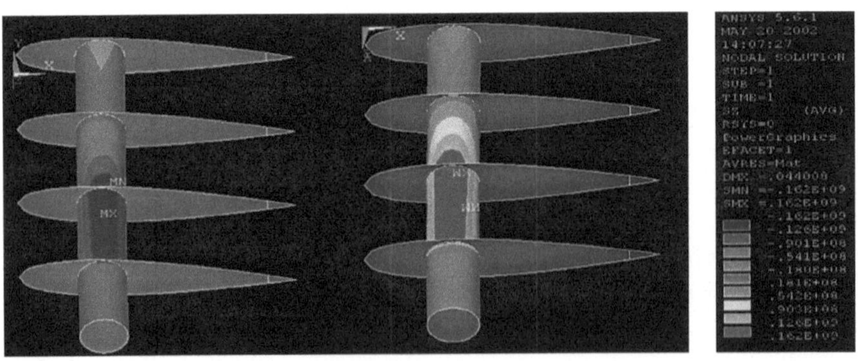

Figure 11.19 *Z*-component stresses in the stock (See Plate 11 of Colour Plate Section)

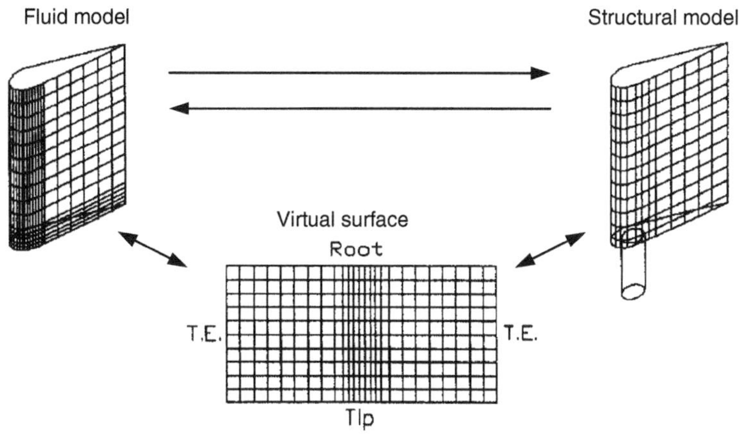

Figure 11.20 Coupled fluid-structure model

parameters can have a significant influence on the relative magnitude of the rudder forces, as described in Section 5.4 and on the propeller performance as described in Section 5.9. The rudder and propeller may be treated as a combination, but the influences of one on the other may in some cases be conflicting. The following discussion identifies the principal changes in rudder and propeller performance as a result of the changes in the geometric parameters *X/D*, *Y/D* and *Z/D*.

(a) *Design implications – single screw vessels:* In the case of single-screw vessels there is some latitude in the choice of longitudinal separation (*X/D*), but the

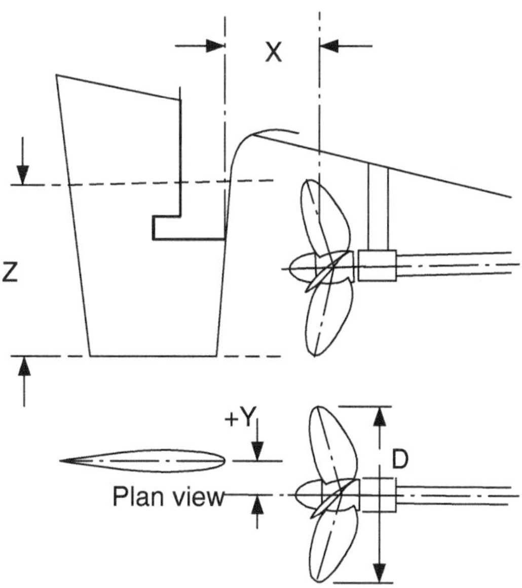

Figure 11.21 Definitions of *x, y, z*

results presented in Sections 5.4 and 5.9 indicate that changes in X/D can have conflicting influences on manoeuvring and propulsion. Decreasing X/D from 0.4 to 0.25 (a typical range for layout purposes) can lead to a net improvement in the propulsive effect of the rudder–propeller combination, but lift curve slope and hence coursekeeping and manoeuvring sideforce production is reduced by about 7% at typical ship operational speeds. Whilst, traditionally, the primary factors dictating X/D tend to be propulsive effects, and these can be derived from model self-propulsion tests with zero rudder incidence, the results presented indicate that the influences on manoeuvring performance can be significant.

On smaller vessels, it is often not practical or possible to design for $Z/D = 1.0$. Values of $Z/D < 1.0$ have to be used and the results indicate that significant penalties can be incurred. The lift curve slope is reduced and there is a zero lift offset of the order of 3°–5° (depending on thrust loading), which requires a rudder angular offset in normal operational conditions. Rudder drag is not a minimum at zero lift and K_T decreases compared with complete rudder coverage of the propeller race. Surhbier [5.135] carried out tests on this type of rudder–propeller layout and similar observations were made. The results indicate that, where possible, this type of arrangement should be avoided and Z/D maximised.

(b) *Design implications – twin screw vessels:* The comments on the effects of longitudinal separation for single-screw vessels also apply to twin-screw vessels. An even greater latitude in the choice of X/D is generally available in twin-screw layouts. It is surprising that large values of X/D in the range 0.50–0.70 are sometimes employed which decrease the benefits of rudder–propeller interaction from the point of view of both manoeuvring and propulsion.

Whilst nonzero values of Y/D in the case of twin-screw vessels might be necessary from a practical operational and maintenance viewpoint, it is found that increasing the value of Y/D can have detrimental effects on both manoeuvring and propulsion. Rudder drag is higher and ΔK_T is lower than at $Y/D = 0$, leading to a net decrease in propulsive thrust. There is a significant shift in zero lift angle, which increases with increase in propeller thrust loading. The results illustrate the need in twin-screw installations to compensate for this and minimise parasitic induced drag by incorporating appropriate rudder toe-in or toe-out. The results show the effects on manoeuvring to be reasonably symmetrical and hence similar for inboard and outboard shift of rudder. This is not necessarily true for the propulsive effects, the results indicating a larger decrease in ΔK_T and increase in C_D for positive Y/D. The decision to site the rudders inboard or outboard of the propeller shaft will depend on whether inboard or outboard turning propellers are employed.

On larger twin-screw vessels the rudder tip will generally coincide with the propeller tip ($Z/D = 1.0$) and Z/D influences will not be present. On smaller twin-screw vessels, values of $Z/D < 1.0$ generally have to be used. The penalties incurred are broadly the same as for smaller single-screw vessels, described earlier, except that in the twin-screw case asymmetry can be compensated by the twin rudders and counter rotating propellers.

(c) *Design implications – overall:* Significant changes in both propulsion and manoeuvring performance can occur when the relative positions of the rudder and propeller are changed. These changes can sometimes be conflicting. The designer generally has some latitude in the choice of longitudinal rudder–propeller separation. Whilst smaller separation can improve overall propulsive efficiency, some gain in manoeuvring performance can be achieved by increasing the separation. Improvements in propulsive efficiency may be achieved by the incorporation of twist across the rudder span (Section 5.8 and Example application 6).

Any lateral offset between the rudder and propeller centrelines tends to have detrimental effects on both propulsion and manoeuvring. The level of these effects depends on whether the rudders are situated inboard or outboard of the propeller centreline for a given direction of rotation. It is apparent that, within practical constraints, lateral offset should be minimised. The relative vertical positions of the rudder and propeller can have detrimental effects on both manoeuvring and propulsion when the rudder is not fully within the propeller race.

(d) *Twin screw vessels – one rudder versus two rudders:* Early twin-screw vessels were equipped with a single rudder, a practice that survived for many years. Current practice is generally to equip twin-screw ships with twin rudders, when the benefits of the accelerated propeller flow can be utilised.

An investigation into the number of rudders suitable for a twin-screw ferry was reported by Avellino and Coan [11.4]. Model and full-scale validation tests were carried out. It was found that the single-rudder configuration performed in a manner equivalent to the twin-rudder solution. Regarding cavitation behaviour and forces induced by the propeller on the hull, in the twin-rudder configuration, the rudder–propeller interaction tends to amplify the forces and makes the rudder more susceptible to cavitation. Conclusions were not clear-cut as decisions, as well as depending on hydrodynamic performance, will depend on cost, layout and required comfort levels.

11.3 Small craft

11.3.1 Example application 10. Cruising yacht: Skeg rudder

Calculation of approximate maximum torque, bending moment and stock diameter:

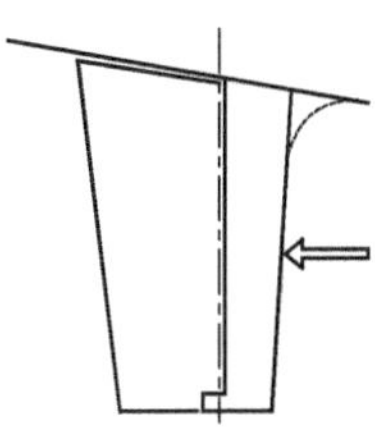

Consider an outline rudder design for a cruising yacht having a waterline length of 9.0 m.

Data from existing craft (see Section 5.6) suggest a rudder area of about 0.50 m^2.

Assume speed for the derivation of maximum forces $V = 8$ knots $= 4.12$ m/s.

The proposed rudder has the following particulars:

Mean chord $\bar{c} = 0.50$ m and span $S = 1.00$ m. Hence area $A = 0.50\,m^2$ and $AR_G = 2.0$

Assume the skeg chord is 30% of total.

For estimates of maximum forces, assume rudder is fitted closely to hull, then $AR_E = 4.0$.

$$Re = V\bar{c}/\nu = 4.12 \times 0.50/1.19) \times 10^6 = 1.73 \times 10^6$$

From Figure 5.34, or Equation (5.18), for $Am/A_T = 0.7$, $Ls/L = 0.85$. This similarly follows by comparing Figure 5.31 with Figure 5.14(a), i.e., lift on the rudder plus skeg combination is 85% of the all-movable equivalent for most of the lift curve. C_{Lmax} for the skeg rudder will however be about the same as for the all-movable case. Hence from the all-movable data, Figure 5.18, assuming C_{Lmax} at $\beta = 0$ is the same as for the all-movable (although α_{stall} will be larger), for $AR_E = 4.0$, $C_{Lmax} = 1.10$ for the all-movable equivalent.

$$\begin{aligned} \text{Maximum lift (rudder + skeg)} &= C_{L\,max} \times \tfrac{1}{2} \times \rho \times AV^2 \\ &= 1.1 \times 0.5 \times 1{,}025 \times 0.5 \times 4.12^2/10^3 \\ &= 4.78\,\text{kN} \end{aligned}$$

From Figure 5.35, or Equation (5.19), for $Cm/c = 0.7$,

Lift on movable part of rudder = 54% of total = $0.54 \times 4.78 = 2.58$ kN

And acts at approximately $0.40\bar{c}$ aft of stock at stall,

i.e., $\bar{x} = 0.4 \times (0.7 \times 0.5) = 0.140$ m aft of the stock.

In the absence of drag data, the maximum force is assumed equivalent to the maximum lift for torque and bending moment calculations.

Hence, torque on stock: $Q_R = N \times \bar{x} = 2.58 \times 0.140 = 0.361$ kNm.

From Equation (7.10) and assuming all the bending moment is carried by the skeg,

$$\text{Equivalent torque } Q_{RE} = Q_R = 0.361\,\text{kNm}.$$

Assume a design stress = 235 N/mm^2 for steel, i.e., about yield stress

Using Equation (7.11),

$$\text{Stock diameter } D = [(0.361 \times 1{,}000 \times 16)/(\pi \times 235 \times 1{,}000^2)]^{1/3} \times 1{,}000 = 20.0\,\text{mm}$$

This assumes all bending is carried by the skeg. If say 30% of bending moment is assumed to be carried by the skeg, the moment carried by the stock is

$$M = 4.78 \times 0.7 \times (1.0 \times 0.48) = 1.61\,\text{kNm}$$

(where 1.0 m is the span and $CPs = 0.48S$ for the equivalent all-movable rudder)

Using Equation (7.9),

$$\text{Equivalent bending moment } BM_E = 1.61/2 + 0.5(1.61^2 + 0.361^2)^{0.5} = 1.63\,\text{kNm}$$

Using Equation (7.11),

$$\text{Stock diameter } D = [(1.63 \times 1{,}000 \times 32)/(\pi \times 235 \times 1{,}000^2)]^{1/3} \times 1{,}000 = 41.3\,\text{mm}$$

(Or using Equation (7.12), an equivalent hollow stock could have an external diameter of 50 mm and internal diameter of 40 mm)

If, as part of the construction, a 8 mm thick side plate is added to each side of the stock, total thickness = 41.3 + 8 + 8 = 57.3 mm and $t/c = 57.3/500 = 0.115$.

Using the properties of the skeg material (e.g. steel or aluminium alloy or FRP) and construction, a check on the skeg root bending moment and hence stress would be required to confirm that this thickness is satisfactory. Bending moments can be derived by direct calculation or approximate simplified formulae provided by the classification societies.

11.3.2 Example application 11. Offshore racing yacht: Spade rudder

Calculation of approximate maximum torque and bending moment:

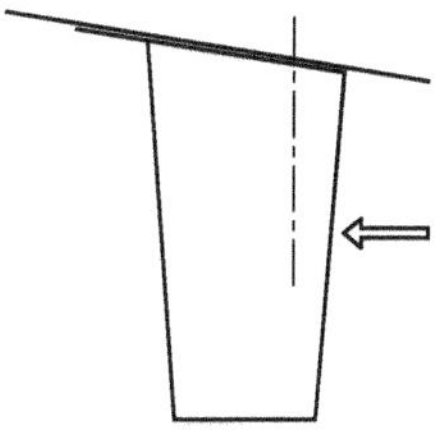

Consider an outline rudder design for an offshore racing yacht having a waterline length $L_{WL} = 14.5$ m and displacement $\Delta = 17$ tonnes.

Data from existing craft (see Section 5.6) suggest a rudder area of about $0.72\,\text{m}^2$.

Assume speed for the derivation of maximum forces $V = 12$ knots $= 6.17$ m/s

The proposed rudder has the following particulars:

Mean chord $\bar{c} = 0.56$ m and span $S = 1.30$ m. Hence, area $A = 0.728\,\text{m}^2$ and $AR_G = 2.32$.

For estimates of maximum forces, assume rudder is closely fitted to hull, then $AR_E = 4.64$

Assume stock is located 18% of mean chord aft of leading edge

Assume 150 mm from lower bearing to top of rudder

$$Re = V\bar{c}/\nu = (6.17 \times 0.56/1.19) \times 10^6 = 2.9 \times 10^6$$

Initially assume $t/c = 0.17$

From Figure 5.17, $\alpha_{stall} = 20°$; Figure 5.18, $C_{Lmax} = 1.20$; Figure 5.14(b), $C_D = 0.17$

From Figure 5.15, CPc at stall $= 0.22\bar{c}$ and from Figure 5.16, CPs at stall $= 0.47S$

$$\text{Maximum lift} = C_{Lmax} \times \tfrac{1}{2}\rho A V^2 = 1.2 \times 0.5 \times 1{,}025 \times 0.728 \times 6.17^2/1{,}000 = 17.04\,\text{kN}$$

Drag $= 0.17 \times 0.5 \times 1{,}025 \times 0.728 \times 6.17^2/1{,}000 = 2.41$ kN

Normal force $N = 17.04 \cos 20° + 2.41 \sin 20° = 16.84$ kN

Resultant force $R = (L^2 + D^2)^{0.5} = (17.04^2 + 2.41^2)^{0.5} = 17.21$ kN

Torque lever $\bar{x} = (0.22 - 0.18) \times 0.56 = 0.022$ m

Torque $= N \times \bar{x} = 16.84 \times 0.022 = 0.370$ kNm

Bending moment lever $y = (CPs \times S) + 0.150 = 0.761$

And bending moment at bearing $M = R \times y = 17.21 \times 0.761 = 13.10$ kNm

Using Equation (7.9),

$$\text{Equivalent bending moment } BM_E = 13.10/2 + 0.5(13.10^2 + 0.370^2)^{0.5} = 13.10\,\text{kNm}$$

Using Equation (7.11),

$$\text{Stock diameter } D = [(13.10 \times 1{,}000 \times 32)/(\pi \times 235 \times 1{,}000^2)]^{1/3} \times 1{,}000 = 82.8\,\text{mm}$$

(where the assumed maximum stress $= 235\,\text{N/mm}^2$, approximately the yield stress for steel).

If, as part of the construction, a 10 mm thick side plate is added to each side of the stock, total thickness $= 82.8 + 10 + 10 = 102.8$ mm and $t/c = 102.8/560 = 0.184$, which is reasonable for the root and can be decreased a little towards, the tip. See the rules of classification societies for permissible stress levels, scantlings of stiffeners and webs and decrease in thickness between root and tip.

Stock diameter using classification society rules: Using the American Bureau of Ships (ABS) rules for offshore racing yachts [7.38],

Rudder force $P = 15.58$ kN

Torque lever = (0.33 − 0.18) × 0.56 = 0.084 m

Torque = 15.58 × 0.084 = 1.31 kNm

Bending moment = 15.58 × 0.761 = 11.86 kNm

Equivalent bending moment $BM_E = 11.86/2 + 0.5(11.86^2 + 1.31^2)^{0.5} = 11.90$ kNm

ABS recommends the use of the minimum yield strength for metals, approximately 235 N/mm^2 for steel.

$$\text{Stock diameter } D = [(11.90 \times 1{,}000 \times 32)/(\pi \times 235 \times 1{,}000^2)]^{1/3} \times 1{,}000$$
$$= 80.2\,\text{mm}$$

This about 3% less than the diameter derived from the above detailed calculations.

11.3.3 Example application 12. Comparison between all-movable and full-skeg rudders

Assume that a lift coefficient of 0.33 is required, based on the same total areas, rudder span = 0.68 m, chord = 0.47 m and inflow speed = 2.5 m/s (4.9 knots).

$AR_G = 0.68/0.47 = 1.45$ and, assuming a small rudder-hull clearance, $AR_E = 2.90$.

Area $A = 0.68 \times 0.47 = 0.32\,\text{m}^2$

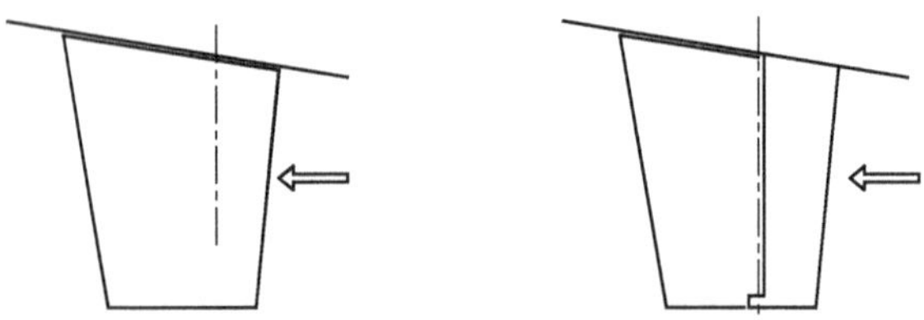

Using Figure 5.32, ($AR = 2.8$) for $Cm/c = 60\%$

	All-movable	*Skeg rudder*
C_L	0.33	0.33
α	7.0°	9.8°
L/D	11.0	6.8
C_D	0.030	0.049

The approximate increase in C_D for the skeg rudder = (0.049 − 0.030)/0.030 = 63%

For a rudder of span 0.68 m and chord 0.47 m at a speed of 2.5 m/s this amounts to a total original drag on the all-movable rudder of

$$d = C_D \times \tfrac{1}{2}\rho \times A \times V^2$$
$$= 0.030 \times (1{,}025/2) \times 0.32 \times 2.5^2 = 30.75\,\text{N}$$

and an increase in drag when using a comparable skeg rudder of approximately,

$$\Delta d = (0.049 - 0.030) \times (1{,}025/2) \times 0.32 \times 2.5^2 = 19.48\,\text{N}$$

If the calculations are repeated with an assumed leeway angle 5° at the rudder, then this requires a weather helm on the all-movable rudder of 2° for an effective α of 7°.

For the skeg rudder, from Figure 5.30(c) at $\beta = +5°$

$$C_L = 0.33 \text{ is achieved at } \alpha = 4.5°$$

and $C_D = 0.045$, hence $L/D = 7.33$.

This amounts to an increase in C_D for the skeg rudder over the all-movable rudder of 50%.

These simple calculations illustrate that the skeg rudder is less efficient than the all-movable rudder when developing the same lift or sideforce. The overall manoeuvring and coursekeeping performance of the skeg rudder will be less efficient due to the occurrence of negative drift on the skeg, Figure 5.38. The influence of negative angle of attack on the skeg is more pronounced as drift angle is developed. For example, at a drift angle β of say $-10°$, the all-movable rudder would require a helm of 17° to give a net effective α of 7° for the development of say $C_L = 0.33$ at maximum $L/D = 11$. However, at $\beta = -10°$ the skeg rudder requires a helm δ of about 23° to develop $C_L = 0.33$ with a C_D of about 0.07 and $L/D = 4.7$.

The calculations indicate that the skeg rudder is less efficient than the all-movable equivalent, even when the skeg rudder is working near its ideal efficiency. Minimising or, better still, effectively sealing the gap between skeg and rudder will show a marked improvement in performance although, as shown in Figure 5.34, it will never achieve the same performance as the all-movable rudder. When manoeuvring, the calculations indicate that the skeg rudder is even less efficient with a much lower rate of response and with the added penalty of higher drag.

The fixed skeg contributes to the directional stability for the rudder-free condition, which may be a desirable feature for some sailing craft, and its performance for sail craft when beating or running can be reasonably satisfactory, providing the skeg length does not exceed about 30% of total mean chord. Further improvements follow if the skeg to rudder gap is sealed with say flexible feathered strips attached to the skeg, Figure 11.22. This technique has also been applied to semi-balanced skeg rudders on small craft. Skeg-rudder manoeuvring characteristics are however much worse and skeg length should, again, not exceed about 30% of total mean chord if satisfactory manoeuvring performance is to be achieved.

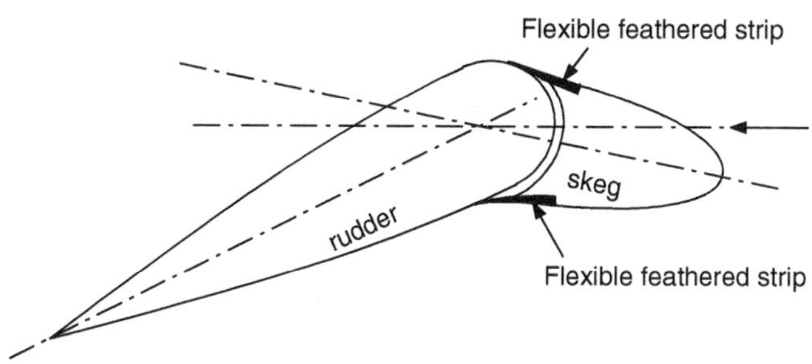

Figure 11.22 Use of flexible feathered strips to seal rudder to skeg gap

11.3.4 Example application 13. Aspect ratio vs. thickness ratio

Relationship between aspect ratio and thickness ratio to meet structural requirements.

Assume a spade rudder fitted close to the hull of a sailing yacht has an area of $0.40\,m^2$, an effective aspect ratio of 3.0, a thickness ratio (t/c) of 0.12 and taper ratio 0.80.

It is proposed to fit a new rudder with an effective aspect ratio of 5.0, and an area necessary to develop the same lift as the original rudder in an assumed design condition of 6° effective angle of attack and a speed of 6 knots. It is required to determine the area of the new proposal and decrease in rudder drag, for the design conditions, and the root bending moment at stall for the original and new proposals.

Original rudder	*New proposal*
$AR_E = 3.0$	$AR_E = 5.0$
Speed $V = 6 \times 0.5144 = 3.08\,m/s$	$V = 3.08\,m/s$
At 6°, from Figure 5.14 (a), $C_L = 0.330$	$C_L = 0.414$

Since $C_L = L/\frac{1}{2}\rho AV^2$, $A\alpha\ 1/C_L$ and for same lift, $A_N = A_0 \times C_{L0}/C_{LN}$

$= 0.40 \times 0.330/0.414$

$= 0.319\,m^2$

Original rudder	*New proposal*
Mean $\bar{c} = (A/AR_G)^{1/2} = (0.40/1.5)^{1/2}$	$= (0.319/2.5)^{1/2}$
$= 0.516\,m$	$= 0.357\,m$
$S = 0.40/0.516 = 0.775\,m$	$= 0.894\,m$
$Re = Vc/\nu = 3.08 \times 0.516 \times 10^6/1.19$	$= 3.08 \times 0.357 \times 10^6/1.19$
$= 1.34 \times 10^6$	$= 0.924 \times 10^6$

From Figures 5.14(b) and 5.19, for $t/c = 0.12$, and Equation (5.10b) with $k_i = 0.37$

Original rudder	*New proposal*
$C_{D0} = 0.011$	$C_{D0} = 0.0114$
$C_{Di} = 0.013$	$C_{Di} = 0.013$
At 6°, $C_D = 0.024$	$C_D = 0.024$
$L/D = 0.330/0.024 = 13.75$	$L/D = 0.414/0.024 = 17.25$
Drag $= C_D \times \frac{1}{2}\rho AV^2$	
Drag $= 0.024 \times 1{,}025/2 \times 0.4 \times 3.08^2$	$= 0.024 \times 1{,}025/2 \times 0.319 \times 3.08^2$
$= 46.67\,N$	$= 37.22\,N$

Hence decrease in drag = 9.47 N, i.e., about 20% of original

From Figures 5.17, 5.18, and 5.14(a)

Original rudder	*New proposal*
$\alpha_{stall} = 22°$, $C_{Lmax} = 0.90$	$\alpha_{stall} = 18.2°$, $C_{Lmax} = 0.935$
$C_D = 0.208$	$C_D = 0.128$
$C_N = C_L \cos\alpha + C_D \sin\alpha$	
$= 0.90 \times 0.927 + 0.208 \times 0.375$	$= 0.935 \times 0.950 + 0.128 \times 0.312$
$= 0.912$	$= 0.928$
$N = C_N \times \frac{1}{2}\rho AV^2$	
$= 0.912 \times 1{,}025/2 \times 0.4 \times 3.08^2$	$= 0.928 \times 1{,}025/2 \times 0.319 \times 3.08^2$
$= 1773.6\,N$	$= 1439.2\,N$

Assuming stock is located 7% $\bar{c}$ forward of CPc at stall in both cases

Lever $= \bar{x} = 0.07 \times 0.516 = 0.036\,\text{m}$
And torque $Q = N \times \bar{x} = 1773.6 \times 0.036$
$= 63.8\,\text{Nm}$

$\bar{x} = 0.07 \times 0.357 = 0.025\,\text{m}$
$= 1439.2 \times 0.025$
$= 36.0\,\text{Nm}$

Resultant force $R = (C_L^2 + C_D^2)^{1/2} \times \frac{1}{2}\rho A V^2$
$= (0.90^2 + 0.208^2)^{1/2} \times 1{,}025/2 \times 0.4 \times 3.08^2$
$= 1796.4\,\text{N}$
From Figure 5.16, at stall $CPs = 49.2\%$ S
$CPs = 0.492 \times 0.775 = 0.381\,\text{m}$
$M = R \times CPs = 1796.4 \times 0.381 = 684.4\,\text{Nm}$

$= (0.935^2 + 0.128^2)^{1/2} \times 1{,}025/2 \times 0.319 \times 3.08^2$
$= 1463.6\,\text{N}$
$= 46.7\%$ S
$CPs = 0.467 \times 0.894 = 0.417\,\text{m}$
$M = 1463.6 \times 0.417 = 610.3\,\text{Nm}$

Equivalent bending moment $BM_E = M/2 + \frac{1}{2}(M^2 + Q^2)^{1/2}$

$BM_E = 684.4/2 + 0.5 \times (684.4^2 + 63.8^2)^{1/2}$
$= 685.9\,\text{Nm}$

$BM_E = 610.3/2 + 0.5 \times (610.3^2 + 36.0^2)^{1/2}$
$= 610.8\,\text{Nm}$

From Equation 7.11, stock diameter $D\alpha(BM_E)^{1/3}$

Original rudder has thickness ratio $t/c = 0.12$, $\bar{c} = 0.516$ and $t = 62\,\text{mm}$ with stock diameter of say 40 mm, allowing $2 \times 11\,\text{mm}$ abreast stock

For new proposal, required thickness $= 40 \times (610.8/685.9)^{1/3} + (62 - 40)$
therefore, $t = 60.5\,\text{mm}$
chord $= 0.357\,m = 357\,\text{mm}$
Hence, $t/c = 0.17$

From Figure 5.19, this increase in thickness ratio leads to an increase in C_{Do} for the new design of about 0.001 which, together with the new area of the new design, leads to a net decrease in minimum drag (as $\alpha \to 0$) of about 10% compared with a decrease of about 17% if t/c had remained at 0.12. Similarly, the 25% increase in L/D ratio at 6° if t/c remained constant at 0.12 is reduced to a 20% increase when the increase in t/c to 0.17 is allowed for.

It is therefore seen that the required significant increase in thickness ratio due to an increase in aspect ratio negates an appreciable amount of the gains in hydrodynamic efficiency expected from the increase in aspect ratio.

11.3.5 Example application 14. Rudder area vs. L/D ratio

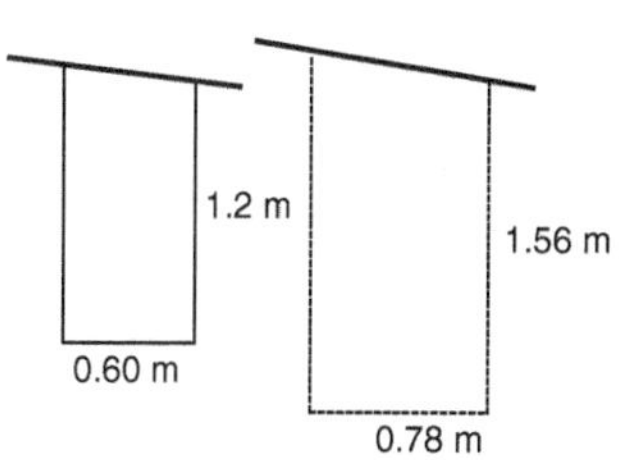

To illustrate possible beneficial effects of an increase in rudder area for a sail craft rudder when, in order to generate a particular sideforce, the original design has to work at an angle of attack somewhat larger than that for L/D_{max}

The existing design has $AR_G = 2.0$ (assumed $AR_E = 4.0$), chord 0.6 m, span 1.2 m (hence area $= 0.72\,\text{m}^2$)

and is required to operate at 10° angle of attack in order to produce the required sideforce at 6 knots (3.08 m/s).

Using Equations (5.6 and 5.10b) or Figure 5.14 (a) and (b), at $\alpha = 10°$ for $AR_E = 4.0$, $C_L = 0.640$ and $L/D = 13.4$, $C_D = 0.048$

Assuming aspect ratio remains unchanged, angle of attack for maximum L/D is 6°, when $C_L = 0.380$ and $L/D = 16.3$, $C_D = 0.023$

$C_L = L/\frac{1}{2}\rho AV^2$ and, for same lift, $A \alpha 1/C_L$

and new area = $0.72 \times 0.640/0.380 = 1.21\,\text{m}^2$
new $\bar{c} = (A/AR_G)^{1/2} = (1.21/2.0)^{1/2} = 0.777\,\text{m}$
new $S = 1.21/0.777 = 1.56\,\text{m}$

$$\text{For original area, at } 10°,\ C_D = 0.048, \text{ and } D = C_D \times \tfrac{1}{2}\rho AV^2 = 0.048 \times 1{,}025/2 \times 0.72 \times 3.08^2 = 168.0\ \text{N}$$

$$\text{For new area, at } 6°,\ C_D = 0.023, \text{ and } D = C_D \times \tfrac{1}{2}\rho AV^2 = 0.023 \times 1{,}025/2 \times 1.21 \times 3.08^2 = 135.3\ \text{N}$$

i.e., a saving in drag of 32.7 N, approximately 19%.

The minimum drag coefficient (as $\alpha \to 0$) for both rudders is approximately the same at 0.010 and the increase in drag for the new design will be proportional to the increase in area.

$$\text{For the original design, drag } D = 0.01 \times 1{,}025/2 \times 0.72 \times 3.08^2 = 35.0\ \text{N}$$

$$\text{For the new design, drag } D = 0.01 \times 1{,}025/2 \times 1.21 \times 3.08^2 = 58.8\ \text{N}$$

an increase of 23.8 N, i.e., approximately 68%.

These simple calculations for a particular case indicate that an increase in rudder area of about 68% with a subsequent decrease in operating angle of attack has led to a decrease in drag of 32.7 N (about 19%) whilst producing the same sideforce. This saving in drag of 32.7 N for the production of a chosen sideforce is made at the cost of an increase in drag of 23.8 N at zero angle of attack. It can be deduced from the data that upward of about 4.5° the larger rudder incurs less drag than the smaller rudder for the production of equal sideforce. This result may be of more significance when related to the rudders on large performance yachts.

11.3.6 Example application 15. Semi-displacement craft: spade rudder downstream of propeller

Forces and stock size are required for a spade rudder acting in the slipstream of a relatively low-revolution propeller on a semi-displacement craft of length 14.0 m, $C_B = 0.55$ and maximum speed 18 knots (9.26 m/s). The hull is assumed to be moderately shaped in way of the rudder and the gap between the rudder and hull is 10 mm.

A geometric aspect ratio $AR_G = 1.5$ and taper ratio $TR = 0.80$ are assumed as satisfactory values for this vessel type.

From Figure 5.109, required area $A = 0.34\,\text{m}^2$

Hence $\bar{c} = (A/AR_G)^{1/2} = (0.34/1.5)^{1/2} = 0.476\,\text{m}$

And Span $S = 0.34/0.476 = 0.714\,\text{m}$

From Section 3.7, an approximate wake fraction $w_T = 0.18$

As the propeller characteristics are not known, assume the correction due to the propeller slipstream as $V = 1.15\ V_A$ (using Equation (3.39).

Hence the propeller inflow speed, corrected for wake and propeller effects is:

$$\begin{aligned} V &= 9.26 \times (1 - 0.18) \times 1.15 \\ &= 8.73\ \text{m/s} \end{aligned}$$

$$\begin{aligned} Re = V\bar{c}/\nu &= 8.73 \times 0.476 \times 10^6/1.19 \\ &= 3.5 \times 10^6 \end{aligned}$$

Assume in the first instance that $t/c = 0.15$ (i.e., maximum thickness = 71.4 mm) and $AR_E = 2 \times AR_G = 2 \times 1.5 = 3.0$

Hence from Figure 5.17, for $t/c = 0.15$, $\alpha_{stall} = 26°$

From Figure 5.18, $C_{Lmax} = 1.32$

Figure 5.14(b), $C_D = 0.290$

Gap $G = 10$ mm, and $G/c = 10/476 = 0.021$

Hence, from Section 5.5.2, there is no significant influence on C_{Lmax} due to gap.

From Figure 5.102, due to gap, increase in $C_D = 0.002$

From Figure 5.19, $C_{D0} = 0.010$ for no gap, hence corrected C_{D0} with gap = 0.012.

From Figure 5.102, induced drag factor $k_i = 0.45$.

From Equation (5.10b), $C_D = C_{D0} + k_i\ C_L^2/AR_E$

Hence drag from Figure 5.14(b) is corrected as follows:

Corrected $C_D = 0.012 + (0.290 - 0.010) \times 0.45/0.37$

Due to hull shape (Section 5.5.3) there is a 10% reduction in C_{Lmax}, 15% increase in k_i and C_{D0} is unchanged.

Hence corrected $C_{Lmax} = 1.32 \times 0.9 = 1.188$

And corrected drag $= 0.012 + (0.290 - 0.010) \times (0.45/0.37) \times 1.15 \times 1.188^2/1.32^2$

Hence $C_D = 0.329$

$$\begin{aligned} \text{Hence at stall, } C_N &= C_L \cos\alpha + C_D \sin\alpha \\ &= 1.188 \cos 26 + 0.329 \sin 26 \\ &= 1.21 \end{aligned}$$

$$\begin{aligned} \text{and normal force } N &= C_N \times \tfrac{1}{2}\rho A V^2 \\ &= 1.21 \times 1{,}025/2 \times 0.34 \times 8.73^2 \\ &= 16068.9\ \text{N} \end{aligned}$$

$$\begin{aligned} \text{Resultant force } R &= (C_L^2 + C_D^2)^{1/2} \times \tfrac{1}{2}\rho A V^2 \\ &= (1.188^2 + 0.329^2)^{1/2} \times 1{,}025/2 \times 0.34 \times 8.73^2 \\ &= 16370.6\ \text{N} \end{aligned}$$

From Figure 5.15, forward most position of $CPc = 19\%$ aft of LE. Hence locate stock at say 14% aft of LE.

CPc at stall = 23.5% aft LE

Hence lever $\bar{x} = (23.5 - 14.0)/100 \times 0.476 = 0.045$ m.

$$\begin{aligned}\text{From Figure 5.16, } CPs \text{ at stall} &= 49.2\% \text{ S from root}\\ &= 0.492 \times 0.714\\ &= 0.351 \text{ m.}\end{aligned}$$

plus say 20 mm from top of rudder to bearing,
giving a total bending moment lever = 0.351 + 0.02 = 0.371 m

$$\begin{aligned}\text{hence at stall, torque } Q = N \times \bar{x} &= 16068.9 \times 0.045\\ &= 723.1 \text{ Nm}\end{aligned}$$

$$\begin{aligned}\text{Bending moment } M = R \times CPs \times S &= 16370.6 \times 0.371\\ &= 6073.5 \text{ Nm}\end{aligned}$$

$$\begin{aligned}\text{Equivalent bending moment } BM_E &= 6073.5/2 + (6073.5^2 + 723.1^2)^{1/2}/2\\ &= 6094.9 \text{ Nm}\end{aligned}$$

$$\begin{aligned}\text{and stock diameter } D &= (6094.9 \times 32/\pi \times 130 \times 1{,}000^2)^{1/3}\\ &= 78.2 \text{ mm}\end{aligned}$$

(assuming a design stress of 130 N/mm^2 for mild steel, i.e., about half yield stress) Allowing say 6 mm for plating each side of stock, overall rudder thickness would amount to 90.2 mm, leading to a thickness ratio $t/c = 0.19$. This would lead to a decrease in C_{Lmax} to a new value of approximately 1.24, Figure 5.18.

Further iterations could be carried out, leading to a small decrease in stock size.

It can be noted that if the above calculation is repeated with the speed corrections included, but neglecting the influences of gap and hull shape, the stock diameter derived is 80.6 mm, leading to an effective design stress of about 119 N/mm^2. This illustrates that whilst corrections for gap and hull shape give useful indications of their influence on lift curve slope and rudder response over the working range, their inclusion is not critical for the derivation of stock size.

11.4 Low speed and manoeuvring

11.4.1 Example application 16. Rudder forces: Low speed and bollard condition

Consider a tug with the following particulars:

$L = 50$ m, $B = 12$ m, $T = 4$ m

Propeller diameter = 2.5 m, propeller revs = 180 rpm = 3 rps.

Rudder span = 3.0 m, chord = 2.0 m, area = 6 m^2

For $V = 0$, $J = 0$

From Figure 5.80, at 45° (stall):

$C_L^* = 1.0$, $C_D^* = 0.7$

Assuming propeller $P/D = 0.95$, similar to test data, then at $J = 0$, $K_T = 0.37$

$$\begin{aligned}\text{Propeller thrust } T &= K_T \times \rho \times n^2 \times D^4 \\ &= 0.37 \times 1{,}025 \times 3^2 \times 2.5^4/1{,}000 = 133.3\,\text{kN}\end{aligned}$$

$$\begin{aligned}\text{Rudder lift } L &= C_L^* \times \tfrac{1}{2}\rho A K_T n^2 D^2 \\ &= 1.0 \times \tfrac{1}{2} \times 1{,}025 \times 6.0 \times 0.37 \times 3^2 \times 2.5^2/1{,}000 = 64.0\,\text{kN}\end{aligned}$$

Rudder drag $d = 0.7 \times \frac{1}{2} \times 1{,}025 \times 6.0 \times 0.37 \times 3^2 \times 2.5^2/1{,}000 = 44.8\,\text{kN}$

It is noted that the lift force (sideforce) is almost half of the propeller thrust.

At say 1.28 m/s (2.5 knots), $J = V/nD = 1.28/3 \times 2.5 = 0.17$

From propeller chart at $P/D = 0.95$, $K_T = 0.33$

From Figure 5.79, stall angle is about 42° and C_L^* and C_D^* remain approximately the same.

Assuming the same revolutions, $L = 57.1\,\text{kN}$, $d = 40.0\,\text{kN}$ and $T = 118.9\,\text{kN}$.

These calculations can be repeated for different rudder angles and a polar plot produced for the net manoeuvring force, as shown schematically in Figure 8.4.

11.5 Control

11.5.1 Example application 17. Fin stabiliser design

The calculations are based on the outline design procedure described in Section 9.1.

Particulars of ship:

L: 100 m × B: 17 m × T: 6.2 m

Displacement (mass) Δ: 5,960 tonnes

Speed V: 20 knots

GM: 0.57 m

R: 9.8 m

Stabilisers situated 50 m from fore end.

Assume a design wave slope $\phi = 4°$.

Assume a rectangular stabiliser fin and, as a starting point, the size of each fin to be of span: 2.2 m and chord: 1.1 m, being typical for this size of ship. This leads to a geometric aspect ratio of 2.2/1.1 = 2.0.

Assume a relatively small gap between the root of the fin and the hull in the zero incidence position, leading to an effective aspect ratio $AR = 4.0$.

Reynolds number and boundary layer thickness at the stabiliser location:

$$\begin{aligned}Re &= VL/\nu \\ &= 20 \times 0.5144 \times 50/1.19 \times 10^{-6} = 4.32 \times 10^{8}\end{aligned}$$

(using $\nu = 1.19 \times 10^{-6}$ from Table 3.1)

Using Equation (3.17), boundary layer thickness $\delta = 50 \times 0.370(4.32 \times 10^8)^{-1/5}$
$= 347\,\text{mm}$
and using Equation (3.20), displacement thickness $\delta^* = 50 \times 0.0463(4.32 \times 10^8)^{-1/5}$
$= 43\,\text{mm}$

A velocity correction can be made across the inboard 347 mm of the span based on a power-law turbulent velocity distribution, Figure 3.11, and using Equation (9.2). An alternative approximate approach is to use the displacement thickness to correct the span, area and aspect ratio. That is, an effective span of (2,200 − 43) = 2,157 mm, an effective area of $2.157 \times 1.1\,\text{m}^2$ and a geometric aspect ratio 2,157/1,100 = 1.96. These corrected values are applied to the data base and coefficients in the following calculations.

Using Equation (9.1), and noting the use of radians and Newtons for consistent units, the required lift force L_F per fin is

$$\begin{aligned} L_F &= \phi \Delta g GM/2R \\ &= (4/57.3)\ (5{,}960 \times 9.81)\ 0.57/2 \times 9.8 \\ &= 118.7\,\text{kN} \end{aligned}$$

Fin lift $L_F = C_L \times \frac{1}{2} \rho A V^2$

A value of $C_L = 0.922$ is required to match the required lift of 118.7 kN.

(Check $L_F = 0.922 \times \frac{1}{2} \times 1{,}025\ (2.157 \times 1.1)\ (20 \times 0.5144)^2 = 118.7\,\text{kN}$)

The effect of the adjacent shaped hull on fin efficiency is described in Section 9.1. This particularly refers to the data for rudders. In the case of stabiliser fins mounted higher and at say 15° to the horizontal, Figure 9.2, it is seen that the effect of hull shape on fin root gap and fin efficiency is relatively small and, for the purposes of this example calculation, is assumed to be offset by the effects of the velocity changes around the hull.

Hence the required $C_L = 0.922$ using a fin with an effective aspect ratio $AR = 1.96 \times 2 = 3.92$. This may be achieved by an all-movable foil or high-lift foils such as with a flap, fishtail, or wedge at the aft end, Figure 5.22. The all-movable, flapped and wedge cases are considered, Figure 9.6.

(a) *All-movable foil:* From Figure 5.14(a) and/or Equation (5.6), $C_L = 0.922$ would be achieved with an incidence of 15°, and with an approximate $C_D = 0.10$.
(b) *Flapped foil:* Using a 20% flap/chord ratio, then from Figure 5.29 and using Equation (5.17) to correct from $AR = 2.8$ to 3.92, $C_L = 0.922$ would be achieved with an incidence of 9.8° (using a flap incidence to rudder incidence ratio = 2.0). At this incidence, approximate $C_D = 0.14$.
(c) *Wedge:* Using a 10% wedge (Table 5.5) and using Equation (5.17) to correct from $AR = 1.0$ to 3.92, $C_L = 0.922$ would be achieved with an incidence of 12.3°, with an approximate $C_D = 0.17$.

Cavitation check: As described in Section 9.1, the cavitation check can be carried out using the principles applied to rudders (Section 5.8).

Assume the depth of immersion $h = 4.7\,\text{m}$ in the worst (minimum immersion) case, including a reduction due to 4° roll angle.

$$\begin{aligned}\text{Cavitation Number } \sigma &= (P_{AT} + \rho g h - P_V)/\tfrac{1}{2}\rho V^2 \\ &= 101 \times 10^3 + 1{,}025 \times 9.81 \times 4.7 - 3 \times 10^3/\tfrac{1}{2} \times 1{,}025\,(20 \times 0.5144)^2 \\ &= 2.68\end{aligned}$$

Using the cavitation inception envelopes for symmetrical all-movable sections, Figure 5.139, it is seen that with a $C_L = 0.922$ at a cavitation number $\sigma = 2.68$, some back-sheet cavitation would be expected. Figure 5.139 would suggest that, at this cavitation number, the lift coefficient should not exceed about 0.80. This could be achieved with a 15% increase in fin area, suggesting fin dimensions of 2.36 m × 1.18 m. The flapped foil can have a slightly more uniform chordwise pressure distribution and cavitation might not occur in this particular situation. Alternative section shapes that might delay the onset of cavitation (Section 5.8) may also be considered.

Centre of pressure, torque and stock diameter: Based on the reasoning in Section 9.1, the fin stock can be placed a little forward of the estimated centre of pressure, thus ensuring that negative torques do not occur. For example, Figure 5.15 would indicate that for an all-movable foil with an aspect ratio of 4.0, the *CPc* travels from about 20% chord aft of leading edge at low angle of attack to about 22%c at stall. Hence if the stock is situated at say 17%c aft of LE of the 1.10 m chord, then the torque lever $\bar{x}$ in Figure 5.1 is

$$\bar{x} = (0.20c - 0.17c) = 0.03 \times 1.10 = 0.033\,\text{m}$$

and at large angles = $(0.22c - 0.17c) = 0.05 \times 1.10 = 0.055\,\text{m}$.
For the all-movable fin with $C_D = 0.1$,

$$\begin{aligned}\text{Drag } d &= C_D \times \tfrac{1}{2}\rho A V^2 \\ &= 0.1 \times \tfrac{1}{2} \times 1{,}025(2.157 \times 1.1)(20 \times 0.5144)^2 = 12.9\,\text{kN}\end{aligned}$$

and using Equation (7.3),

$$\begin{aligned}\text{Normal force } N &= L\cos\alpha + d\sin\alpha \\ &= 118.7\cos 15 + 12.9\sin 15 = 118.0\,\text{kN}\end{aligned}$$

Hence torque on the fin $Q = N \times \bar{x} = 118.0 \times 0.055 = 6.5\,\text{kNm}$

From Figure 5.16, $CPs = 0.49$ $S = 0.49 \times 2.2 = 1.08\,\text{m}$ and bending moment $BM = 118.0 \times 1.08 = 127.44\,\text{kNm}$.

Using Equation (7.9),

$$\begin{aligned}\text{Equivalent bending moment } BM_E &= 127.44/2 + 0.5\,(127.44^2 + 6.5^2)^{1/2} \\ &= 127.52\,\text{kNm}\end{aligned}$$

and using Equation (7.11),

$$\begin{aligned}\text{stock diameter } D &= (127.52 \times 1{,}000 \times 32/\pi \times 235 \times 1{,}000 \times 1{,}000)^{1/3} \times 1{,}000 \\ &= 177\,\text{mm}.\end{aligned}$$

The flapped and trailing edge wedge fins have larger drag coefficients of 0.14 and 0.17 for $C_L = 0.922$. Using the same reasoning and form of calculation leads to a stock diameter of 178 mm for these cases.

The above calculations assume that the incidence and forces will be limited to these values. Whilst this may be possible using the fin controller, a calculation providing a higher degree of reliability in operation would use the maximum possible forces, as follows for the all-movable case.

For the all-movable fin, and using Figures 5.17, 5.18, 5.14(b), stall angle $\alpha_{stall} = 22°$, $C_{Lmax} = 1.2$ for a $Re = 9.5 \times 10^6$ based on fin chord, and $C_D = 0.19$. Following the same form as the above calculations this leads to a torque of 8.4 kNm and a fin stock diameter of 188 mm. The flapped fin and T.E. wedge fin will each have a C_{Lmax} of about 1.4 and, based on the above reasoning, would lead to a stock diameter of 198 mm.

If a thickness/chord ratio $t/c = 0.25$ is assumed, then the maximum thickness would be $1.1 \times 1{,}000 \times 0.25 = 275$ mm, which would accommodate these stock diameters.

A further factor of safety might be incorporated by using a lower design stress than the 235 N/mm^2 assumed for steel.

11.5.2 Example application 18. Heel angle due to rudder

The calculations are based on Equations (4.2) and (4.4) derived in Chapter 4, and the notation shown in Figure 4.3.
Assume the same ship as example 17.
Particulars of ship:
L: 100 m $\times$ B: 17 m $\times$ T: 6.2 m
Displacement (mass) Δ: 5,960 tonnes
Speed V: 20 knots
Wake fraction = 0.250
$KG = 7.4$ m, $GM = 0.57$ m
Rudder area A is assumed to be $= 0.02 \times L \times T = 0.02 \times 100 \times 6.2 = 12.4\,\text{m}^2$
Rudder span say = 4.5 m
Then chord = 12.4/4.5 = 2.76 m
$AR_G = 1.63$, and assume $AR_E = 3.26$
In service speed condition, $K_T/J^2 = 0.5$
From lift/drag data for $K_T/J^2 = 0.5$, interpolating in Figure 5.58, for $AR_E = 3.0$, at 25° rudder incidence, $C_L = 1.79$.
Using Equation 5.17 to correct to $AR_E = 3.26$,

$$C_L = 1.79\,[(1 + 3/3.0)/(1 + 3/3.26)] = 1.86$$

Lift (sideforce on ship body axis) $F_{R2} = C_L \times \frac{1}{2}\rho A V^2$

$$= 1.86 \times 0.5 \times 1{,}025 \times 12.4\,[20 \times 0.5144\,(1 - 0.25)]^2$$
$$= 703.7\,\text{kN}$$

Assume $CPs = 0.48 \times S$, Figure 5.58, $= 0.48 \times 4.5 = 2.16$ m

Bottom of rudder is 100 mm above base,
and F_{R2} acts at approximately 4.5 − 2.16 + 0.10 = 2.44 m above base.
CLR is assumed to be 4.0 m above base.
Then, in Figure 4.3:

$$y_1 = 7.4 - 2.44 = 4.96\,\text{m},$$
$$y_2 = 7.4 - 4.0 = 3.4\,\text{m}$$
$$y_3 = 4.0 - 2.44 = 1.56\,\text{m}$$

Assume diameter of turning circle to be six ship lengths = 6 × 100 = 600 m and radius of turning circle $R = 300$ m

Initial angle of heel (inward): Using Equation (4.2), and noting the use of Newtons for consistent units,

$$\begin{aligned}\text{Heel angle } \phi &= \sin^{-1}[F_{R2} \times y_1/\Delta g GM] \\ &= \sin^{-1}[703.7 \times 1{,}000 \times 4.96/5{,}960 \times 1{,}000 \times 9.81 \times 0.57] \\ &= 6.0^\circ\end{aligned}$$

This heel angle may not be achieved in practice with 25° helm as the *effective* rudder incidence will decrease as drift develops.

On the turn, steady angle of heel (outward): Using Equation (4.4),

$$\begin{aligned}\text{Heel angle } \phi &= \sin^{-1}\{[(\Delta V^2/R) \times y_2 - F_{R2} \times y_3]/\Delta g GM\} \\ &= \sin^{-1}\{[5{,}960 \times 1{,}000\,[20 \times 0.5144\,(1 - 0.25)]^2/300) \times 3.4 \\ &\quad - 703.7 \times 1{,}000 \times 1.56]/5{,}960 \times 1{,}000 \times 9.81 \times 0.57\} \\ &= 5.0^\circ\end{aligned}$$

This is the heel angle developed on a steady turn with an *effective* rudder incidence of 25° and it is not unusual for this level of heel to occur in practice.

In both of the above cases, the importance of *GM* can be noted. For example, a doubling of *GM* would lead to approximately half the values of the predicted heel angles. As noted in Section 4.1.4, these predicted heel angles are based on a number of approximations and should therefore be treated only as approximate estimates.

It is interesting to note the value of the rudder heeling moment ($= F_{R2} \times y_1 =$ 703.2 × 4.96 = 3,490 kNm) and to compare this with the heeling/righting moment of the fin stabiliser (example 17), ($= 2 \times L_F \times R =$ 2,327 kNm). This sizeable rudder heeling moment does indicate the reasons why the rudder has been considered for use as a roll stabiliser (Section 9.1.7).

References

11.1 Molland, A.F. and Turnock, S.R. A compact computational method for predicting forces on a rudder in a propeller slipstream. *Transactions of the Royal Institution of Naval Architects*, Vol. 138, 1996, pp. 227–244.

11.2 Harding, R. *The performance of twisted and cambered rudders.* Part III Individual Project Report, School of Engineering Sciences, Ship Science, University of Southampton, 2000.

11.3 Turnock, S.R. and Wright, A.M. Directly coupled fluid structure model of a ship rudder behind a propeller. *Marine Structures* Vol. 13, 2000, pp. 53–72.

11.4 Avellino, G. and Coan, A. One or two rudders? A dilemma in fast twin screw vessel design. Proceedings of *Seventh International Conference on Fast Sea Transportation, FAST2003*, Ischia, Italy, October, 2003.

Appendix 1

Tabulated test data

Section 5.4 describes the extensive investigations into rudder–propeller interaction carried out in a wind tunnel at the University of Southampton, UK. The rudder–propeller test rig is shown in Figure 5.52 and the coordinate system in Figure 5.54. Alternative configurations for the tests are shown in Figure 5.56. A NACA 0020 section is used for all the models and normalised offsets for the NACA 00 sections are given in Table 5.4.

An extensive database of the test results for various parametric changes has been developed. Most of the data are available in the form of Ship Reports, ISSN 0140 3818, published by the University of Southampton [5.37,5.38,5.75–5.80], and published papers [5.23,5.68–5.74]. A number of the results are used in Figures 5.58–5.74 to illustrate the influences on rudder performance of parametric changes in rudder–propeller arrangements.

Most of the wind tunnel test data have been brought together in an electronic database. This database is available through the publisher's web site.

Table A1.1 lists the various test cases for which the force and moment data are available.

Table A1.2 lists the tests for which rudder and hull pressure data are available.

Tables A1.3–A1.14 show examples of the tabulated test data, for the all-movable spade Rudder No. 2 and the semi-balanced skeg Rudder No. 0.

Table A1.1 Force and moment data
Propeller data:
Propeller open water data: modified Wageningen B4.40
Rudder data: as follows

Rudder No. 0	*Skeg rudder*	*Skeg angle $\beta = 0$*	*Movable rudder plus skeg*	
Z/D	*Y/D*	*X/D*	*V (m/s)*	*rpm*
Free	stream	data	10.0	0
0.75	−0.25	0.39	10.0	2,100
0.75	−0.25	0.39	10.0	1,460
0.75	−0.25	0.39	10.0	800

(Continued)

Table A1.1 Continued

Rudder No. 0	*Skeg rudder*	*Skeg angle* $\beta = 0$	*Movable rudder plus skeg*	
Z/D	*Y/D*	*X/D*	*V (m/s)*	*rpm*
0.75	−0.125	0.39	10.0	2,100
0.75	−0.125	0.39	10.0	1,460
0.75	0.00	0.39	3.3	1,460
0.75	0.00	0.39	10.0	2,100
0.75	0.00	0.39	10.0	1,460
0.75	0.00	0.39	10.0	800
0.75	0.125	0.39	10.0	2,100
0.75	0.125	0.39	10.0	1,460
0.75	0.25	0.39	10.0	2,100
0.75	0.25	0.39	10.0	1,460
0.75	0.375	0.39	0.0	1,460
0.75	0.000	0.46	0.0	1,460
0.75	0.000	0.52	0.0	1,460
0.75	−0.375	0.39	0.0	1,460
0.75	−0.25	0.39	0.0	1,460
0.75	−0.125	0.39	0.0	1,460
0.75	0.00	0.39	0.0	1,460
0.75	0.125	0.39	0.0	1,460
0.75	0.25	0.39	0.0	1,460
Rudder No. 0	*Skeg rudder*	*Skeg angle* $\beta = 0$	*Movable rudder alone*	
Z/D	*Y/D*	*X/D*	*V (m/s)*	*rpm*
Free	stream	data	10.0	0
0.75	0.00	0.39	3.3	1,460
0.75	0.00	0.39	10.0	2,100
0.75	0.00	0.39	10.0	1,460
0.75	0.00	0.39	10.0	800
0.75	0.25	0.39	10.0	2,100
0.75	0.25	0.39	10.0	1,460
Rudder No. 1				
0.75	0.00	0.34	10.0	2,750
0.75	0.00	0.34	10.0	2,130
0.75	0.00	0.34	10.0	1,420
0.75	0.00	0.34	10.0	779
Rudder No. 2				
Free	stream	data	10.0	0
Free	stream	data	25.0	0
Free	stream	data	32.0	0

Table A1.1 Continued

Z/D	Y/D	X/D	V (m/s)	rpm
0.75	0.00	0.30	10.0	2,078
0.75	0.00	0.30	10.0	1,433
0.75	0.00	0.30	10.0	792
0.75	−0.25	0.39	10.0	2,158
0.75	−0.25	0.39	10.0	1,479
0.75	−0.25	0.39	10.0	781
0.75	0.00	0.39	10.0	2,103
0.75	0.00	0.39	10.0	1,460
0.75	0.00	0.39	10.0	792
0.75	0.25	0.39	10.0	2,159
0.75	0.25	0.39	10.0	1,489
0.75	0.25	0.39	10.0	782
0.75	0.00	0.52	10.0	2,079
0.75	0.00	0.52	10.0	1,433
0.75	0.00	0.52	10.0	791
1.125	0.00	0.39	10.0	2,165
1.125	0.00	0.39	10.0	1,460
1.125	0.00	0.39	10.0	784
0.75	0.00	0.30	0.0	1,460
0.75	0.00	0.36	0.0	1,460
0.75	−0.375	0.39	0.0	1,460
0.75	−0.25	0.39	0.0	1,460
0.75	−0.125	0.39	0.0	1,460
0.75	0.00	0.39	0.0	1,460
0.75	0.125	0.39	0.0	1,460
0.75	0.25	0.39	0.0	1,460
0.75	0.375	0.39	0.0	1,460
0.75	0.00	0.46	0.0	1,460
0.75	−0.25	0.52	0.0	1,460
0.75	−0.125	0.52	0.0	1,460
0.75	0.00	0.52	0.0	1,460
0.75	0.125	0.52	0.0	1,460
0.75	0.25	0.52	0.0	1,460
Rudder No. 2, propeller $P/D = 0.69$				
0.75	0.00	0.39	10.0	1,790
0.75	0.00	0.39	10.0	1,719
0.75	0.00	0.39	10.0	1,072
Rudder No. 2, propeller $P/D = 1.34$				
0.75	0.00	0.39	10.0	2,165
0.75	0.00	0.39	10.0	1,220
0.75	0.00	0.39	10.0	574

(Continued)

Table A1.1 Continued

Z/D	*Y/D*	*X/D*	*V (m/s)*	*rpm*
Rudder No. 3				
Free	stream	data	10.0	0
Free	stream	data	25.0	0
0.75	0.00	0.30	10.0	790
0.75	0.00	0.39	10.0	2,079
0.75	0.00	0.39	10.0	1,434
0.75	0.00	0.39	10.0	791
0.75	0.00	0.52	10.0	2,076
0.75	0.00	0.52	10.0	1,432
0.75	0.00	0.52	10.0	792
Rudder No. 4				
Free	stream	data	10.0	0
1.125	0.00	0.39	10.0	779
1.125	0.00	0.39	10.0	1,460
1.125	0.00	0.39	10.0	2,163
Rudder No. 5				
0.75	0.00	0.39	10.0	785
0.75	0.00	0.39	10.0	1,460
0.75	0.00	0.39	10.0	2,160
Rudder No. 6				
0.75	0.00	0.39	10.0	784
0.75	0.00	0.39	10.0	1,462
0.75	0.00	0.39	10.0	2,163
Rudder No. 2, four-quadrant data				
0.75	0.00	0.39	0.0	750
0.75	0.00	0.39	0.0	1,400
0.75	0.00	0.39	10.0	2,780
0.75	0.00	0.39	10.0	2,100
0.75	0.00	0.39	10.0	1,460
0.75	0.00	0.39	10.0	805
0.75	0.00	0.39	10.0	735
0.75	0.00	0.39	10.0	400
0.75	0.00	0.39	10.0	200
0.75	0.00	0.39	10.0	0
0.75	0.00	0.39	10.0	0
0.75	0.00	0.39	10.0	−200
0.75	0.00	0.39	10.0	−400
0.75	0.00	0.39	10.0	−800
0.75	0.00	0.39	10.0	−1,453
0.75	0.00	0.39	−10.0	−2,105
0.75	0.00	0.39	−10.0	−1,460

Table A1.1 Continued

Z/D	*Y/D*	*X/D*	*V (m/s)*	*rpm*
0.75	0.00	0.39	−10.0	−800
0.75	0.00	0.39	−10.0	−400
0.75	0.00	0.39	−10.0	0
0.75	0.00	0.39	−10.0	0
0.75	0.00	0.39	−10.0	400
0.75	0.00	0.39	−10.0	800
0.75	0.00	0.39	−10.0	1,460
0.75	0.00	0.39	0.0	750
0.75	0.00	0.39	0.0	1,400

Table A1.2 Pressure data

Free stream rudder data
Rudder Nos. 0, 2 and 4
Pressures for the 25 chordwise tappings at each span position

Test cases
Rudder angles $\delta = 0°, 10°, 20°, 25°$ and $30°$

Rudder plus propeller data
Rudder No. 2
Pressures for the 25 chordwise tappings at each span position

Test cases
$J = 0.51$, $K_T/J^2 = 0.88$
Drift angles $\beta = -15°, -7.5, +7.5°, +15°$
Rudder angles $\delta = -30°, -20°, -10°, 0°, 10°, 20°, 30°$

Hull pressure data
Rudder No. 0
Dimensions/offsets for the hull, and sitings of the hull pressure tappings are given in the database

Test cases
$\beta = 0$; free stream
Rudder angles $\delta = -30°, -20°, -10°, 0°, 10°, 20°, 30°$

$\beta = 0$; $J = 0.35, 0.51, 0.94$; $X/D = 0.39$
Rudder angles $\delta = -30°, -20°, -10°, 0°, 10°, 20°, 30°$

$\beta = 0$; $J = 0.17, 0.35, 0.51, 0.94$; $X/D = 0.64$
Rudder angles $\delta = -30°, -20°, -10°, 0°, 10°, 20°, 30°$

$\beta = -7.5°$; $J = 0.35, 0.51, 0.94$; $X/D = 0.0.39$
Rudder angles $\delta = -30°, -20°, -10°, 0°, 10°, 20°, 30°$

Note: A NACA 0020 section is used for all the models and normalised offsets for the NACA 00 sections are given in Table 5.4. Sitings of pressure tappings are given in Tables 5.11(a) and (b).

Table A1.3–A1.12 Force and moment data, Rudder No. 2

Table A1.3

Free stream		*Rudder No. 2*			*10 m/s*		
Angle	C_L	C_D	*CPc*	*CPs*	*Cmz*	*Cmx*	*Cmy*
−40	−0.561	0.559	43.3	41.2	−0.07	−0.387	0.259
−35	−0.527	0.459	41	43.4	−0.051	−0.363	0.219
−30	−0.493	0.382	39.7	44.8	−0.04	−0.334	0.191
−25	−0.85	0.239	26.5	50.5	0.02	−0.596	0.123
−20	−0.926	0.167	21.1	48	0.055	−0.616	0.084
−15	−0.751	0.099	18.2	47	0.059	−0.488	0.051
−10	−0.498	0.056	15.9	45.5	0.047	−0.315	0.028
−5	−0.237	0.031	13	45.2	0.027	−0.149	0.015
0	0	0.018	–	–	–	–	–
5	0.237	0.031	13	45.2	−0.027	0.149	0.015
10	0.498	0.056	15.9	45.5	−0.047	0.315	0.028
15	0.751	0.099	18.2	47	−0.059	0.488	0.051
20	0.926	0.167	21.1	48	−0.055	0.616	0.084
25	0.85	0.239	26.5	50.5	−0.02	0.596	0.123
30	0.493	0.382	39.7	44.8	0.04	0.334	0.191
35	0.527	0.459	41	43.4	0.051	0.363	0.219
40	0.561	0.559	43.3	41.2	0.07	0.387	0.259

Table A1.4

800 rpm		*Rudder No. 2*			*10 m/s*		
Angle	C_L	C_D	*CPc*	*CPs*	*Cmz*	*Cmx*	*Cmy*
−30.8	−1.245	0.481	35.726	49.5091	−0.076	−0.85	0.296
−25.8	−1.484	0.329	27.982	49.1283	0.03	−0.988	0.221
−20.8	−1.299	0.208	23.43	46.7435	0.084	−0.831	0.143
−15.8	−0.973	0.119	21.525	45.6152	0.082	−0.61	0.089
−10.8	−0.657	0.066	20.554	43.938	0.062	−0.401	0.055
−5.8	−0.349	0.042	19.867	40.4896	0.036	−0.201	0.037
−3.8	−0.238	0.032	21.612	37.1357	0.02	−0.129	0.032
−2.8	−0.175	0.033	–	–	0.011	−0.09	0.03
−1.8	−0.12	0.034	–	–	–	–	–
−0.8	−0.065	0.033	–	–	–	–	–
0.2	−0.003	0.036	–	–	–	–	–
1.2	0.052	0.037	–	–	–	–	–
2.2	0.107	0.038	–	–	–	–	–
4.2	0.508	0.116	–	–	0.033	0.082	−0.008
9.2	0.514	0.087	15.799	52.0777	−0.074	0.359	0.052
14.2	0.818	0.146	19.015	50.6333	−0.091	0.561	0.083
19.2	1.12	0.228	21.466	50.7516	−0.097	0.772	0.134
24.2	1.333	0.357	24.789	52.4144	−0.071	0.951	0.208
29.2	1.482	0.501	29.798	52.7407	−0.003	1.069	0.303
34.2	1.568	0.675	33.868	52.9234	0.065	1.154	0.402

Table A1.5

1,460 rpm		*Rudder No. 2*			*10 m/s*		
Angle	C_L	C_D	*CPc*	*CPs*	*Cmz*	*Cmx*	*Cmy*
−30.8	−2.668	0.903	31.912	45.2373	−0.053	−1.644	0.617
−25.8	−2.535	0.552	23.149	47.2165	0.172	−1.622	0.395
−20.8	−2.062	0.348	20.766	46.065	0.189	−1.298	0.254
−15.8	−1.56	0.198	19.145	44.28	0.169	−0.955	0.154
−10.8	−1.068	0.101	18.789	40.8925	0.12	−0.619	0.084
−5.8	−0.579	0.05	19.665	34.2841	0.06	−0.297	0.051
−3.8	−0.39	0.048	20.307	27.9107	0.038	−0.176	0.043
−2.8	−0.288	0.044	–	–	0.025	−0.108	0.041
−1.8	−0.193	0.022	–	–	–	–	–
−0.8	−0.099	0.033667	–	–	–	–	–
0.2	−0.005	0.032	–	–	–	–	–
1.2	0.097	0.034	–	–	–	–	–
2.2	0.191	0.041	–	–	−0.057	0.221	0.043
4.2	0.379	0.051	–	–	−0.083	0.351	0.049
9.2	0.848	0.114	13.422	62.9583	−0.142	0.683	0.084
14.2	1.321	0.217	16.827	59.7085	−0.176	1.025	0.149
19.2	1.8	0.368	19.355	57.7826	−0.194	1.364	0.251
24.2	2.244	0.577	22.1	57.0307	−0.18	1.687	0.397
29.2	2.619	0.903	27.239	56.0493	−0.075	1.955	0.612

Table A1.6

2,100 rpm		*Rudder No. 2*			*10 m/s*		
Angle	C_L	C_D	*CPc*	*CPs*	*Cmz*	*Cmx*	*Cmy*
−30.8	−4.326	1.206	22.607	47.9057	0.32	−2.798	0.841
−25.8	−3.665	0.81	20.441	46.9988	0.349	−2.331	0.591
−20.8	−2.945	0.486	18.879	45.6327	0.325	−1.829	0.386
−15.8	−2.235	0.251	17.884	43.4158	0.268	−1.339	0.23
−10.8	−1.093	0.189	42.852	36.6013	−0.143	−0.612	−0.007
−5.8	−0.808	0.012	20.901	30.6188	0.073	−0.384	0.055
−3.8	−0.54	0	–	–	0.031	−0.2	0.038
−2.8	−0.406	0.002	–	–	–	–	–
−1.8	−0.265	−0.001	–	–	–	–	–
−0.8	−0.13967	−0.00933	–	–	–	–	–
1.2	0.143	−0.004	–	–	–	–	–
4.2	0.558	0.023	–	–	−0.171	0.607	0.046
9.2	1.22	0.113	7.803	71.9401	−0.271	1.09	0.107
14.2	1.879	0.269	9.987	66.4443	−0.377	1.581	0.213
19.2	2.526	0.483	12.825	64.1801	−0.437	2.074	0.363
24.2	3.208	0.792	15.779	62.4522	−0.462	2.589	0.579
29.2	3.853	1.177	18.989	61.565	−0.433	3.086	0.859

Table A1.7

Rudder plus skeg forces							
800 rpm		*Skeg-Rudder No. 0*			*10 m/s*		
Angle	C_L	C_D	*CPc*	*CPs*	*Cmz*	*Cmx*	*Cmy*
−35.2	−1.15	0.517	36.922	18.1247	−0.019	−0.339	0.284
−30.2	−1.089	0.395	35.149	18.1203	0.003	−0.341	0.222
−25.2	−0.9	0.294	32.038	17.8358	0.031	−0.284	0.176
−20.2	−0.752	0.216	30.55	17.9001	0.038	−0.243	0.139
−15.2	−0.669	0.15	27.305	19.185	0.055	−0.232	0.104
−10.2	−0.488	0.079	19.737	19.2989	0.077	−0.173	0.064
−5.2	−0.451	0.072	49.983	12.1086	−0.067	−0.146	−0.116
−3.2	−0.1725	0.0465	–	–	0.0555	−0.084	0.001
−2.2	−0.13	0.041	–	–	0.036	−0.042	0.008
−1.2	−0.083	0.035	–	–	–	–	–
−0.2	−0.0086	0.0354	–	–	–	–	–
0.8	0.029	0.03	–	–	–	–	–
1.8	0.062	0.036	–	–	–	–	–
2.8	0.129	0.039	–	–	0.005	0.071	0.031
4.8	0.214	0.044	–	–	0.01	0.078	0.05
9.8	0.462	0.076	33.659	20.9504	−0.008	0.171	0.068
14.8	0.644	0.15	39.369	21.6377	0.026	0.24	0.105
19.8	0.771	0.218	40.699	19.0913	0.043	0.264	0.131
24.8	0.671	–	–	–	0.492	0.032	0.226
29.8	0.99	0.364	40.452	20.6054	0.053	0.328	0.224
34.8	1.155	0.473	42.369	20.2706	0.085	0.361	0.287
39.8	1.241	0.563	43.624	19.0536	0.108	0.341	0.34

Table A1.8

Rudder plus skeg forces							
1,460 rpm		*Skeg-Rudder No. 0*			*10 m/s*		
Angle	C_L	C_D	*CPc*	*CPs*	*Cmz*	*Cmx*	*Cmy*
−35.2	−2.369	1.051	36.721	32.5593	−0.034	−1.098	0.65
−30.2	−2.319	0.814	33.779	33.1514	0.039	−1.114	0.517
−25.2	−2.133	0.598	29.855	33.4838	0.121	−1.045	0.395
−20.2	−1.763	0.391	25.933	33.1052	0.169	−0.866	0.269
−15.2	−1.331	0.241	21.721	30.8103	0.184	−0.628	0.17
−10.2	−0.868	0.139	16.81	28.3	0.163	−0.391	0.099
−5.2	−0.702	0.124	–	–	–	–	–
−2.2	−0.22	0.077	–	–	–	–	–
−1.2	−0.125	0.072	–	–	–	–	–
−0.2	0.009	0.065	–	–	–	–	–
1.8	0.164	0.069	–	–	0.075	0.146	0.039
4.8	0.427	0.079	47.18	49.1102	0.051	0.284	0.062
9.8	0.857	0.14	36.93	42.9985	0.013	0.516	0.1
14.8	1.302	0.249	35.639	41.5385	0.003	0.764	0.165
19.8	1.728	0.387	36.43	40.8383	0.018	0.999	0.25
24.8	2.037	0.602	40.118	41.7592	0.1	1.192	0.391
29.8	2.134	0.789	43.205	40.0651	0.175	1.197	0.508
34.8	2.252	0.971	44.637	41.1599	0.222	1.272	0.64

Table A1.9

Rudder plus skeg forces							
2,100 rpm		*Skeg-Rudder No. 0*			*10 m/s*		
Angle	C_L	C_D	*CPc*	*CPs*	*Cmz*	*Cmx*	*Cmy*
−35.2	−4.006	1.789	36.601	38.4294	−0.052	−2.137	1.147
−30.2	−3.718	1.271	30.69	38.9111	0.181	−2.032	0.829
−25.2	−3.4	0.948	26.746	39.0065	0.301	−1.874	0.635
−20.2	−2.661	0.555	22.027	37.2834	0.359	−1.43	0.38
−15.2	−1.993	0.319	18.446	34.74	0.34	−1.026	0.22
−10.2	−1.319	0.161	13.227	30.5941	0.294	−0.629	0.11
−5.2	−0.888	0.121	–	–	0.073	−0.317	−0.114
−2.2	−0.285	0.065	–	–	–	–	–
−0.2	−0.02467	0.066	–	–	–	–	–
1.8	0.234	0.078	–	–	0.137	0.264	0.032
4.8	0.632	0.103	–	–	0.094	0.508	0.065
9.8	1.256	0.2125	38.673	52.8355	0.042	0.885	0.139
14.8	1.892	0.377	34.97	50.6426	−0.008	1.292	0.249
19.8	2.522	0.598	34.91	49.4865	−0.012	1.692	0.392
24.8	3.122	0.903	36.303	49.3417	0.03	2.086	0.604
29.8	3.652	1.302	39.817	49.6353	0.169	2.445	0.887
34.8	3.933	1.739	44.032	49.5199	0.365	2.623	1.183

Table A1.10

Rudder alone forces							
800 rpm		*Skeg-Rudder No. 0*			*10 m/s*		
Angle	C_L	C_D	*CPc*	*CPs*	*Cmz*	*Cmx*	*Cmy*
−35.2	−0.853	0.472	49.213	14.8033	−0.134	−0.192	0.271
−30.2	−0.817	0.382	46.921	14.0861	−0.104	−0.2	0.221
−25.2	−0.638	0.27	43.404	13.8176	−0.055	−0.156	0.176
−20.2	−0.531	0.193	41.754	12.4719	−0.036	−0.128	0.141
−15.2	−0.496	0.115	38.645	14.784	−0.017	−0.142	0.104
−10.2	−0.325	0.05	32.465	12.9271	0.01	−0.09	0.066
−5.2	−0.093	0.006	–	–	0.042	−0.035	−0.046
−2.2	−0.095	0.018	–	–	−0.029	−0.008	−0.019
−0.2	0.017	0.008	–	–	–	–	–
1.8	0.086	0.007	–	–	0.005	0.047	0.025
4.8	0.173	0.007	27.987	29.3012	−0.013	0.077	0.044
9.8	0.347	0.039	31.168	24.4815	−0.015	0.138	0.064
14.8	0.528	0.1	36.461	24.2179	0.006	0.206	0.097
19.8	0.701	0.196	42.667	22.4377	0.053	0.261	0.13
24.8	0.77	0.261	43.079	20.6907	0.062	0.26	0.174
29.8	0.807	0.338	44.904	19.7441	0.083	0.248	0.218
34.8	0.892	0.421	47.143	19.971	0.114	0.256	0.27

Table A1.11

Rudder alone forces							
1,460 rpm		*Skeg-Rudder No. 0*			*10 m/s*		
Angle	C_L	C_D	*CPc*	*CPs*	*Cmz*	*Cmx*	*Cmy*
−35.2	−1.63	0.939	48.882	33.2839	−0.253	−0.728	0.618
−30.2	−1.622	0.747	45.706	34.2727	−0.183	−0.77	0.507
−25.2	−1.454	0.528	41.917	33.5232	−0.101	−0.689	0.381
−20.2	−1.158	0.341	38.132	32.9551	−0.033	−0.551	0.264
−15.2	−0.833	0.197	35.727	29.599	−0.003	−0.372	0.168
−10.2	−0.495	0.101	34.301	22.126	0.005	−0.184	0.105
−5.2	−0.232	0.072	–	–	−0.051	−0.033	−0.009
−2.2	0.006	0.032	–	–	–	–	–
−0.2	0.171	0.018	–	–	–	–	–
1.8	0.282	0.018	–	–	−0.026	0.236	0.036
4.8	0.488	0.026	–	–	−0.036	0.35	0.057
9.8	0.82	0.067	29.398	49.219	−0.049	0.539	0.089
14.8	1.157	0.161	32.207	47.0797	−0.037	0.735	0.15
19.8	1.524	0.303	35.966	45.5262	0.009	0.946	0.232
24.8	1.807	0.521	40.153	45.6857	0.089	1.123	0.37
29.8	1.98	0.717	43.591	44.5756	0.17	1.194	0.507
34.8	2.026	0.889	45.853	45.0339	0.227	1.213	0.634

Table A1.12

Rudder alone forces							
2,100 rpm		*Skeg-Rudder No. 0*			*10 m/s*		
Angle	C_L	C_D	*CPc*	*CPs*	*Cmz*	*Cmx*	*Cmy*
−35.2	−2.852	1.528	45.489	41.902	−0.325	−1.575	1.076
−30.2	−2.595	1.149	41.492	42.4508	−0.172	−1.481	0.817
−25.2	−2.197	0.774	37.375	41.5256	−0.046	−1.24	0.578
−20.2	−1.69	0.467	33.691	38.9305	0.03	−0.914	0.372
−15.2	−1.155	0.262	32.047	34.8653	0.039	−0.581	0.225
−10.2	−0.64	0.135	32.821	25.1979	0.017	−0.26	0.131
−5.2	−0.202	0.073	–	–	−0.062	0.023	0.008
−2.2	0.148	0.024	–	–	−0.057	0.226	0.006
−0.2	0.28	0.00725	–	–	–	–	–
1.8	0.567	0.019	–	–	−0.072	0.497	0.029
4.8	0.553	−0.319	–	–	0.662	0.538	−0.206
9.8	1.379	0.116	26.459	57.3398	−0.123	1.026	0.12
14.8	1.911	0.261	28.363	54.5969	−0.134	1.367	0.227
19.8	2.432	0.472	30.808	53.7941	−0.112	1.724	0.365
24.8	2.956	0.754	33.681	54.1685	−0.051	2.106	0.568
29.8	3.452	1.088	36.866	53.8886	0.052	2.443	0.815
34.8	3.769	1.531	41.732	53.6531	0.252	2.653	1.13

Table A1.13 Pressure data, Cp, Rudder No. 2, Free stream, $\delta = 20°$

Rudder No. 2				*Free stream*				
%c	S1	S2	S3	S4	S5	S6	S7	S8
0	−1.985	−2.350	−1.760	−1.740	−1.734	−1.340	−0.295	−0.240
2.5	0.875	0.974	0.985	0.980	1.004	1.013	0.951	0.859
5	0.910	1.020	0.980	0.978	0.936	0.519	0.772	0.677
10	0.713	0.771	0.723	0.729	0.065	0.601	0.415	0.312
20	0.414	0.433	0.393	0.391	0.318	0.259	0.118	0.029
30	0.257	0.284	0.198	0.203	0.142	0.081	−0.017	−0.097
40	0.144	0.143	0.084	0.099	0.047	0.007	−0.065	−0.141
50	0.093	0.077	0.042	0.046	0.009	−0.015	−0.093	−0.167
60	0.052	0.013	0.007	0.013	−0.032	−0.046	−0.120	−0.196
70	−0.009	0.007	−0.024	0.011	−0.037	−0.022	−0.110	−0.163
80	−0.032	0.002	−0.035	0.029	−0.026	−0.022	−0.099	−0.130
90	−0.080	−0.068	−0.077	0.016	0.004	0.011	−0.097	−0.112
95	−0.073	−0.075	−0.048	0.004	0.052	0.039	−0.050	−0.141
95	−0.233	−0.189	−0.264	−0.046	0.000	0.020	−0.600	−0.598
90	−0.289	−0.284	−0.270	−0.082	−0.049	−0.057	−0.899	−0.831
80	−0.304	−0.270	0.360	−0.119	−0.150	−0.207	−1.138	−1.088
70	−0.459	−0.336	−0.453	−0.223	−0.298	−0.327	−0.972	−0.941
60	−0.507	−0.334	−0.444	−0.340	−0.260	−0.450	−0.880	−1.189
50	−0.457	−0.323	−0.492	−0.490	−0.524	−0.534	−0.725	−1.666
40	−0.483	−0.404	−0.536	−0.720	−0.702	−0.682	−0.750	−0.965
30	−0.588	−0.879	−0.607	−1.068	−0.936	−0.922	−0.798	−0.978
20	−1.013	−1.444	−1.440	−1.55	−1.436	−1.257	−1.069	−0.967
10	−1.940	−2.303	−2.415	−2.358	−2.111	−1.952	−1.277	−1.165
5	−2.716	−3.426	−3.130	−3.263	−2.959	−2.645	−1.942	−1.640
2.5	−3.392	−3.950	−3.615	−3.764	−3.365	−3.028	−2.226	−1.908

Note: Pressure tappings chordwise are numbered anticlockwise from LE, starting with face. Location of span positions S1–S8 are given in Table 5.11.

Table A1.14 Pressure data, Cp, Rudder No. 2, $\beta_R = -7.5°$, $\delta = 20°$, $J = 0.51$ ($K_T/J^2 = 0.88$)

%c	S1	S2	S3	S4	S5	S6	S7	S8
0	0.222	1.685	3.746	−0.521	−2.557	−1.689	0.014	−0.028
2.5	0.833	1.658	2.014	1.875	3.771	4.784	3.676	2.222
5	0.681	0.918	1.042	1.611	3.643	4.243	2.845	1.736
10	0.389	0.164	0.085	0.889	2.900	2.946	1.563	0.819
20	0.153	−0.247	−0.225	−0.167	1.771	1.811	0.493	0.153
30	0.056	−0.301	−0.282	−0.903	1.143	1.000	0.113	−0.125
40	0.028	−0.274	−0.239	−1.056	0.843	0.757	0.000	−0.111
50	0.028	−0.178	−0.239	−0.917	0.600	0.568	−0.113	−0.208
60	0.014	−0.096	−0.183	−0.639	0.343	0.378	−0.169	−0.278
70	0.056	−0.014	−0.113	−0.444	0.229	0.284	−0.155	−0.250

(*Continued*)

Table A1.14 Continued

%c	S1	S2	S3	S4	S5	S6	S7	S8
80	0.097	0.110	0.085	−0.139	0.200	0.108	−0.056	−0.222
90	0.069	0.260	0.352	0.125	0.343	0.203	0.000	−0.153
95	0.111	0.493	0.535	0.319	0.586	0.365	0.070	−0.083
95	0.097	0.425	0.507	0.528	0.629	0.378	−0.451	−0.653
90	0.000	0.164	0.310	0.431	0.429	0.203	−0.803	−0.861
80	−0.153	−0.151	−0.014	0.097	0.157	0.027	−0.915	−1.139
70	−0.236	−0.356	−0.296	−0.208	−0.029	−0.122	−0.775	−1.389
60	−0.361	−0.603	−0.634	−0.569	−0.243	−0.270	−0.704	−1.750
50	−0.556	−0.877	−1.028	−1.056	−0.586	−0.446	−0.761	−1.625
40	−0.722	−1.192	−1.493	−1.569	−1.029	−0.757	−0.887	−1.097
30	−0.889	−1.575	−2.324	−2.347	−1.900	−1.378	−1.268	−1.236
20	−1.139	−1.767	−2.732	−2.889	−2.714	−1.973	−1.549	−1.194
10	−1.389	−1.918	−3.577	−3.750	−4.500	−3.811	−2.225	−1.736
5	−1.528	−2.616	−3.268	−3.528	−6.300	−6.700	−3.282	−2.181
2.5	−1.417	−2.014	−2.690	−2.917	−7.471	−7.203	−4.775	−2.722

Note: Pressure tappings chordwise are numbered anticlockwise from LE, starting with face. Location of span positions S1–S8 are given in Table 5.11.

Appendix 2

Rudder and propeller design software

A theoretical model of rudder–propeller interaction has been developed using lifting line theory to model the rudder and blade element-momentum theory to model the propeller. The theoretical methods and basis for the approach are described in Chapter 6.

A computer program has been written, based on the theoretical model, which is useful for preliminary design purposes and parametric studies. The main components of the lifting line program, together with a summary of the blade element- momentum theory, are described in Chapter 6 and reference [A2.1].

The program exe file, together with operating instructions, is available through the publisher's web site.

Operation of the program

The input parameters and output data are summarised as follows:

INPUT:
Propeller parameters:
J, *P/D*, BAR, No. of blades
Rudder parameters:
S, *TR*, *D/S*, *Y1/S*, *X/D*, rudder angle

OUTPUT:
Spanwise distribution of K_T, K_Q and η_0 for propeller,
Spanwise propeller induced flow velocities and angles,
Integration of propeller spanwise distributions to derive total K_T, K_Q and η_0,
Rudder spanwise distribution of local inflow velocity and angle and local normal force coefficient C_N,
Integration of rudder spanwise distribution to derive the total normal force coefficient C_N and spanwise centre of pressure *CPs*,
Rudder induced drag coefficient C_{Di} and total drag coefficient C_D,
Approximate rudder stall angle, based on experimental data.

The overall flow path of the theoretical analysis is shown in Figure A2.1.

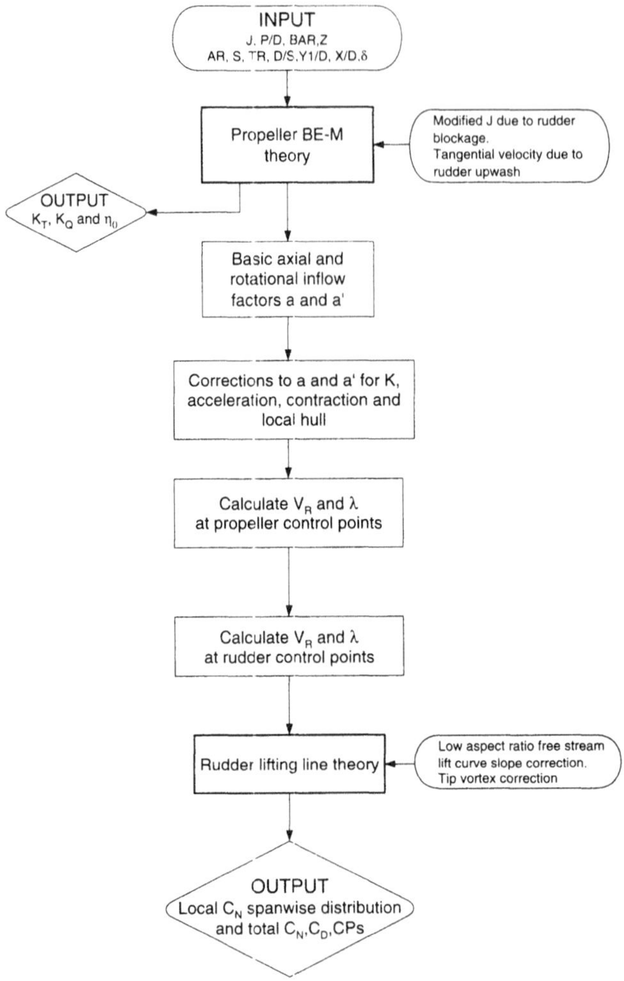

Figure A2.1 Lifting line–blade element flow path

Limits of operation of the program:

It is (currently) set up for all-movable rudders only
It is not suitable for zero and very low speeds or astern operation.
It does not include estimates of chordwise centre of pressure or rudder torque.

Reference

A2.1 Molland, A.F. and Turnock, S.R. A compact computational method for predicting forces on a rudder in a propeller slipstream. *Transactions of the Royal Institution of Naval Architects*, Vol. 138, 1996, pp. 227–244.

Index

Kolibri 360 GmbH
Ferdinand-Jühlke-Straße 7
99095 Erfurt, Deutschland
produktsicherheit@kolibri360.de